T0338271

JUNCTIONLESS FIELD-EFFECT TRANSISTORS

IEEE Press
445 Hoes Lane
Piscataway, NJ 08854

JUNCTIONLESS FIELD-EFFECT TRANSISTORS

Design, Modeling, and Simulation

SHUBHAM SAHAY
MAMIDALA JAGADESH KUMAR

IEEE Press Series on Microelectronic Systems

Published by John Wiley & Sons, Inc., Hoboken, New Jersey.
Published simultaneously in Canada.

Library of Congress Cataloging-in-Publication Data is available.

ISBN 978-1-119-52353-6

Printed in the United States of America.

V10007872_012819

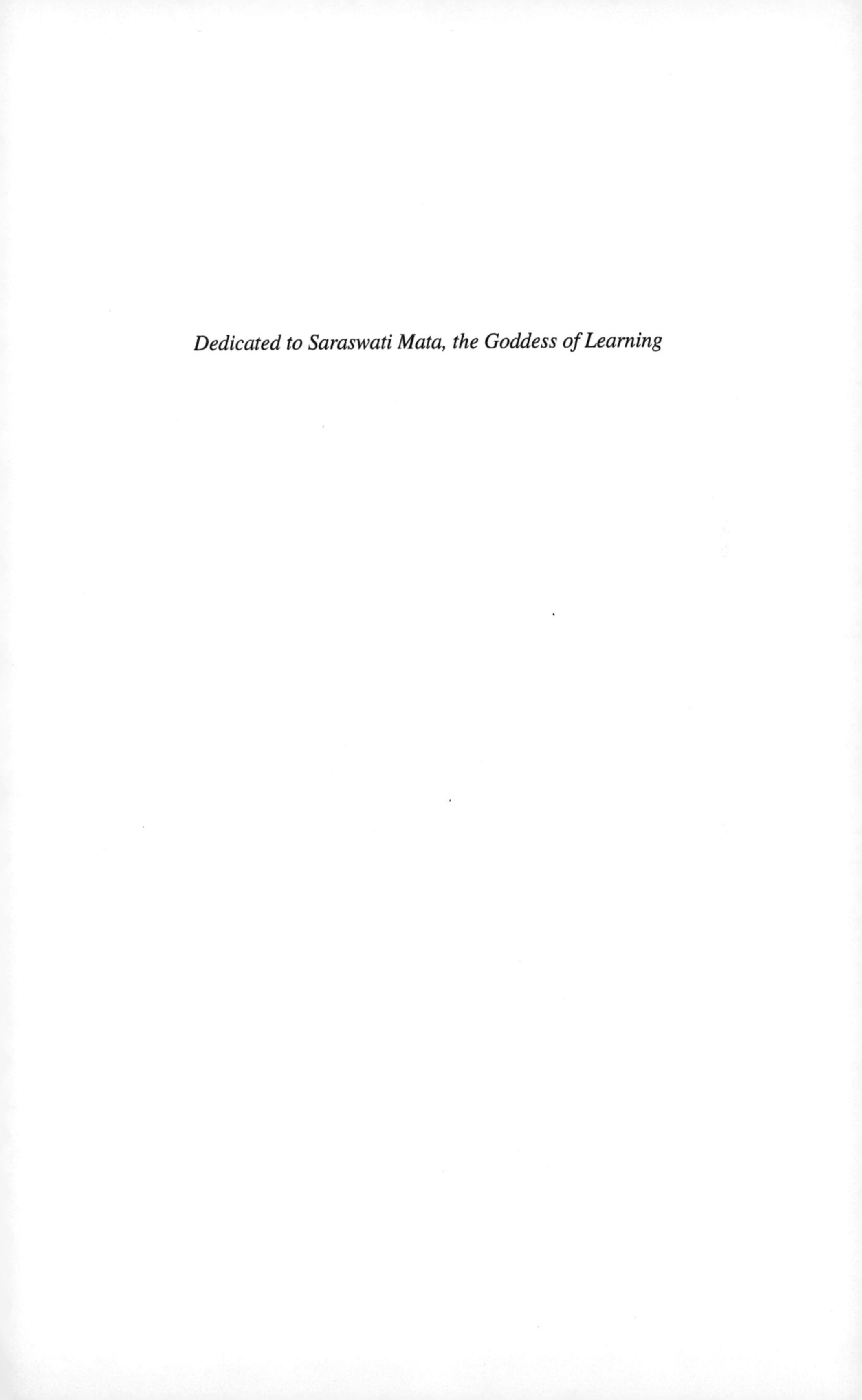

Dedicated to Saraswati Mata, the Goddess of Learning

CONTENTS

PREFACE

We are living in an era of supercomputing where smartphones, smartwatches, and smart technology have become an inevitable part of our daily life. The research and development in the field of transistors, which forms the basic building block of the computing devices, has driven this "smart" revolution. The dimensions of the transistors have been incessantly scaled down to increase the number of transistors per chip, which has not only reduced the chip area enabling hand-held devices but also increased the functionality and operating frequency and decreased the power dissipation. However, all the modern-day transistors such as metal-oxide-semiconductor field-effect transistors (MOSFETs) (or tunnel field-effect transistors or ferroelectric field-effect transistors, etc.) contain two metallurgical junctions: one at the source–channel interface and other at the channel–drain interface. To further scale down the modern transistors to the sub-10 nm regime and exploit the performance improvements brought by the scaling process, the doping must change abruptly from a high value (typically $\sim 10^{20}$ cm^{-3}) in the source and drain regions to a low value (typically $\sim 10^{14}$–10^{16} cm^{-3}) with complementary dopants in the channel region within a span of a few nanometers ($\sim$1–2 nm). Experimental realization of such an ultrasteep doping profile is extremely difficult even with the industry-standard ion-implantation process. To add to this misery, achieving high dopant activation in the heavily doped source/drain regions requires a high-temperature annealing. The annealing process, in turn, leads to a thermally assisted lateral diffusion of dopant atoms from source/drain regions into the channel region. This further restricts the possibility of realizing ultrasteep doping profiles in MOSFETs. As lateral diffusion is inevitable while annealing, the simultaneous requirement of a high dopant activation and an ultrasteep doping profile puts a complex constraint on the thermal budget. Our lives as device designers would have been much easier if there have been no metallurgical junctions. Therefore, to alleviate the need for ultrasteep doping profiles, field-effect transistors without any metallurgical junction were proposed to facilitate the scaling down of the conventional MOSFETs. These junctionless FETs (JLFETs) utilize an ultrathin semiconductor film with a gate stack to control its resistance and modulate the current flowing through it. The absence of a metallurgical junction leads to an altogether new conduction mechanism and device properties, which are different from conventional MOSFETs.

Surprisingly, the working principle of the JLFET was conceptualized and patented by Austrian-Hungarian physicist Julius Edgar Lilienfield in 1930 even before the discovery of the point-contact transistor by Shockley, Brattain, and Bardein in 1947. But it was only with the recent advancements in the fabrication technology that nanowire JLFETs were experimentally realized in 2010, inspired by Lilienfield's work. An exhaustive research has been carried out on JLFETs since then. The number of research papers on JLFETs has increased exponentially, and our understanding of JLFETs has also improved significantly over the years. The junctionless architecture, owing to its low cost, low fabrication complexity, and lower thermal budget, has opened up a new domain of exciting possibilities whereby JLFETs could be employed as sensors, memories, such as capacitor-less DRAM, NAND flash memory, display devices, and for biocompatible, optoelectronic, and three-dimensional (3D) sequential integrated circuit applications apart from logic applications. The enormous possibilities offered by the junctionless transistor architecture are exciting opportunities to the researchers to explore and invent novel device structures for a variety of applications ranging from logic circuits to memories, sensors, 3D integration, and display technology. However, due to the lack of a comprehensive textbook, research papers are currently the primary source of knowledge on JLFETs. With a plethora of research papers appearing on JLFETs, gaining a basic understanding of a JLFET and keeping track of the latest research is a challenge.

This book endeavors to be a comprehensive guide for those who are about to begin their study (and research) or have already started working on JLFETs. It provides a one-stop volume for studying JLFETs for someone having a basic knowledge of device physics. The book covers the fundamental physics behind the operation of JLFETs and provides a comparative analysis of different performance metrics of the JLFETs with respect to the MOSFETs. The book unfolds the challenges for JLFETs if they were to replace MOSFETs and incorporates a comprehensive study of the device architectures and designs proposed in the literature to mitigate the challenges and improve the performance of JLFETs. The book also includes a detailed analysis of the junctionless devices realized without the need for conventional chemical doping. In addition, it discusses in detail the different approaches used for analytical or compact modeling of JLFETs for the purpose of circuit design and circuit simulation. Therefore, this book is the first attempt to encompass the research reported on JLFETs on aspects spanning from device architectures and simulations to analytical modeling. Also, every aspect of the JLFET has been compared to the MOSFET so that the material presented in the book allows the entire semiconductor device fraternity to evaluate the potential of JLFETs and take informed decisions regarding its integration with the prevailing technology in the industry. Another unique feature of this book is that it describes the process of carrying out numerical simulations of JLFETs using the technology computer-aided design (TCAD) tool Sentaurus S-device. TCAD simulations are helpful for studying the behavior of any semiconductor device without getting into the complex process of fabrication and characterization, thus reducing the time to market. The calibrated simulation setup provided in the book would definitely aid the researchers especially the beginners in the field and provide them with an effective tool to analyze, evaluate, and invent new junctionless architectures for different applications, which may serve as a stepping-stone in the early stage of their work. We hope that this book covering the fundamentals of the JLFET along with their analytical modeling and simulation using TCAD would encourage the beginners to pursue research on JLFETs and augment the efforts of the existing researchers to realize a power-efficient JLFET for "green" electronics, which would eventually lead to a better society.

1

INTRODUCTION TO FIELD-EFFECT TRANSISTORS

We are living in an era of information technology where smartphones, smart watches, and smart technology have become an inevitable part of our lives. You might have observed a drastic improvement in the performance of these smart devices. For instance, the shift from single core processors to multicore processors, the increase in CPU's frequency from few MHz to several GHz, the increase in the RAM from few MB to several GB, and so on. All these factors have led to a tremendous increase in the performance of these computing devices. The smart devices found in every household nowadays have a performance metric comparable to the earlier supercomputers. For instance, the Apple watch has twice the processing power of a 1985 Cray-2 supercomputer [1]. In addition, the device size has also shrunk significantly and the focus in the research and development of computing devices has shifted toward mobile devices. Moreover, the functionality per device has also increased considerably. For instance, the present day smartphones not only have processing capabilities of a supercomputer but can also perform the functions of a good quality camera, a Wi-Fi dongle, an X-BOX gaming system, and so on. To summarize, every other person in this modern era has access to low-cost, high-performance gadgets.

Have you ever wondered what drives the "smartness" and the supercomputing capabilities of all the smart technology gadgets? Let us try to understand this from a human body–gadget analogy. Just like the human body is composed of cells as the building block, the electronic gadgets are made up of transistors. In human body, the

Junctionless Field-Effect Transistors: Design, Modeling, and Simulation, First Edition.
Shubham Sahay and Mamidala Jagadesh Kumar.

cells are grouped together to perform a particular function and form an organ. Therefore, the efficiency and the number of different functions that can be performed by the body depends exclusively on these cells. Similarly, the transistors act like a switch and are wired together in a chip (which is similar to the organ from body–gadget analogy) in a specific manner to enable a particular function. The larger the number of transistors in a gadget, the more the number of functions it can perform. The research and development in the field of transistors has driven this "smart" revolution. It is indeed very interesting how such small chunks of silicon chips drive our lives.

1.1 TRANSISTOR ACTION

But what exactly is a transistor? The word transistor was given by its first inventors: Shockley, Brattain, and Bardeen in 1947 [2–5]. At that time, no one would have wondered that this discovery (which actually was an accident) would be driving the lives of common people for generations to come. The transistors are often conceived as a device where the resistance between two terminals may be controlled by the current/voltage at the third terminal. Therefore, transistor refers to any three-terminal device where the current (or voltage) between two terminals may be controlled by the action of voltage (or current) at the third terminal.

In the subsequent sections, we shall see how the most common transistors work from both a qualitative approach and an energy band diagram perspective. The bipolar junction transistors (BJTs) dominated the semiconductor industry until late 1970s. Although BJTs are still used in the high-frequency circuits such as in radio frequency circuits, the throne is captured by the metal-oxide-semiconductor field-effect transistors (MOSFETs) and they continue to drive the semiconductor industry even today. Therefore, we shall discuss the MOSFETs in detail in the next section.

Transistors such as MOSFETs act as switches in the integrated circuits. However, it may be noted that the MOSFETs are not ideal switches (which are expected to consume no power when switched-OFF and deliver a high current instantaneously when switched-ON). The MOSFETs exhibit a small leakage current and, therefore, consume power from the supply even when they are switched-OFF. This power consumption is termed as the static power dissipation (P_s) given as

$$P_S = V_{DD} \cdot I_{OFF} \tag{1.1}$$

where V_{DD} is the supply voltage and I_{OFF} is the leakage current that flows through the transistor when the switch is turned off. Furthermore, the MOSFETs also consume a significant power when switched from the ON-state to OFF-state or vice versa. This power consumption also depends on the frequency of switching of the MOSFETs and is termed as the dynamic power dissipation (P_d) given as

$$P_D = V_{DD}^2 C_L f \alpha \tag{1.2}$$

where V_{DD} is the supply voltage, f is the frequency of operation and α is the switching probability, which simply tells us that the MOSFET is not switched in each cycle, and C_L is the load capacitance. In a wired network of MOSFETs, a MOSFET drives another MOSFET. Therefore, in most cases C_L is the input capacitance of the MOSFET. The interested readers are requested to refer [5] for more details.

Until recent past, the focus of the researchers all over the world was to miniaturize the dimensions of the MOSFETs so as to increase the number of MOSFETs per chip, which would not only reduce the area enabling mobile devices but also increase the number of operations that may be performed by a single chip. Scaling the MOSFET dimensions also reduces the input capacitance and increases its capability (current) to drive another MOSFET in the wired chip network and helps to achieve large frequency of operation due to fast charging of C_L. Although the drive current of MOSFET increases with scaling, the OFF-state current also increases drastically due to the short-channel effects that are triggered by MOSFET gate length scaling. The increase in the OFF-state current results in a significant static power dissipation. While the dynamic power dissipation was a major concern for the researchers until recent past, the scaling trends suggest that the static power dissipation would eventually surpass the dynamic power dissipation if the conventional MOSFETs are scaled aggressively.

A high static power consumption means that the MOSFETs would draw a significantly large power from the supply even when it is switched-OFF. Therefore, the chip would drain the battery or the power source even when the functionality provided by the chip is not being utilized. This is detrimental to the performance of computing devices especially for the hand-held devices like smartphones, which have a limited supply available in the form of a battery. Furthermore, the static power dissipation also heats up the chip and degrades the performance of the gadgets which are designed for room temperature operation. Of course, every consumer wants to have a smart device with an unlimited battery or power supply with no heating effects. To reduce the power dissipation, we can reduce the supply voltage as evident from equations (1.1) and (1.2). However, there lies a fundamental limitation on the MOSFETs which is inherent to the very physics of the device. The current in a MOSFET cannot increase by more than a ten-fold when the input voltage is raised by 60 mV. This limitation is due to the Maxwell–Boltzmann distribution of electrons in matter and is often referred to as the "Boltzmann tyranny." The application of MOSFET as a switch requires that the ON-state to OFF-state current ratio be high so that these states are easily distinguishable ($\sim 10^4$ to 10^6). To achieve an ON-state to OFF-state current ratio of a million, the variation of the input voltage, and therefore the supply voltage, needs to be at least equal to $60 \times \log(10^6) = 360$ mV. This limitation simply implies that if we have an extremely scaled supply voltage, the ratio of the ON-state current to the OFF-state current of the transistor would be very low and the MOSFET would cease to act like a switch. Therefore, the Boltzmann limit hinders the use of the conventional MOSFETs as a switch for ultralow supply voltages.

As a result, the conventional MOSFETs cannot cater the need of yielding an area and power-efficient chip with multiple functionalities. Moreover, scaling the conventional MOSFETs also requires a large investment from the manufacturing point of view. Therefore, present day research focuses on design of a low-cost and highly scalable MOSFET with minimum power dissipation. As you would have noted, the research and development in this context has gradually shifted from an area-driven perspective to a power-driven scenario.

This chapter will help to develop a basic understanding of the conventional MOSFETs. After a subtle discussion of the various modes of operation of these devices, Section 1.3 describes how basic circuits can be formed using MOSFETs. Section 1.3.2 focuses on different types of power dissipations reported earlier in the introduction.

1.2 METAL-OXIDE-SEMICONDUCTOR FIELD-EFFECT TRANSISTORS

To understand a MOSFET, we shall first get an in-depth understanding of a MOS capacitor (Fig. 1.1) which is the heart of a MOSFET, grasp the concept of "field–effect," and then discuss operation of MOSFETs. The MOS capacitor consists

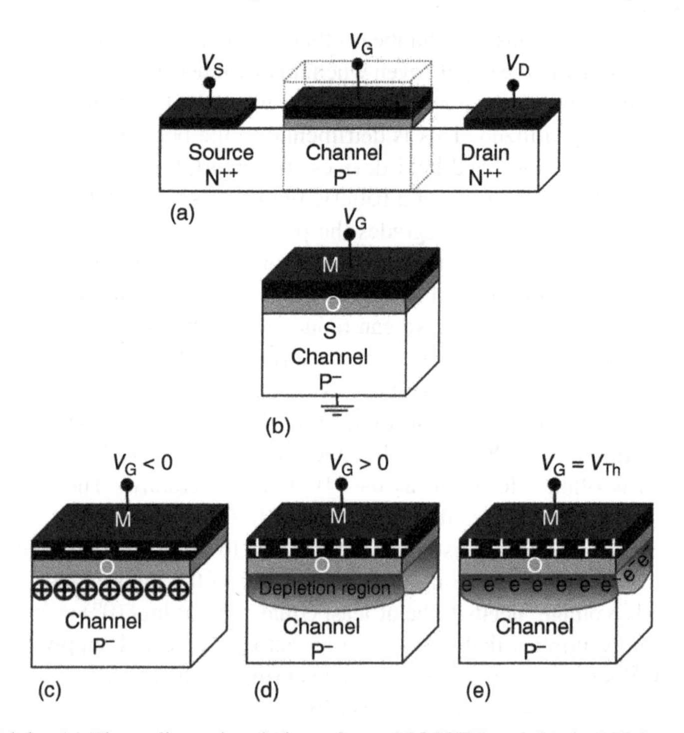

FIGURE 1.1 (a) Three-dimensional view of an n-MOSFET and (b) the MOS capacitor and the operation mode of a MOS capacitor in (c) accumulation regime, (d) depletion regime, and (e) inversion regime.

of three layers as the name suggests: metal-oxide-semiconductor. A thin insulating oxide layer is sandwiched between a metal and a semiconductor. Since the structure has a dielectric inserted between two conducting plates (assuming that the semiconductor is doped or at room temperature), the MOS structure is essentially a capacitor. The MOS capacitor with the p-type doped semiconductor is called a p-type MOS capacitor, whereas the MOS capacitor with an n-type doped semiconductor is called an n-type MOS capacitor.

The capacitance of the MOS structure can be controlled by the gate voltage just like the capacitance of a p–n junction is controlled by the applied bias. However, the range of capacitances exhibited by the MOSFET is large compared to the p–n junction capacitance.

At this point, we would also like to mention that the property of bulk atoms and surface atoms are different. Indeed, in the words of W. Pauli (who gave the Pauli exclusion principle), "God made the bulk; interfaces were invented by the devil" [7]. In MOS devices, all the charge dynamics occur at the surface. Therefore, silicon is the most preferred material for MOS devices as the Si–SiO_2 interface constitutes the best quality semiconductor–insulator interface. Silicon is also abundant on earth in the form of sand (silica).

1.2.1 "Field-Effect" and Operation Modes

To understand the different modes of operation of a MOS capacitor, it is essential to understand the concept of "field-effect" applied to the MOS devices. The field-effect simply means controlling the charge dynamics with the aid of an electric field. Now, let us look at how electric field controls charges in case of a p-type MOS capacitor shown in Fig. 1.1(b).

If we apply a negative voltage on the gate terminal with respect to silicon, an electric field will be generated across the insulator with a direction from the semiconductor to the metal. The applied negative potential on the gate can be conceptualized as depositing negative charges on the gate which attract the majority holes in the p–Si toward the Si–SiO_2 interface. Therefore, the holes would be accumulated at the Si–SiO_2 surface due to the application of a negative bias on the gate. From an electric field perspective, the holes move in the direction of the electric field and accumulate at the Si–SiO_2 interface. As a result, the effective carrier concentration at the interface is increased in the accumulation mode as shown in Fig. 1.1(c).

Now, if a positive voltage is applied to the gate terminal, the electric field direction across the insulator is reversed and points toward the semiconductor from the gate. A positive potential at the gate can also be conceptualized as depositing positive charges on the gate which repel the majority holes close to the Si–SiO_2 interface. The repelled holes move into the bulk leaving behind uncovered negative acceptor ions. Therefore, a depletion region is formed in the semiconductor in the vicinity of the Si–SiO_2 interface. In other words, the electric field pushes holes away from the interface to the bulk, increasing the depletion region width. This region of operation is called the depletion mode.

Now, what happens if the positive voltage is increased even further? One may expect that the depletion region would continue to expand until it spans the entire semiconductor. However, this is not what happens since the minority electrons are also there in the bulk of the p-type semiconductor, which may provide negative charge for the electric field lines from the gate to terminate. Therefore, when we continue to increase the positive voltage, the depletion region increases until a maximum value and then the minority electrons move to the surface from the bulk and start accumulating to facilitate the termination of the electric field lines. From electric field perspective, the field becomes so strong that it pulls the minority electrons to the surface. The value of gate voltage at which the electron concentration at the surface becomes equal to the bulk doping concentration of the p-type semiconductor is called the threshold voltage. At the threshold voltage, the majority carriers change from holes to electrons at the surface. This phenomenon is called inversion, and the electron layer is called the inversion layer.

For the n-type MOS capacitor, accumulation of electrons takes place for positive gate voltages. Upon application of a negative gate voltage, the semiconductor is depleted first and then the depletion region reaches its maximum value. As the magnitude of the negative gate voltage is increased further, the minority holes start moving to the surface and, eventually, an inversion layer of holes is formed. Therefore, the characteristics of a p-type MOS capacitor are just complementary to the n-type MOS capacitor and hence do not require a detailed discussion.

Now, with this background, we shall discuss the structure of a MOSFET.

1.2.2 MOSFET as a Switch

A MOSFET consists of the MOS capacitor appended by the source and drain regions, which makes it a three-terminal device as shown in Fig. 1.1(a). The three terminals of the MOSFET are source (acts as a source of carriers), gate (controls the amount/concentration of carriers), and drain (acts as the sink for carriers). The source and drain are heavily doped, whereas the channel is lightly doped with a polarity opposite to that of source and drain. MOSFETs also utilize a fourth terminal called the body terminal. The body terminal is connected to the channel and is used to manipulate the electron conduction (and the threshold voltage) under special circumstances. Otherwise, it is normally grounded.

As discussed in Section 1.2.1, if a positive bias greater than the threshold voltage is applied, an inversion layer of electrons is formed at the surface of p-type silicon. Now, if a positive voltage is applied at the drain terminal, the electrons of the source would find a low-resistance conduction path via the inversion layer and flow into the drain. Therefore, the electrons would flow from the source to the drain region and MOSFET would act as a closed switch. However, when the applied bias is lower than the threshold voltage, the channel region remains depleted and offers a high-resistance path. Therefore, the electrons in the source find it difficult to reach the drain via the channel and the MOSFET behaves like a open switch. Therefore, the MOSFET acts like a switch which can be switched ON or OFF depending on

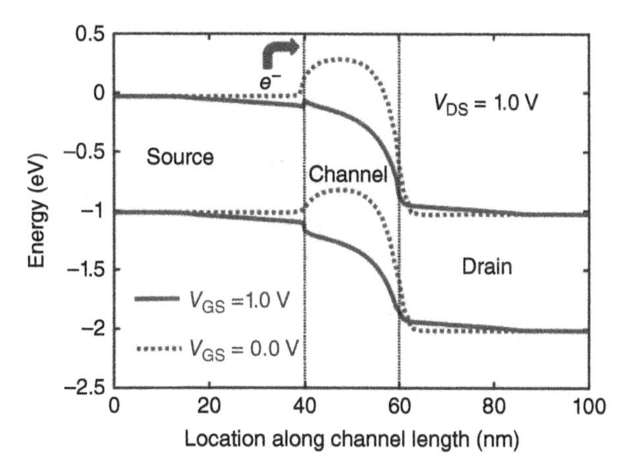

FIGURE 1.2 Energy band profiles of the MOSFET at ON-state ($V_{GS} = 1.0$ V) and OFF-state ($V_{GS} = 0.0$ V) showing that the gate voltage modulates the barrier height for source electrons.

the voltage applied to the gate. The drain to source current can, therefore, be controlled by the voltage applied to the gate. Since current through two terminals is being controlled by the voltage at the third terminal, the MOSFET is called a transistor, i.e. a resistor whose resistance may be controlled by the gate. As shown in Fig. 1.2, the gate voltage simply modulates the effective barrier height seen by the source electrons to move into the drain region through the channel.

1.2.3 Transfer Characteristics and Output Characteristics

At this point, we would like to introduce the concept of transfer characteristics and the output characteristics. The output characteristics refer to the relation between the output current (drain current in the case of a MOSFET) with the output voltage (drain voltage). Therefore, the output characteristics in a MOSFET are simply a plot between the drain current versus the drain voltage for a particular gate voltage as shown in Fig. 1.3(a). The drain current first increases linearly with the drain voltage and then gets saturated owing to the pinch-off of the channel region at the drain end. The inversion layer charge increases with increasing gate voltage leading to a larger drain current.

The relationship between the output current (drain current) and the input voltage (gate voltage) for a particular drain voltage is called the transfer characteristics. The drain current is very low below the threshold voltage (subthreshold regime) and is governed by diffusion of carriers from the source to the drain region. As the gate voltage increases above the threshold voltage, an inversion layer forms and the drain current increases significantly (Fig. 1.3(b)). If the transfer characteristics are plotted on a linear scale, the threshold voltage can be extracted by extrapolating the drain current after which the current starts increasing dramatically (Fig. 1.3(b)).

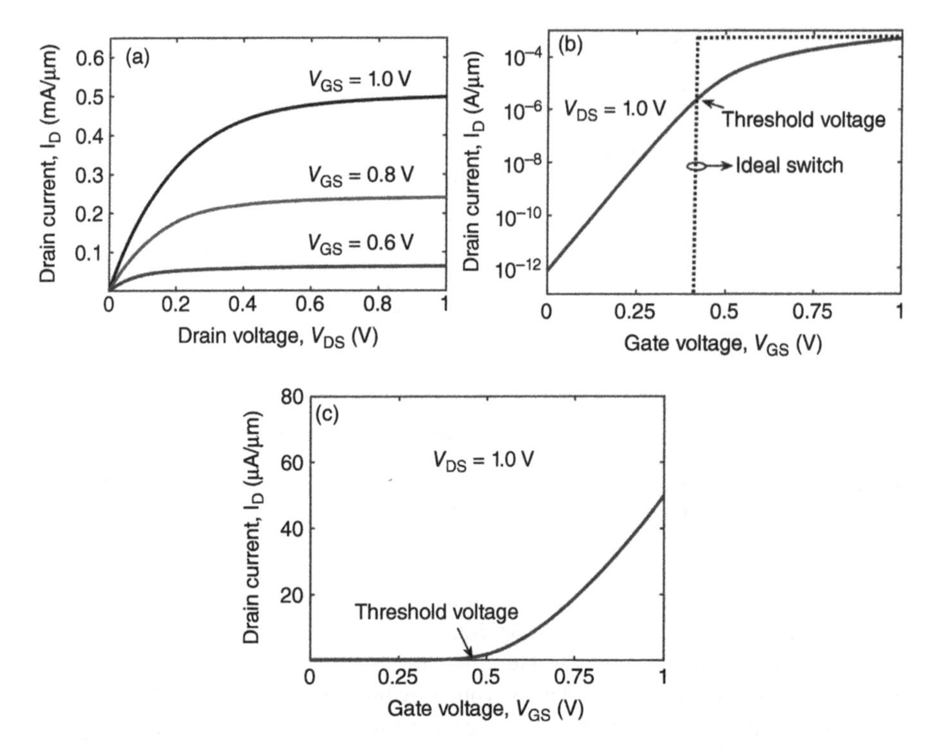

FIGURE 1.3 (a) Output characteristics and transfer characteristics of the MOSFET in (b) log scale and (c) linear scale.

Several methods have been proposed for determination of the threshold voltage. Some of these methods include finding the zeroes of the double derivative of the transfer characteristics (which is equivalent to finding maxima of the derivative), whereas others rely on finding the exact surface potential-gate voltage relationship by solving the Poisson equation in the channel region and equating the surface potential equal to twice the Fermi potential at threshold condition [8]. The Poisson equation simply relates the potential to the charge contained in any region and is defined as

$$\nabla^2 \varphi = -\frac{\rho}{\varepsilon_{\mathrm{Si}}} \tag{1.3}$$

Though equation (1.3) appears very simple, it is very difficult to solve analytically and requires numerical solvers or approximations as discussed in Chapter 7. The simplest approach to find the threshold voltage is the constant current method, which defines threshold voltage as the gate voltage for a particular constant drain current, generally $(W/L_g) \times 10^{-7}$ A [8]. This approach is simplest as it does not involve any modulation or extrapolation or numerical solver.

In addition, other important parameters can also be extracted from the transfer characteristics at a drain voltage of V_{DD}. The drain current corresponding to the $V_{GS} = V_{DS} = V_{DD}$ is termed as the ON-state current, and the drain current at

$V_{GS} = 0$, $V_{DS} = V_{DD}$ is termed as the OFF-state current. Furthermore, the subthreshold swing can also be extracted from the transfer characteristics. There are two subthreshold swings for each transfer characteristics: a point subthreshold slope and an average subthreshold slope. The point subthreshold slope is simply the derivative of the transfer characteristics at a given gate voltage, whereas the average subthreshold slope is calculated by taking the mean of the point subthreshold slopes at different gate voltages ranging from the OFF-state voltage ($V_{GS} = 0$) to the threshold voltage ($V_{GS} = V_{Th}$) and is given as

$$SS_{avg} = \frac{V_{Th}}{\log{(I_D)_{V_{Th}}} - \log(I_{OFF})} \tag{1.4}$$

These are the parameters that are given as the design specification to a device designer. In general, there is a minimum ON-state current to OFF-state current ratio ($I_{ON}/I_{OFF} \sim 10^4$–10^6) and a maximum subthreshold swing, which a MOSFET must satisfy to be used in circuits.

With a background of essential physics of MOSFETs, we can now discuss the implementation of circuits with the help of MOSFET as a switch.

1.3 MOSFET CIRCUITS: THE NEED FOR COMPLEMENTARY MOS

If you remember from our discussion in Section 1.1, it is indeed these MOSFETs in the form of switches which are wired together in the integrated circuits to perform particular functions. Nearly every MOSFET in a circuit has to drive a load capacitance, which may correspond to input capacitance of other MOSFETs or an external load. Therefore, for circuit representation, we connect a load capacitance at the output terminal of the MOSFET. We chose the simplest circuit, i.e. an inverter, to give an insight into the digital circuit implementation using MOSFETs.

An inverter is a NOT gate which essentially inverts the input logic “0” into output logic “1” and vice versa. How can a MOSFET perform this action? Let us consider the case of an n-MOSFET with a load capacitance connected to the drain end (Fig. 1.4(a)), which represents the output load. If the output is initially at logic “1”, i.e. if the load capacitance is initially charged, then the voltage across this capacitor is essentially the drain voltage of the n-MOSFET. Now, if an input logic “0” (V_G = 0) is applied, since $V_{GS} < V_{Th}$ (assuming $V_{Th} \sim 0.2\ V_{DD}$), the n-MOSFET remains OFF and, hence, the output logic remains at “1”. However, if an input logic “1” ($V_G = V_{DD}$) is applied to the n-MOSFET, since $V_{GS} > V_{Th}$, the n-MOSFET turns on and current flows from the drain terminal to the source terminal. This current would drain the charges on the output load capacitance to the ground. This can also be viewed as discharging of the output load capacitance through the resistance of the n-MOSFET in the ON-state. Therefore, the load capacitance is discharged to the ground potential which corresponds to the output logic “0”. Therefore, the application of an input logic “0” leads to an output logic “1” and vice versa and the n-MOSFET acts like an inverter.

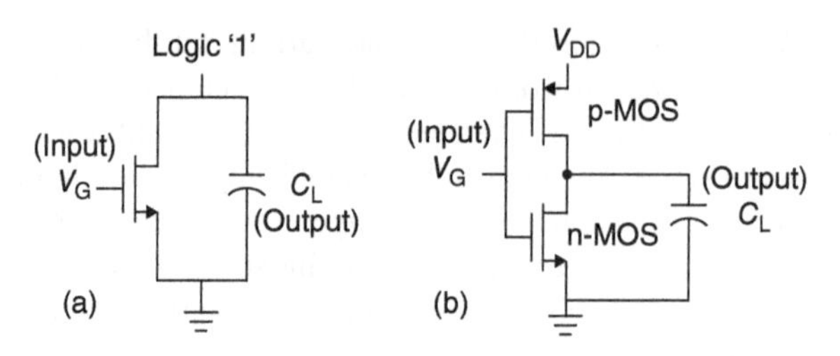

FIGURE 1.4 (a) n-MOSFET connected to a load capacitance, which is initially charged to logic "1", i.e., V_{DD} and (b) schematic view of a complementary metal-oxide-semiconductor (CMOS) inverter.

At this juncture, you may wonder what would happen in case the load capacitance was not charged initially? If the load capacitance is discharged, regardless of whether the input logic is "0" or "1" (n-MOSFET is OFF or ON), the output logic state remains at "0" because there is no means to charge the output load capacitance. Therefore, an n-MOSFET can only perform the inverter operation when the output logic is "1" and fails when the output logic is "0". Therefore, an n-MOSFET satisfies only half logic function and cannot be used alone for making even a simple inverter. Similarly, the p-MOSFET can only charge the load capacitance to logic "1" and works fine if output logic is initially "0" but fails when the output logic is initially "1". Therefore, even the p-MOSFET also satisfies only half logic function.

How can we make a complete inverter logic from MOSFETs if the n-MOSFET and p-MOSFET cannot individually perform the inverter operation? This calls for the need of complementary metal-oxide-semiconductor (CMOS) process. Since the p-MOSFET can charge the output to logic "1" whereas n-MOSFET can discharge the output to logic "0", both charging and discharging paths can be realized if they are used together as shown in Fig. 1.4(b). The p-MOSFET and n-MOSFET complement each other's logical function and can perform complete logic implementation only if used together. The circuit implementations in which both n-MOS and p-MOS are used together are known as CMOS circuits.

Now that we have recognized the importance of the CMOS process, let us analyze the working of a CMOS inverter and how a complete NOT operation is performed.

1.3.1 CMOS Inverter

The schematic of a CMOS inverter is shown in Fig. 1.4(b). When the input voltage is close to 0 V and low, for n-MOSFET, $V_{GS} < V_{Th}$ and it does not conduct. However, for the p-MOSFET, $|V_{GS}| > |V_{Th}|$. Therefore, the p-MOSFET turns ON and conducts. The load capacitance gets charged to V_{DD} via the p-MOSFET. Therefore, the output logic becomes "1". Similarly, when the input voltage becomes high, i.e. close to V_{DD}, for the n-MOSFET, $V_{GS} > V_{Th}$ and it conducts while for the p-MOSFET, $|V_{GS}| < |V_{Th}|$ and it remains switched OFF. Therefore, the load capacitance discharges to output logic "0" via the n-MOSFET. Hence, the inverter action is realized.

1.3.2 Power Dissipation in CMOS Inverter

A CMOS inverter takes in current from the supply only when both n-MOSFET and p-MOSFET are ON simultaneously, resulting in a path from the supply to the ground. Therefore, the current flows in a CMOS inverter only when the input voltage is close to 0.5 V_{DD} [6].

Ideally, we assume that the transistors do not consume any current when they are in the OFF-state. However, from our discussion in Section 1.1, we know that even below threshold voltage, the current is not equal to zero and a finite subthreshold leakage current flows through the MOSFETs. This leads to a power dissipation even when the input and output states of the CMOS inverter remain idle. This is called static power dissipation given by equation (1.1).

Also, every time the CMOS inverter output switches from "0" to "1", the load capacitor gets charged by V_{DD}. Therefore, the charge deposited on the load capacitance is $V_{DD}C_L$. Now, the energy taken from the supply is simply $C_L V_{DD}^2$. However, when a capacitor is charged with a voltage V_{DD}, the energy stored in the capacitor is only 0.5 $C_L V_{DD}^2$. Therefore, out of the total $C_L V_{DD}^2$ energy taken from the supply, only half is stored in the load capacitance. Since the law of conservation of energy explicitly says that energy can neither be created nor be destroyed, where did the half of the energy go? Actually, the capacitor gets charged via the p-MOSFET, which acts as a resistor, and this half energy is dissipated as heat across this resistor. Now, when the inverter output switches from "1" to "0", capacitor discharges through the n-MOSFET and the energy stored in the capacitor is dissipated through the n-MOSFET as heat. Hence, in every cycle of switching from "1" to "0" and back from "0" to "1", a power equal to $C_L V_{DD}^2$ is dissipated in the CMOS inverter. Since power is dissipated only when the inverter switches, this power dissipation is called the dynamic power dissipation. The dynamic power dissipation may be generalized for any CMOS circuit as shown in equation (1.2).

1.4 THE NEED FOR CMOS SCALING

In Section 1.3.2, we analyzed the different power consumption mechanisms in the CMOS inverter. You may wonder what can be done to reduce the power consumption? The static power dissipation depends on the supply voltage V_{DD} and the OFF-state leakage current. Therefore, a reduction in the supply voltage or the leakage current can reduce static power dissipation. Since the digital circuits, for example, the microprocessor runs at a dramatically high frequency (~GHz range), the contribution from the dynamic power dissipation is most significant in the total power dissipation. The dynamic power dissipation depends on α, f, V_{DD}, and C_L. The parameter α depends on the functionality of the digital circuit and cannot be altered. The frequency of operation needs to be increased for faster computation speed. As a result, the power dissipation can only be lowered by reducing V_{DD} and C_L.

However, reducing the supply voltage reduces the ON-state to OFF-state current ratio (I_{ON}/I_{OFF}) due to the fundamental Boltzmann limit on subthreshold swing. For feasible switching operation, the I_{ON}/I_{OFF} has to be at least 10^4. Therefore, the supply voltage cannot be reduced significantly.

The load capacitance C_L is essentially the input capacitance of similar logic circuits. C_L can be minimized by scaling down the area of MOSFETs. This calls for the need of CMOS scaling. Scaling the length of the MOSFET not only reduces the input capacitance but also increases the speed of the transistor as the ON-state current varies inversely with the gate length. Since the output capacitances are essentially charged and discharged, a higher drive current will increase the rate of charging or discharging. Therefore, scaling the MOSFET gate length leads to a reduced capacitance and facilitates high-frequency operation while increasing the number of MOSFETs and functionalities in a given chip area. CMOS scaling seems to be the best method to achieve a power efficient and high-speed multifunctionality device.

There are two ways in which CMOS scaling can be performed. These scaling techniques are categorized depending upon whether the supply voltage is scaled along with the channel length and width. A constant electric field scaling rule or the full scaling rule implies that the supply voltage is scaled by the same ratio as that of the length and the width. In the fixed-voltage scaling rule, the voltage is not scaled along with the length and width. The constant electric field scaling rule, which is also referred to as the Dennard's scaling rule, is followed in the industry, and the impact of the scaling factor (S) on various parameters is summarized in Table 1.1 [9].

TABLE 1.1 Scaling Factor for Different Parameters Utilizing Dennard's Scaling Rule [9]

Parameter	Scaling factor	Scaled value considering $S = \sqrt{2}$	Relative change
Channel length (L_g) Channel width (W) Gate oxide thickness (t_{OX}) Supply voltage (V_{DD})	$1/S$	0.7	30% ↓ 😊
Gate capacitance (C_{gg}) ~ $\left(\frac{WL_g}{t_{ox}}\right)$ Depletion region thickness (x_j) Intrinsic delay (τ) ~ $\left(\frac{C_{gg}V_{DD}}{I_{eff}}\right)$	$1/S$	0.7	30% ↓ 😊
Power dissipation (P_D) ~ (V_{DD}•I_{eff}) Area (A) ~ (W•L_g)	$1/S^2$	0.5	50% ↓ 😃
Power-delay product ~ (P_D• τ)	$1/S^3$	0.35	65% ↓ 😁
Electric field ~ $\left(\frac{V_{DD}}{t_{ox}}\right)$ Power density ~ $\left(\frac{P_D}{A}\right)$	1	1	0% = 😐
Frequency (f) ~ $\left(\frac{1}{\tau}\right)$	S	1.4	40% ↑ 😊

1.5 MOORE'S LAW

At this point, we would also introduce the famous law of CMOS scaling introduced by Gordon Moore, which says that the number of transistors in a chip will double after every one and a half years (18 months). The CMOS industry followed this famous Moore's law for more than 40 years with the help of CMOS scaling [10]. However, the dimensions of the scaled MOSFETs gradually became comparable to the depletion region widths at the source–channel and channel–drain interfaces. Such MOSFETs in which the channel length approaches the source–channel or channel–drain depletion region width are known as short-channel MOSFETs. The MOSFET electrostatics, which we have discussed until now, can be extended to the short-channel MOSFETs with slight modifications, which arise due to the effects that originate only when devices are scaled to the short-channel regime. These short-channel effects are discussed in Section 1.7.1.

1.6 KOOMEY'S LAW

Another parameter for estimating the efficiency of the microprocessors apart from the number of transistors in a chip is the number of computations it performs per unit power consumption. This is a more fundamental property since it relates to the energy efficiency of the microprocessors. This parameter is evaluated as the number of computations performed by the microprocessor every kilo-watt-hour of power consumed by it when it is operating at its peak output frequency.

Interestingly, because of a reduction in the power dissipation and improvement in the operating speed owing to scaling, even the number of computations performed per unit energy consumption follows the same trend as the number of transistors per chip. Even this parameter has nearly doubled every 18 months [11]. This observation was first made by Jonathan G. Koomey and hence came to be known as the Koomey's law. However, unlike the deviation from the Moore's law owing to the short-channel effects, the Koomey's law has remained intact even in the post 2010 scenario and the computing efficiency with respect to power has been doubling every 18 months. Since the physical basis for Koomey's law is more centric to today's energy-efficient computing systems including Internet of things (IoT), servers, and big data systems, it is expected to last longer than the Moore's law.

Now, in the subsequent section, we will discuss about the challenges while scaling the MOSFETs.

1.7 CHALLENGES IN SCALING THE MOSFET

1.7.1 Short-Channel Effects

In Section 1.5, we discussed that if the channel length of a MOSFET is comparable to the length of the source–channel and channel–drain depletion regions, it is termed as the short-channel MOSFET [12]. You may wonder how much exactly is the depletion region width at the source–channel or channel–drain interface? Typically, the doping concentration of source/drain region is $N_D = 10^{20}$ cm^{-3} and that of the channel is $N_A = 10^{16}$ cm^{-3}. The expression for depletion region width (x_{dep}) for a one-sided p–n junction with doping levels similar to MOSFET is

$$x_{dep} = \frac{2\varepsilon_{Si} V_{bi}}{qN_A} \tag{1.5}$$

$$\text{where} \quad V_{bi} = \frac{kT}{q} \ln\left(\frac{N_A \cdot N_D}{n_i^2}\right), \tag{1.6}$$

Solving this expression, the typical depletion region width at the source–channel or channel–drain interface comes out to be ~350 nm. As the channel lengths are scaled in this regime, several new physical phenomena arise and degrade the performance of MOSFETs. Therefore, in this section, we would give a brief overview of few short-channel effects, which dominate the performance of the MOSFETs in this ultrashort-channel length regime. Interested readers are directed to [13–15] for more detailed analysis of the short-channel effects.

1.7.1.A Threshold Voltage Roll-Off In a MOSFET, the depletion region of the heavily doped source/drain region protrudes into the p-channel and depletes it at the source–channel and channel–drain interface. As the gate length of the MOSFET is scaled to the short-channel regime, the source/drain-induced depletion region widths become a significant proportion of the overall channel length. The charge dynamics of the source/drain induced depletion region in the channel is no longer controlled solely by the "field-effect" of the gate electrode. Therefore, the effective channel charge that may be controlled by the gate electrode reduces significantly. As a result, the amount of gate voltage required to invert the channel region reduces considerably as compared to an undepleted MOS capacitor of similar dimensions. This reduction in the channel charge and the threshold voltage required to invert the channel with gate length scaling is known as threshold voltage roll-off [16, 17].

We already know that the subthreshold current, which contributes to the OFF-state leakage current, is due to diffusion of carriers from source to drain region and varies exponentially with the gate overdrive voltage, i.e., $V_{GS} - V_{Th}$. A lower V_{Th} simply means that the subthreshold leakage current would increase significantly. Therefore, the threshold voltage roll-off due to gate length scaling increases the OFF-state leakage current exponentially. It may be noted that while the increase in the ON-state current due to channel length scaling is linear, the OFF-state current increases exponentially. This leads to a significant reduction in the I_{ON}/I_{OFF} with channel length scaling.

1.7.1.B Drain-Induced Barrier Lowering As shown in Fig. 1.2, the application of a gate voltage simply modulates the source to channel barrier height and alters the injection of electrons from source to drain region in a MOSFET. The source to channel barrier height in an ideal "long" channel MOSFET is controlled exclusively by the gate voltage. However, when the channel length is scaled to the short-channel regime, the channel–drain depletion region may interact with the source–channel depletion region. As a result, in the short-channel MOSFETs, the application of a drain voltage not only reduces the electron energy level in the drain region but the drain electric field also couples through the depletion regions and reduces the source to channel barrier height. This reduction in the source–channel barrier height with the drain voltage is not pronounced in the long-channel MOSFETs due to the absence of interaction between the depletion regions. However, in the short-channel MOSFETs, this drain-induced barrier lowering (DIBL) reduces the source to channel barrier height considerably and results in an increased leakage current [18–21]. Nowadays, one of the major challenges is to design a MOSFET with minimum drain–source coupling such that the source to channel barrier height is controlled exclusively by the gate voltage.

1.7.2 Hot Electron Effect

When the channel length is scaled, the lateral electric field estimated as V_{DS}/L_g increases significantly. You may wonder how much is the average lateral electric field in a typical MOSFET? For a MOSFET with gate length $L_g = 1\ \mu m$ and $V_{DS} = 5.0\ V$, the average lateral electric field is 5 MV/m whereas for a MOSFET with $L_g = 100\ nm$ and $V_{DS} = 2.0\ V$, the lateral electric field is 20 MV/m. As you can see, the magnitude of the electric field is very high for the short-channel MOSFETs. The channel electrons gain high momentum and kinetic energy due to this lateral field and collide with the atoms exchanging momentum and energy. The collision may result in generation of an electron–hole pair due to the impact ionization mechanism if the electrons transfer sufficient energy to the atoms.

The application of a gate voltage also creates a longitudinal field in the channel region. The impact-generated electrons are, therefore, attracted by the longitudinal electric field and may enter the oxide and knock out electrons even in the oxide region [22]. However, the energy required by these hot electrons to enter the oxide region is quite high (~3.1 eV). The hot electrons entering the oxide layer degrade the insulating capability of the oxide and lead to a current through the gate. This gate leakage current results in a reduced input impedance and is undesirable for a good transistor action. The hot electrons may accumulate inside the oxide layer as charges and affect the threshold voltage.

1.7.3 Gate-Induced Drain Leakage

For understanding the gate-induced drain leakage (GIDL), we would like to introduce the concept of tunneling at this juncture. Tunneling is a quantum-mechanical phenomenon. The fact that light not only behaves as a particle but also as a wave was established by the classical experiments of Thomas Young [23]. Electrons show a similar behavior and undergo diffraction as conceived by Davisson and

Germer during their famous experiment, which gained them the noble prize for physics [23].

Now, as you may recall from freshman course on physics, the waves penetrate through the objects in their path unlike particles. Upon impinging on any object, some part of the wave is reflected while the remaining portion is transmitted. Depending on the object, there exists a finite probability that the wave would be transmitted through it. You would also remember that the larger the thickness of the object, the more the interaction between the wave and the object and the consequent attenuation (a decrease in the amplitude) in the output wave that comes out after penetrating through the object would also be significantly higher. Therefore, the transmission probability depends on the thickness of the object which poses itself as a barrier for the wave propagation.

For finding the exact solution for the transmission probability, one needs to solve the Schrödinger equation:

$$\left[-\frac{\hbar^2}{2m^*}\nabla^2 - q\varphi\right]\psi_i = E_i\psi_i \tag{1.7}$$

which gives the wave function (ψ_i). The probability of finding electrons is simply obtained by multiplying the wave function with its complex conjugate yielding the square of the magnitude of the wave function. The probability of finding an electron outside a potential barrier of a given height and length can be easily found by solving the Schrödinger equation in the different regions. This famous potential well problem is available in all standard text books [13–15].

Now, we shall talk about the implication of the wave nature of electrons in a reverse biased p–n junction. Let us take the case when the semiconductor is lightly doped on the p-side and the n-side. Figure 1.5(a) shows the energy band profiles of a p–n junction when the p and n sides are symmetrically doped to 10^{15} cm^{-3}. Let us

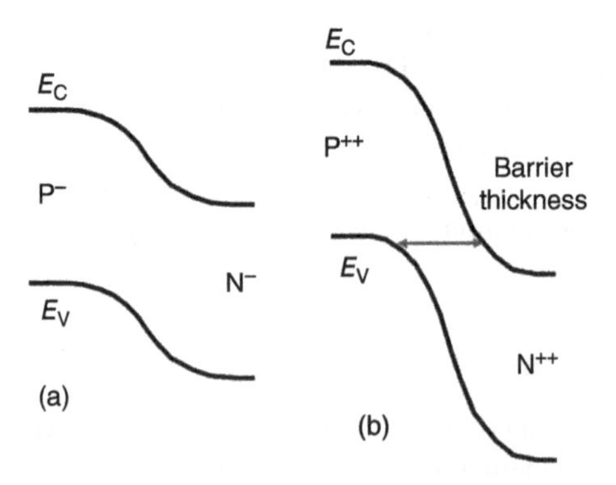

FIGURE 1.5 Energy band profile of symmetrically doped p–n junctions with (a) p and n doping concentration = 10^{15} cm^{-3} and (b) p and n doping concentration = 10^{20} cm^{-3}.

focus our attention on an electron in the valence band of the p-side. To become a free electron, jump to the conduction band and move to the n-side, the electron sees a potential barrier height equal to the band gap.

Now, we shall take the case of a heavily doped symmetrical p–n junction with doping equal to 10^{20} cm^{-3}. In this case also, let us focus our attention to an electron in the valence band on p-side. For this electron to become free and contribute to current conduction, it sees a potential barrier height equal to the band gap. However, the thickness of the potential barrier created (see Fig. 1.5(b)) is extremely thin as the depletion region width is significantly low for heavily doped p–n junctions.

Although this electron in the valence band may not get enough energy to surmount the potential barrier and move into the conduction band on the p-side, considering the wave nature of this electron, there is a finite probability that it may tunnel through the barrier since the barrier width is extremely small. The extremely thin barrier width offered by the ultrathin depletion region facilitates propagation of this electron wave directly into the conduction band on the n-side and become free. This phenomenon of transmission of electrons through potential barriers is called band-to-band tunneling (BTBT). However, BTBT is not significant in the first case since the potential barrier thickness is large. Therefore, it is not only the potential barrier height, which governs the conduction of electrons due to the drift and diffusion mechanism, but also the potential barrier width, which governs the conduction through the BTBT mechanism. Both these phenomena must be accounted for while analyzing the transport in any device.

However, the above analysis may arise a question in the minds of the readers: Will BTBT occur whenever the potential barrier is thin? To the dismay of readers, the answer is no. The quantum mechanical perspective of electron motion is quite different from the classical Newtonian perspective. According to the Bohr's theory, the electrons may occupy only specific energy "states" at discrete energy "levels." Without getting into details of the theory, we would like to mention that the electrons traverse only when they find an empty energy state corresponding to the energy level of that particular electron. This is quite intuitive and simply states that the electrons would tunnel only when they find empty energy states at that particular energy level. In case II shown earlier, the valence band electrons of p-side can easily tunnel into the conduction band of n-side since the conduction band contains energy states which are largely empty at room temperature. However, had this band alignment been between valence band of n- and p-side, no tunneling would have taken place since the valence band is nearly filled and hardly contains any empty states.

Let us focus our attention on the GIDL phenomenon in the MOSFETs. The process of fabrication of MOSFETs induces an inherent gate-on drain overlap region. The source/drain ion implantation process is performed after gate polysilicon (gate metal) deposition in the self-aligned gate CMOS process. In addition to the lateral straggle of ion-implanted dopant atoms, the thermal annealing process to reduce the defects and activate the dopants in the heavily doped source/drain regions also leads to lateral diffusion of dopant atoms. Therefore, the dopant atoms from source and drain

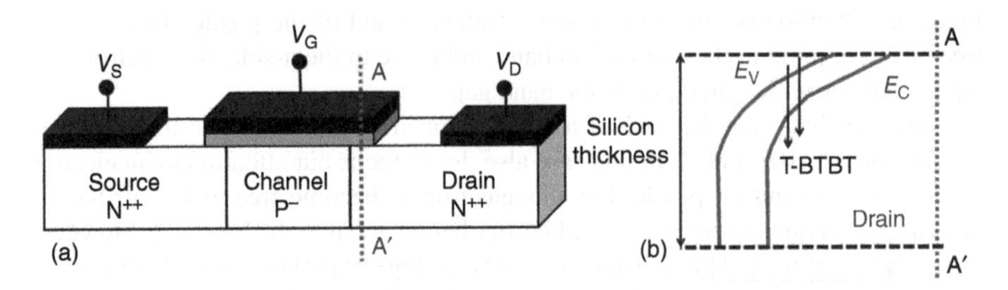

FIGURE 1.6 (a) Three-dimensional view of the fabricated MOSFET after the thermal annealing process and (b) energy band profiles along the cutline A-A′.

regions diffuse under the gate region. This leads to the gate-on source/drain overlap architecture, which exhibits a reduced channel length as compared to the defined gate length.

Now, let us consider the case of an n-MOSFET as shown in Fig. 1.6(a). When a positive drain voltage is applied and the gate voltage is negative ($V_{GS} \leq 0.0$ V), the drain region in the gate-on drain overlap area is depleted. If we analyze the band diagram along a vertical cutline as shown in Fig. 1.6(b), we can clearly observe that there is a large band bending in the drain region under the gate. Also, the band bending is sufficient to align the filled valence band of the drain region close to the surface with the empty conduction band of the drain region away from the surface. This band alignment facilitates BTBT of electrons within the drain region from the depleted surface to the bulk. This phenomenon is named transverse BTBT since this tunneling takes place in a direction perpendicular to the electron flow which is from the source to the drain. This BTBT significantly increases the leakage current of MOSFETs and is detrimental to the device performance [24–38].

Had drift-diffusion been the only mechanism governing the drain current of MOSFETs, the drain current would reduce with negative gate voltage due to the absence of carriers available in the channel region for current conduction. However, GIDL is the reason for an increase in the current in conventional MOSFETs for negative gate voltages [24–38].

Since GIDL originates due to gate-on drain overlap, it may be mitigated by properly designing the drain profile so that there is an exact alignment of the gate with the drain. Furthermore, the use of a lightly doped drain (LDD) close to the channel region may be beneficial for reduction of GIDL since the depletion region width would be larger in the LDD region [39]. A larger depletion region width would increase the tunneling width reducing the BTBT.

1.7.4 Direct Source to Drain Tunneling

In the preceding section, we discussed about GIDL in MOSFETs which arises due to BTBT. The tunneling takes place between the valence band and the conduction band in the gate-on drain overlap region. This kind of tunneling, which takes place

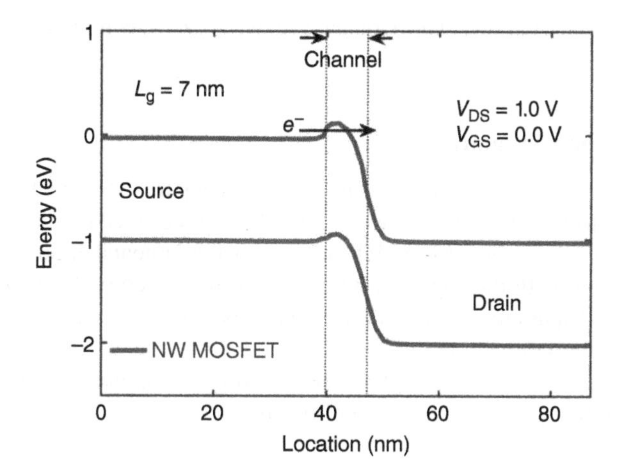

FIGURE 1.7 Energy band profile of a nanowire MOSFET with a gate length of 7 nm.

between two different energy bands (conduction to valence or vice versa), is termed as interband tunneling. However, there is yet another tunneling process that may take place between same type of bands, i.e. between conduction bands or valence bands. This tunneling phenomenon is termed as intraband tunneling.

If we carefully observe the band diagram of an ultrashort-channel ($L_g < 10$ nm) MOSFET in the OFF-state as shown in Fig. 1.7, we will notice that although the source–channel potential barrier is sufficient to prevent electrons from surmounting the barrier, the potential barrier width is significantly small owing to the ultrashort-channel length. The density of states increases with the energy level in the conduction band. However, the energy states at higher energy level in the conduction band have a negligible probability of occupancy. As a result, the conduction band electrons on the source side see empty energy states in the conduction band of the drain region. The ultrathin potential barrier, therefore, facilitates a tunneling of source conduction band electrons into the conduction band of the drain since both the conditions listed for BTBT are satisfied. This source to drain intraband tunneling is referred to as direct source to drain tunneling (DSDT) and is the most severe short-channel effect increasing the OFF-state current in ultrashort-channel ($L_g < 10$ nm) MOSFETs. The DSDT does not allow the ultrashort-channel MOSFETs to turn off. Reducing DSDT is a major challenge while designing FETs for the sub-10-nm regime [40–42].

One of the approaches proposed to mitigate DSDT is to use a different crystal orientation of silicon for active device layer. For instance, <211> silicon has a high effective carrier mass. Therefore, if <211> silicon is used in the active device layer of the MOSFETs, the tunneling probability and the consequent tunneling current would be reduced significantly. Therefore, DSDT can be mitigated by appropriately selecting the orientation or the channel material. Architectural-level designs that somehow increase the attenuation of the energy bands at the source–channel and channel–drain interface like gate sidewall spacer or gate underlap architectures may also reduce the DSDT.

1.7.5 Boltzmann Tyranny

For any device to work as a switch, we need a very steep transition in the current from the OFF-state to the ON-state. However, the thermionic injection over the source–channel barrier height in a MOSFET imposes a fundamental limit on the rate of change of current with the applied gate voltage as discussed in the beginning of the chapter. Let us discuss the inversion layer charge below the threshold voltage to gain a better insight into this limitation.

The inversion layer charge (Q_n) can be found by integrating the inversion layer charge density (n), which is a function of position:

$$Q_n = \int n(y)dy \tag{1.8}$$

Now, $n(y)$ may be obtained from the surface potential as

$$n(y) = N_\mathrm{C} e^{\frac{(E_f - E_\mathrm{C}(y))}{kT}} \tag{1.9}$$

where N_C is the density of states in the conduction band. The conduction band energy may be expressed in terms of the surface potential as

$$E_\mathrm{C}(y) = E_\mathrm{C}(\mathrm{bulk}) - q\varphi_\mathrm{S}(y) \tag{1.10}$$

where φ_S is the surface potential. Utilizing equations (1.10) in (1.9), we obtain

$$n(y) = N_\mathrm{C} e^{\frac{(E_f - E_\mathrm{C}(\mathrm{bulk}))}{kT}} e^{\frac{q\varphi_\mathrm{S}(y)}{kT}} = n_\mathrm{B} e^{\frac{q\varphi_\mathrm{S}(y)}{kT}} \tag{1.11}$$

where n_B is the minority carrier concentration given by n_i^2/N_A.

Now, we may change the variable of integration in equation (1.8) from position to surface potential as

$$Q_n = \int n_\mathrm{B} e^{\frac{q\varphi_\mathrm{S}(y)}{kT}} d\varphi_\mathrm{S} \left(\frac{dy}{d\varphi_\mathrm{S}}\right) \tag{1.12}$$

Now, we assume that the electric field (E_S) within the inversion layer is constant because the entire inversion layer is located at the surface, we obtain:

$$Q_n = \frac{qn_\mathrm{B}}{E_\mathrm{S}} \int e^{\frac{q\varphi_\mathrm{S}(y)}{kT}} d\varphi_\mathrm{S} \tag{1.13}$$

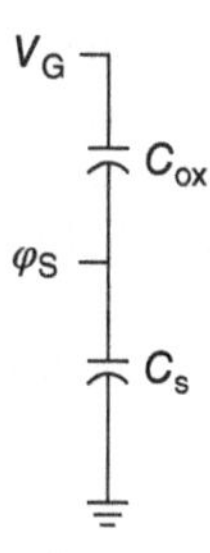

FIGURE 1.8 The effective capacitor divider network representation of the gate electrode, gate oxide, and the semiconductor channel.

Without getting into detailed mathematical analysis, we simply predict that the inversion layer charge depends exponentially on the surface potential. Now, how can we relate surface potential to the gate voltage? We introduce a simple capacitor divider principle to relate the gate voltage with the surface potential.

As shown in Fig. 1.8, the surface potential is simply the voltage that falls across the semiconductor capacitance (C_S). Therefore, the relation between φ_S and V_G can be given as

$$\varphi_S = V_G \left(\frac{C_{ox}}{C_{ox} + C_S} \right) = \frac{V_G}{m} \tag{1.14}$$

In the subthreshold regime, the semiconductor is essentially in the depletion mode. Therefore, the semiconductor capacitance can be approximated by a depletion capacitance. Using this simple relationship between φ_S and V_G, we can clearly see that the inversion layer charge density and, hence, the drain current varies exponentially with the gate voltage as

$$I_D \sim Q_n \sim e^{\frac{qV_G}{mkT}} \tag{1.15}$$

Now, we introduce the concept of the subthreshold swing, which is defined as the amount of voltage required to change the subthreshold current by one order of magnitude (one decade). For a conventional MOSFET, using equation (1.15), the subthreshold slope can be given as

$$\frac{dV_G}{d(\log_{10} I_D)} = \frac{2.3mkT}{q} \approx 60m \text{ mV/decade at } T = 300 \text{ K} \tag{1.16}$$

Now, since $m = 1 + C_S/C_{ox}$, and both C_S and C_{ox} are positive, the value of subthreshold swing is always more than 60 mV/decade. This limitation on the minimum value of the attainable subthreshold swing is called the Boltzmann tyranny. For changing the drain current by one order of magnitude, a voltage equal to 60 mV must be supplied at the gate even if C_S/C_{ox} is assumed to be negligible. This restricts a steep transition from the OFF-state to the ON-state in a MOSFET. The lower the value of subthreshold swing, the steeper is the transition from the OFF-state to the ON-state, and the better is the switch. Therefore, most of the present-day research on MOSFETs is focused on reducing the subthreshold swing below 60 mV/decade. It may be noted that this limit arises out of the fundamental thermionic emission mechanism inherent to the MOSFETs as the current is inversely proportional to the exponential of the barrier height.

The lower limit of 60 mV/decade on the subthreshold swing of MOSFETs also hinders the scaling of their supply voltage. For effective operation of MOSFETs as a switch, the ON-state current to OFF-state current ratio should be sufficient (so that the states are easily distinguishable), at least 10^4. However, if a low supply voltage is utilized, the current swing between the ON-state and the OFF-state would be compromised owing to the limitation on the minimum attainable subthreshold swing. Therefore, the Boltzmann tyranny limits the efficiency of the conventional MOSFETs as switch at low supply voltages. To obtain high I_{ON}/I_{OFF} even at lower supply voltages which is the prime goal of the device designers, the subthreshold slope should be as low as possible. It should be close to 0 mV/decade for abrupt turning on of the FETs. Therefore, the quest for ultralow power steep subthreshold swing devices is underway.

1.7.6 Ultrasteep Doping Profile

The MOSFET contains two p–n junctions: one at the source–channel interface and the other at the channel–drain interface. The doping at the source–channel and channel–drain junction should change abruptly from a high value (typically $\sim 10^{20}$ cm^{-3}) at the source and drain regions to a low value (typically $\sim 10^{15}$ to 10^{17} cm^{-3}) with complimentary dopants in the channel region. Otherwise, the effective channel length would be significantly reduced as compared to the drawn channel length and this would increase the short-channel effects considerably. However, realizing such an ultrasteep doping profile is extremely difficult.

The ion-implantation process employed for doping the source/drain regions inherently leads to a stochastic distribution of dopant atoms (leading to lateral straggle) and creates defect centers as the dopant ions are bombarded onto the semiconductor film [43–45]. The typical value of the doping gradient between source/drain region and the channel region in MOSFETs doped using ion implantation ranges between 2 and 3 nm/decade [43–45]. This simply means that the doping can be changed, for instance, from 10^{20} cm^{-3} in the source/drain region to 10^{17} cm^{-3} in the channel region within 5–10 nm. Therefore, in any case 5–10 nm of the channel region would be consumed by the source/drain doping and unintentionally doped higher than the channel doping concentration. The effective channel length reduces by 5–10 nm from the drawn channel length due to the inability to form abrupt source/drain doping profiles.

Also, the dopant activation in the heavily doped source/drain regions requires a high-temperature annealing which in turn leads to a thermally assisted lateral diffusion of dopant atoms from source/drain regions into the channel region, minimizing the possibility of realizing ultrasteep doping profiles [46]. This puts a complex constraint on the thermal budget as lateral diffusion is inevitable while annealing. Therefore, development of alternative doping techniques and ultrafast annealing systems is essentially required for realizing ultrasteep doping profiles.

1.8 CONCLUSION

In this chapter, we discussed the fundamentals of a MOSFET. We saw how MOSFETs work as switch and can be wired together to form circuits. We also analyzed the power dissipation in circuits made from CMOS. The need for scaling and the scaling rules were also discussed in detail. We saw how MOSFETs continued to shrink in size following the Moore's law. However, it is indeed the Koomey's law which is more fundamental. The various short-channel effects such as DIBL, threshold voltage roll-off, hot electron effects, DSDT, GIDL, and so on, which degrade the performance of the MOSFETs upon scaling were also examined carefully. A brief overview of the Boltzmann tyranny in MOSFETs was provided. This chapter lays the basis for the discussions presented in rest of the book. In the next chapter, we look at the different device architectures proposed to mitigate the challenges faced by the conventional MOSFETs. We also discuss the alternate conduction mechanisms, which may be useful to overcome the fundamental Boltzmann limit of the conventional MOSFETs.

REFERENCES

[1] Apple discussion forum [online]. Available: https://apple.stackexchange.com/questions/194367/does-the-apple-watch-have-more-processing-power-than-a-cray-2-supercomputer, Accessed Dec. 23, 2017.

[2] W. H. Brattain and B. John, "Three-electrode circuit element utilizing semi conductive materials," U.S. Patent 2524035, Oct. 1950.

[3] J. Bardeen and W. H. Brattain, "The transistor, a semiconductor triode," *Proc. IEEE*, vol. 86, no. 1, pp. 29–30, Jan. 1998.

[4] I. M. Ross, "The invention of the transistor," *Proc. IEEE*, vol. 86, no. 1, pp. 7–28, Jan. 1998.

[5] M. Riordan, L. Hoddeson, and C. Herring, "The invention of the transistor," *Rev. Mod. Phys.*, vol. 71, no. 2, pp. S336, Mar. 1999.

[6] J. M. Rabaey, A. P. Chandrakasan, and B. Nikolic, *Digital Integrated Circuits*, Englewood Cliffs, NJ: Prentice-Hall, 2002.

[7] B. Jamtveit and P. Meakin, "Growth, dissolution and pattern formation in geosystems," in *Growth, Dissolution and Pattern Formation in Geosystems*, pp. 1–19, Springer, Dordrecht, the Netherlands, 1999.

[8] A. Ortiz-Conde, F. G. Sánchez, J. J. Liou, A. Cerdeira, M. Estrada, and Y. Yue, "A review of recent MOSFET threshold voltage extraction methods," *Microelectron. Rel.*, vol. 42, no. 4, pp. 583–596, Apr. 2002.

[9] S. Saurabh and M. J. Kumar, *Fundamentals of Tunnel Field-Effect Transistors*, CRC Press (Taylor & Francis Group), Boca Raton, FL, 2016.

[10] Moore's law [online]. Available: https://en.wikipedia.org/wiki/Moore%27s_law#/media/File:Moore%27s_Law_Transistor_Count_1971-2016.png, Accessed Dec. 23, 2017.

[11] Intel Newsroom [online]. Available: http://download.intel.com/pressroom/pdf/computertrendsrelease.pdf, Accessed Dec. 23, 2017.

[12] A. Chaudhry and M. J. Kumar, "Controlling short-channel effects in deep submicron SOI MOSFETs for improved reliability: A review," *IEEE Trans. Dev. Mater. Rel.*, vol. 4, pp. 99–109, Mar. 2004.

[13] Y. Taur and T. H. Ning, *Fundamentals of Modern VLSI Devices*, Cambridge University Press, New York, 2013.

[14] S. M. Sze and K. K. Ng, *Physics of Semiconductor Devices*, John Wiley & Sons, Inc., Hoboken, NJ, 2007.

[15] B. G. Streetman and S. K. Banerjee, *Solid State Electronic Devices*, Pearson Education, 2016.

[16] P. K. Chatterjee and J. E. Leiss, "An analytic charge-sharing predictor model for submicron MOSFETs," in *Proc. IEDM*, pp. 28–33, 1980.

[17] P. K. Chatterjee, W. R. Hunter, T. C. Holloway, and Y. T. Lin, "The impact of scaling laws on the choice of n-channel or p-channel for MOS VLSI," *IEEE Electron Dev. Lett.*, vol. 1, no. 10, pp. 220–223, Oct. 1980.

[18] R. R. Troutman, "VLSI limitations from drain-induced barrier lowering", *IEEE J. Solid-State Circuits*, vol. 14, no. 2, pp. 383–391, Apr. 1979.

[19] T. Toyabe and S. Asai, "Analytical models of threshold voltage and breakdown voltage of short-channel MOSFET's derived from two-dimensional analysis," *IEEE Trans. Electron Devices*, vol. 26, no. 4, pp. 453–461, Apr. 1979.

[20] J. J. Barnes, K. Shimohigashi, and R. W. Dutton, "Short-channel MOSFETs in the punch through current mode," *IEEE J. Solid-State Circuits*, vol. 14, no. 2, pp. 368–375, Apr. 1979.

[21] R. T. Jerdonek, W. R. Bandy, and J. Birnbaum, "A model for the submicrometer n-channel deep-depletion SOS/MOSFET," *IEEE Trans. Electron Devices*, vol. 27, no. 8, pp. 1566–1570, Aug. 1980.

[22] S. Tam, F. C. Hsu, P. K. Ko, C. Hu, and R. S. Muller, "Hot-electron induced excess carriers in MOSFET's," *IEEE Electron Device Lett.*, vol. 3, no. 12, pp. 376–378, 1982.

[23] M. J. Kumar, R. Vishnoi, and P. Pandey, *Tunnel Field-effect Transistors (TFET): Modelling and Simulation*, John Wiley and Sons Ltd, West Sussex, UK, 2016.

[24] J. Fan, M. Li, X. Xu, Y. Yang, H. Xuan, and R. Huang, "Insight into gate-induced drain leakage in silicon nanowire transistors," *IEEE Trans. Electron Devices*, vol. 62, no. 1, pp. 213–219, Jan. 2015.

[25] J. Hur, B.–H. Lee, M.–H. Kang, D.–C. Ahn, T. Bang, S.–B. Jeon, and Y.–K. Choi, "Comprehensive analysis of gate-induced drain leakage in vertically stacked nanowire FETs: Inversion-mode vs. junctionless mode," *IEEE Electron Device Lett.*, vol. 37, no. 5, pp. 541–544, May 2016.

[26] S. Sahay and M. J. Kumar, "Physical insights into the nature of gate-induced drain leakage in ultrashort channel nanowire FETs," *IEEE Trans. Electron Devices*, vol. 64, no. 6, pp. 2604–2610, June 2017.

[27] S. Sahay and M. J. Kumar, "A novel gate-stack-engineered nanowire FET for scaling to the sub-10-nm regime," *IEEE Trans. Electron Devices*, vol. 63, no. 12, pp. 5055–5059, Dec. 2016.

[28] S. Sahay and M. J. Kumar, "Spacer design guidelines for nanowire FETs from gate-induced drain leakage perspective," *IEEE Trans. Electron Devices*, vol. 64, no. 7, pp. 3007–3015, July 2017.

[29] S. Sahay and M. J. Kumar, "Insight into lateral band-to-band-tunneling in nanowire junctionless FETs," *IEEE Trans. Electron Devices*, vol. 63, no. 10, pp. 4138–4142, Oct. 2016.

[30] S. Sahay and M. J. Kumar, "Controlling L-BTBT and volume depletion in nanowire JLFETs using core-shell architecture," *IEEE Trans. Electron Devices*, vol. 63, no. 9, pp. 3790–3794, Sept. 2016.

[31] S. Sahay and M. J. Kumar, "Diameter dependency of leakage current in nanowire junctionless field-effect transistors," *IEEE Trans. Electron Devices*, vol. 64, no. 3, pp. 1330–1335, Mar. 2017.

[32] S. Sahay and M. J. Kumar, "Nanotube junctionless FET: Proposal, design, and investigation," *IEEE Trans. Electron Devices*, vol. 64, no. 4, pp. 1851–1856, Apr. 2017.

[33] M. J. Kumar and S. Sahay, "Controlling BTBT induced parasitic BJT action in junctionless FETs using a hybrid channel," *IEEE Trans. Electron Devices*, vol. 63, no. 8, pp. 3350–3353, Aug. 2016.

[34] S. Sahay, and M. J. Kumar, "Realizing efficient volume depletion in SOI junctionless FETs," *IEEE J. Electron Devices Soc.*, vol. 4, no. 3, pp. 110–115, May 2016.

[35] S. Sahay and M. J. Kumar, "Symmetric operation in an extended back gate JLFET for scaling to the 65 nm regime considering quantum confinement effects," *IEEE Trans. Electron Devices*, vol. 64, no. 1, pp. 21–27, Jan. 2017.

[36] A. K. Jain, S. Sahay, and M. J. Kumar, "Controlling L-BTBT in emerging nanotube FETs using dual-material gate," *IEEE J. Electron Dev. Soc.*, vol. 6, pp. 611–621, June 2018.

[37] V. Nathan and N. C. Das, "Gate-induced drain leakage currents in MOS devices," *IEEE Trans. Electron Devices*, vol. 40, no. 10, pp. 1888–1890, Oct. 1993.

[38] T. Hoffmann, G. Doornbos, I. Ferain, N. Collaert, P. Zimmerman, M. Goodwin, R. Rooyackers, A. Kottantharayil, Y. Yim, A. Dixit, K. De Meyer, M. Jurczak, and S. Biesemans, "GIDL (gate induced drain leakage) and parasitic Schottky barrier leakage elimination in aggressively scaled HfO_2/TiN FinFET devices," in *IEDM Tech. Dig.*, pp. 725–729, 2005.

[39] S. Ogura, P. J. Tsang, W. W. Walker, D. L. Critchlow, and J. F. Shepard, "Design and characteristics of the lightly doped drain–source (LDD) insulated gate field-effect transistor," *IEEE Trans. Electron Devices*, vol. 27, no. 8, pp. 1359–1367, Aug. 1980.

[40] H. Kawaura, T. Sakamoto, and T. Baba, "Observation of source-to-drain direct tunneling current in 8 nm gate electrically variable shallow junction metal–oxide–semiconductor field-effect transistors," *Appl. Phys. Lett.*, vol. 76, no. 25, pp. 3810–3812, June 2000.

[41] R. A. Vega and T. J. K. Liu, "Dopant-segregated Schottky source/drain double-gate MOSFET design in the direct source-to-drain tunneling regime," *IEEE Trans. Electron Devices*, vol. 56, no. 9, pp. 2016–2026, Sept. 2009.

[42] W. S. Cho and K. Roy, "The effects of direct source-to-drain tunneling and variation in the body thickness on (100) and (110) sub-10-nm Si double-gate transistors," *IEEE Electron Device Lett.*, vol. 36, no. 5, pp. 427–429, May 2015.

[43] J. F. Gibbons, "Ion implantation in semiconductors—Part I: Range distribution theory and experiments," *Proc. IEEE*, vol. 56, no. 3, pp. 295–319, Mar. 1968.

[44] J. F. Gibbons, "Ion implantation in semiconductors—Part II: Damage production and annealing," *Proc. IEEE*, vol. 60, no. 9, pp. 1062–1096, Sept. 1972.

[45] S. Furukawa, H. Matsumura, and H. Ishiwara, "Theoretical considerations on lateral spread of implanted ions," *Jap. J. Appl. Phys.*, vol. 11, no. 2, pp. 134, Feb. 1972.

[46] J.-P. Colinge, C.-W. Lee, A. Afzalian, N. D. Akhavan, R. Yan, I. Ferain, P. Razavi, B. O'Neill, A. Blake, M. White, A.-M. Kelleher, B. McCarthy, and R. Murphy, "Nanowire transistors without junctions," *Nature Nanotechnol.*, vol. 5, no. 3, pp. 225–229, Mar. 2010.

2

EMERGING FET ARCHITECTURES

In the preceding chapter, we discussed about the fundamentals of the metal-oxide-semiconductor field-effect transistors (MOSFET). We saw how the various short-channel effects, the inherent thermionic emission mechanism, and the inability to realize ultrasharp doping profiles degrade the performance of the conventional MOSFETs. These limitations hinder the scaling of MOSFETs to the sub-10 nm regime. Several device architectures have been proposed to mitigate these challenges. Some of these emerging field-effect transistor (FET) architectures rely on alternate conduction mechanisms such as band-to-band tunneling (BTBT), impact ionization, and so forth to achieve steep switching, whereas others tend to improve the electrostatic control of the channel region. In this chapter, we discuss these emerging FET architectures proposed to replace the conventional MOSFETs in detail. We begin our discussion by analyzing tunnel FETs (TFETs), impact ionization MOSFETs (I-MOS), and ferroelectric FETs, which exhibit a sub-60 mV/decade subthreshold swing followed by a discussion on the architectural design improvements such as two-dimensional (2D) materials for channel, nanowires, and nanotube architectures to sustain the scaling of the conventional MOSFETs beyond the 10-nm technology node.

Junctionless Field-Effect Transistors: Design, Modeling, and Simulation, First Edition.
Shubham Sahay and Mamidala Jagadesh Kumar.

2.1 TUNNEL FETs

The life of the device designers would be greatly simplified if a complementary metal-oxide-semiconductor (CMOS) compatible novel device with steep switching characteristics could be realized. TFETs emerged as a result of the efforts of the researchers in this direction. Since then, TFETs have attracted much attention as an alternative to the MOSFETs for low-power applications.

2.1.1 Structure

The three-dimensional (3D) view of a TFET is shown in Fig. 2.1. The TFET is essentially a reverse biased gated p–i–n diode [1–40]. The n-TFET consists of a p^+ source, an intrinsic channel with a gate to modulate its potential, and n^+ drain. The source region needs to be heavily doped ($>10^{20}$ cm^{-3}) to achieve a high tunneling efficiency. Also, the doping profile at the source–channel interface must be abrupt to achieve a large band bending and increase the tunneling probability. Realizing an abrupt junction at the source–channel interface requires ultrafast annealing techniques such as spike anneal or microwave annealing.

2.1.2 Operation

The TFETs work on the principle of interband tunneling. The gate effectively modulates the tunneling efficiency at the source–channel interface as shown in Fig. 2.2. In TFETs, as the gate voltage is increased, the intrinsic channel enters the strong inversion regime in the ON-state ($V_{GS} = 1.0$ V). Under this condition, the band profiles of the TFET are shown in Fig. 2.2(a). As can be observed from Fig. 2.2(a), there exists a considerable spatial proximity between the valence band of the source region and the conduction band of the channel region. This facilitates a BTBT of electrons from source to the channel region leading to a drain current. The gate modulates the band bending of the channel region and dictates the tunneling length (λ).

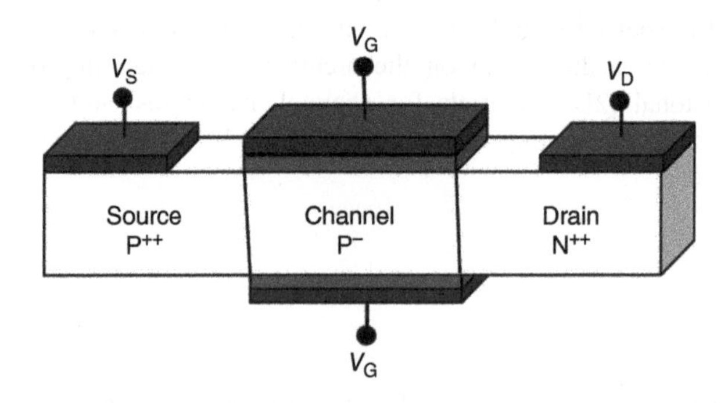

FIGURE 2.1 Three-dimensional view of a double gate n-TFET.

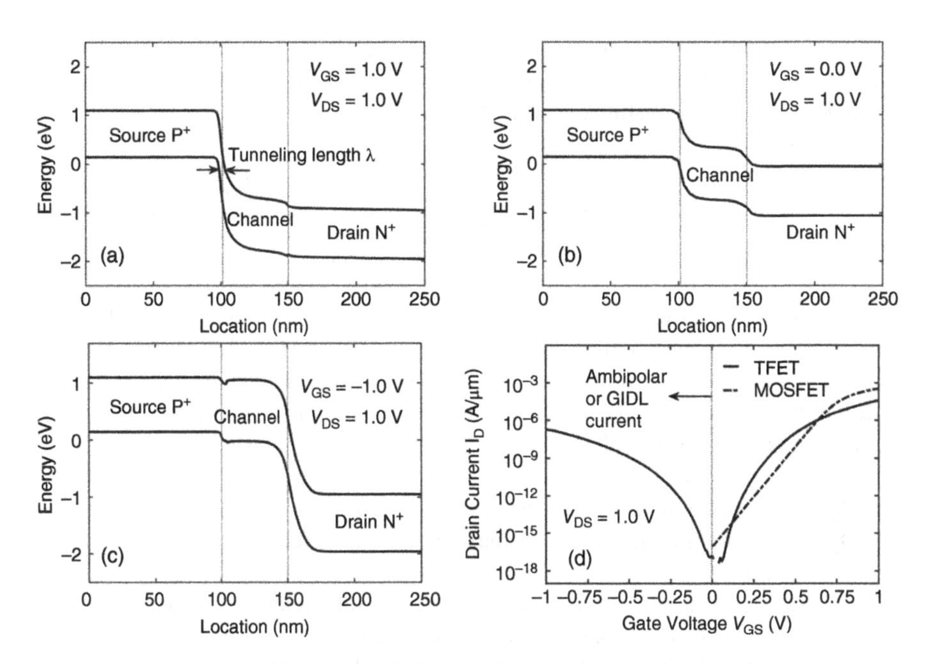

FIGURE 2.2 Energy band profiles of the double-gate tunnel field-effect transistor at a cutline 1 nm below the Si–SiO_2 interface in (a) ON-state ($V_{GS} = 1.0$ V), (b) OFF-state ($V_{GS} = 0.0$ V), (c) ambipolar state ($V_{GS} = -1.0$ V), and (d) transfer characteristics of the TFET and MOSFET.

In the OFF-state ($V_{GS} = 0.0$ V), the valence band of the source region overlaps the forbidden gap of the channel region. Therefore, the possibility of tunneling is eliminated as shown in Fig. 2.2(b). Since TFET is essentially a reverse biased p–i–n diode, in the OFF-state, the drain current is simply the leakage current of a p–i–n diode, which is extremely low as shown in Fig. 2.2(d) ($\sim 10^{-15}$ to 10^{-17} A/μm). It is also evident from Fig. 2.2(b) that the energy band profile of TFETs in the OFF-state increase monotonically. As a result, there is lesser impact of short-channel effects such as drain-induced barrier lowering (DIBL) on TFETs.

From a potential perspective, in the OFF-state, the whole drain voltage falls across the intrinsic channel region (offering the maximum series resistance) and across the source–channel and channel–drain depletion region widths. However, in the ON-state due to the formation of an inversion layer, the series resistance of the channel region is reduced significantly and the entire drain voltage falls across the source–channel and channel–drain depletion region width. This results in a large lateral electric field at the source–channel interface leading to a large band bending. The large band bending leads to the spatial proximity between the source valence band and the channel conduction band, resulting in injection of electrons into the channel region as shown in Fig. 2(a). The tunneled electrons then flow via drift–diffusion through the channel to the drain.

The tunneling efficiency at the source–channel junction (approximated as a triangular potential barrier) can be calculated using Wentzel–Kramers–Brillouin (WKB) approximation. The probability of tunneling is given as [3–7]

$$I_{DS} \approx T_{\mathrm{WKB}} \approx \exp\left(\frac{-4\sqrt{2m^*}E_g^{3/2}}{3q\hbar E}\right) \approx \exp\left(\frac{-4\lambda\sqrt{2m^*}E_g^{3/2}}{3\hbar(\Delta\varphi + E_g)}\right) \tag{2.1}$$

where m^* is the effective carrier mass, E_g is the band gap, E is the electric field at the source–channel tunnel junction, λ is the tunneling length (Fig. 2.2(a)) and depends significantly on the electric field as it describes the spatial extent of the transition region at the source–channel interface, $\Delta\phi$ is the tunneling window over which the tunneling takes place, $\hbar$ is the reduced Planck's constant, and q is the elementary charge. We can already make few inferences from equation (2.1). The ON-state current can be increased by using a material with low effective carrier mass or low band gap or by utilizing architectures that may increase the electric field at the source–channel interface.

To understand how TFETs can exhibit a subthreshold swing less than 60 mV/decade, let us focus our attention on Fig. 2.3. As discussed in Section 1.7.5, in case of MOSFETs, the electrons with an energy higher than the source–channel barrier height, which can be identified as the electrons at the tail of the Fermi–Dirac distribution in Fig. 2.3(a) easily surmount the barrier and contribute to the leakage current. These high-energy electrons at the tail of the Fermi–Dirac distribution restrict the subthreshold swing of the MOSFETs to 60 mV/decade. However, in case of TFETs, as shown in Fig. 2.3(b), the energy level of the electrons at the tail of the Fermi–Dirac distribution lies in the forbidden gap of the source where no energy states are present. As a result, these highly energetic electrons in the source do not contribute to current conduction. Similarly, the electrons with low-energy level in the valence band of the source region (with probability of occupancy close to 1)

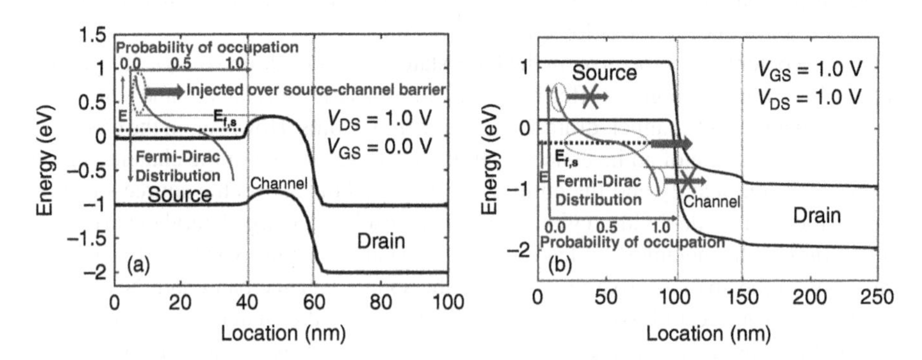

FIGURE 2.3 Energy band profiles at a cutline 1 nm below the Si–SiO_2 interface in (a) MOSFET and (b) TFET.

cannot tunnel into the channel region since there are no available energy states in the forbidden gap of the channel region (refer Fig. 2.3(b)). Only the electrons with energy level close to the Fermi-level in the source can tunnel into the channel region and contribute to the current conduction. Therefore, there is a band pass filtering of source electrons in the energy domain in TFETs. Moreover, the probability of occupancy of electrons changes dramatically near the Fermi-level in the Fermi–Dirac distribution leading to a steep change in the current and a subthreshold swing below 60 mV/decade in TFETs. Therefore, subthreshold swing below 60 mV/decade has been demonstrated experimentally for TFETs.

2.1.3 Challenges

Although TFETs have attracted much attention of the researchers owing to the low leakage current, steep subthreshold swing, better immunity to short-channel effects, and compatibility with the conventional CMOS process, their application is limited due the factors discussed below.

2.1.3.A Asymmetric Structure As discussed in Section 2.1.1, a TFET is a p–i–n structure unlike n–i–n structure of a MOSFET. Since the MOSFET architecture is symmetric, it offers flexibility while selecting the n^+ region as the source or drain region and the electrodes can be interchanged without affecting the operation of the device as n-MOSFET. This unique property of MOSFETs allows the circuit layout designers to reduce the layout area using simple techniques such as contact sharing. However, such a provision cannot be exploited in TFETs. Therefore, the circuits utilizing TFETs would consume a large area as compared to MOSFETs in the layout. This poses a technological challenge while scaling the size of the integrated circuits made from TFETs.

2.1.3.B Low ON-State Current The ON-state current is limited in the TFETs by the tunneling efficiency at the source–channel junction. The ON-state current of the TFETs is smaller than the MOSFETs as shown in Fig. 2.2(d). As can be observed from equation (2.1), the ON-state current may be improved by reducing the tunneling distance. This may be achieved by increasing the electric field at the tunnel junction. An abrupt doping profile with a high source doping concentration is required for a higher electric field at the source–channel interface. This puts a stringent constraint on the annealing techniques used for dopant activation since thermal annealing would lead to lateral diffusion and a consequent graded doping profile instead of the much needed abrupt profile. Techniques such as spike annealing or microwave annealing were proposed for realizing ultrasteep doping profiles.

Also, different architecture-level techniques have been proposed in the literature to increase the ON-state current including the use of a high-κ gate dielectric [7], an n^+ pocket at the source–channel interface yielding a p–n–p–n TFET [22–24], low band gap material at the source side [25], and so on. The presence of a high-κ gate dielectric increases the gate coupling and hence the vertical electric field leading to

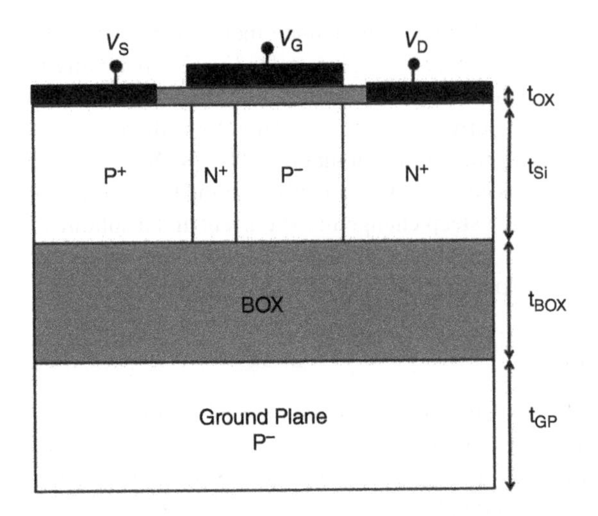

FIGURE 2.4 Schematic view of a N^+-pocket (p–n–p–n) TFET.

an increased band bending and a consequent lower tunneling width increasing the ON-state current (see equation 2.1). A reduction in the band gap leads to a higher tunneling probability as can be observed from equation (2.1) leading to a larger ON-state current.

The application of an n^+ pocket at the source–channel tunneling interface leads to a larger electric field at the tunneling junction. This can be simply understood from the fact that the peak electric field at the heavily doped p^+–n^+ junction is larger than the peak electric field at the p^+–n junction. As a result, inclusion of an n^+ pocket of appropriate thickness such that it is depleted completely in the ON-state leads to a larger band bending owing to the higher electric field and a consequent higher tunneling probability in the ON-state. PNPN TFETs (Fig. 2.4) have been reported to have a steeper subthreshold slope, a reduced operating voltage, and an improved reliability compared to the p–i–n TFETs in addition to increased ON-state current [22–24]. However, even after utilizing these techniques, the ON-state current of TFET remains lower than the conventional MOSFETs leading to a higher intrinsic delay and a lower speed. Therefore, III–V heterojunction TFETs were also explored for increased ON-state current [26–31]. The III–V materials offer a smaller effective tunneling mass and a smaller band gap which increases the tunneling probability. In addition, III–V materials are direct band gap materials, and the momentum of the electrons at the bottom of the conduction band is same as the electron momentum at the top of the valence band. On the other hand, silicon and germanium are indirect band gap materials implying that the momentum at the bottom of the conduction band is not the same as the top of the valence band. As a result, the momentum needs to be changed for transition of electrons from the top of the valence band to the bottom of the conduction band. The extra energy required for this momentum change is obtained from collisions with the crystal lattice and reduces the tunneling probability in silicon and germanium TFETs.

The III–V TFETs with a direct band gap do not require any momentum change or additional energy and, therefore, offer a higher ON-state current. However, integration of III–V materials with conventional CMOS is a technological challenge. Furthermore, in the fabricated III–V TFETs, a subthreshold slope below 60 mV/decade is achieved only for a low range of drain currents [26–31].

2.1.3.C Ambipolar Conduction Until now, we have focused only on two states of TFET, i.e., the ON-state and the OFF-state. Interestingly, in a TFET, when a negative gate voltage is applied, there is a significant band bending at the channel–drain interface, which leads to a spatial overlap between the valence band of the channel region and the conduction band of the drain region. This band alignment favors the tunneling at the channel–drain interface as shown in Fig. 2.2(c). This leads to a gate-controlled drain current even when the polarity of gate voltage is reversed as shown in Fig. 2.2(d). This unique property of TFETs allows them to conduct even for negative gate voltage, and the behavior is termed as an ambipolar conduction. The ambipolar conduction limits the utility of TFETs for digital circuit applications [32–39]. To alleviate the ambipolar behavior, several device architectures including gate–drain underlap, low drain doping, lateral heterostructure with high bandgap material at the drain side, low-κ spacers, and back gate engineering have been proposed [32–39].

A reduction in the drain doping leads to an extension of the channel–drain depletion region width even on the drain side increasing the tunneling width. Therefore, the tunneling probability and hence the ambipolar current is suppressed. However, a lower drain doping is undesirable for realizing ohmic contacts and low drain series resistance [25]. Since the tunneling probability reduces with an increase in the band gap according to equation (2.1), utilizing a material with a large band gap in the drain region leads to a suppressed ambipolar behavior. However, the experimental realization of such heterostructures increases the process complexity.

Since the ambipolar conduction essentially arises due to the gate-controlled band bending at the channel–drain interface, gate–drain underlap architecture can significantly suppress the ambipolar conduction. However, it increases the device footprint as the gate–drain distance is usually larger than the electrostatic isolation length required for the contacts [35]. The ON-state current is also reduced due to the increased channel series resistance, and current saturation is not achieved in the output characteristics. Another method to minimize the impact of gate on the channel–drain interface is to replace the gate-dielectric toward the drain side of the channel by a low-κ material while utilizing a high-κ material on the source side [34]. This reduces the tunneling length and, hence, the ambipolar conduction.

The extension of the gate region over the drain effectively depletes the drain region at the channel–drain interface leading to a larger tunneling width. Therefore, the ambipolar conduction is reduced in TFETs with a gate-on drain overlap. However, the increase in the Miller capacitance and the large overlap length required to mitigate ambipolar conduction reduce the scaling flexibility of this approach [38]. Another technique to deplete the drain region at the channel drain interface is by utilizing a heterodielectric buried oxide (BOX) in the silicon-on-insulator (SOI) architecture with a heavily doped ground plane. The heavily doped ground plane acts like a metal with a high work function and coupled with the lower effective oxide thickness of the high-κ dielectric BOX below the drain region; it effectively depletes the drain region at the channel–drain interface increasing the tunneling width on the drain side [39]. This leads to a suppressed ambipolar conduction.

2.1.3.D Improved Subthreshold Swing for Limited Current Range In most of the experimentally reported TFETs, the optimum subthreshold swing below 60 mV/decade is obtained only for low current values and that too for only 3–4 decades of drain current [26–31]. The subthreshold swing degrades with increasing drain current values. The TFETs exhibit the lowest subthreshold swing at the voltage where the valence band of the source just overlaps the conduction band of the channel region. As the gate voltage is increased above this point, the tunneling phenomenon is not limited by the tunneling distance (or the tunneling width modulation) but by the limited density of states (DOS) available in the channel region for electrons to tunnel. Therefore, to obtain low subthreshold slope at high current values, alternate materials or systems with large DOS should be chosen.

2.1.3.E Trap-Assisted Tunneling In many fabricated TFETs, subthreshold swing higher than 60 mV/decade has also been observed although negative differential resistance, which indicates BTBT as the conduction mechanism, was observed in the output characteristics [29]. This is attributed to the trap-assisted tunneling (TAT) mechanism [29] or Auger generation [40]. The electrons may tunnel into the traps and subsequently get injected in the channel region via thermal processes. TAT is responsible for poor subthreshold swing behavior of TFETs. TAT is a major challenge for TFETs, and TFETs may be viable for low-power application only once TAT is eliminated.

2.2 IMPACT IONIZATION MOSFET

In Section 2.1.2, we discussed how changing the conduction mechanism from a thermionic injection in MOSFETs to interband tunneling in TFETs enables realization of sub-60 mV/decade subthreshold swing. Apart from BTBT, impact ionization is yet another conduction mechanism which may facilitate abrupt transition from OFF-state to ON-state. The avalanche breakdown of the p–n diodes due to impact ionization–induced positive feedback leads to a sharp increase in the current within a short span of applied voltage close to the breakdown voltage. If somehow the breakdown phenomenon may be controlled using a gate electrode, abrupt transitions between ON-state and OFF-state may be obtained in FETs.

2.2.1 Structure

The impact I-MOS is also a reverse biased p–i–n diode with a gated intrinsic region [41–53]. However, as shown in Fig. 2.5, the gate electrode does not span over the

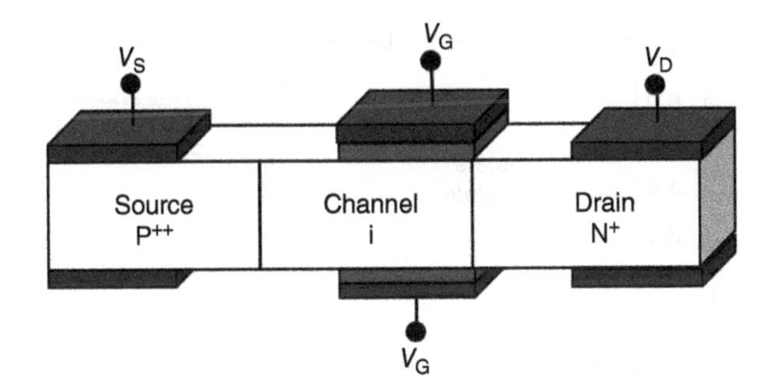

FIGURE 2.5 Three-dimensional view of a double gate I-MOS.

entire intrinsic region in an I-MOS. In n-channel I-MOS, a portion of the intrinsic region close to the source–channel interface is intentionally not covered by the gate electrode, i.e. the gate electrode is underlapped from the source end. As a result, the intrinsic region is divided into two distinct regions: a gated channel region and an ungated intrinsic region. The p–i–n architecture is chosen since the electric field magnitude and the consequent voltage required for avalanche multiplication induced breakdown is smaller for p–i–n diodes. In p–n diodes, owing to the narrow depletion region width, the electric field is very high at a breakdown condition and the device breaks down mainly due to BTBT-enhanced impact ionization.

2.2.2 Operation and Characteristics

The gate electrode in the I-MOS device modulates the effective length of the device over which the large drain voltage is sustained. For instance, in the OFF-state, the entire intrinsic region (both gated + ungated) acts like the channel region. This increases the effective channel length in the OFF-state, increasing the immunity of I-MOS against short-channel effects. In addition, the monotonically increasing energy band profiles similar to the TFETs in I-MOS devices provide immunity against DIBL.

However, if we carefully observe the energy band profiles of I-MOS in the OFF-state (Fig. 2.6), we will notice that there exists a significant overlap between the valence band of the channel region and the conduction band of the drain region in the intrinsic region. Therefore, BTBT is inevitable in the OFF-state of I-MOS and the OFF-state current is dictated by the BTBT leakage.

In the OFF-state, the applied drain voltage is sustained by the entire intrinsic region and by the source–channel and channel–drain depletion regions. However, as the gate voltage is increased, the gated intrinsic region is inverted. This reduces the series resistance of the gated channel region significantly. Consequently, the entire drain voltage falls across the small ungated intrinsic region. The large electric field owing to the high drain voltage is greater than the critical field required to initiate impact ionization–induced breakdown. Therefore, the application of the gate voltage effectively modulates the effective length over which the drain voltage is sustained. Once

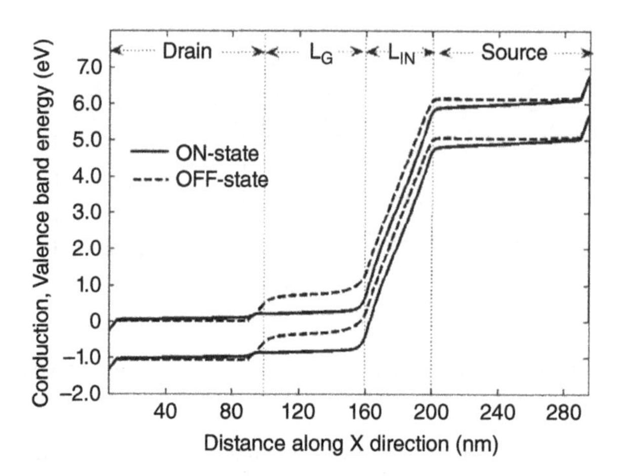

FIGURE 2.6 Energy band profiles of the I-MOS in ON-state ($V_{GS} = 1.0$ V) and OFF-state ($V_{GS} = 0.0$ V) [48].

the critical voltage (corresponding to the critical electric field required to initiate the impact ionization) is reached, the current abruptly increases from the leakage current to a large value.

Subthreshold swing values below 5 mV/decade and the ON-state to OFF-state current ratio above 5 orders of magnitude have been experimentally demonstrated in I-MOS devices as shown in Fig. 2.7. However, the output current does not saturate in I-MOS [48].

At this point, you may be tempted to think that scaling the length of the ungated intrinsic region may reduce the voltage required for impact ionization since the electric field increases with scaling. However, for impact ionization to produce a positive feedback, the generated carriers must move appropriate distance to attain sufficient kinetic energy to generate another electron–hole pair. Scaling the length of the ungated region hinders the impact-generated carriers to attain the kinetic energy level required for sustaining the positive feedback. In addition, BTBT in the OFF-state would increase significantly with incessant scaling of the ungated intrinsic region. Therefore, the length of the ungated region should be appropriately chosen.

2.2.3 Challenges

2.2.3.A Large Operating Voltage Since I-MOS operates on the principle of avalanche multiplication–induced breakdown, a large drain voltage in excess of the material band gap (E_g/q) is required, which makes I-MOS unsuitable for low-power applications [49]. Moreover, the breakdown in I-MOS takes place at a larger voltage as compared to the conventional p–i–n diode of same dimension. While the current flows through the surface inversion layer in I-MOS, it flows through the bulk in the

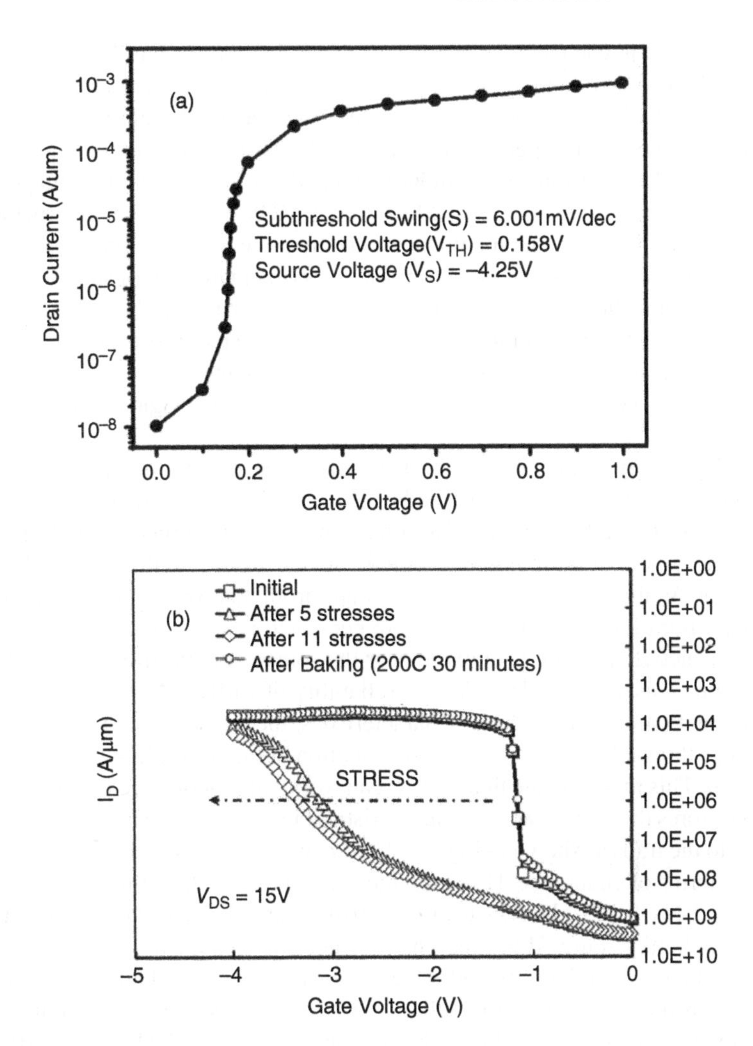

FIGURE 2.7 (a) Transfer characteristics of a virgin I-MOS [47] and (b) after subsequent stress and anneal [41].

p–i–n diode. An increase in the breakdown voltage is attributed to the lower impact ionization coefficients at the surface as compared to the bulk.

Several device architectures have been proposed to reduce the breakdown voltage of I-MOS. A depletion mode I-MOS with a doped gated channel region as opposed to an intrinsic channel region in the conventional inversion-mode I-MOS was proposed [43]. The avalanche multiplication–induced breakdown occurs in the bulk of the channel region in a depletion mode I-MOS. This not only increases the area over which impact ionization takes place but also helps to exploit the higher impact ionization coefficients of the bulk as compared to the surface. A reduction in the breakdown voltage is attributed to these factors. A double spacer I-MOS with a shallow source junction to increase the electric field (which increases impact ionization) and a lightly doped drain to reduce the impact of drain voltage was also proposed to improve the performance of I-MOS [44]. Materials with lower band gap to achieve impact ionization–induced breakdown at low supply voltages were also explored in I-MOS [46, 51]. Moreover, I-MOS utilizing heterostructures to increase the electric field and the impact ionization rate were also proposed [47]. An I-MOS with a Schottky source junction was experimentally demonstrated [45]. The increased electric field and the tunneling generated carriers at the Schottky source junction enhance the impact ionization rate and lead to a reduced breakdown voltage. Although these architectures reduce the breakdown voltage by ~1–2 V, the operating voltage is still very high making these architectures unfit for logic applications.

2.2.3.B Reliability Concern The impact ionization process requires a very high operating voltage (in excess of 5 V) leading to a large lateral electric field. This results in a carrier heating effect and hot electron generation. The vertical gate electric field aids the movement of these highly energetic hot electrons into the gate oxide on the source side. While the hot electron effects are prominent on the drain side in conventional MOSFETs, the hot electrons are generated near the source side in I-MOS and may penetrate into the gate oxide near the source side in I-MOS while flowing to the drain. This severely degrades the reliability of I-MOS. As a result, with each measurement (stress) of the device characteristics, the oxide quality becomes poorer resulting in threshold voltage shift and degradation in subthreshold swing as shown in Fig. 2.7(b). This may indicate that I-MOS can work only for one cycle. Annealing the I-MOS restores the virgin transfer characteristics but with an increased OFF-state current due to the traps as shown in Fig. 2.7(b). It may appear that I-MOS performs very poorly in terms of endurance. However, the use of novel architectures such as depletion mode I-MOS [43], I-MOS using heterostructures [47], vertical impact region [44, 46], Schottky source/drain regions [45] not only reduce the operating voltage but also enhance the reliability and the endurance of I-MOS. The impact-generated hot electrons are produced in the bulk of the depletion mode I-MOS, limiting their interaction with the gate oxide [43]. The electric field peaks at the heterojunction [47] or at the Schottky source junction [45]. Therefore, the hot electrons are generated far from the gate oxide, which leads to an enhancement in reliability and endurance.

2.2.3.C Circuit Design Using IMOS Since I-MOS also possesses an asymmetric p–i–n structure like TFETs, the circuits made from I-MOS are expected to occupy larger area in layout since contact sharing and other area reduction techniques are not applicable.

Also, as the drain voltage is reduced from a high value to a lower value below breakdown voltage, the drain current drops abruptly to the OFF-state leakage current of a p–i–n diode. Therefore, if I-MOS is utilized for digital circuit applications, it may not be able to offer rail-to-rail swing at the output. To understand this, let us consider the simple case of an n-channel I-MOS used for discharging a load capacitor. As the capacitor discharges, the output voltage across the capacitor reduces. This output voltage is the drain voltage of the I-MOS. Once this output voltage becomes less than the breakdown voltage of the I-MOS device, the current through the I-MOS device would abruptly drop. As a result, the capacitor would discharge slowly once the output voltage becomes less than the breakdown voltage of the I-MOS. This clearly shows the inefficiency of I-MOS in exhibiting rail-to-rail swing when used in high-frequency digital circuits. Therefore, it would be extremely difficult, if not impossible, to design complex logic circuits using I-MOS. Also, the drain current does not saturate in I-MOS and continues to increase with the drain voltage [50]. The poor saturation characteristics may hinder the application of I-MOS for analog applications. Moreover, impact ionization–based device would rarely be used for radio frequency applications because noise immunity is reduced when impact ionization is the dominant mechanism.

A novel symmetric I-MOS with reduced operating voltage and suppressed hot electron effect exploiting the bipolar junction transistor (BJT) in open base configuration has been proposed to overcome the challenges faced by the conventional I-MOS. We discuss this bipolar I-MOS in the subsequent sections.

2.3 BIPOLAR I-MOS

2.3.1 Structure

The structure of the bipolar I-MOS is shown in Fig. 2.8. It can be viewed as a n–p–n MOSFET with the gate partially covering the channel region, i.e. the gate does not overlap the channel region on the drain side. The main difference between the bipolar I-MOS and the conventional I-MOS is that bipolar I-MOS utilizes symmetric n–p–n or p–n–p doping unlike asymmetric p–i–n in conventional I-MOS. Second, while the channel region close to the source side is not covered by the gate in conventional I-MOS leading to impact ionization in the source side, the channel region on the drain side is ungated in the bipolar I-MOS which facilitates impact ionization on the drain side.

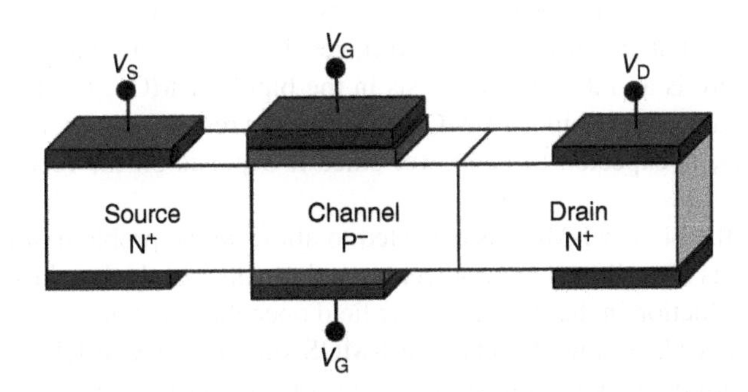

FIGURE 2.8 Three-dimensional view of a bipolar I-MOS.

Bipolar I-MOS exploits the current gain of the parasitic BJT present in a MOSFET with source, channel, and drain acting as the emitter, base, and collector, respectively, of the parasitic BJT [54]. Therefore, the silicon channel needs to be of high resistivity (lightly doped) to have an increased bipolar action. Also, to avoid hysteresis, a fully depleted silicon body should be used in bipolar I-MOS.

2.3.2 Operation and Characteristics

The bipolar I-MOS exploits the internal amplification utilizing the current gain of the parasitic BJT that exists in a MOSFET in an open base configuration since there is no body contact. Upon application of a high collector to emitter voltage (drain to source voltage), impact ionization is triggered in the channel region, which acts as the base current. In an open base configuration, the collector current, which is essentially the drain current in a bipolar I-MOS, is given as

$$I_{\mathrm{C}} = \frac{\beta \times M \times I_{\mathrm{B}}}{1 - (M - 1)\beta} \tag{2.2}$$

where I_{C} is the collector current, I_{B} is the base current, M is the avalanche multiplication factor, and β is the current gain of the parasitic BJT. Therefore, to obtain an avalanche effect in the bipolar I-MOS, a collector to emitter (drain to source voltage) must be sufficient to obtain $M = 1 + 1/\beta$. However, in a conventional I-MOS, M is given as

$$M = \frac{1}{1 - \int \alpha \, dx} \tag{2.3}$$

where α is the impact ionization coefficient. Realizing avalanche multiplication in conventional I-MOS demands that M tends to infinity, i.e. M to be very large. Since a lower voltage is needed to obtain the condition for avalanche multiplication in bipolar I-MOS than conventional I-MOS, the bipolar I-MOS exhibits a lower breakdown voltage (~2.8 V) as compared to the conventional I-MOS (~5.5 V) as shown in Fig. 2.9 [54].

Also, since a bipolar positive feedback is utilized in the bipolar I-MOS, the breakdown is initiated at different gate voltages during forward and reverse sweeps. Therefore, there is an inherent hysteresis in the bipolar I-MOS. Furthermore, since the body is floating in the bipolar I-MOS, the transient response, especially the turnoff characteristics, is expected to be slower since it depends on the recombination of the carriers.

However, the bipolar I-MOS is expected to alleviate the problem of the hot electron effect. This is attributed to (a) lower operating voltage which reduces the carrier heating, (b) reduction in the lateral electric field once the gate voltage is applied (vertical field reduces lateral field) in bipolar I-MOS similar to MOSFETs as opposed to increased lateral field upon application of gate voltage (vertical field aids lateral field) in conventional I-MOS, and (c) reduction in interaction between the gate oxide and

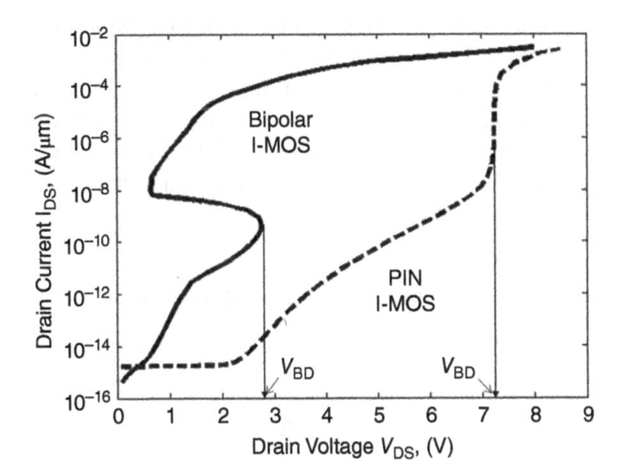

FIGURE 2.9 Output characteristics of bipolar I-MOS [54].

the impact generated carriers in the bipolar I-MOS [54]. Since in a bipolar I-MOS, the impact ionization occurs near the drain side, the impact-generated carriers easily get carried away by the drain and do not flow toward the gate oxide. However, in the conventional I-MOS, the impact-generated carriers on the source side flow through the gated region toward the drain, resulting in increased possibility of gate oxide damage due to the hot electron effect.

2.3.3 Challenges

Although the bipolar I-MOS reduces the operating voltage, suppresses the hot carrier effects, and yields a symmetric design, the contact sharing and other circuit techniques are still not viable due to the stringent requirement of a gate–drain underlap. Even though the breakdown voltage is lower in bipolar I-MOS, it is not going to exhibit a rail-to-rail swing when utilized in digital circuits and it may not even completely discharge a capacitive load as is the case with conventional I-MOS. The hysteresis in the transfer characteristics due to the bipolar mechanism further restricts its utility in the digital circuits. But the bipolar I-MOS is expected to yield a larger output swing owing to the lower breakdown voltage. In addition, the drain current does not saturate even for the bipolar I-MOS hindering its application in the analog circuits.

2.4 NEGATIVE CAPACITANCE FETs

Although TFET and I-MOS devices exhibit a sub-60 mV/decade subthreshold slope, they are asymmetric p–i–n structures. Therefore, the existing circuits and layouts would have to be redesigned for implementing circuits with TFET and I-MOS.

If somehow, the Boltzmann tyranny could be surpassed exploiting the symmetric n–p–n structure, the life of the circuit designers would be easier. The use of negative capacitance in FETs (NCFET) enables a sub-60 mV/decade subthreshold swing while utilizing the symmetric conventional n–p–n structure. Therefore, the NCFETs could be a lucrative alternative to the conventional MOSFETs. In the subsequent sections, we discuss more about the negative capacitance and how it may be exploited for obtaining steep subthreshold swing devices.

2.4.1 Negative Capacitance in Ferroelectric Materials

At first instance, you may wonder whether capacitance can be negative. Theoretically, capacitance (C) is defined as

$$C = \frac{\partial Q}{\partial V} \tag{2.4}$$

where Q is the charge stored in the capacitor and V is the voltage applied across its terminals. A negative capacitance simply means that the charge stored in the capacitor reduces as the voltage across it is increased. Furthermore, the free energy (U) stored in a linear capacitor is simply given as

$$U = \frac{Q^2}{2C} \tag{2.5}$$

In a more generalized form, the capacitance can also be defined utilizing the free energy landscape as

$$C = \left(\frac{\partial^2 U}{\partial Q^2}\right)^{-1} \tag{2.6}$$

While the energy varies as a parabola with the charge in linear capacitors, the energy follows a different curvature with charge for some materials. The ferroelectric materials form a special class of materials, which show a unique charge–energy relationship. The ferroelectric materials are piezoelectric and are known to have a noncentrosymmetric structure [55–72]. In symmetric lattice structures, the net dipole moment is zero as the dipoles are aligned randomly. However, a noncentrosymmetric structure simply implies that ferroelectric materials would be inherently polarized as the net dipole moment is not negligible. Typically, the charge density in a material is given as $Q = \varepsilon E + P$, where E is the applied electric field, P is the net polarization, and ε is the permittivity of the material. However, for typical ferroelectric materials, the polarization is significantly larger. Therefore, the charge is dominated by the polarization, i.e., $P \cong Q$.

The ferroelectric materials form a class of insulators, which consist of two distinct stable or metastable polarization states. Furthermore, the state of the ferroelectric materials may be reversibly switched from one stable state to the other by the

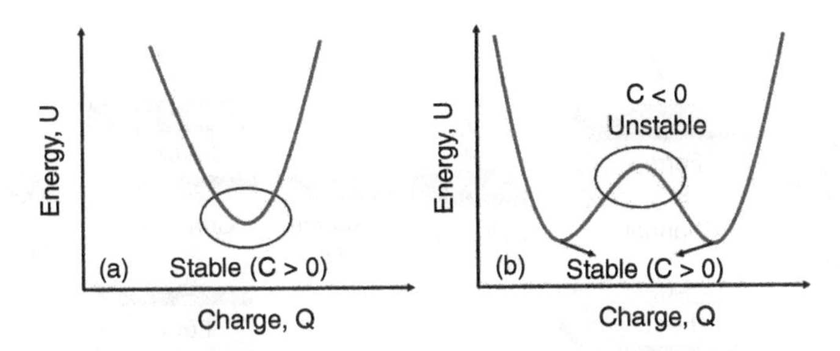

FIGURE 2.10 Energy landscape of (a) normal dielectric capacitor and (b) ferroelectric dielectric capacitor.

application of an electric field larger than the coercive field which is the characteristic of the ferroelectric material [55–72]. Therefore, the energy landscape of a typical ferroelectric material has two energy minima corresponding to the two stable states and a transition region as shown in Fig. 2.10. From Fig. 2.10(b), you may observe that in the transition region of the energy curvature, $\partial^2 U/\partial Q^2$ is negative, which clearly indicates that the capacitance is negative in this regime of operation. However, the transition region is not a stable state and is encountered only when the state of the ferroelectric film is switched and, therefore, the negative capacitance cannot be measured in a regular experimental setup [56]. The negative capacitance can only be harnessed in a ferroelectric material during the polarization switching operation.

Since the charge is dominated by polarization in the ferroelectric materials, the polarization-free energy relationship for a ferroelectric material can be described by the same energy landscape and can be expressed mathematically according to the Landau's theory as [55]

$$U = \alpha P^2 + \beta P^4 + \gamma P^6 - EP \tag{2.7}$$

where α, β, and γ are material anisotropic constants. At equilibrium, the free energy is minimum which implies $dU/dP = 0$. Therefore, $E = 2\alpha P + 4\beta P^3 + 6\gamma P^5$. This relationship yields the polarization-electric field (voltage) relationship of the ferroelectric materials and clearly indicates the presence of a negative capacitance [55].

2.4.2 Structure

The schematic view of the NCFET is shown in Fig. 2.11. The NCFETs may be obtained from conventional MOSFETs by simply replacing the gate oxide with a ferroelectric material. Also, NCFETs retain the symmetric nature of MOSFETs with only difference in the gate oxide. Therefore, from fabrication perspective, NCFETs are not expected to pose severe challenge as the MOSFET technology is already mature. However, the commonly used ferroelectric materials with large polarization such as lead titanate ($PbTiO_3$) or lead zirconate ($PbZrO_3$) contain elements such as

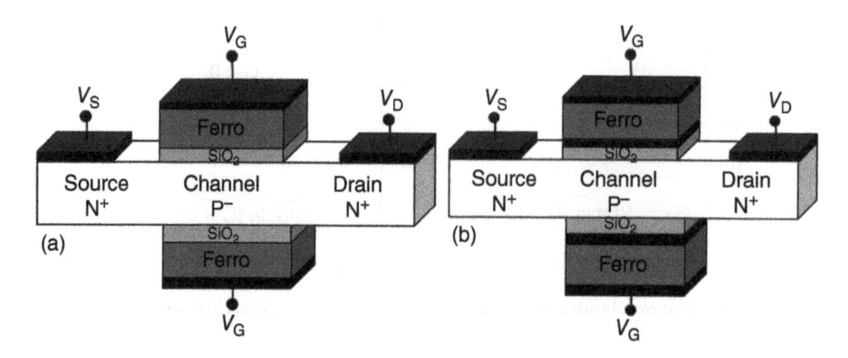

FIGURE 2.11 The 3D view of two different configurations of NCFETs: (a) metal–ferroelectric–insulator semiconductor and (b) metal–ferroelectric–metal–insulator semiconductor.

lead, which may contaminate the CMOS process flow. Therefore, integration of ferroelectric material with the existing CMOS is a technological challenge. With the introduction of aluminum- or zirconium-doped hafnium oxide–based ferroelectric materials, the CMOS integration challenges may be overcome [59, 60]. Also, the interface of ferroelectric material with silicon is expected to have a lot of traps and defects due to lattice mismatch.

Two versions of NCFETs have been proposed: (a) metal–ferroelectric–insulator semiconductor (Fig. 2.11(a)), where an ultrathin layer of SiO_2 is sandwiched between the semiconductor and the ferroelectric material to reduce the defects and traps, and (b) metal–ferroelectric–metal–insulator semiconductor (MFMIS) as shown in Fig. 2.11(b), where a layer of ferroelectric material and top metal layer is deposited on a conventional MOSFET. The MFMIS configuration is better from fabrication perspective as the metal layer eliminates interaction between the ferroelectric material and the gate oxide insulator and at the same time prevents the percolation of contaminants. It also leads to a uniform electric field distribution within the ferroelectric film [64].

2.4.3 Operation and Characteristics

No matter how efficiently the MOSFET is designed, since the Boltzmann tyranny arises due to the fundamental thermionic injection mechanism, the best possible subthreshold slope that may be obtained would still be limited to 60 mV/decade. A large chunk of the gate voltage is coupled via the gate oxide capacitance to the internal node in the form of the surface potential as shown in Fig. 1.8. The surface potential–drain current relationship follows an exponential behavior ($I_D \sim e^{q\varphi s/KT}$), which is attributed to the thermionic injection mechanism. This relationship can be altered only when an alternate injection mechanism like BTBT or impact ionization is chosen.

The subthreshold swing (*S*) of the conventional MOSFETs is defined as

$$S = \frac{\partial V_G}{\partial \log_{10} I_D} = \left(\frac{\partial V_G}{\partial \varphi_S}\right)\left(\frac{\partial \varphi_S}{\partial \log_{10} I_D}\right) \tag{2.8}$$

The second term owing to the thermionic injection mechanism is restricted to 60 mV/decade as discussed in Section 1.7.5. The first term, on the other hand, is generally referred to as the body factor (m) and can be expressed as $m = \partial V_G / \partial \varphi_S = 1 + C_S / C_{ox}$ from the simple potential divider rule in a series capacitor network as shown in Fig. 1.8. The body factor would always be larger than unity since the gate capacitance and the semiconductor capacitance in a MOSFET are always positive forcing the subthreshold swing to be higher than 60 mV/decade.

However, by utilizing the ferroelectric material as the gate oxide, negative capacitance may be harnessed. This allows to achieve a body factor lower than unity and a surface potential which is larger than the applied gate voltage. This inherent voltage amplification due to the negative capacitance is the operating principle of NCFETs [55]. The negative capacitance is utilized to reduce the body factor below unity and attain sub-60 mV/decade subthreshold swing without altering the inherent conduction mechanism or introducing asymmetry in the device architecture.

The transfer characteristics of the NCFETs are shown in Fig. 2.12. It may be noted that negative capacitance may only be observed when the polarization of the ferroelectric material switches from one state to another and that too for a limited range of voltages. Therefore, steep switching is observed only for a limited range of gate voltage [55]. Also, steep switching is observed during both OFF-state to ON-state transition and ON-state to OFF-state transition. However, the threshold voltage is different for OFF-state to ON-state transition and ON-state to OFF-state transition as shown in Fig. 2.12. This hysteresis in the transfer characteristics is due to the capacitance mismatch effect, which makes the body factor negative and hence unstable [62]. This phenomenon is discussed in detail in the subsequent section.

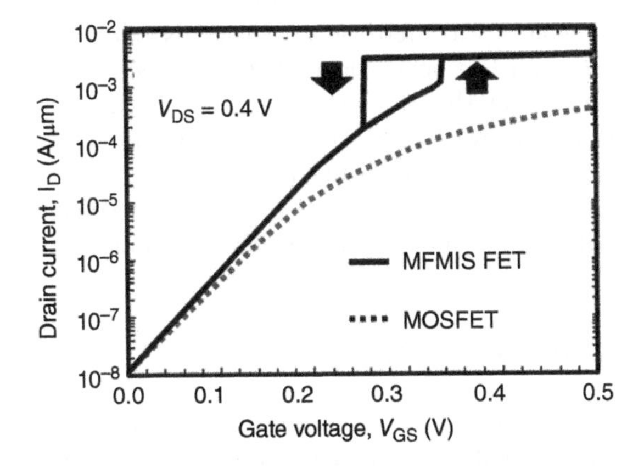

FIGURE 2.12 Transfer characteristics of the MFMIS FET [109].

2.4.4 Challenges

Although the NCFETs may appear as the best alternative to the conventional MOSFETs, they suffer from several challenges, which are discussed in the subsequent sections.

2.4.4.A Hysteresis The body factor may be reduced below unity utilizing negative capacitance. However, if the NCFET is not designed properly, i.e. if the semiconductor capacitance C_S happens to be larger than the negative oxide capacitance, a negative body factor may be obtained. A negative body factor implies unstable operation resulting in hysteresis in the transfer characteristics [57, 62, 64] as shown in Fig. 2.12. A hysteresis in the transfer characteristics means different threshold voltages for turn-ON and turn-OFF characteristics. This may hinder the application of NCFETs in the digital circuits. Therefore, appropriate capacitance matching techniques must be utilized to ensure a positive body factor during all regimes of operation of a NCFET [64].

2.4.4.B Ferroelectric Material As also discussed in Section 2.4.2, most ferroelectric materials are incompatible with the CMOS process and may contaminate the process flow. They also exhibit a large dielectric constant (in excess of 1000) which may increase the gate capacitance by several orders, degrading the dynamic performance.

With the introduction of aluminum- and zirconium-doped hafnium oxide ($HfAlO_x$/$HfZrO_x$)-based ferroelectric materials, CMOS compatibility and low dielectric constants (<30) may be achieved [69, 70]. However, the polarization of the hafnium oxide–based ferroelectric materials is lower as compared to the lead titanate or lead zirconate–based materials. A lower value of remnant polarization results in a lower ON-state current [67].

2.4.4.C Frequency of Operation The negative capacitance phenomenon is observed only during the polarization switching of the ferroelectric materials. Therefore, the speed of NCFETs is limited by the time for the polarization to switch from one stable state to another. Most ferroelectric materials exhibit a large polarization switching time. However, zirconium-doped hafnium oxide ($HfZrO_x$)-based ferroelectric materials have been shown to exhibit negative capacitance and polarization switching even at a frequency of 1 MHz [63]. The use of doped hafnium oxide–based NCFETs may pave the way for its utilization in the low-power Internet of things (IoT) applications. Alternate materials with lower switching time would be required to target the GHz frequency range of operations, which is common in digital circuits.

2.5 TWO-DIMENSIONAL FETs

Recently, the discovery of a single atom thin layer of carbon allotrope graphite, known as graphene, created a sensation in the research community. Prof. Andre Gem and Konstantin Novoselov from University of Manchester not only won the

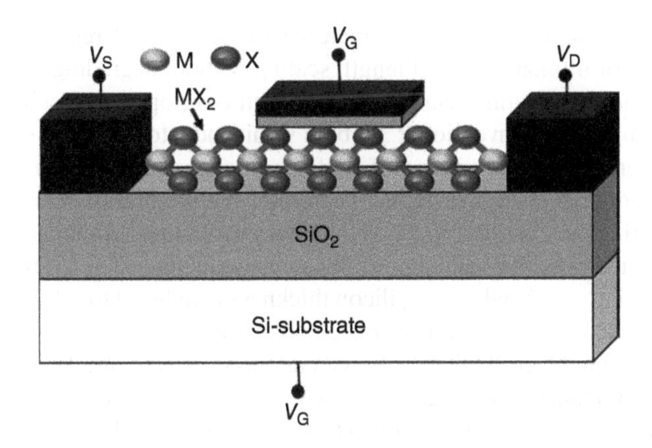

FIGURE 2.13 Three-dimensional view of a 2D FET.

Nobel prize for physics in 2010 for their ground-breaking experiments on graphene, but the experimental realization of a 2D material also opened up a new horizon for research in material sciences. Subsequently, other monolayer chalcogenide 2D materials (transition metal dichalcogenide [TMDs]) such as MoS_2, WSe_2, WTe_2, single-atom layer of elements such as phosphorus called phosphorene, silicon called silicene, and germanium called germanene, and so forth were also experimentally realized. The 2D material technology is at the cusp of next semiconductor revolution after silicon, considering its enormous potential for FETs, optoelectronics devices, and flexible electronics [73–93].

2.5.1 Structure

The 3D view of a 2D FET is shown in Fig. 2.13. The 2D FETs employ the 2D materials such as TMD, graphene, phosphorene, silicone, germanene, and so on as the active device layer. A heavily doped substrate at the bottom of the active 2D layer acts as the gate, and the source/drain contacts are formed at the top of the 2D layer. The device is passivated at the top to avoid interaction of the 2D materials with the ambient to maintain its stability.

2.5.2 Operation

The 2D materials represent ultimate miniaturization of the vertical dimension. The minimum channel length for maintaining the electrostatic integrity in FETs depends significantly on the channel thickness (t_{ch}) [93] and is given as

$$L_g \geq \beta(t_{ch} t_{ox} \varepsilon_{ch} \varepsilon_{ox})^{1/2} \tag{2.9}$$

where L_g is the channel length, t_{ch} is the channel thickness, and ε_{ch} and ε_{ox} are the dielectric constants of the channel material and oxide material, respectively.

Therefore, the application of 2D material in the channel region of FETs may pave the way for ultimate channel length scaling without degrading the electrostatic integrity. Although several techniques have been developed for realizing ultrathin layers (<5 nm) of the conventional 3D bulk semiconductors such as silicon, the 2D materials offer inherent advantage of uniform thickness and ultrasmooth dangling bond-free atomic layer thin surfaces. Therefore, the possibility of trap generation, carrier scattering, and process-induced thickness variation is eliminated in 2D materials [93]. In addition, the band gap of 3D semiconductors such as silicon becomes significantly high ~3 eV when the silicon thickness is reduced to ~1 nm. This makes the ultrathin silicon devices unfit for logic applications.

The TMD monolayer 2D materials such as MoS_2, WSe_2, and WTe_2 exhibit high carrier mobility, enhanced band gap (~1–2 eV), high effective carrier mass, and a consequent higher DOS. The enhanced band gap offered by the TMDs reduces the tunneling leakage current and significantly lowers the power dissipation. Furthermore, the high effective carrier mass suppresses the intraband direct source to drain tunneling current, which is the major short-channel effect hindering the scaling of FETs beyond the sub-10 nm regime [95]. Moreover, the ultimately thin channel leads to the best possible gate control of the channel region and reduces the influence of the drain electrostatics on the channel region. Consequently, DIBL is also significantly reduced in 2D FETs. Therefore, the TMDs are most promising materials for ultimate scaling of the FETs to the sub-10 nm regime.

2.5.3 Challenges

Fabrication of 2D FETs is the major technological challenge. The methods available for realizing atomically thin layers are limited and crude. Mechanical exfoliation of monolayer from the bulk sample using scotch tape was the most feasible technique until the development of CVD technique for growth of 2D materials. Interested readers are advised to watch the video for mechanical exfoliation of MoS_2 in [94].

In addition, limited doping techniques are available for the TMDs and other 2D materials [81–86]. Realizing high doping concentration in the source/drain regions in 2D FETs is a technological challenge. Moreover, the experimentally observed channel mobility of monolayer 2D materials is significantly low [95]. Therefore, 2D FETs suffer from a low ON-state current owing to the inefficient source/drain doping and low carrier mobility.

The absence of dangling bonds on the pristine surface of TMDs leads to weak bond formation with the contacting metals [87, 88]. Therefore, the contact resistance is quite large in 2D FETs, which is further worsened by the inefficient source/drain doping. In addition, it is also difficult to realize a defect-free interface between the 2D material and the gate dielectric in 2D FETs [89–93]. These factors degrade the performance of 2D FETs significantly. As a result, the 2D FETs cannot outperform the conventional FETs in the sub-10 nm regime unless these technological challenges are overcome.

2.6 NANOWIRE FETs

Thc gatc-all-around nanowire has been hailed as the most promising architecture for ultimate scaling of the MOSFETs [95]. Although silicon nanowires were conceptualized for application in FETs in the early 2000s itself [96], with Intel, IBM, Global Foundaries, and Interuniversitair Micro-Electronica Centrum (IMEC)'s announcement of shifting to lateral NWFETs for the 5-nm technology node and vertical NWFETs for the 3-nm technology node [97], the interest in nanowire (NW) architecture has increased manifold.

2.6.1 Structure and Characteristics

As the name suggests, in the gate-all-around nanowire FETs, the gate electrode wraps around the entire silicon channel as shown in Fig. 2.14. The nanowire FETs may be fabricated using the popular bottom-up techniques such as vapor–liquid–solid (VLS) growth, noncatalytic growth, solution-processed growth, and suspended nanowire growth. Recently, arrays of nanowire and standalone nanowire FETs have also been experimentally demonstrated utilizing the top-down approach such as self-limiting oxidation [95] and inductive plasma etching along with stress-limited oxidation [96].

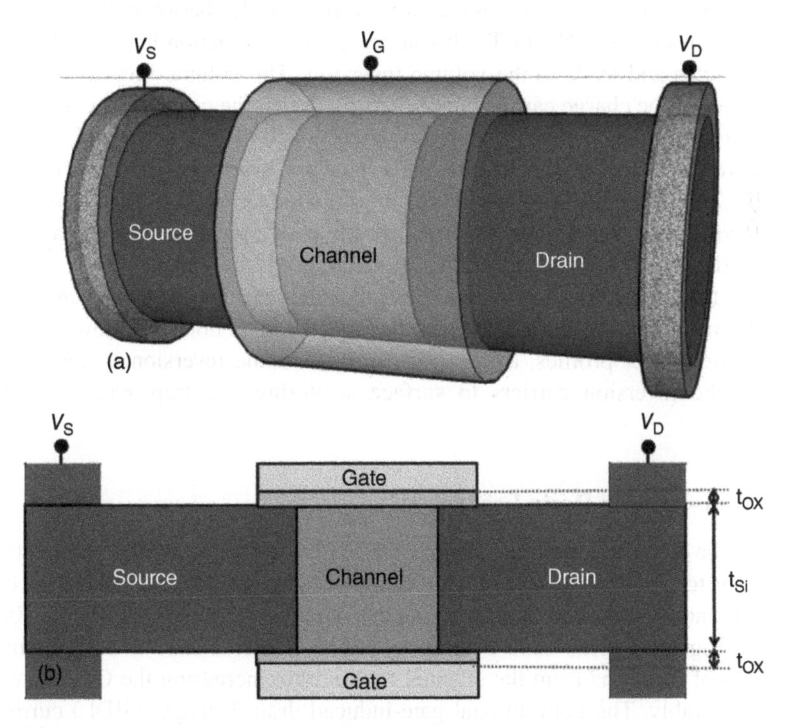

FIGURE 2.14 (a) Three-dimensional view and (b) cross-sectional view of a gate all around nanowire.

The gate-all-around architecture facilitates an efficient gate control of the channel charge and reduces the influence of the drain region on the electrostatics of the channel region [98–112]. As a result, the short-channel effects such as DIBL, threshold voltage roll-off, and so on are significantly suppressed in the NWFETs [95, 98–108]. NWFETs exhibit a close to ideal 60 mV/decade subthreshold swing due to the enhanced gate control [95].

Another salient feature of the nanowire architecture is volume inversion, which occurs when the nanowire diameter is small. A reduction in the nanowire diameter below the 10-nm regime results in a significant energy quantization due to the quantum confinement effects. This increases the number of energy states available at the center of the nanowire as compared to the energy states available at the surface of nanowire. Therefore, the electrons confined in a very narrow inversion region at the nanowire surface ($Si–SiO_2$ interface) move to the center (bulk) of the nanowire (with a large number of available energy states). This leads to confinement of electrons at the center of the nanowire rather than at the surface, and the phenomenon is called volume inversion. It may be noted that the total DOS in a nanowire is still lower than the bulk MOSFETs.

In other words, as can be observed from Fig. 2.14(b), the cross section of the NWFET appears like a double gate FET (DGFET). Reducing the diameter increases the proximity between the inversion charge carrier profiles between the two gates in the cross section of the NWFET. This increases the interaction between the carrier charge profiles and leads to the volume inversion. The volume inversion increases the mobility of the charge carriers since their motion at the center of the nanowire is expected to be less perturbed by surface scattering and the traps.

However, the limited dimensions of the nanowires restrict their current driving capability. Therefore, the effective drive current, which may be extracted from a single nanowire, is extremely low [98–108]. To obtain a large drain current, nanowires are stacked in parallel. However, vertical stacking limits the area efficiency advantage offered by the nanowires. A larger nanowire diameter can also be used to enhance the ON-state current in NWFETs. However, the large diameter does not allow interaction between the carrier profiles, and, consequently, volume inversion is not achieved exposing the inversion carriers to surface scattering and trap-induced mobility degradation.

2.6.2 Gate-Induced Drain Leakage

Although the effective gate control in the NWFETs reduces the short-channel effects, it also leads to a considerable spatial proximity between the valence band of the channel region and the conduction band of the drain region in the OFF-state [98–108] as shown in Fig. 2.15. This band overlap facilitates a lateral band-to-band tunneling (L-BTBT) of electrons from the channel to the drain increasing the OFF-state current considerably. The conventional gate-induced drain leakage (GIDL) current in MOSFETs is mainly attributed to electron–hole pair generation by the vertical tunneling (T-BTBT) of valance band electrons into the conduction band in gate-to-drain

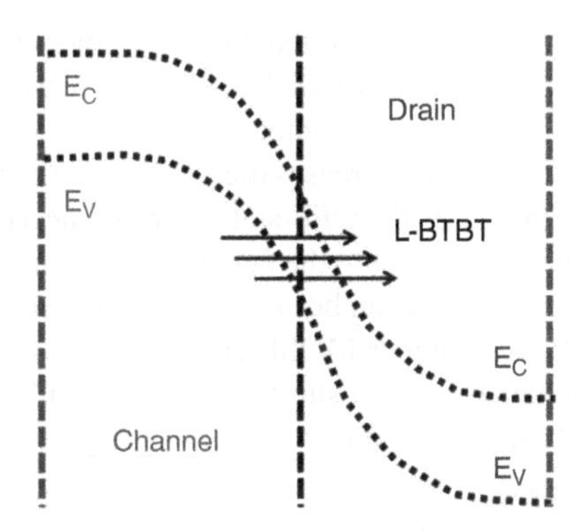

FIGURE 2.15 Lateral BTBT component of GIDL in NWFETs.

overlapped region via the BTBT and TAT mechanism [113, 114]. However, the conventional GIDL due to gate–drain overlap is dominant only at large and negative gate voltages ($V_{GS} \ll 0$ V) in NWMOSFETs [98–108]. L-BTBT is the dominant leakage mechanism in the OFF-state of NWMOSFETs and dictates the leakage current of NWMOSFETs. Therefore, L-BTBT significantly degrades the performance of the NWMOSFETs and hinders their scaling to the sub-10 nm regime.

Considering the importance of the GIDL phenomenon in NWFETs for scaling to the sub-10 nm regime, we have dedicated an entire Section 4.4.7 to provide physical insights into the nature of GIDL in different NWFET configurations.

2.6.3 Challenges

Although NWFETs appear to be the most promising candidate for the ultimate scaling of MOSFETs, the fabrication of NWFETs using a top-bottom approach is a technological challenge. Several bottom-top techniques like VLS growth, noncatalytic growth, solution-processed growth, and suspended nanowire growth have been developed [95]. However, a simple and dedicated top-bottom fabrication flow for the fabrication of NWFETs is required for mass production and commercial viability. Furthermore, electrical shorting and charge trapping is common in most bottom-top approaches due to the contamination from residual noble metal catalyst seed [96].

The drive current, which can be extracted from a single NWFET, is low owing to the limited dimension (width) of the nanowire. As a result, the NWFETs must be vertically stacked to enhance the ON-state current. Although vertical stacking increases the ON-state current linearly, the leakage current also increases linearly [98]. Therefore, vertical stacking from the perspective of power dissipation is detrimental.

2.7 NANOTUBE FETs

From our discussion in Section 2.6, we infer that although NWFETs exhibit a reduced short-channel effect owing to the proficient gate control, they offer an extremely low current driving capability due to their limited dimension. For extracting larger current, NWFETs need to be stacked vertically which leads to increased fabrication complexity and area inefficiency considering 3D integration. Moreover, the recently observed new L-BTBT component of GIDL increases their OFF-state current significantly. The vertical stacking further increases the OFF-state current linearly and restricts the viability of such an approach in improving the performance of NWFETs. Therefore, the incessant scaling of the MOSFETs can be continued only if somehow a larger current driving capability similar to a planar MOSFET is obtained along with an enhanced immunity to the short-channel effects similar to the NWFETs. The nanotube FETs (NTFETs) offer such a unique blend of high ON-state current with extremely suppressed short-channel effects.

2.7.1 Structure

The 3D view and 2D schematic view of a NTFET are shown in Fig. 2.16. It may be simply viewed as a nanowire with a core gate [115–123]. It consists of a hollow cylindrical (tubular) channel, which is controlled by two gates: an inner core gate and

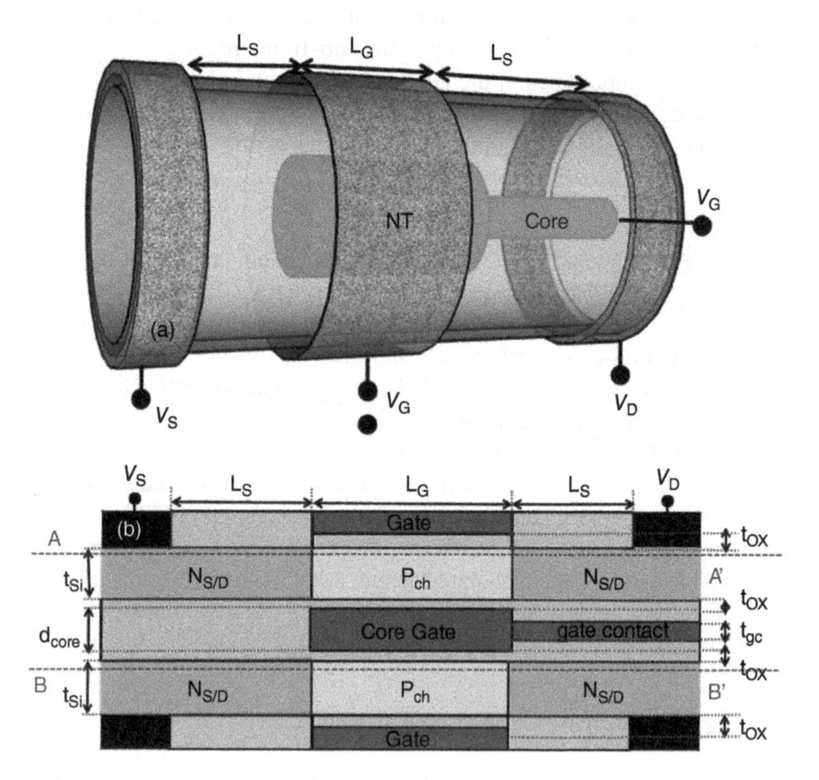

FIGURE 2.16 (a) Three-dimensional view and (b) cross-sectional view of the nanotube FETs.

an outer gate in a gate-all-around fashion. The nanotube (NT) architecture provides a better gate control of the channel region than the NWFET due to the presence of the core gate. Since the contact to the core gate is taken at the drain end, a portion of the core gate also extends below the drain region. The diameter of the core gate below the drain must be taken as small as possible to minimize the miller capacitance owing to gate-on drain overlap, which leads to overshoot in the inverter voltage-transfer characteristics [122].

The core gate may be tied to the outer gate or may operate independently. This independent operation of the core gate and the outer gate may enable the NTFETs to exhibit multithreshold FET characteristics by simply adjusting the core gate voltage [118]. Therefore, by exploiting the independent core gate, the same NTFET may be tuned for either high performance (HP) or for low standby power (LSTP) applications. This would be highly beneficial for circuit design using NTFETs as it would eliminate the need for two different processes for obtaining HP and LSTP corners.

NTFETs may be fabricated using selective epitaxial growth following the process flow outlined in [115–118]. The channel region is undoped to reduce random dopant fluctuations. The source and drain regions can be *in situ* doped during the bottom-top approach used for selective epitaxy, and ultrasteep doping profile may be realized [115]. Since NTFET is essentially a vertical structure, the gate length is dictated by the metal electrode thickness and may be easily controlled. The footprint of the NTFETs is governed by the core gate diameter and the silicon film thickness along with the oxide thickness and outer gate thickness. This provides flexibility in selecting the spacer thickness and the source/drain extension region lengths since the device footprint is not dependent on these parameters. Furthermore, hollow silicon nanorods have already been experimentally demonstrated using plasma-treated nanosphere lithography [124].

2.7.2 Operation and Characteristics

If we carefully look at the cross section of the NTFETs in Fig. 2.16(b), we will observe two silicon channels surrounded by four gates. The cross section of NTFETs appears as if two DGFETs are stacked on top of each other unlike NWFETs whose cross section seems like a single DGFET. Now, we may recall from our discussion in Section 2.6.1 that reducing the diameter increases the proximity between the inversion charge carrier profiles between the two gates in the cross section of the NWFET and leads to volume inversion.

A similar phenomenon is expected in the NTFETs with two DGFETs stacked on top of each other in the cross section. A silicon thickness of 10 nm is sufficient in NTFETs to have a significant interaction between the carrier charge profiles. Therefore, there is a surge in the DOS at the middle of the silicon channel even in the NTFETs which leads to volume inversion [115, 116]. The volume inversion phenomenon occurs in both the DGFETs in the cross section of a NTFET. Volume inversion enhances the mobility of carriers in the NTFETs.

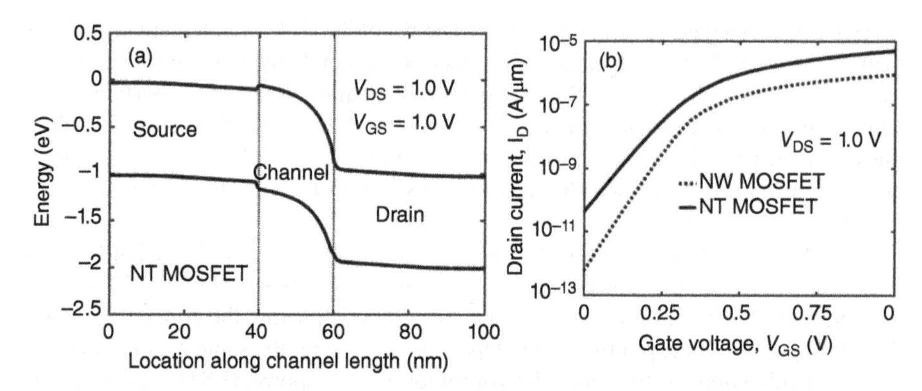

FIGURE 2.17 (a) Energy band profiles in the ON-state of NTMOSFET and (b) the transfer characteristics of NWMOSFET and NTMOSFET.

The energy band profiles of the NTMOSFET in the ON-state are shown in Fig. 2.17(a). We observe that the source-to-channel barrier height is significantly lower in the NTMOSFET in the ON-state. A lower source–channel barrier height in the ON-state not only increases the injection of electrons from source to drain but also reduces the back scattering of electrons. At this juncture, we would like to discuss about the ballistic transport, which occurs when the channel length is reduced below the mean free path of the semiconductors. The mean free path represents the average distance that an electron moves without encountering any collision or scattering event. When the channel length is reduced below the mean free path, the electrons from source move unimpeded and unscattered like a bullet fired from the gun. This phenomenon is called ballistic transport.

The presence of a large source-to-channel barrier height would increase the backscattering events at the source–channel interface. However, a lower source to channel barrier height in the NTMOSFET reduces the probability of backscattering and enables enhanced injection of electrons even in the ballistic regime. Therefore, NTFETs may serve as an effective tool for understanding the ballistic properties of the semiconductors. Furthermore, reducing the semiconductor channel thickness in NTFETs may enable the researchers to study the atomic-scale properties of the semiconductors [118].

Owing to larger volume inversion and a lower source-to-channel barrier height, the ON-state current is expected to be larger in NTFETs as compared to the NWFETs. The transfer characteristics of the NTMOSFET and NWMOSFET are compared in Fig. 2.17(b) For a fair comparison, the drain current values have been normalized with respect to the width of the active channel region (through which current flows) which is circumference for the NW and the average circumference for the NT, i.e. the average of the outer gate and the core gate circumferences. The ON-state current is larger in NTMOSFETs as compared to NWMOSFETs. This is attributed to the volume inversion–induced improved mobility and the enhanced carrier ballisticity owing to lower source–channel barrier height.

The NTMOSFETs also offer an inherent area advantage. For instance, the absolute value of the drain current, which may be extracted from a single NT, is ~1 mA whereas that obtained from a single nanowire is only ~10 μA. Thus, a single NT is equivalent to ~100 stacked NWs in terms of current drivability [116]. Also, the area of a NT with a core diameter of 100 nm and a silicon thickness of 10 nm (~3500 nm^2) is ~45 times more than the area of a NW with a diameter of 10 nm (~80 nm^2) [116]. Therefore, the effective drive current per unit area offered by the NT is also more than two times larger than the NW. The drive current per unit area is a more fundamental parameter for comparison since the current-carrying capacity depends on the number of modes, which is proportional to the cross-sectional area of the active silicon through which the current flows. The modes can be simply thought of as the effective number of channels through which the electrons flows from the source to the drain.

2.7.3 Gate-Induced Drain Leakage

From our discussion in Section 2.6.2, we may conclude that L-BTBT arises due to the efficient gate control in NWFETs. Although the NT architecture seems to offer the best possible gate control (better than the NW architecture), unfortunately, this proficient gate control may not be favorable from an L-BTBT GIDL perspective. The impact of L-BTBT GIDL on the performance of NTFETs was not studied in [115–121], and even the results shown in Fig. 2.17 were obtained without using BTBT models to account for L-BTBT. As discussed in Section 2.6.2, L-BTBT significantly degrades the performance of the NWFETs and hinders their scaling to the sub-10 nm regime. Therefore, it becomes imperative to analyze the performance of NTFETs in the presence of L-BTBT GIDL, which is discussed in [122, 123].

The transfer characteristics of the NW and the NTMOSFET including BTBT models to account for L-BTBT are compared in Fig. 2.18(a). Although the NTMOSFET exhibits a higher ON-state current as compared to the NWMOSFET, it also offers a significantly high OFF-state current by an order of magnitude.

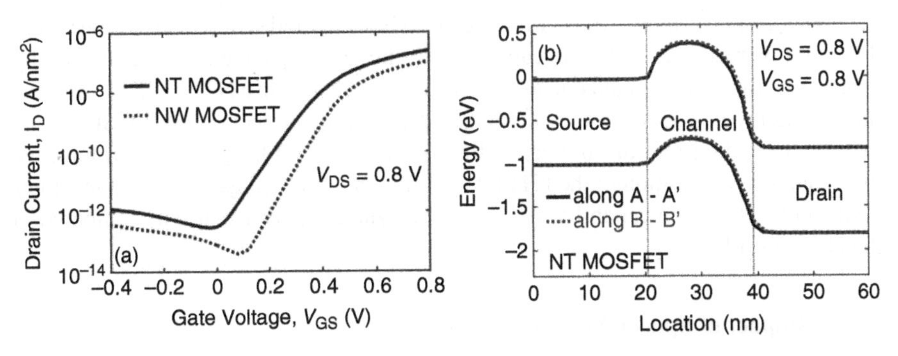

FIGURE 2.18 (a) Transfer characteristics and (b) the energy band profiles of the NTMOSFET.

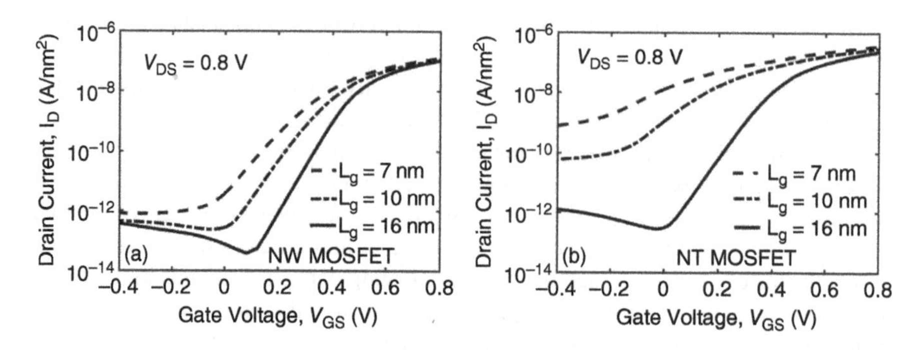

FIGURE 2.19 Impact of gate length scaling on NTMOSFETs and NWMOSFETs.

In NWFETs, the L-BTBT occurs only due to the outer surrounding gate. However, in NTFETs, the core gate also leads to significant band overlap between the valence band of the channel region and the conduction band of the drain region. This facilitates the tunneling of electrons from the channel to the drain at the Si–SiO_2 interface adjacent to the core gate as shown in Fig. 2.18(b). Therefore, in the NTMOSFETs, in addition to the L-BTBT induced by the outer surrounding gate, there is an additional L-BTBT component of GIDL due to the core gate, which further increases the OFF-state current and degrades the performance of the NTMOSFET as compared to the NWMOSFET.

As shown in Fig. 2.19, the OFF-state current of both NW and NTMOSFETs increases significantly with gate length scaling. However, the performance of the NTMOSFET degrades severely as compared to the NWMOSFET with gate length scaling. This clearly indicates that the NT architecture is more susceptible to gate length scaling induced increase in the OFF-state current. Therefore, the presence of core gate from L-BTBT GIDL perspective is detrimental.

2.7.4 Dynamic Performance

The dynamic performance of NTFETs needs to be evaluated since the core gate would definitely increase the total gate capacitance. The gate capacitance C_{gg} increases by nearly two times as expected due to the additional core gate in a NTFET. However, the intrinsic delay is smaller in the NTMOSFET as compared to the NWMOSFET due to an increased current driving capability of the NTMOSFET [122]. Therefore, we may conclude that although the NTFETs exhibit an increased gate capacitance, their higher current driving capability improves the dynamic performance as compared to the NWMOSFET.

2.7.5 Impact of Spacer Material

Since a gate–sidewall spacer is inevitable for realizing NTFETs, the spacer material may be used as an efficient tool for tuning the performance of the NTFETs. The application of a high-κ spacer leads to a reduced OFF-state current owing to the attenuation in the transition of the energy bands due to the fringing fields [122]. However, the fringing fields through the spacer also lead to an increased outer fringe capacitance component, which increases the total gate capacitance degrading the dynamic performance. Therefore, there is a trade-off between the dynamic performance and the standby performance while selecting the spacer dielectric.

2.7.6 Impact of Core Diameter

As discussed in Section 2.7.2, it is the core diameter and the silicon thickness that dictate the footprint of the NTFETs. Therefore, the impact of core diameter on the NTMOSFET has also been analyzed in [122]. An increase in the core diameter leads to a reduction in the flux of electric field lines. A reduction in the electric field intensity reduces the band bending and leads to a lower L-BTBT. However, a large core diameter is not favorable from area scaling perspective. Therefore, there is a trade-off between the area and the standby performance of the NTMOSFET with respect to the core diameter.

2.7.7 Challenges

Although silicon nanotubes may pave the way for ultimate scaling of NWFETs, realizing a core gate is very difficult. The formation of a good quality core gate dielectric interface to the silicon film is the major challenge. Apart from this, formation of silicon nanotubes with precise dimensions is also a fabrication challenge. Therefore, NFETs can be realized only when these fabrication challenges are overcome.

Also, the additional L-BTBT GIDL induced by the core gate increases the OFF-state current significantly and restricts the scaling of the NTFETs to the sub-10 nm regime. Therefore, the presence of the core gate from scaling perspective is detrimental. The L-BTBT must be mitigated in NTFETs to exploit its higher current driving capability for practical applications.

2.8 CONCLUSION

In this chapter, a detailed overview of the different emerging FET architectures proposed as an alternative to the conventional MOSFETs was provided. We discussed the operation of TFET, I-MOS, and NCFETs in detail and saw how they exhibit a sub-60 mV/decade subthreshold swing. We also looked at the challenges that these devices need to overcome to replace the conventional MOSFETs for low-power applications. We also discussed the prospects of the 2D material FETs and the technological challenges for utilizing 2D channel materials in FETs. The NWFETs, which are hailed as the most promising architecture for ultimate scaling of silicon FETs below the 7-nm technology node, were also analyzed in detail. The salient features of the NTFETs that are essentially NWFET with a core gate were also discussed. Although these emerging FETs appear promising, they still require an ultrasteep doping profile at the source–channel and channel–drain interface which is difficult to realize. The life of the device designers would be greatly simplified had there been no metallurgical junctions. As we see in the next chapter, it is indeed possible to realize FETs without any metallurgical junctions. These FETs, known as the junctionless FETs are the simplest transistor architecture that exists. In the next chapter, we specifically analyze the junctionless FETs in detail, which is the main theme of this book.

REFERENCES

[1] S. Saurabh and M. J. Kumar, *Fundamentals of Tunnel Field-Effect Transistors*, CRC Press (Taylor & Francis Group), Boca Raton, FL, 2016.

[2] M. J. Kumar, R. Vishnoi, and P. Pandey, *Tunnel Field-effect Transistors (TFET): Modelling and Simulation*, John Wiley and Sons, Ltd, West Sussex, UK, 2016.

[3] J. Knoch and J. Appenzeller, "A novel concept for field-effect transistors—The tunneling carbon nanotube FET," in *Proc. Device Res. Conf. Dig.*, 2005, pp. 153–156.

[4] W. Y. Choi, B.-G. Park, J.D. Lee and T.-J. K. Liu "Tunneling field-effect transistors (TFETs) with subthreshold swing (SS) less than 60 mV/dec," *IEEE Electron Device Lett.*, vol. 28, no. 8, pp. 743–745, Aug. 2007.

[5] A. C. Seabaugh and Q. Zhang, "Low-voltage tunnel transistors for beyond CMOS logic," *Proc. IEEE*, vol. 98, no. 12, pp. 2095–2110, Dec. 2010.

[6] A. M. Ionescu and H. Riel, "Tunnel field-effect transistors as energy efficient electronic switches," *Nature*, vol. 479, no. 7373, pp. 329–337, Nov. 2011.

[7] K. Boucart and A. M. Ionescu, "Double-gate tunnel FET with high-κ gate dielectric," *IEEE Trans. Electron Devices*, vol. 54, no. 7, pp. 1725–1733, July 2007.

[8] R. Vishnoi and M. J. Kumar, "A compact analytical model for the drain current of gate all around nanowire tunnel FET accurate from sub-threshold to ON-state," *IEEE Trans. Nanotechnol.*, vol. 14, no. 3, pp. 358–362, Mar. 2015.

[9] R. Vishnoi and M. J. Kumar, "An accurate compact analytical model for the drain current of a TFET from sub-threshold to strong inversion," *IEEE Trans. Electron Devices*, vol. 62, pp. 478–484, Feb. 2015.

[10] R. Vishnoi and M. J. Kumar, "Two-dimensional analytical model for the threshold voltage of a tunneling FET with localized charges," *IEEE Trans. Electron Devices*, vol. 61, pp. 3054–3059, Sept. 2014.

[11] R. Vishnoi and M. J. Kumar, "A pseudo 2D-analytical model of dual material gate all-around nanowire tunneling FET," *IEEE Trans. Electron Devices*, vol. 61, pp. 2264–2270, July 2014.

[12] R. Vishnoi and M. J. Kumar, "Compact analytical drain current model of gate-all-around nanowire tunneling FET," *IEEE Trans. Electron Devices*, vol. 61, pp. 2599–2603, July 2014.

[13] R. Vishnoi and M. J. Kumar, "Compact analytical model of dual material gate tunneling field-effect transistor using interband tunneling and channel transport," *IEEE Trans. Electron Devices*, vol. 61, no. 6, pp. 1936–1942, June 2014.

[14] D. B. Abdi and M. J. Kumar, "Two-dimensional threshold voltage model for the double gate pnpn TFET with localized charges," *IEEE Trans. Electron Devices*, vol. 63, no. 9, pp. 3663–3668, Sept. 2016.

[15] D. B. Abdi and M. J. Kumar, "PNPN tunnel FET with controllable drain side tunnel barrier width: proposal and analysis," *Superlattices Microstruct.*, vol. 86, pp. 121–125, Oct. 2015.

[16] D. B. Abdi and M. J. Kumar, "Dielectric modulated overlapping gate-on-drain tunnel-FET as a label-free biosensor," *Superlattices Microstruct.*, vol. 86, pp. 198–202, Oct. 2015.

[17] M. J. Kumar and S. Janardhanan, "Doping-less tunnel field-effect transistor: Design and investigation," *IEEE Trans. Electron Devices*, vol. 60, pp. 3285–3290, Oct. 2013.

[18] S. Saurabh and M. J. Kumar, "Investigation of the novel attributes of a dual material gate nanoscale tunnel field-effect transistor," *IEEE Trans. Electron Devices*, vol. 58, pp. 404–410, Feb. 2011.

[19] S. Saurabh and M. J. Kumar, "Estimation and compensation of process induced variations in nanoscale tunnel field-effect transistors (TFETs) for improved reliability," *IEEE Trans. Dev. Mater. Rel.*, vol. 10, pp. 390–395, Sept. 2010.

[20] P. Chaturvedi and M. J. Kumar, "Impact of gate leakage considerations in tunnel field-effect transistor design," *Jap. J. Appl. Phys.*, vol. 53, 074201, June 2014.

[21] M. S. Ram and D. B. Abdi, "Single grain boundary dopingless PNPN tunnel FET on recrystallized polysilicon: Proposal and theoretical analysis," *IEEE J. Electron Dev. Soc.*, vol. 3, no. 3, pp. 291–296, 2015.

[22] N. V. Nagavarapu, R. Jhaveri, and J. C. S. Woo, "The tunnel source (PNPN) n-MOSFET: A novel high performance transistor," *IEEE Trans. Electron Devices*, vol. 55, no. 4, pp. 1013–1019, Apr. 2008.

[23] W. Cao, C. J. Yao, G. F. Jiao, D. Huang, H. Y. Yu, and M.-F. Li, "Improvement in reliability of tunneling field-effect transistor with p-n-i-n structure," *IEEE Trans. Electron Devices*, vol. 58, no. 7, pp. 2122–2126, July 2011.

[24] D. B. Abdi and M. J. Kumar, "In-built N+ pocket p-n-p-n tunnel field-effect transistor," *IEEE Electron Device Lett.*, vol. 35, no. 12, pp. 1170–1172, Dec. 2014.

[25] T. Krishnamohan, D. Kim, S. Raghunathan, and K. C. Saraswat, "Double-gate strained-Ge heterostructure tunneling FET (TFET) with record high drive currents and <60 mV/dec subthreshold slope," in *IEDM Tech. Dig.*, pp. 947–949, 2008.

[26] S. Mookerjea, D. Mohata, R. Krishnan, J. Singh, A. Vallett, A. Ali, T. Mayer, V. Narayanan, D. Schlom, A. Liu, and S. Datta, "Experimental demonstration of 100 nm channel length $In_{0.53}Ga_{0.47}As$-based vertical inter-band tunnel field-effect transistors (TFETs) for ultra low-power logic and SRAM applications," in *IEDM Tech. Dig.*, pp. 949–951, Dec. 2009.

[27] D. Mohata, S. Mookerjea, A. Agrawal, Y. Li, T. Mayer, V. Narayanan, A. Liu, and S. Datta, "Experimental staggered-source and N+ pocket-doped channel III-V tunnel field-effect transistors and their scalabilities," *Appl. Phys. Expr.*, vol. 4, 024105, Feb. 2011.

[28] D. K. Mohata, R. Bijesh, S. Mujumdar, C. Eaton, R. Engel-Herbert, T. Mayer, V. Narayanan, J. Fastenau, D. Loubychev, A. Liu, and S. Datta, "Demonstration of MOSFET-like on-current performance in arsenide/antimonide tunnel FETs with staggered hetero-junctions for 300 mV logic applications," in *IEDM*, Washington, DC, Dec. 5–7, 2011.

[29] S. Mookerjea, D. Mohata, T. Mayer, V. Narayanan, and S. Datta, "Temperature-dependent I-V characteristics of a vertical $In_{0.53}Ga_{0.47}As$ tunnel FET," *IEEE Electron Device Lett.*, vol. 31, no. 6, pp. 564–567, June 2010.

[30] R. Li, Y. Lu, G. Zhou, Q. Liu, S. D. Chae, T. Vasen, W. S. Hwang, Q. Zhang, P. Fay, T. Kosel, M. Wistey, H. Xing, and A. Seabaugh, "AlGaSb/InAs tunnel field-effect transistor with on-current of 78 μA/μm at 0.5 V," *IEEE Electron Device Lett.*, 33, pp. 363–365, 2012.

[31] G. Zhou, Y. Lu, R. Li, Q. Zhang, Q. Liu, T. Vasen, H. Zhu, J.-M. Kuo, T. Kosel, M. Wistey, P. Fay, A. Seabaugh, and H. Xing, "InGaAs/InP tunnel FETs with a subthreshold swing of 93 mV/dec and I_{ON}/I_{OFF} ratio near 10^6," *IEEE Electron Dev. Lett.*, 33, 6, pp. 782–784, 2012.

[32] A. Hraziia, C. Andrei, A. Vladimirescu, A. Amara, and C. Anghel, "An analysis on the ambipolar current in Si double-gate tunnel FETs," *Solid-State Electron.*, vol. 70, pp. 67–72, Apr. 2012.

[33] A. Hraziia, A. Gupta, A. Vladimirescu, A. Amara, and C. Anghel, "30-nm Tunnel FET with improved performance and reduced ambipolar current," *IEEE Trans. Electron Devices*, vol. 58, pp. 1649–1654, June 2011.

[34] W. Y. Choi and W. Lee, "Hetero-gate-dielectric tunnelling field-effect transistors," *IEEE Trans. Electron Devices*, vol. 57, no. 9, pp. 2317–2319, Sept. 2010.

[35] A. S. Verhulst, W. G. Vandenberghe, K. Maex, and G. Groeseneken, "Tunnel field-effect transistor without gate–drain overlap," *Appl. Phys. Lett.*, vol. 91, no. 5, 053102, July 2007.

[36] J. Wan, C. Le Royer, A. Zaslavsky, and S. Cristoloveanu, "SOI TFETs: Suppression of ambipolar leakage and low-frequency noise behavior," in *Proc. European Solid-State Device Research Conference (ESSDERC)*, 2010, pp. 341–344.

[37] J. Wana, C. L. Royer, A. Zaslavsky, and S. Cristoloveanu, "Tunneling FETs on SOI: Suppression of ambipolar leakage, low-frequency noise behavior, and modeling," *Solid-State Electron.*, vol. 65, pp. 226–233, Nov. 2011.

[38] D. B. Abdi and M. J. Kumar, "Controlling ambipolar current in tunneling FETs using overlapping gate-on-drain," *IEEE J. Electron Devices Soc.*, vol. 2, no. 6, pp. 187–190, Nov. 2014.

[39] S. Sahay and M. J. Kumar, "Controlling the drain side tunneling width to reduce ambipolar current in tunnel FETs using heterodielectric BOX," *IEEE Trans. Electron Devices*, vol. 62, no. 11, pp. 3882–3886, Nov. 2015.

[40] J. T. Teherani, S. Agarwal, W. Chern, P. M. Solomon, E. Yablonovitch, and D. A. Antoniadis, "Auger generation as an intrinsic limit to tunneling field-effect transistor performance," *Jour. Appl. Phys.*, vol. 120, no. 8, 084507, Aug. 2016.

[41] K. Gopalakrishnan, P. B. Griffin, and J. D. Plummer, "Impact ionization MOS (I-MOS)—Part I: Device and circuit simulations," *IEEE Trans. Electron Devices*, vol. 52, no. 1, pp. 69–76, Jan. 2005.

[42] K. Gopalakrishnan, R. Woo, C. Jungemann, P. B. Griffin, and J. D. Plummer, "Impact ionization MOS (I-MOS)—Part II: Experimental results," *IEEE Trans. Electron Devices*, vol. 52, no. 1, pp. 77–84, Jan. 2005.

[43] C. Onal, R. Woo, H.-Y. S. Koh, P. B. Griffin, and J. D. Plummer, "A novel depletion-IMOS (DIMOS) device with improved reliability and reduced operating voltage," *IEEE Electron Device Lett.*, vol. 30, no. 1, pp. 64–67, Jan. 2009.

[44] E.-H. Toh, G. H. Wang, L. Chan, G. Samudra, and Y.-C. Yeo, "A double spacer I-MOS transistor with shallow source junction and lightly doped drain for reduced operating

voltage and enhanced device performance," *IEEE Electron Device Lett.*, vol. 29, no. 2, pp. 189–191, Feb. 2008.

[45] Q. Huang, R. Huang, Z. Wang, Z. Zhan, and Y. Wang, "Schottky barrier impact-ionization metal–oxide–semiconductor device with reduced operating voltage," *Appl. Phys. Lett.*, vol. 99, no. 8, 083507, Aug. 2011.

[46] E.-H. Toh, G. H. Wang, L. Chan, G.-Q. Lo, G. Samudra, and Y.-C. Yeo, "Strain and materials engineering for the I-MOS transistor with an elevated impact-ionization region," *IEEE Trans. Electron Devices*, vol. 54, no. 10, pp. 2778–2785, Oct. 2007.

[47] D. Sarkar, N. Singh, and K. Banerjee, "A novel enhanced electric-field impact-ionization MOS transistor," *IEEE Elec. Dev. Lett.*, vol. 31, no. 11, pp. 1175–1177, Nov. 2010.

[48] S. Ramaswamy and M. J. Kumar, "Junctionless impact ionization MOS: Proposal and investigation," *IEEE Trans. Electron Devices*, vol. 61, no. 12, pp. 4295–4298, Dec. 2014.

[49] C. W. Lee, A. N. Nazarov, I. Ferain, N. D. Akhavan, R. Yan, P. Razavi, R. Yu, R. T. Doria, and J. P. Colinge, "Low subthreshold slope in junctionless multigate transistors," *Appl. Phys. Lett.*, vol. 96, no. 10, 102106, 2010.

[50] Z. Wang and R. Huang, "Investigations on the performance limits of the IMOS transistor," in *Proc. IEEE ICSICT*, pp. 72–75, Oct. 2008.

[51] T. V. Dinh, R. Kraus, and C. Jungemann, "Investigation of the performance of strained-SiGe vertical IMOS-transistors," *Solid-State Electron.*, vol. 54, no. 9, pp. 942–949, Sept. 2010.

[52] F. Mayer, C. Le Royer, G. Le Carval, L. Clavelier, and S. Deleonibus, "Static and dynamic TCAD analysis of IMOS performance: From the single device to the circuit," *IEEE Trans. Electron Devices*, vol. 53, no. 8, pp. 1852–1857, Aug. 2006.

[53] F. Mayer, C. Le Royer, G. Le Carval, C. Tabone, L. Clavelier, and S. Deleonibus, "Comparative study of the fabricated and simulated impact ionization MOS (IMOS)," *Solid-State Electron.*, vol. 51, no. 4, pp. 579–584, Apr. 2007.

[54] M. J. Kumar, M. Maheedhar, and P. P. Varma, "Bipolar I-MOS—An impact-ionization MOS with reduced operating voltage using the open-base BJT configuration," *IEEE Trans. Electron Devices*, vol. 62, no. 12, pp. 4345–4348, Dec. 2015.

[55] S. Salahuddin and S. Datta, "Use of negative capacitance to provide voltage amplification for low power nanoscale devices," *Nano Lett.*, vol. 8, no. 2, pp. 405–410, 2008.

[56] A. I. Khan, K. Chatterjee, B. Wang, S. Drapcho, L. You, C. Serrao, S. R. Bakaul, R. Ramesh, and S. Salahuddin, "Negative capacitance in a ferroelectric capacitor," *Nature Mater.*, vol. 14, no. 2, pp. 182–186, 2014.

[57] D. J. Appleby, N. K. Ponon, K. S. Kwa, B. Zou, P. K. Petrov, T. Wang, N. M. Alford, and A. O'Neill, "Experimental observation of negative capacitance in ferroelectrics at room temperature," *Nano Lett.*, vol. 14, no. 7, pp. 3864–3868, 2014.

[58] A. I. Khan, C. W. Yeung, C. Hu, and S. Salahuddin, "Ferroelectric negative capacitance MOSFET: Capacitance tuning & antiferroelectric operation," in *IEDM*, pp. 11–13, 2011.

[59] K. Majumdar, S. Datta, and S. P. Rao, "Revisiting the theory of ferroelectric negative capacitance," *IEEE Trans. Electron Devices*, vol. 63, no. 5, pp. 2043–2049, 2017.

[60] T. Rollo, and D. Esseni, "Energy minimization and Kirchhoff's laws in negative capacitance ferroelectric capacitors and MOSFETs," *IEEE Electron Device Lett.*, vol. 38, no. 6, pp. 814–817, 2017.

[61] Z. Dong and J. Guo, "A simple model of negative capacitance FET with electrostatic short-channel effects," *IEEE Trans. Electron Devices*, vol. 64, no. 7, pp. 2927–2934, July 2017.

[62] E. Ko, J. W. Lee, and C. Shin, "Negative capacitance FinFET with sub-20-mV/decade subthreshold slope and minimal hysteresis of 0.48 V," *IEEE Electron Device Lett.*, vol. 38, no. 4, pp. 418–421, 2017.

[63] Z. C. Yuan, S. Rizwan, M. Wong, K. Holland, S. Anderson, T. B. Hook, D. Kienle, S. Gadelrab, P. S. Gudem, and M. Vaidyanathan, "Switching-speed limitations of ferroelectric negative-capacitance FETs," *IEEE Trans. Electron Devices*, vol. 63, no. 10, pp. 4046–4052, 2017.

[64] S. Khandelwal, J. P. Duarte, A. I. Khan, S. Salahuddin, and C. Hu, "Impact of parasitic capacitance and ferroelectric parameters on negative capacitance FinFET characteristics," *IEEE Electron Device Lett.*, vol. 38, no. 2, pp. 142–144, 2017.

[65] J. Seo, J. Lee, and M. Shin, "Analysis of drain-induced barrier rising in short-channel negative-capacitance FETs and Its applications," *IEEE Trans. Electron Devices*, vol. 64, no. 4, pp. 1793–1798, 2017.

[66] A. Aziz, S. Ghosh, S. K. Gupta, and S. Datta, "Polarization charge and coercive field dependent performance of negative capacitance FETs," in *Dev. Res. Conf.*, pp. 1–2, 2016.

[67] M. Kobayashi and T. Hiramoto, "On device design for steep-slope negative-capacitance field-effect-transistor operating at sub-0.2 V supply voltage with ferroelectric HfO_2 thin film," *AIP Adv.*, vol. 6, no. 2, 025113, Feb. 2016.

[68] S. Dasgupta, A. Rajashekhar, K. Majumdar, N. Agrawal, A. Razavieh, S. Trolier-Mckinstry, and S. Datta, "Sub-kT/q switching in strong inversion in $PbZr_{0.52}Ti_{0.48}O_3$ gated negative capacitance FETs," *IEEE J. Explor. Solid-State Comput. Devices Circuits*, vol. 1, pp. 43–48, 2015.

[69] M. Hoffmann, M. Pešić, K. Chatterjee, A. I. Khan, S. Salahuddin, S. Slesazeck, U. Schroeder, and T. Mikolajick, "Direct observation of negative capacitance in polycrystalline ferroelectric HfO_2," *Adv. Func. Mater.*, vol. 26, no. 47, pp. 8643–8649, 2016.

[70] S. Mueller, J. Mueller, A. Singh, S. Riedel, J. Sundqvist, U. Schroeder, and T. Mikolajick, "Incipient ferroelectricity in Al-doped HfO_2 thin films," *Adv. Func. Mater.*, vol. 22, no. 11, pp. 2412–2417, 2012.

[71] F. A. McGuire, Y. C. Lin, K. M. Price, G. B. Rayner, S. Khandelwal, S. Salahuddin, and A. D. Franklin, "Sustained sub-60 mV/decade switching via the negative capacitance effect in MoS_2 transistors," *Nano Lett.*, vol. 17, no. 8, pp. 4801–4806, 2017.

[72] A. Nourbakhsh, A. Zubair, S. Joglekar, M. Dresselhaus, and T. Palacios, "Subthreshold swing improvement in MoS_2 transistors by the negative-capacitance effect in a ferroelectric Al-doped-HfO_2/HfO_2 gate dielectric stack," *Nanoscale*, vol. 9, no. 18, pp. 6122–6127, 2017.

[73] F. Schwierz, J. Pezoldt, and R. Granzner, "Two-dimensional materials and their prospects in transistor electronics," *Nanoscale*, vol. 7, no. 18, pp. 8261–8283, 2015.

[74] Z. Lin, A. McCreary, N. Briggs, S. Subramanian, K. Zhang, Y. Sun, X. Li, N. J. Borys, H. Yuan, S. K. Fullerton-Shirey, and A. Chernikov, "2D materials advances: From large scale synthesis and controlled heterostructures to improved characterization techniques, defects and applications," *2D Mater.*, vol. 3, no. 4, 042001, 2016.

[75] P. Ajayan, P. Kim, and K. Banerjee, "Two-dimensional van der Waals materials," *Phys. Today*, vol. 69, no. 9, pp. 38–44, 2016.

[76] G. R. Bhimanapati, Z. Lin, V. Meunier, Y. Jung, J. Cha, S. Das, D. Xiao, Y. Son, M. S. Strano, V. R. Cooper, and L. Liang, "Recent advances in two-dimensional materials beyond graphene," *ACS Nano*, vol. 9, no. 12, pp. 11509–11539, 2015.

[77] B. W. Baugher, H. O. Churchill, Y. Yang, and P. Jarillo-Herrero, "Optoelectronic devices based on electrically tunable p-n diodes in a monolayer dichalcogenide," *Nature Nanotechnol.*, vol. 9, no. 4, pp. 262–267, 2015.

[78] G. H. Lee, Y. J. Yu, X. Cui, N. Petrone, C. H. Lee, M. S. Choi, D. Y. Lee, C. Lee, W. J. Yoo, K. Watanabe, and T. Taniguchi, "Flexible and transparent MoS_2 field-effect transistors on hexagonal boron nitride-graphene heterostructures," *ACS nano*, vol. 7, no. 9, pp. 7931–7936, 2013.

[79] M. Y. Li, C. H. Chen, Y. Shi, and L. J. Li, "Heterostructures based on two-dimensional layered materials and their potential applications," *Mater. Today*, vol. 19, no. 6, pp. 322–335, 2016.

[80] D. Jariwala, T. J. Marks, and M. C. Hersam, "Mixed-dimensional van der Waals heterostructures," *Nature Mater.*, vol. 16, no. 2, pp. 170–181, 2016.

[81] B. Radisavljevic, A. Radenovic, J. Brivio, I. V. Giacometti, and A. Kis, "Single-layer MoS2 transistors," *Nature Nanotechnol.*, vol. 6, no. 3, pp. 147–150, 2013.

[82] D. Lembke, S. Bertolazzi, and A. Kis, "Single-layer MoS_2 electronics," *Acc. Chem. Res.*, vol. 48, no. 1, pp. 100–110, 2015.

[83] T. Roy, M. Tosun, J. S. Kang, A. B. Sachid, S. B. Desai, M. Hettick, C. Hu, and A. Javey, "Field-effect transistors built from all two-dimensional material components," *ACS Nano*, vol. 8, no. 6, pp. 6259–6264, 2014.

[84] P. Zhao, S. Desai, M. Tosun, T. Roy, H. Fang, A. Sachid, M. Amani, C. Hu, and A. Javey, "2D layered materials: From materials properties to device applications," in *IEDM*, pp. 27–33, 2015.

[85] W. Cao, J. Kang, D. Sarkar, W. Liu, and K. Banerjee, "2D semiconductor FETs—Projections and design for sub-10 nm VLSI," *IEEE Trans. Electron Devices*, vol. 62, no. 11, pp. 3459–3469, 2015.

[86] K. Alam, and R. K. Lake, "Monolayer MOS_2 transistors beyond the technology road map," *IEEE Trans. Electron Devices*, vol. 59, no. 12, pp. 3250–3254, 2012.

[87] A. Allain, J. Kang, K. Banerjee, and A. Kis, "Electrical contacts to two-dimensional semiconductors," *Nature Mater.*, vol. 14, no. 12, pp. 1195–1205, 2015.

[88] Y. Xu, C. Cheng, S. Du, J. Yang, B. Yu, J. Luo, W. Yin, E. Li, S. Dong, P. Ye, and X. Duan, "Contacts between two-and three-dimensional materials: Ohmic, Schottky, and p–n heterojunctions," *ACS Nano*, vol. 10, no. 5, pp. 4895–4919, 2016.

[89] D. Jiménez, "Drift-diffusion model for single layer transition metal dichalcogenide field-effect transistors," *Appl. Phys. Lett.*, vol. 101, no. 24, 243501, 2012.

[90] D. Krasnozhon, S. Dutta, C. Nyffeler, Y. Leblebici, and A. Kis, "High-frequency, scaled MoS_2 transistors," in *IEDM*, pp. 24–27, 2015.

[91] H. Liu, K. Xu, X. Zhang, and P. D. Ye, "The integration of high-κ dielectric on two-dimensional crystals by atomic layer deposition," *Appl. Phys. Lett.*, vol. 100, no. 15, 152115, 2015.

[92] X. W. Jiang and S. S. Li, "Performance limits of tunnel transistors based on monolayer transition-metal dichalcogenides," *Appl. Phys. Lett.*, vol. 104, no. 19, 193510, 2014.

[93] W. Cao, J. Kang, W. Liu, and K. Banerjee, "A compact current–voltage model for 2D semiconductor based field-effect transistors considering interface traps, mobility degradation, and inefficient doping effect," *IEEE Trans. Electron Devices*, vol. 61, no. 12, pp. 4282–4290, 2014.

[94] Fabrication of TMD heterostructures [online]. Available: https://www.youtube.com/watch?v=5T0uihls40U, Accessed: Dec. 23, 2017.

[95] K. J. Kuhn, "Considerations for ultimate CMOS scaling," *IEEE Trans. Electron Devices*, vol. 59, no. 7, pp. 1813–1828, July 2012.

[96] M. S. Gudiksen et. al., "Growth of nanowire superlattice structures for nanoscale photonics and electronics," *Nature*, vol. 415, no. 6872, pp. 617–620, Feb. 2002.

[97] IBM unveils world's first 5 nm chip [online]. Available: https://arstechnica.co.uk/gadgets/2017/06/ibm-5nm-chip, Accessed Dec. 23, 2017.

[98] J. Fan, M. Li, X. Xu, Y. Yang, H. Xuan, and R. Huang, "Insight into gate-induced drain leakage in silicon nanowire transistors," *IEEE Trans. Electron Devices*, vol. 62, no. 1, pp. 213–219, Jan. 2015.

[99] J. Hur, B.-H. Lee, M.-H. Kang, D.-C. Ahn, T. Bang, S.-B Jeon, and Y.-K. Choi, "Comprehensive analysis of gate-induced drain leakage in vertically stacked nanowire FETs: Inversion-mode vs. junctionless mode," *IEEE Electron Device Lett.*, vol. 37, no. 5, pp. 541–544, May 2016.

[100] S. Sahay and M. J. Kumar, "Physical insights into the nature of gate-induced drain leakage in ultrashort channel nanowire FETs," *IEEE Trans. Electron Devices*, vol. 64, no. 6, pp. 2604–2610, June 2017.

[101] S. Sahay and M. J. Kumar, "A novel gate-stack-engineered nanowire FET for scaling to the sub-10-nm regime," *IEEE Trans. Electron Devices*, vol. 63, no. 12, pp. 5055–5059, Dec. 2016.

[102] S. Sahay and M. J. Kumar, "Spacer design guidelines for nanowire FETs from gate-induced drain leakage perspective," *IEEE Trans. Electron Devices*, vol. 64, no. 7, pp. 3007–3015, July 2017.

[103] S. Sahay and M. J. Kumar, "Insight into lateral band-to-band-tunneling in nanowire junctionless FETs," *IEEE Trans. Electron Devices*, vol. 63, no. 10, pp. 4138–4142, Oct. 2016.

[104] S. Sahay and M. J. Kumar, "Controlling L-BTBT and volume depletion in nanowire JLFETs using core-shell architecture," *IEEE Trans. Electron Devices*, vol. 63, no. 9, pp. 3790–3794, Sept. 2016.

[105] S. Sahay and M. J. Kumar, "Diameter dependency of leakage current in nanowire junctionless field-effect transistors," *IEEE Trans. Electron Devices*, vol. 64, no. 3, pp. 1330–1335, Mar. 2017.

[106] S. Sahay and M. J. Kumar, "Nanotube junctionless FET: proposal, design, and investigation," *IEEE Trans. Electron Devices*, vol. 64, no. 4, pp. 1851–1856, Apr. 2017.

[107] M. J. Kumar and S. Sahay, "Controlling BTBT induced parasitic BJT action in junctionless FETs using a hybrid channel," *IEEE Trans. Electron Devices*, vol. 63, no. 8, pp. 3350–3353, Aug. 2016.

[108] S. Sahay, and M. J. Kumar, "Realizing efficient volume depletion in SOI junctionless FETs," *IEEE J. Electron Devices Soc.*, vol. 4, no. 3, pp. 110–115, May 2016.

[109] S. Sahay and M. J. Kumar, "Symmetric operation in an extended back gate JLFET for scaling to the 5 nm regime considering quantum confinement effects," *IEEE Trans. Electron Devices*, vol. 64, no. 1, pp. 21–27, Jan. 2017.

[110] M. Li, K. H. Yeo, S. D. Suk, Y. Y. Yeoh, D.-W. Kim, T. Y. Chung, K. S. Oh, and W.-S. Lee., "Sub-10 nm gate-all-around CMOS nanowire transistors on bulk Si substrate," in *Proc. VLSI Tech. Symp.*, pp. 94–95, June 2009.

[111] N. Singh, F. Y. Lim, W. W. Fang, S. C. Rustagi, L. K. Bera, A. Agarwal, C. H. Tung, K. M. Hoe, S. R. Omampuliyur, D. Tripathi, A. O. Adeyeye, G. Q. Lo, N. Balasubramanian, and D. L. Kwong, "Ultra-narrow silicon nanowire gate-all-around CMOS devices: Impact of diameter, channel-orientation and low temperature on device performance," in *IEDM Tech. Dig.*, pp. 548–551, Dec. 2006.

[112] S. Bangsaruntip, G. M. Cohen, A. Majumdar, Y. Zhang, S. U. Engelmann, N. C. M. Fuller, L. M. Gignac, S. Mittal, J. S. Newbury, M. Guillorn, T. Barwicz, L. Sekaric, M. M. Frank, and J. W. Sleight, "High performance and highly uniform gate-all-around silicon nanowire MOSFETs with wire size dependent scaling," in *IEEE IEDM Tech. Dig.*, pp. 297–300, Dec. 2009.

[113] V. Nathan and N. C. Das, "Gate-induced drain leakage currents in MOS devices," *IEEE Trans. Electron Devices*, vol. 40, no. 10, pp. 1888–1890, Oct. 1993.

[114] T. Hoffmann, G. Doornbos, I. Ferain, N. Collaert, P. Zimmerman, M. Goodwin, R. Rooyackers, A. Kottantharayil, Y. Yim, A. Dixit, K. De Meyer, M. Jurczak, and S. Biesemans, "GIDL (gate induced drain leakage) and parasitic Schottky barrier leakage elimination in aggressively scaled HfO_2/TiN FinFET devices," in *IEDM Tech. Dig.*, pp. 725–729, 2005.

[115] H. M. Fahad, C. E. Smith, J. P. Rojas, and M. M. Hussain, "Silicon nanotube field-effect transistor with core–shell gate stacks for enhanced high performance operation and area scaling benefits," *Nano Lett.*, vol. 11, no. 10, pp. 4393–4399, Oct. 2011.

[116] H. M. Fahad and M. M. Hussain, "Are nanotube architectures advantageous than nanowire architectures for field-effect transistor applications?" *Sci. Rep.*, vol. 2, no. 2, 475, June 2012.

[117] D. Tekleab, H. H. Tran, J. W. Sleight, and D. Chidambarrao, "Silicon nanotube MOSFET," *U.S. Patent* 0 217 468, Aug. 30, 2012.

[118] D. Tekleab, "Device performance of silicon nanotube field-effect transistor," *IEEE Electron Device Lett.*, vol. 35, no. 5, pp. 506–508, May 2014.

[119] A. N. Hanna, H. M. Fahad, and M. M. Hussain, "InAs/Si hetero-junction nanotube tunnel transistors" *Sci. Rep.*, vol. 9, 9843, Apr. 2015.

[120] H. M. Fahad and M. M. Hussain, "High-performance silicon nanotube tunneling FET for ultralow-power logic applications," *IEEE Trans. Electron Devices*, vol. 60, no. 3, pp. 1034–1039, Mar. 2013.

[121] A. N. Hanna, and M. M. Hussain, "Si/Ge hetero-structure nanotube tunnel field-effect transistor," *J. Appl. Phys.*, vol. 117, no. 1, 014310, Jan. 2015.

[122] S. Sahay and M. J. Kumar, "Comprehensive analysis of gate-induced drain leakage in emerging FET architectures: Nanotube FETs vs. nanowire FETs," *IEEE Access*, vol. 5, pp. 18918–18926, Dec. 2017.

[123] A. K. Jain, S. Sahay, and M. J. Kumar, "Controlling L-BTBT in emerging nanotube FETs using dual-material gate," *IEEE J. Electron Dev. Soc.*, vol. 6, pp. 611–621, June 2018.

[124] S. S. Amiri, A. Gholizadeh, S. Rajabali, Z. Sanaee, and S. Mohajerzadeh, "Formation of Si nanorods and hollow nanostructures using high precision plasma-treated nanosphere lithography," *RSC Adv.*, vol. 4, pp. 12701–12709, Feb. 2014.

3

FUNDAMENTALS OF JUNCTIONLESS FIELD-EFFECT TRANSISTORS

From our discussions in Chapter 1, you may have started comprehending why the incessant scaling of the conventional metal-oxide-semiconductor field-effect transistors (MOSFETs) is extremely difficult and not favorable from a performance perspective. The MOSFET's structure consists of p–n junctions at the source–channel and the channel–drain interface. Scaling the MOSFETs to the sub-10 nm regime requires that the doping at the source–channel and channel–drain junction changes abruptly from a high value (typically $\sim 10^{20}$ cm^{-3}) at the source and drain regions to a low value (typically $\sim 10^{15}$–10^{17} cm^{-3}) with complimentary dopants in the channel region. However, realizing such an ultrasteep doping profile is extremely difficult.

The ion implantation process employed for doping the source/drain regions inherently leads to a stochastic distribution of dopant atoms as the dopant ions are bombarded onto the semiconductor film [1–3]. The typical value of the doping gradient between source/drain region and the channel region in MOSFETs doped using ion implantation ranges between 2 and 3 nm/decade [1–3]. This simply means that the doping can be changed, for an instance, from 10^{20} cm^{-3} in the source/drain region to 10^{17} cm^{-3} in the channel region within 5–10 nm, which is effectively the channel length in sub-10 nm regime. In addition, the dopant activation in the heavily doped source/drain regions requires a high-temperature annealing, which in turn leads to a thermally assisted lateral diffusion of dopant atoms from source/drain regions into the channel region minimizing the possibility of realizing ultrasteep doping profiles

Junctionless Field-Effect Transistors: Design, Modeling, and Simulation, First Edition.
Shubham Sahay and Mamidala Jagadesh Kumar.

[2]. This puts a complex constraint on the thermal budget as lateral diffusion is inevitable while annealing. Therefore, development of alternative doping techniques and ultrafast annealing systems are essentially required for realizing ultrasteep doping profiles.

In the last chapter, we discussed how the nanowire architecture, tunnel FETs (TFET), impact ionization metal-oxide-semiconductor (IMOS), negative capacitance FETs (NCFETs), and alternative materials like graphene and other two-dimensional semiconductors, and so on can overcome the fundamental performance limits of the conventional MOSFETs. However, all these alternatives to the MOSFET essentially require the formation of p–n junctions at the source–channel and channel–drain interface. The requirement of ultrasteep doping profile is even more stringent for devices such as TFETs where the tunneling process significantly depends on the steepness of the doping profile at the source–channel tunneling junction [4–8]. Therefore, as device designers, our lives would have been easier had there been no metallurgical junctions in transistors.

Surprisingly, even before the discovery of transistor by Shockley, Brattain, and Bardein in 1947 [9–12], the working principle of the transistor was conceptualized and patented by Austrian-Hungarian physicist Julius Edgar Lilienfield in 1930 with the title “Method and apparatus for controlling electrical current” [13]. The apparatus consisted of two thick metal electrodes on a glass substrate with an extremely thin metal electrode positioned in between these electrodes. An ultrathin layer of a compound of copper and sulfur was deposited over the three metal electrodes and formed the conducting channel in between the two thick metal electrodes as shown in Fig. 3.1. The layer thickness was so small that the electric field due to the thin metal electrode could penetrate the entire thickness of the conducting channel and modulate the conductivity [13] much like today’s junction FETs. Therefore, the current through the two terminals could be modulated by the (electric-field) voltage at the third terminal, which is essentially a transistor action. Subsequently, Lilienfield also conceptualized the conductivity modulation by the aid of electric field in a metal-oxide-semiconductor configuration and patented it in 1933 with the title “Device for controlling electrical current” [14]. The proposed device consisted of a conducting layer of copper sulfide over an ultrathin insulator, preferably a compound of the metal over which insulating layer is deposited (aluminum oxide and aluminum being the choice in this case). The significantly high-electric field through the insulator can be used to modulate the resistance and the current flowing through the conducting film between two metal electrodes using the metal-oxide-semiconductor (MOS)-like stack. The Lilienfield transistor does not contain any metallurgical junction and is essentially a gate-controlled resistor. Although the Lilienfield transistors appear promising and relevant even in the present scenario, unfortunately, the fabrication technology during 1930s was not matured enough to realize working transistors.

However, taking inspiration from Lilienfield’s work, the concept of transistors without any metallurgical junction was revived in 2010 to alleviate the need for ultrasteep doping profiles and nanowire field-effect transistors without any junction

Jan. 28, 1930. J. E. LILIENFELD 1,745,175

METHOD AND APPARATUS FOR CONTROLLING ELECTRIC CURRENTS

Filed Oct. 8, 1926

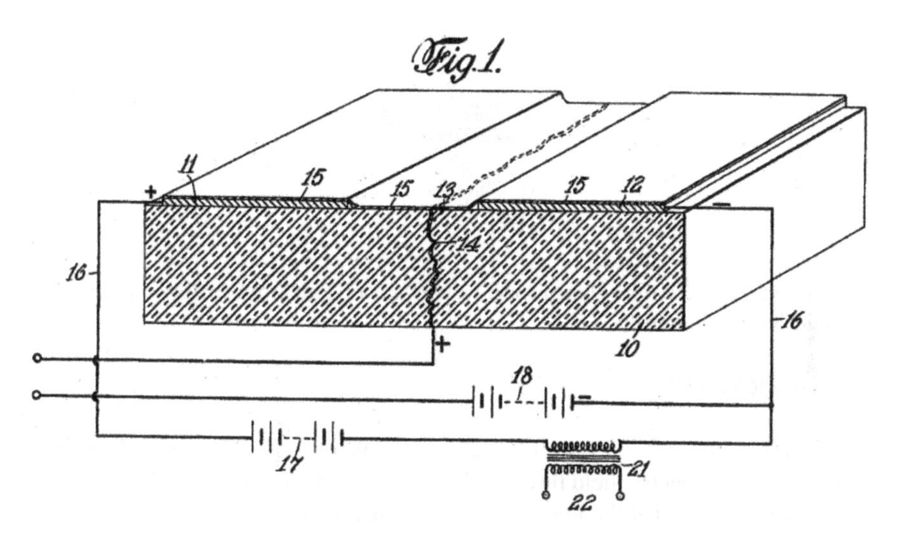

FIGURE 3.1 The apparatus proposed by Lilienfield to control electric current between two metal films using the voltage on the third metal electrode, which essentially resembles transistor action [13].

were experimentally demonstrated [15]. These devices, known as the junctionless FETs (JLFETs), utilize a semiconductor film with a gate to control its resistance and, hence, the current flowing through it. Therefore, JLFETs are also known as gated resistors [16].

3.1 DEVICE STRUCTURE

The schematic structure of a JLFET is shown in Fig. 3.2(a). It consists of a heavily doped semiconductor film with a gate to modulate the channel carrier concentration and, hence, the resistivity of the silicon film in the channel [15]. A high doping is required to achieve ohmic contacts at the source and drain electrodes. Since the whole semiconductor film from source through the channel toward the drain is doped with the same type and same concentration of impurities, there are no metallurgical junctions in this device unlike the other semiconductor devices such as MOSFETs [15]. The absence of a source/drain junction in JLFETs leads to an altogether different current conduction mechanism in JLFETs compared to the MOSFETs.

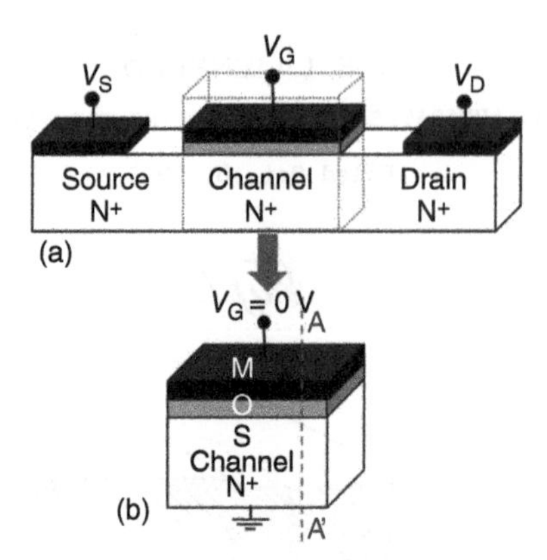

FIGURE 3.2 (a) The three-dimensional view of a JLFET and (b) the MOS capacitor which forms the heart of a JLFET.

3.2 OPERATION

The gated resistor or JLFET works on the principle of modulation of carriers by the application of an electric field through the gate [13–15]. To understand the operation of JLFETs, let us revisit the impact of a work function difference between the metal and the semiconductor in a MOS capacitor. Consider the case of a MOS capacitor with an n-substrate (Fig. 3.2(b)) with the metal–semiconductor work function difference ϕ_{MS} $(= \phi_M - \phi_S) > 0$. The energy band profiles of this MOS capacitor along the cutline A–A′ are shown in Fig. 3.3(a). Since the Fermi level of the metal is less than the Fermi level of the semiconductor, the electrons would flow from the semiconductor interface to the metal as both are shorted via ground (Fig. 3.3(a)). The ionized donor atoms at the semiconductor interface are uncovered. A positive sheet charge is created on the semiconductor, and the electrons from the semiconductor create a negative sheet charge on the metal. Therefore, when $\phi_{MS} > 0$, it leads to an in-built electric field from the semiconductor to the metal and consequent depletion of the

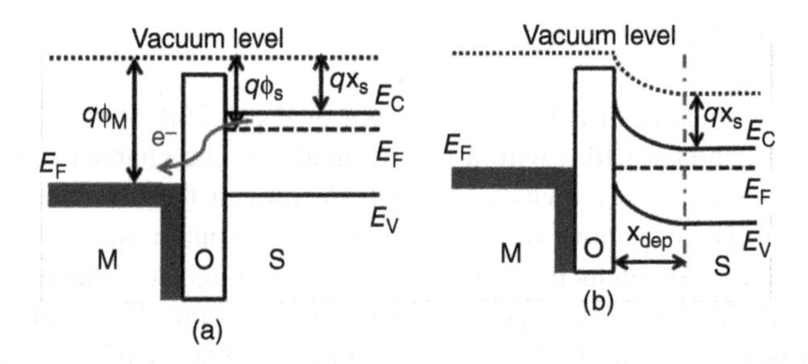

FIGURE 3.3 Energy band profiles of the MOS capacitor along cutline A–A′ (a) before thermal equilibrium and (b) at thermal equilibrium. χ_S represents the electron affinity of the semiconductor, and X_{dep} is the length of the depletion region in the semiconductor.

semiconductor at the interface since the electrons at the interface flow into the metal as shown in Fig. 3.3(b). If the semiconductor film is thin enough, it is possible to deplete it completely. This situation is called volume depletion in which the entire semiconductor film under the influence of the gate is depleted [15–20].

As shown in Fig. 3.2(b), this n-MOS capacitor forms the heart of the JLFET. Therefore, the JLFET can be simply viewed as an n-MOS capacitor with n-type source/drain regions appended on both sides of the channel. The gate electrode work function is used as an effective tool to switch off the JLFETs by achieving volume depletion in the OFF-state. The operation of a gate voltage drives the gated resistor from volume depletion into partial depletion and subsequently into a flat-band condition and accumulation region [15–20].

The different operation regimes of JLFETs are discussed in the subsequent sections.

3.2.1 Full Depletion

In JLFETs, when the gate voltage (V_{GS}) is 0.0 V, it is essential to achieve volume depletion under the gate in the channel region to completely inhibit majority carrier current flow from drain to source. The gate electrode work function needs to be adjusted to achieve volume depletion when $V_{GS} = 0.0$ V. Since the semiconductor film is heavily doped in JLFETs, to achieve volume depletion, a gate electrode with a high work function (≥ 5.1 eV) is needed in n-JLFETs whereas a low work function (≤ 4.1 eV) is required in p-JLFETs [17]. The depletion of majority carriers leads to a very low conductivity (large resistance) during the OFF-state operation of JLFET. Therefore, in JLFETs, the OFF-state leakage current is determined by the gate-induced depletion of the channel region rather than by the reverse biased p–n junction leakage current as in the case of a MOSFET. Achieving volume depletion would be easier if the gate control is enhanced [17]. This implies that multigate architectures [21] such as double-gate (DG) [22–44] (Fig. 3.4(a)), tri-gate (TG) [45–53] (Fig. 3.4(b)), and gate-all-around nanowire configurations [54–72] (Fig. 3.4(c)) would be more efficient in achieving volume depletion compared to silicon-on-insulator (SOI) JLFETs with a top gate [19, 20] as demonstrated in [17]. In these architectures, since the channel region is under the influence of the gate electrode from more than one side, the gate-induced depletion is efficient in achieving volume depletion.

Figure 3.5(a) shows the energy band diagram along the channel thickness at the middle of the channel region for a DG n-JLFET. As can be observed from Fig. 3.5(a), the intrinsic level is less than the Fermi level throughout the channel thickness. This implies that the channel is volume depleted. This is also evident from the electron concentration contour plots of DGJLFETs shown in Fig. 3.5(b).

3.2.2 Partial Depletion

The application of a positive potential on the gate electrode leads to a reduction in the depletion layer width and uncovers a neutral (undepleted) region at the middle

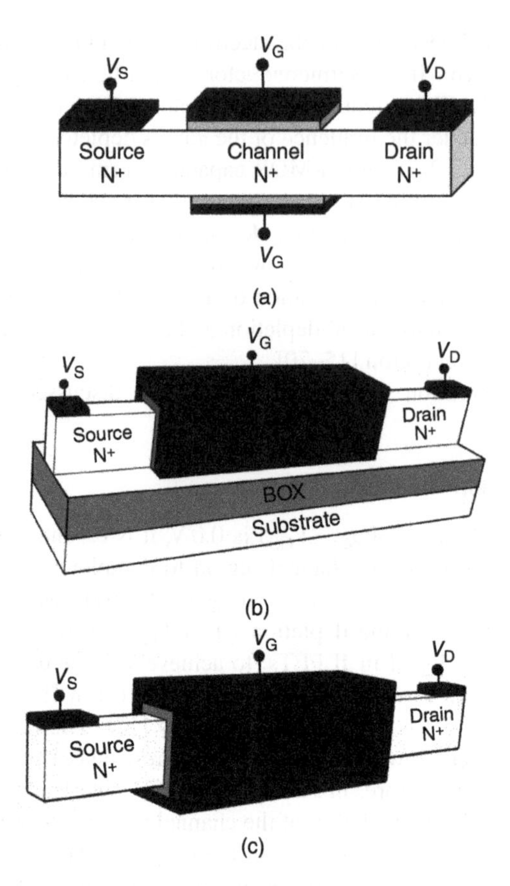

FIGURE 3.4 The three-dimensional view of (a) double-gate, (b) trigate, and (c) gate all-around nanowire JLFET.

of the channel region. This mode of operation is called partial depletion as a part of the channel region remains depleted, while the remaining neutral region, which is heavily doped, easily allows the current to flow from the source to the drain region. The neutral region, which contributes to the current flow, is squeezed in the center of the channel in between the depletion regions due to the gate [15]. The gate voltage at which the neutral region at the center of the channel disappears due to the merging of depletion region widths contributed by the gate electrode is defined as the threshold voltage for the JLFETs (Fig. 3.6). This is different from MOSFETs where the threshold voltage is defined at the onset of inversion (when the inversion layer concentration is same as the channel doping concentration).

Let us formulate an analytical model for the threshold voltage in JLFETs based on this definition. For simplicity, let us first take the case of a single-gate JLFET (as shown in Fig. 3.6(c)) and then extend the derived model for extraction of

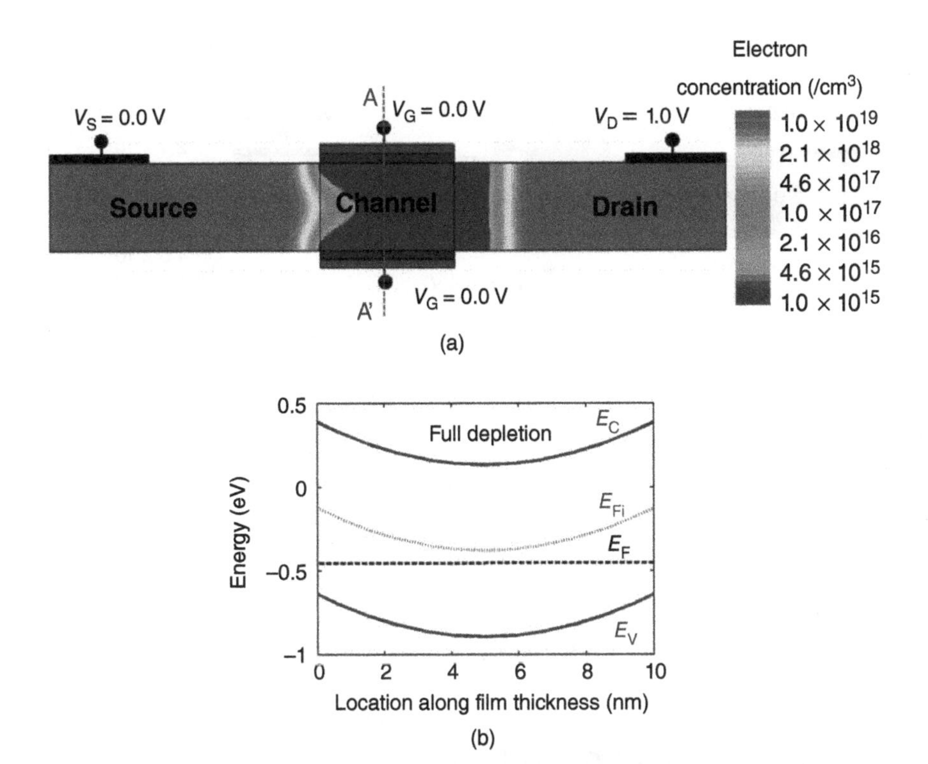

FIGURE 3.5 (a) Electron concentration contour plot of DGJLFET in full depletion mode and (b) energy band profiles taken at a cutline at the center of the channel region along the silicon film thickness ($N_D = 1 \times 10^{19}$ cm^{-3}) (cutline A–A′) in full depletion mode.

threshold voltage for the DGJLFETs. From the basics on semiconductor devices and MOSFETs, we already know that the applied gate voltage (V_G) changes the voltage drop across the gate oxide (φ_{ox}) and the surface potential (φ_S), i.e.

$$V_G = V_{FB} + \varphi_{ox} + \varphi_S \tag{3.1}$$

where V_{FB} is the flat-band voltage, which accounts for the charges in the gate oxide and the work function difference between the gate electrode and the semiconductor. To find the surface potential, we use the depletion approximation representing the depletion charge in the silicon channel as a charge box at the Si–SiO_2 interface as shown in Fig. 3.6(d). Now, to relate the charge and potential, we use the Poisson equation given by

$$\frac{\partial^2 \varphi(x)}{\partial x^2} = -\frac{\rho}{\varepsilon_{Si}} \tag{3.2}$$

where $\varphi(x)$ is the potential distribution, ρ is the charge density in the silicon film, and ε_{Si} is the permittivity of silicon. The charge density in the silicon channel can be

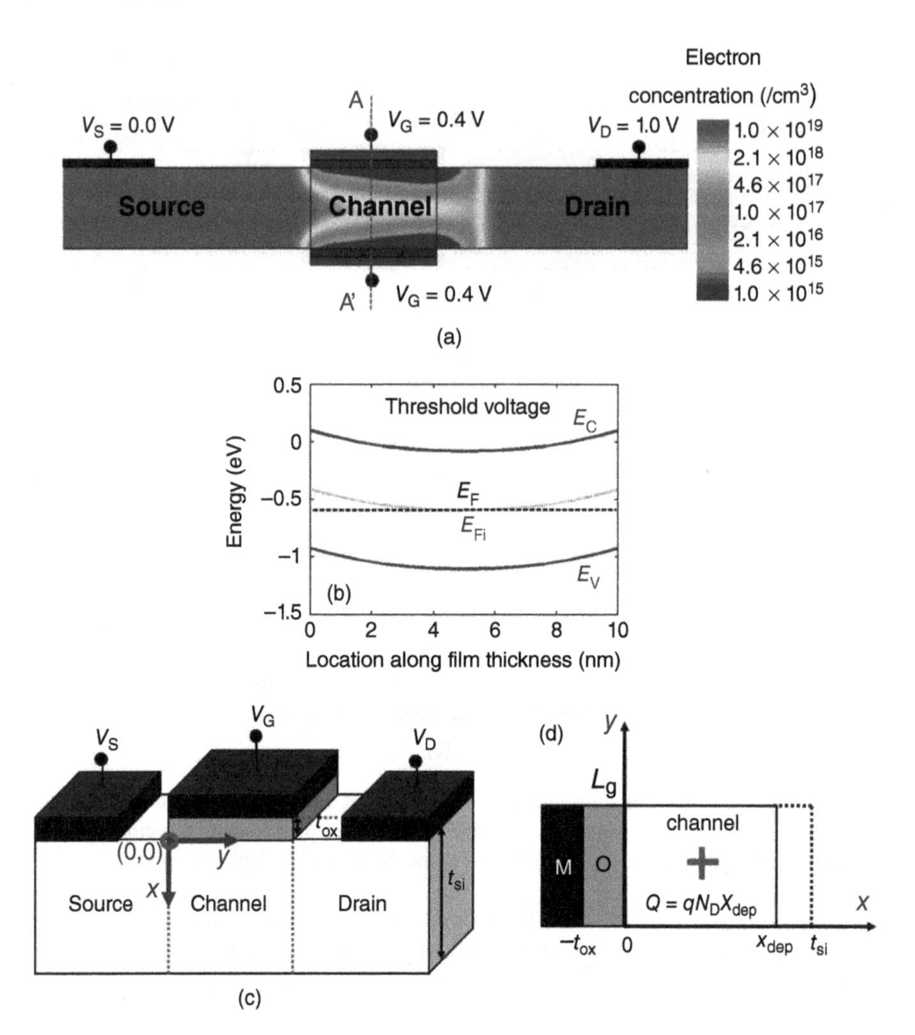

FIGURE 3.6 (a) Electron concentration contour plot of DGJLFET at the threshold voltage and (b) energy band profiles taken at a cutline at the center of the channel region along the silicon film thickness ($N_D = 1 \times 10^{19}$ cm^{-3}) (cutline A–A′) at the threshold voltage, (c) schematic view of a single-gate JLFET, and (d) the depletion charge profile assumed in the silicon film.

simply given as $\rho = qN_D$, where q is the electronic charge and N_D is the donor doping concentration. Now, integrating equation (3.2) with respect to x, we have,

$$E(x) = \frac{qN_D x}{\varepsilon_{Si}} + c \tag{3.3}$$

where $E(x)$ is the electric field distribution within the silicon film and c is the constant of integration. Now, with the assumption that the electric field emanating from the gate diminishes to 0 at $x = x_{dep}$, where x_{dep} is the depletion region width in the silicon film, we can obtain the constant as

$$c = -\frac{qN_D x_{dep}}{\varepsilon_{Si}} \tag{3.4}$$

The surface potential can be found by integrating equation (3.3) with respect to x from $x = 0$ ($Si-SiO_2$) interface to $x = x_{dep}$, which yields

$$\varphi(x_{dep}) - \varphi(0) = \frac{qN_D x_{dep}^2}{2\varepsilon_{Si}} \tag{3.5}$$

Now, assuming that the potential also ceases to be 0 at $x = x_{dep}$ (which is a valid assumption for the volume depletion case at threshold voltage), we can obtain an expression for the surface potential at the $Si-SiO_2$ interface as

$$\varphi_s = -\frac{qN_D x_{dep}^2}{2\varepsilon_{Si}} \tag{3.6}$$

The electric field (E_S) at the surface can be obtained by putting $x = 0$ in equation (3.3) as

$$E_S = -\frac{qN_D x_{dep}}{\varepsilon_{Si}}. \tag{3.7}$$

We can find the electric field (E_{ox}) inside the gate oxide by applying the boundary condition that the electric displacement vector must be continuous at the $Si-SiO_2$ interface, which yields $\varepsilon_{ox}E_{ox} = \varepsilon_{Si}E_S$. Under the assumption of an ideal gate oxide, there is no charge inside the gate oxide implying that the electric field should be constant. Therefore, the voltage drop across the oxide can be easily obtained as

$$\varphi_{ox} = E_{ox}t_{ox} = -\frac{qN_D x_{dep} t_{ox}}{\varepsilon_{ox}} \tag{3.8}$$

where t_{ox} is the gate oxide thickness and ε_{ox} is the permittivity of SiO_2

Using the values obtained for φ_{ox} and φ_S in equation (3.1), we have

$$V_G = V_{FB} - \frac{qN_D x_{dep}^2}{2\varepsilon_{Si}} - \frac{qN_D x_{dep} t_{ox}}{\varepsilon_{ox}} \tag{3.9}$$

Now, as per the definition, at the threshold voltage, the entire silicon film is depleted, i.e. the depletion region width (x_{dep}) spans the entire silicon film thickness (t_{Si}). Replacing V_G by V_{Th}, where V_{Th} is the threshold voltage and x_{dep} by t_{Si} in equation (3.9), we can obtain the final analytical expression for threshold voltage in JLFETs as

$$V_{\mathrm{Th}} = V_{\mathrm{FB}} - \frac{qN_{\mathrm{D}}t_{\mathrm{Si}}^2}{2\varepsilon_{\mathrm{Si}}} - \frac{qN_{\mathrm{D}}t_{\mathrm{Si}}t_{\mathrm{ox}}}{\varepsilon_{\mathrm{ox}}} \tag{3.10}$$

This simple analytical model for threshold voltage in single-gate JLFETs can be easily extended to DGJLFETs. In a DGJLFET, since two gates are used to deplete the entire silicon thickness, one gate has to deplete only half of the silicon film, i.e. $t_{\mathrm{Si}}/2$. Therefore, replacing x_{dep} in equation (3.9) by $t_{\mathrm{Si}}/2$, we obtain the threshold voltage for a DGJLFET as

$$V_{\mathrm{Th}} = V_{\mathrm{FB}} - \frac{qN_{\mathrm{D}}t_{\mathrm{Si}}^2}{8\varepsilon_{\mathrm{Si}}} - \frac{qN_{\mathrm{D}}t_{\mathrm{Si}}t_{\mathrm{ox}}}{2\varepsilon_{\mathrm{ox}}} \tag{3.11}$$

The threshold voltage for a single-gate MOSFET is given as

$$V_{\mathrm{Th}} = V_{\mathrm{FB}} + \frac{\sqrt{4qV_t\varepsilon_{\mathrm{Si}}N_{\mathrm{A}}\ln\left(\frac{N_{\mathrm{A}}}{n_i}\right)}}{\varepsilon_{\mathrm{ox}}}t_{\mathrm{ox}} + 2V_t\ln\left(\frac{N_{\mathrm{A}}}{n_i}\right) \tag{3.12}$$

where n_i is the intrinsic carrier concentration, V_t is the thermal voltage, and N_{A} is the channel doping in the MOSFET. By careful observation of the threshold voltage expressions for the JLFET and MOSFET, i.e. equations (3.10) and (3.12), we notice that while the threshold voltage depends on only the silicon film doping and the gate oxide thickness in MOSFETs, it is also a function of silicon film thickness in JLFETs. This leads to serious design constraints for JLFETs as discussed in Section 3.3.

The electron concentration contour plot of JLFET in partial depletion mode (Fig. 3.7(a)) clearly indicates the presence of an undepleted highly doped semiconductor region in the middle of the channel region, which contributes to the current conduction via majority carriers. The energy band profile of the JLFET at the middle of the channel region in a partial depletion regime is also shown in Fig. 3.7(b). We observe that as compared to the full depletion mode (Fig. 3.5(b), under the partial depletion mode, the Fermi level (E_{F}) moves above the intrinsic Fermi level (E_{Fi}) at the center of the channel. As a result, a narrow part of the semiconductor film at the center of the channel is undepleted and converted into a neutral conduction path. It may also be noted that above threshold voltage ($V_{\mathrm{GS}} > V_{\mathrm{th}}$), the majority carriers contribute to the current which flows in the bulk of the channel through the neutral region in a JLFET unlike MOSFETs where the current flows at the surface through the inversion layer formed by the minority carriers. Therefore, JLFETs are essentially majority carrier devices unlike MOSFETs [15].

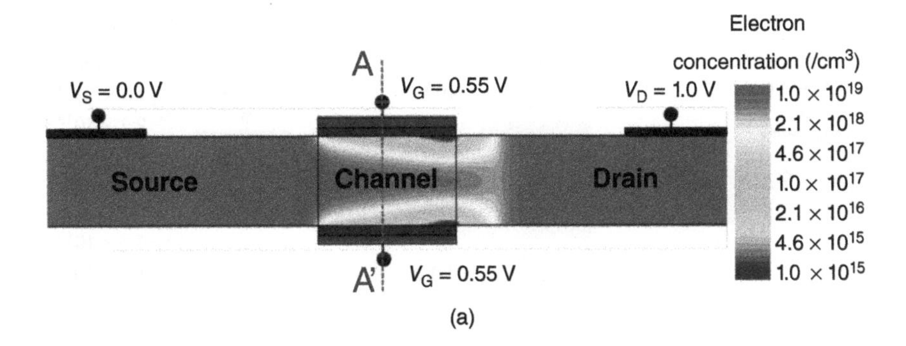

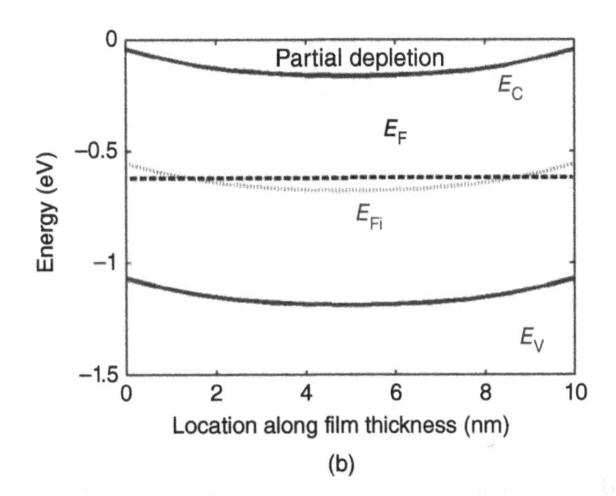

FIGURE 3.7 (a) Electron concentration contour plot of DGJLFET in partial depletion mode and (b) energy band profiles taken at a cutline at the center of the channel region along the silicon film thickness ($N_D = 1 \times 10^{19}$ cm^{-3}) (cutline A–A′) in partial depletion mode.

3.2.3 Flat Band Condition

If we continue to increase the voltage on the gate electrode, the depletion region width owing to the gate would reduce uncovering a larger neutral region. At a particular gate voltage, the gate-induced depletion regions would disappear and the entire semiconductor film would be undepleted and neutral. The gate voltage, which corresponds to a net zero depletion region width in the channel region, is defined as the flat band voltage for JLFETs. Under flat band conditions, the entire channel thickness becomes neutral and actively provides majority carriers for current conduction [15]. The energy bands of the channel region remain flat at this voltage as shown in Fig. 3.8(a), and the whole channel becomes neutral as shown in the electron concentration contour plot in Fig. 3.8(b).

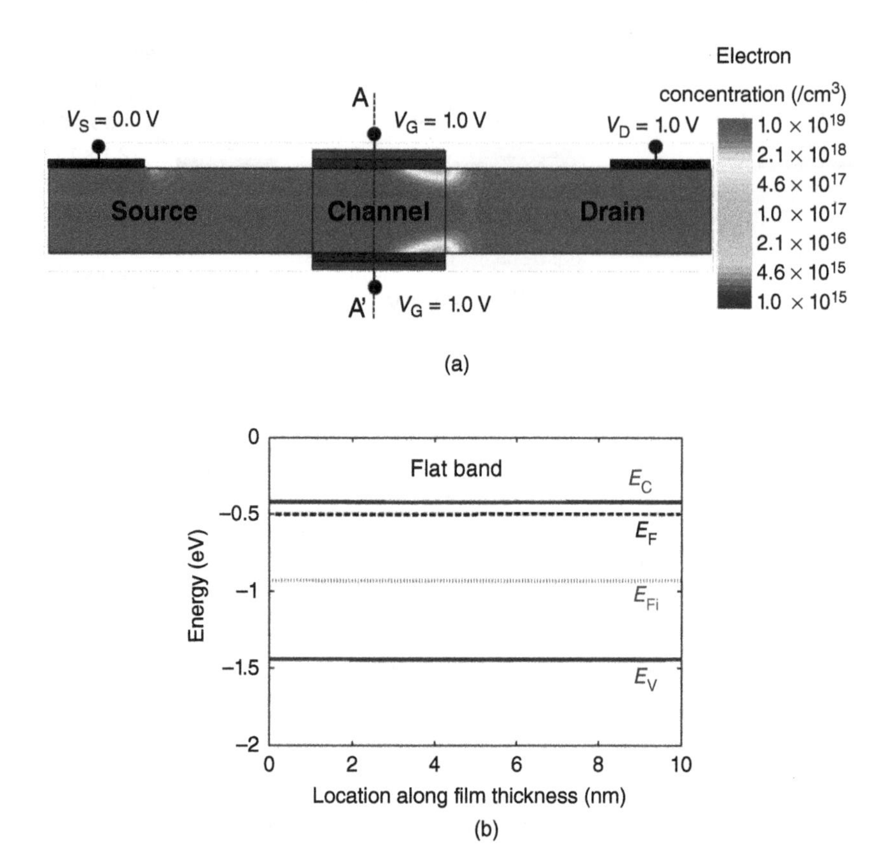

FIGURE 3.8 (a) Electron concentration contour plot of DGJLFET in flat band condition and (b) energy band profiles taken at a cutline at the center of the channel region along the silicon film thickness ($N_D = 1 \times 10^{19}$ cm^{-3}) (cutline A–A′) in flat band conditions.

3.2.4 Accumulation

If the gate voltage is further increased, the electrons would be attracted to the surface of the channel (at the Si–SiO_2 interface) and an accumulation layer of electrons will form. The energy band profiles and the electron concentration contour plot of JLFET in accumulation mode are shown in Fig. 3.9. An accumulation layer of electrons is formed, and the electric field at the surface of the channel region also increases from its minimum value at flat band as can be observed from Fig. 3.9.

The ON-state of the semiconductor devices corresponds to the condition when the gate voltage and drain voltage are maximum and equal to the supply voltage, i.e. $V_{GS} = V_{DS} = V_{DD}$. We would like to point here that the JLFETs are normally designed so that they operate in flat band condition in the ON-state [15]. However, you may wonder that if accumulation layer of electrons may increase the current through the JLFETs, why are the JLFETs designed to have flat band operation in the ON-state?

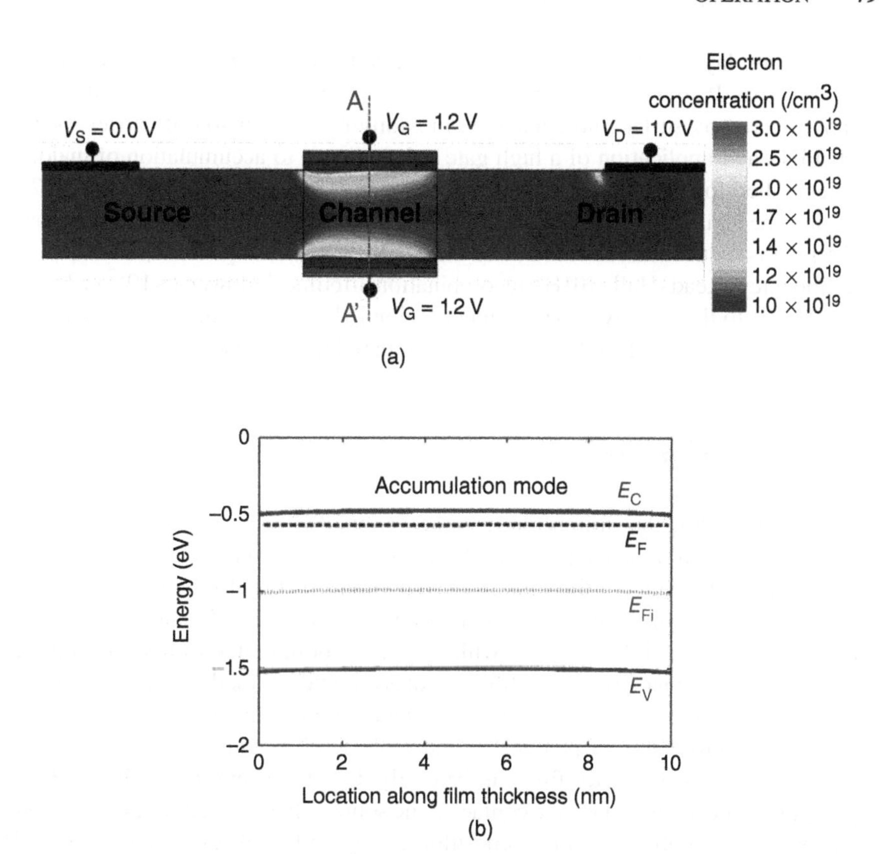

FIGURE 3.9 (a) Electron concentration contour plot of DGJLFET in accumulation-mode and (b) Energy band profiles taken at a cutline at the center of the channel region along the silicon film thickness ($N_D = 1 \times 10^{19}$ cm^{-3}) (cutline A–A′) in accumulation-mode.

This is because the scattering events at the surface reduce the mobility of electrons and degrade the performance as observed in MOSFETs. In addition, the carriers are more susceptible to interactions with the interface states and traps when they flow near the surface. A combined result of these effects is the reduction in the transconductance value with increasing gate voltages which is generally observed in a MOSFET [73]. When operating near the flat band conditions, a significant number of carriers flow in the bulk of the channel and are less affected by surface scattering and interface traps. Therefore, the JLFETs are normally designed so that the ON-state overlaps with the flat band voltage.

At this juncture, we would also like to discuss the key differences between the operation of JLFETs and MOSFETs, which can be inferred from our previous discussions. In an n-channel MOSFET, the flat band conditions occur below the threshold voltage whereas in n-channel JLFETs flat band condition is achieved above threshold voltage. Also, the current conduction is at the surface in a MOSFET through the inversion layer carriers which are susceptible to surface scattering and interface traps whereas JLFETs employ majority carrier conduction through the bulk, which is not affected by surface nonidealities. In addition, no inversion of carriers occurs in JLFETs and application of a high gate voltage leads to accumulation of majority carriers. Therefore, the JLFETs resemble the accumulation mode MOSFETs [74–76].

Since the channel is heavily doped in JLFETs, the effective band gap of the channel region is reduced in JLFETs due to the band gap narrowing effect [77–80]. Also, the Shockley–Read–Hall (SRH) recombination lifetime is lower ($\sim 10^{-7}$ s) for the JLFETs due to the heavily doped channel region [81–83]. This implies that the SRH recombination rate in JLFETs is larger as compared to the MOSFETs.

3.3 DESIGN PARAMETERS

After understanding the various modes of operation of JLFETs, let us now look at the device design requirements. First of all, to realize ohmic contacts at the source and drain contacts, the silicon film must be heavily doped. This also helps to reduce the series resistance of the source/drain regions and increase the ON-state current. However, the electron mobility reduces with increasing doping due to ionized impurity scattering [15]. Second, the silicon film thickness should be small so that the gate electric field may penetrate the entire thickness and volume depletion may be achieved using gate electrodes with practically realizable work functions (3.1–5.93 eV). A lightly doped semiconductor film can be easily depleted using lower electric fields but degrades the quality of ohmic contact at the source/channel electrodes. Therefore, an appropriate combination of silicon film doping and thickness is required while designing JLFETs at a fixed gate oxide thickness. This is also evident from equation (3.10), which simply conveys that the threshold voltage in a JLFET is a strong function of both the silicon film thickness and silicon film doping unlike MOSFETs where the threshold voltage is a function of only doping (equation 3.12).

3.3.1 Fabrication Flow

In this section, we look at a simple fabrication flow to realize JLFETs. As discussed in Section 2.6, since nanowires are hailed as the most promising architecture for ultimate scaling of FETs, we discuss the process steps involved in realizing a gate-all-around nanowire junctionless field-effect transistor (NWJLFET) shown in Fig. 3.4(c). The schematic process flow to realize a NWJLFET is shown in Fig. 3.10. To fabricate a NWJLFET, one can start with a SOI wafer with a p-type top silicon layer. Ion implantation is then performed with an appropriate dopant dose to achieve uniformly doped source/drain and channel regions as shown in Fig. 3.10(a) followed by annealing in nitrogen ambient. It may be noted that uniformly doped SOI wafers with a doping of 10^{19} cm^{-3} are also commercially available, which may directly be processed for fabricating JLFETs.

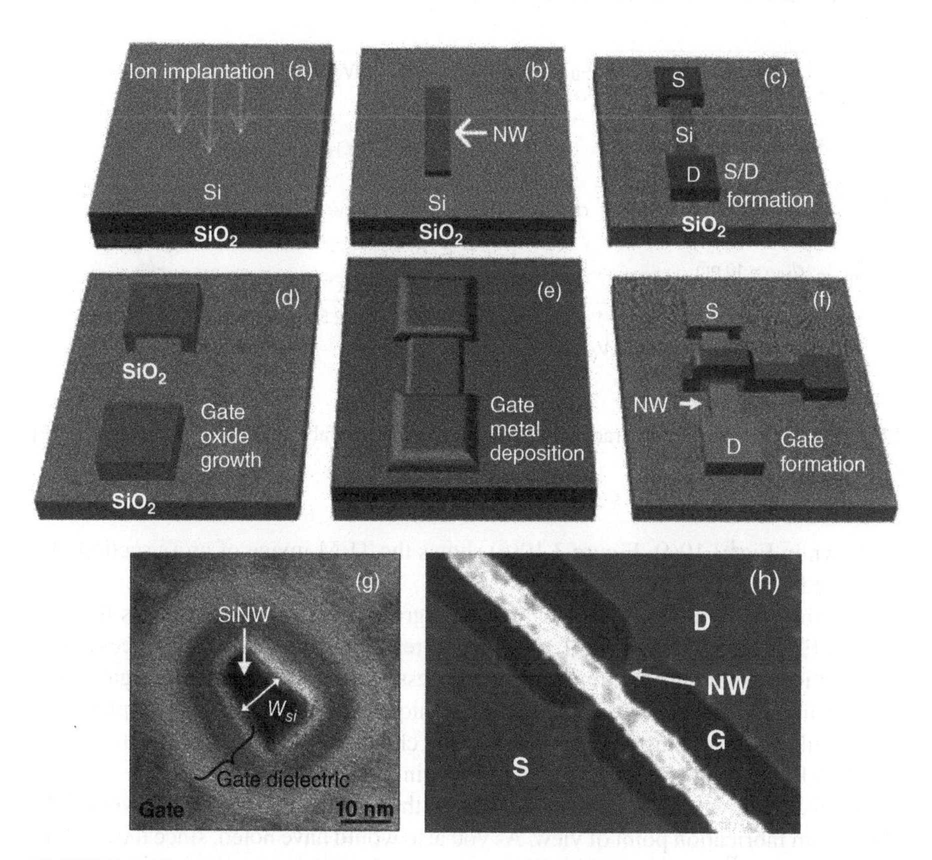

FIGURE 3.10 Schematic process flow to realize NWJLFETs: (a) ion implantation of SOI film, (b) nanowire channel definition, (c) source/drain contact pad formation, (d) gate oxide growth/deposition, (e) gate metal deposition, (f) gate electrode definition [61], (g) cross-sectional TEM image of a fabricated NWJLFET [186], and (h) bird's view SEM image of a fabricated nanowire metal-oxide-semiconductor field-effect transistor (NWMOSFET) [71].

Once the SOI layer is doped, the active channel of the NWJLFET can be defined on the wafer using E-beam lithography followed by a dry etch as shown in Fig. 3.10(b). The width of the channel of NWJLFET is defined in this step. The source and drain contact pads can then be defined in the next step using photolithography or E-beam lithography and dry etching. It may be noted that the thickness and width of the source/drain contact pads is more than the channel width to reduce the contact resistance. After the contact pads and nanowire channel are defined, the remaining silicon film is etched using dry etching as shown in Fig. 3.10(c). A gate oxide of appropriate thickness can then be thermally grown using rapid thermal oxidation or deposited using atomic layer deposition (Fig. 3.10(d)). A gate metal layer such as TiN or even polysilicon may then be deposited for a gate electrode as shown in Fig. 3.10(e). Finally, the gate length may be defined using E-beam lithography and dry etching

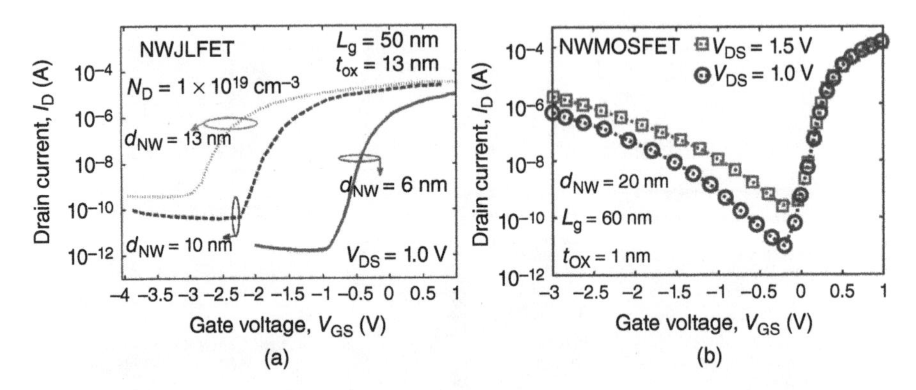

FIGURE 3.11 Transfer characteristics of experimentally fabricated (a) NWJLFETs [186] and (b) NWMOSFET [71].

as shown in Fig. 3.10(f). Figure 3.10(g) shows the TEM image of cross section of a fabricated NWJLFET.

The absence of any metallurgical junction greatly simplifies the process flow for NWJLFETs. The NW MOSFET can also be realized using a similar process flow. An additional source/drain implantation process would be required after gate electrode definition, and ultrafast annealing techniques would have to be employed to achieve ultrasteep doping profiles. The transfer characteristics of experimentally realized NWJLFET and NWMOSFET are shown in Fig. 3.11.

At this point, we would also like to discuss the advantages of JLFETs over MOSFETs from fabrication point of view. As you also would have noted, since the JLFETs do not have any metallurgical junctions, there are no stringent constraints on realizing ultrasteep doping profiles. As a result, JLFETs do not require rapid annealing processes and complex thermal budgets. The lower thermal budget for JLFETs gives the device designers flexibility while selecting the gate stack materials. Furthermore, a relaxation in the fabrication constraints also leads to a reduced cost of the process flow for the JLFETs compared with the MOSFETs [84]. Therefore, JLFETs are advantageous over MOSFETs or other architectures such as TFET, IMOS, and NCFETs, which require formation of metallurgical junctions with ultrasteep doping profiles from fabrication and cost of manufacturing perspective.

3.4 PARAMETERS THAT AFFECT THE PERFORMANCE

In the last section, we saw how the absence of metallurgical junction affects the performance and leads to an altogether new conduction mechanism in the JLFETs. We discussed in detail the various differences in the current conduction mechanisms between the JLFETs and the MOSFETs. These differences have serious implications on mobility, temperature dependence, carrier ballisticity, reliability, and short-channel effects. In the following section, we discuss how each of these parameters affects the performance of JLFETs.

3.4.1 Mobility

The mobility of carriers is a crucial parameter governing the current through the FETs. As discussed earlier, since the JLFETs have a different conduction mechanism (bulk conduction as opposed to surface conduction in MOSFETs), they are expected to have a different mobility dependence compared to the MOSFETs. In a MOSFET, the current flows through the inversion layer close to the surface where the electrons are subjected to high field as well as scattering events due to surface roughness. Both these factors lead to a degradation in the mobility.

However, in a JLFET, the carriers flow in the bulk far away from the interface. In JLFETs, the electric field in the direction transverse to the carrier flow, i.e. along the channel thickness (cutline A–A′ in Fig. 3.9), is minimum when it operates under flat band condition and increases when operated in both depletion and accumulation regimes as shown in Fig. 3.12. It may be noted that the electric field reduces to zero at the center of the silicon film because of the symmetric structure of DGJLFET where the electric fields due to the two gates are in opposite direction.

The JLFETs are normally designed to operate in flat band condition at the ON-state where the longitudinal electric field is nearly zero and the entire heavily doped channel conducts [15–20]. Consequently, surface roughness and electric field do not degrade the mobility of JLFETs unlike MOSFETs [15–20]. Moreover, the major drawback in utilizing alternate high mobility materials such as germanium, indium gallium arsenide (InGaAs), and so on in MOSFETs is that these materials do not have a native oxide of their own and the interface between oxide stack and these materials has a large number of defects and charges which severely degrade the mobility and the threshold voltage control. However, the bulk conduction mechanism in JLFETs

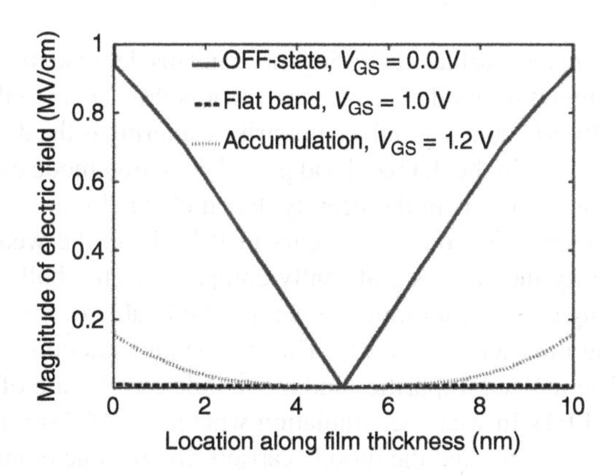

FIGURE 3.12 Vertical electric field along the silicon film thickness (cutline A–A′ in Fig. 3.9) of DGJLFET in different operating regimes.

may pave the way for these alternative materials to be used in JLFETs where the impact of surface defects and charges is lower on the current conduction [85].

From experimental measurements, it was observed that the mobility of JLFETs at flat band is nearly same as that for the MOSFETs in the ON-state [86–89]. Therefore, there must be some alternate mobility degradation mechanism which dominates in the JLFETs as compared to the MOSFETs. As the channel region in JLFETs is heavily doped, the ionized impurities form scattering centers and distort the motion of carriers flowing nearby. This impurity scattering (Coulomb scattering), which is not present in the modern MOSFETs due to the use of undoped channel region, is the dominating mobility degradation mechanism in JLFETs and restricts the mobility of the JLFETs to the bulk mobility at high doping densities. Although surface roughness scattering is low for JLFETs operating close to flat band conditions, it increases significantly as the gate voltage increases above flat band and leads to accumulation of carriers at the surface. The surface roughness scattering degrades the accumulation layer mobility in JLFETs and the inversion layer mobility in MOSFETs in a similar fashion.

The mobility degradation in MOSFETs and JLFETs is related to the interplay of three mechanisms: (a) phonon scattering, which arises due to the vibration of the atoms owing to the thermal energy which changes the periodic potential for current conducting carriers; (b) surface roughness scattering, which originates due to the surface impurities and roughness, and (c) ionized impurity scattering (Coulomb scattering) arising out of the ionized impurity atoms, which act as scattering centers and affect the motion of current carrying carriers. The total mobility is given by the Matthiessen's rule:

$$\frac{1}{\mu_{\text{total}}} = \frac{1}{\mu_{\text{ph}}} + \frac{1}{\mu_{\text{sr}}} + \frac{1}{\mu_{\text{im}}} \tag{3.13}$$

where μ_{total} is the total mobility and μ_{ph}, μ_{sr}, and μ_{im} are the mobility due to phonon scattering, surface roughness, and impurity scattering, respectively, which simply implies that the lowest mobility component would govern the overall mobility which is quite obvious.

Phonon scattering and surface scattering mechanisms degrade mobility of JLFETs in the strong accumulation mode because the carriers flow close to the surface (well above flat band) [86–91], and the ionized impurity scattering is the dominant mobility degradation mechanism in the flat band and partial depletion mode owing to the presence of ionized dopant atoms in the heavily doped channel region of JLFETs [90]. However, an interesting phenomenon occurs in JLFETs in the weak accumulation regime. The mobility increases significantly compared to the bulk silicon mobility for the same doping concentration indicating a considerable decrease in the ionized impurity scattering mechanism [86–91]. This can be understood from the screening effect [86–92]. The ionized impurities and the mobile carriers are of opposite polarities in case of JLFETs in weak accumulation whereas in MOSFETs, due to inversion, the ionized impurities and the mobile carriers are of same polarity [89, 90]. As a result, in JLFETs, the mobile carriers are attracted by the ionized impurities. The mobile carriers encircle the ionized dopant atoms and shield the other carriers from

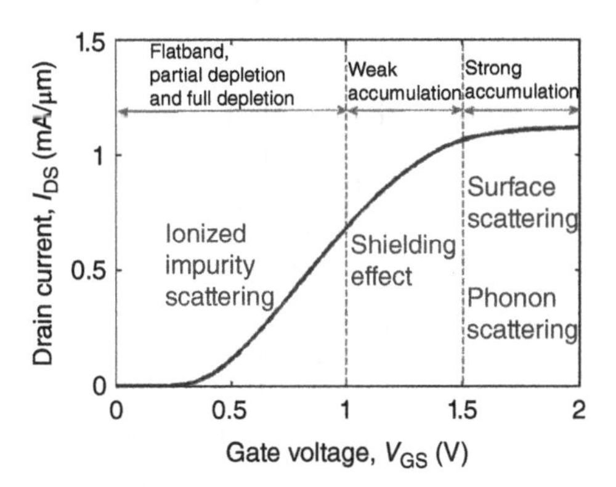

FIGURE 3.13 The dominant scattering mechanisms, which affect the mobility of DGJLFET in different operating regimes.

being scattered [86–92]. This shielding effect reduces the scattering cross section of the ionized impurity and leads to a higher mobility in the weak accumulation mode in JLFETs compared to the bulk silicon mobility. Also, forward scattering mechanisms contribute to enhancement in mobility under weak accumulation regime. Consequently, the mobility in the weak accumulation mode in JLFETs is comparable (or even larger) to the inversion layer mobility in MOSFETs [86–92]. Figure 3.13 summarizes the dominant mobility degradation mechanisms which affect the mobility of JLFETs in different operating regimes.

In addition, the mobility degradation mechanisms also depend on silicon thickness and doping concentration [91]. As the silicon thickness is reduced, (a) the number of dopant atoms, which contribute to impurity scattering reduces [91], and (b) volume accumulation occurs, which increases the screening effect [86–92], leading to an increase in the mobility with a reduction in the silicon thickness. However, as the silicon film thickness is reduced below 4 nm, the impact of surface roughness increases even though volume accumulation occurs in the device [91, 92]. As a result, the mobility degrades significantly.

The mobility is expected to reduce with an increase in the electron concentration or doping. However, screening increases with electron concentration improving the mobility [86–92]. Moreover, the mobility in JLFETs is also found to increase with an increase in the doping concentration for ultrathin NWJLFETs [87]. This is attributed to the close spacing of the donor atoms at the center of JLFETs which not only reduces the potential barrier for electrons but also reduces the tunneling distance between the neighboring ionized impurities.

To understand how mobility is affected under these conditions, let us examine a phenomenon known as resonant tunneling in a potential well surrounded by two potential barriers as shown in Fig. 3.14. In a potential well, owing to the spatial confinement, the electron energy states are significantly discretized and quasi-stationary energy states with unity transmission coefficient (where the ratio of the output

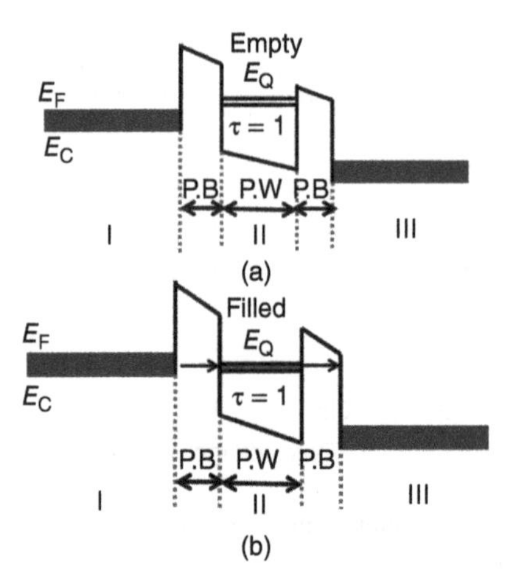

FIGURE 3.14 Energy band profiles of a system consisting of a potential well (P.W) confined between two potential barriers (P.B) when quasistationary state with a unity transmission coefficient ($\tau = 1$) is (a) empty and (b) filled (resonant tunneling).

electron density of electrons to the input electron density is one) are created within the potential well due to the quantum confinement effects. If the Fermi level of the electrons in region II coincides with these quasi-stationary energy levels within the potential well with unity transmission coefficient, a resonant transmission of electrons takes place from region I to region III. Therefore, the presence of an energy level in region II due to quantum confinement induced by the surrounding potential barriers may lead to a significant increase in the current. This phenomenon is called resonant tunneling, which requires the presence of energy levels within the potential well.

In JLFETs, this potential well can be formed in the channel region by an appropriate alignment of the ionized donor atoms. The donor atoms are placed close enough in the channel region in JLFETs since the semiconductor film is heavily doped. As a result, the channel region between the closely spaced ionized donor atoms would be significantly quantized due to the quantum confinement effects. The quantization of the energy levels within the channel region may provide the necessary quasi-stationary energy states for ionized donor atom induced resonant tunneling in JLFETs. This leads to a significant increase in the drain current indicating a mobility enhancement in JLFETs [87].

At this point, we would also like to mention that since the conduction mechanism in JLFETs is significantly different from the MOSFETs, the methods used to extract the mobility in MOSFETs may not be accurate for JLFETs. For instance, the Y-method [95] to extract the mobility in MOSFET assumes that the mobility has a gate voltage dependence given by

$$\mu(V_{GS}) = \frac{\mu_0}{1 + \theta(V_{GS} - V_{th})} \tag{3.14}$$

where μ_0 is the low field mobility, θ is the mobility reduction coefficient, and V_{th} is the threshold voltage. The mobility is extracted from the slope of $I_{Ds}/\sqrt{g_m}$ versus V_{GS}, which is linear as long as equation (3.14) is valid. Although the inversion layer mobility in MOSFETs satisfies equation (3.14), the accumulation layer mobility in JLFETs does not yield a linear relationship between $I_{Ds}/\sqrt{g_m}$ and V_{GS} [96, 97]. Therefore, the accumulation layer mobility cannot be extracted accurately with the help of Y-method [96]. Hence, alternate methods for extracting the mobility using $1/g_m^2(V_{GS})$ [98] or a difference of two current values at different V_{DS} [96] were reported for JLFETs.

The split capacitance–gate voltage (C–V) measurement [97] is one of the popular methods to extract mobility in MOSFETs. It involves the estimation of the different charge components (i.e., mobile charges and bulk charges) by extracting two different C–V relations: (a) gate-channel capacitance-gate voltage (C_{GC}–V_{GS}) characteristics for mobile charges and (b) gate-bulk substrate capacitance-gate voltage characteristics (C_{GB}–V_{GS}) for bulk charges. The C_{GC}–V_{GS} characteristics are then used to extract the magnitude of the mobile charges (Q_m) given by

$$Q_m(V_{GS}) = \int_{-\infty}^{V_G} C_{GC}(V_{GS})dV_{GS} \tag{3.15}$$

The value of $Q_m(V_{GS})$ is then used to extract the mobility using the following simple equation:

$$\mu_{eff} = \frac{L}{W}\frac{I_D(V_{GS})}{Q_m(V_{GS})V_{DS}} \tag{3.16}$$

The split C-V method is found to be accurate even for measuring mobility in JLFETs since it depends on the C_{GC}–V_{GS} characteristics rather than relying on any assumption as in the case of the Y-method.

3.4.2 Impact of Strain on Mobility

Strain engineering is widely used to enhance the mobility of inversion layer in the MOSFETs [99–105]. The application of strain leads to a modification in the energy band structure and alters the effective carrier transport mass, mobility, and, therefore, the conductivity of the silicon film. However, since the conduction mechanism is different in JLFETs, the impact of strain is expected to be different. The impact of strain is measured in terms of piezoresistance coefficient (π) defined as

$$\pi = \frac{\Delta\rho}{\rho \cdot \sigma} \tag{3.17}$$

where σ is the applied stress and ρ is the resistivity [106]. The piezoresistance coefficient decreases for bulk silicon when the doping concentration is increased. This is attributed to a reduced impact of stress on mobility enhancement as the

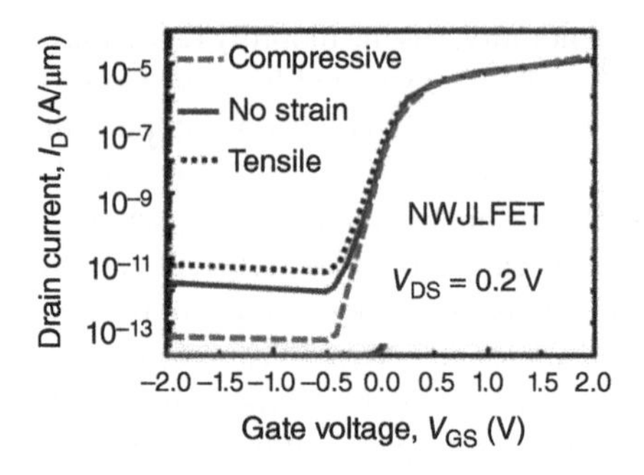

FIGURE 3.15 Transfer characteristics of an n-NWJLFET under compressive strain, tensile strain, and without any external strain [107].

doping concentration increases [106]. Since the current flows in the bulk in JLFETs, the piezoresistance coefficient is expected to be nearly equal to the bulk silicon values at that particular doping concentration. The four probe bending analysis shows that drain current increases with a uniaxial compressive stress in p-JLFETs and uniaxial tensile stress in n-JLFETs (Fig. 3.15) due to the increase in mobility similar to the MOSFETs [106, 107]. The comprehensive analysis of impact of strain in NWJLFETs [108] illustrates that the tensile strain leads to an increase in the interface states, whereas the compressive strain reduces the activation energy of the traps [108]. The interface electron trapping mechanism [107] also plays a major role in determining the impact of strain. Moreover, the NWFET with a thicker oxide was found to be more sensitive to the application of strain.

Also, as expected, the extracted piezoresistance coefficient is nearly same as that of the bulk silicon [107] but less than that of MOSFETs [106]. However, the ratio of increase in the drain current (ΔI_{DS}) to the total drain current (I_{DS}), i.e., $\Delta I_{DS}/I_{DS}$ remains independent of the gate overdrive voltage ($V_{GS} - V_{th}$) in JLFETs, indicating that strain enhances the mobility for all the gate voltages [107]. This is contrary to the MOSFETs where the strain-enhanced mobility degrades with large gate overdrive owing to the mobility degradation of the carriers flowing through the thin inversion layer due to the increased gate field. Since carriers flow in the bulk under very low gate field, no such mobility degradation is observed in JLFETs. Therefore, strain engineering can be effectively utilized to improve the mobility in JLFETs.

3.4.3 Carrier Ballisticity

In the MOSFETs, the electrons move from the source to the drain via the channel under the influence of the drain electric field and continuously collide with each other during the motion. The distance between successive collisions is called the mean free path. The scaling of MOSFETs in the sub-10 nm regime leads to a new kind of transport mechanism. When the channel length becomes smaller than the mean free path, the electrons can move from source to the drain without any collision. The electrons

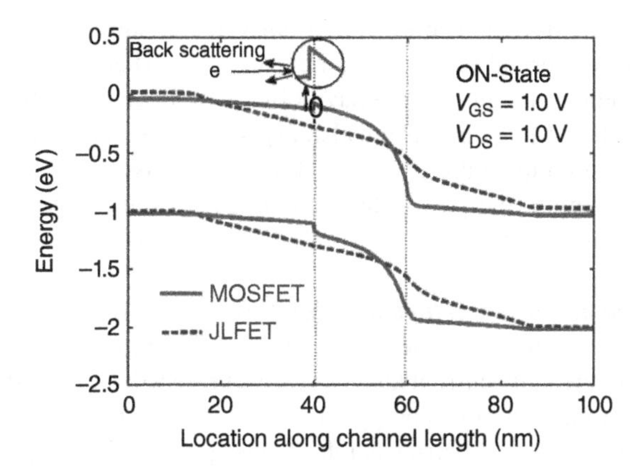

FIGURE 3.16 Energy band profiles of the MOSFET and JLFET in the ON-state ($V_{GS} = V_{DS} = 1.0\,V$) taken at a cutline 1-nm below the Si–SiO_2 interface.

would reach the drain from the source without encountering any scattering event. This motion of electrons where they move unimpeded like a bullet fired from the source reaching the drain is known as ballistic transport [109–111]. The mobility is significantly increased due to the ballistic transport, and this phenomenon is observed only when the channel length becomes smaller than the mean free path [109–111]. Since the JLFETs exhibit a different conduction mechanism than MOSFETs, the nature of ballistic transport would be different in JLFETs compared to the MOSFETs.

The JLFETs present a smooth transition from source to the drain region without any source-to-channel barrier height owing to the volume conduction. However, the gate voltage presents a source-to-channel barrier height even in the ON-state of MOSFETs. This leads to back scattering of carrier injected from the source by the optical phonons [112]. The absence of source-to-channel barrier height in JLFETs eliminates the possibility of back scattering via optical phonons as shown in Fig. 3.16. Therefore, the carrier ballisticity enhanced in the JLFETs compared to the MOSFETs in the ON-state and the ON-state current of JLFETs is expected to be more than that of MOSFETs under the ballistic regime. The carrier ballisticity reduces as the carriers are confined in a narrow silicon thickness due to enhanced electron–phonon scattering [112]. The carriers in the subthreshold regime are confined in a narrow region at the center of the channel in the JLFETs. As a result, the JLFETs exhibit a lower degree of ballisticity compared to the MOSFETs in the subthreshold regime. This enables JLFETs to have a lower OFF-state current compared to the MOSFETs in the ballistic mode of operation [112]. Therefore, the JLFETs are expected to outperform the MOSFETs as the channel length is scaled to the sub-10 nm regime.

3.4.4 Temperature Dependence

The temperature dependence study of nano-scale devices not only provides us with a deeper insight into the intrinsic performance of the device but also helps to investigate the reliability of the device. The traps and other defect sites are inactivated at low temperatures since thermal energy becomes very low. Therefore, low-temperature measurements indicate the intrinsic behavior of devices free from effects of interface traps. Similarly, the CMOS circuits are also expected to operate at high temperatures since the operating temperature in a typical microprocessor rises to ~500 K due to heating effects owing to the power dissipation. As a result, it becomes essential to understand the temperature performance of devices for reliable circuit performance.

3.4.4.A Zero Temperature Coefficient in MOSFETs In MOSFETs, the subthreshold current (I_{sub}) depends exponentially on the temperature. As the temperature increases, the subthreshold current also increases. From a device physics perspective, the electrons gain more thermal energy when temperature increases and easily overcome the source-to channel barrier height. Consequently, the subthreshold current increases with temperature [113]. However, the electrons with high thermal energy undergo more collisions with the lattice atoms and lead to a larger phonon scattering. As a result, the mobility degrades as the temperature is increased and the ON-state current reduces with an increase in the temperature [113–115]. This complex temperature dependence in a MOSFET leads to a zero-temperature coefficient (ZTC) point in the transfer characteristics. The ZTC point corresponds to the gate voltage at which the impact of the current enhancement due to the increased thermal energy of the electrons is nullified by the mobility degradation. Therefore, ZTC point remains unaffected by temperature changes in MOSFETs. The ZTC point is of significant interest to the device designers as the performance of a MOSFET circuit biased at this point remains unaffected by the temperature fluctuations.

3.4.4.B Zero Temperature Coefficient in JLFETs Let us now focus on the temperature dependence of the drain current in JLFETs. Initial studies on the temperature-dependent behavior of long-channel JLFETs revealed that both subthreshold current and ON-state current in JLFETs increase with temperature [114]. The threshold voltage reduction with temperature in JLFETs was found to be more than MOSFETs [114, 115]. As a result, MOSFETs exhibit a lower leakage current and perform better than JLFETs at high temperature. Also, the mobility degradation in JLFETs is mainly due to the impurity scattering effects owing to the heavily doped channel which varies as $T^{3/2}$ whereas the phonon scattering varies as $T^{-3/2}$. The two scattering effects try to compensate each other leading to a smaller degradation in mobility of JLFETs compared to the MOSFETs. Therefore, the current increases with temperature in JLFETs as the threshold voltage reduction with temperature dominates over the mobility degradation and, hence, no ZTC is observed [114, 115].

However, the JLFETs studied in [114, 115] used a high gate oxide thickness (~10 nm) leading to a poor gate control of the channel region. The threshold voltage dependence of a TGJLFET (Fig. 3.17) on the temperature can be expressed as [116]

$$\frac{dV_{Th}}{dT} = \frac{dV_{FB}}{dT} - \frac{dN_D}{dT}\left[\left\{\frac{q}{\varepsilon_{Si}}\left(\frac{WH}{2H+W}\right)^2\right\} + \left(\frac{qWHt_{ox}}{\varepsilon_{ox}}\right)\right] \tag{3.18}$$

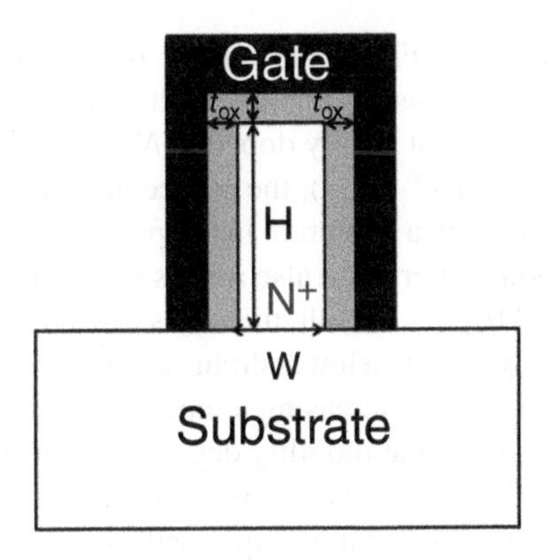

FIGURE 3.17 Cross-sectional view of a trigate JLFET.

where V_{Th} is the threshold voltage, V_{FB} is the flat-band voltage, t_{ox} is the gate oxide thickness, W is the width, and H is the height of the silicon fin. Equation (3.18) implies that the impact of the second term would be significant on the variation on threshold voltage with temperature if a large oxide thickness is used. Therefore, the use of a large oxide thickness overestimates the threshold voltage variation with temperature and also provides a poor gate control which is highly undesirable for achieving volume depletion in JLFETs.

The temperature dependence of JLFETs was revisited using a lower effective oxide thickness (EOT) of 1.5 nm [116, 119]. In these studies, the ZTC point was observed for both bulk and SOI short-channel and long-channel JLFETs as shown in Fig. 3.18. The mobility also shows a considerable dependence on the temperature in the long-channel JLFETs owing to the phonon scattering and the neutral defect scattering [115, 119]. The channel presents a large resistance in long-channel JLFETs compared with the source/drain series resistances. The impact of source/drain series resistance can be neglected while analyzing current in long-channel JLFETs.

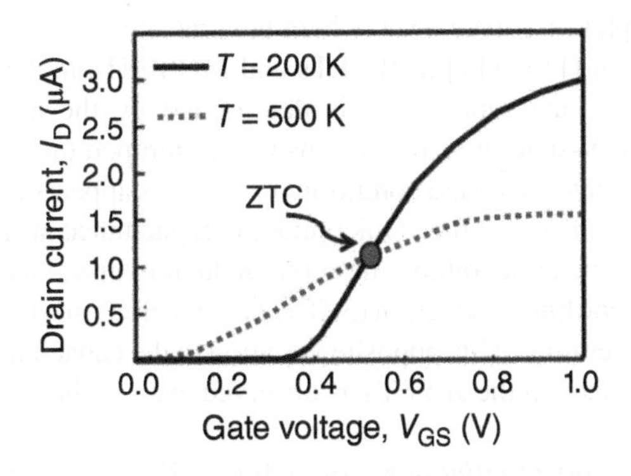

FIGURE 3.18 Transfer characteristics of the JLFET for different temperatures [117].

However, for short-channel JLFETs, the impact of source/drain series resistance is significant and the series resistance governs the current. Since the source/drain regions in JLFETs are somewhat lightly doped ($\sim N_D = 1 \times 10^{19}$ cm^{-3}) compared to the MOSFETs ($\sim N_D = 1 \times 10^{20}$ cm^{-3}), the source and drain present a larger series resistance in JLFETs. Although a reduction in temperature increases the mobility due to a lower acoustic phonon scattering, it also results in incomplete impurity ionization and dopant inactivation [116]. As a result, the series resistance increases with a reduction in the temperature and leads to a lower drain current. This effect complements the threshold voltage reduction with increasing temperature and leads to a ZTC in short-channel JLFETs. Furthermore, the mobility degradation in short-channel JLFETs is dominated by impurity (Coulomb) scattering and neutral impurity defect scattering.

The gate capacitance of a FET can be expressed as

$$\frac{1}{C_{GG}} = \frac{1}{C_{ox}} + \frac{1}{C_S} \tag{3.19}$$

where C_{GG} is the gate capacitance, C_{ox} is the gate oxide capacitance given as ϵ_{ox}/t_{ox}, and C_S is the semiconductor capacitance. C_S is the depletion capacitance (C_d) given as ϵ_{Si}/x_d, where x_d is the depletion region width, when the JLFET or MOSFET is operated in the depletion regime. However, under flat band conditions, C_S is equal to the flat band capacitance, which is given as

$$C_S(\text{flat band conditions}) = \sqrt{q\varepsilon_{Si}N_{ch}\frac{q}{kT}} \tag{3.20}$$

where N_{ch} is the channel doping.

The gate capacitance also exhibits anomalous temperature dependence in the JLFETs as compared to the MOSFETs. The gate capacitance in the subthreshold region of both MOSFETs and JLFETs is dominated by the depletion capacitance. The surface potential decreases with increasing temperature due to a shift in the threshold voltage. A lower surface potential means a reduced depletion region width. As a result, the depletion capacitance, which governs the gate capacitance, increases with the temperature [113, 117] in JLFETs and MOSFETs in the depletion regime. However, while the gate capacitance is determined by the oxide capacitance in MOSFETs in the ON-state, which is insensitive to temperature variations, the gate capacitance in ON-state (flat band condition) of JLFETs appears as a series combination of oxide capacitance and the semiconductor capacitance at flat band condition. With an increase in the temperature, the semiconductor capacitance in flat band condition decreases according to equation (3.20) [117] whereas the depletion capacitance increases with temperature. This opposite variation in the capacitances leads to a ZTC for the gate capacitance in the JLFETs as observed even in the case of drain current.

3.4.4.C Drain Current Oscillations In addition, JLFETs at extremely low temperatures (< 77 K) exhibit oscillations in the transconductance characteristics at low

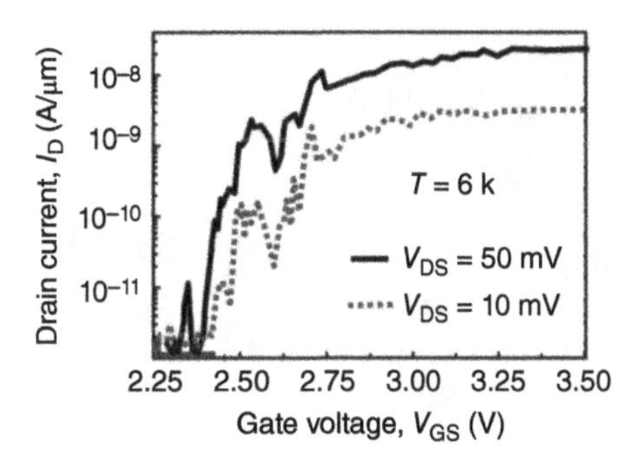

FIGURE 3.19 Transfer characteristics of the JLFET for different drain voltages at low temperature [121].

drain voltages (Fig. 3.19). Since the current flows in a constricted neutral region at the center of the JLFET, the effective area through which the current flows is quite small compared to the silicon thickness. Recently, it was observed that the ionized dopant atoms provide localized states and can act as quantum dots and perturb the potential of the charge carriers [120–124]. The random positioning of the dopant atoms through the conductive channel in JLFETs due to the random dopant fluctuation (RDF) effect [121–123] leads to the tunneling of electrons through these donor-induced quantum dots. The transconductance oscillations are attributed to this phenomenon, which is similar to the Coulomb blockade mechanism [121]. Furthermore, the trap densities within the band gap may also lead to oscillations [122]. Recently, such oscillations were also observed at room temperature for the Ge JLFETs with ultrathin channel leading to significant discretization of the subbands [125]. This is due to the quantum confinement effects and the oscillation peaks correspond to the location of the subbands [121–123, 125].

Moreover, the drain conductance (g_D) also exhibits an anomalous behavior at cryogenic temperatures (4.2 K). The drain conductance normally reduces with the drain voltage. However, at cryogenic temperatures, the drain conductance first increases and then starts decreasing [124]. This is because of dopant inactivation due to reduced ionized impurities at such low temperatures which lead to Schottky barrier and inhibit the formation of ohmic source/drain contacts [124].

The mobility at such low temperatures also exhibits an anomalous behavior. The mobility reduces until 15 K and increases again [123]. This effect can be understood from the interplay of impurity ionization and thermal energy. The ionized impurity scattering dominates the mobility degradation mechanism until 15 K [123]. However, the increase in the mobility with temperature is due to the increased thermal energy of the carriers which may pass the ionized impurity scattering centers without getting significantly deflected. However, as the temperature is reduced below 15 K, the ionization rate reduces significantly and the mobility degradation via the Coulomb scattering mechanism reduces. As a result, the mobility increases while the temperature is lowered from 15 to 6 K [123].

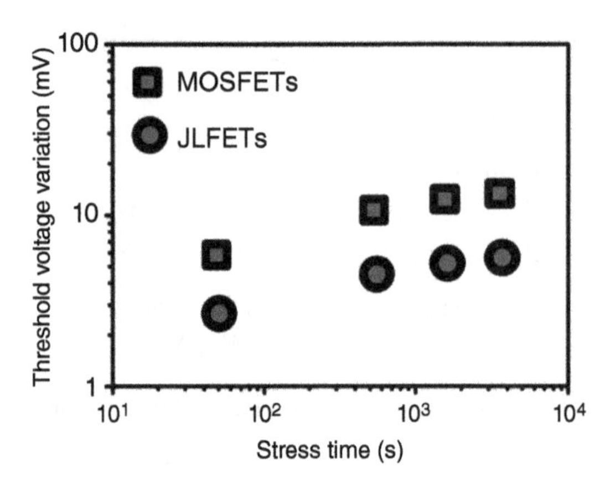

FIGURE 3.20 Threshold voltage variation in the JLFETs and MOSFETs under applied voltage stress for different stress durations [85].

3.4.5 Bias Temperature Instability

The incessant scaling of the MOSFETs and the need to enhance the gate control over the channel region has led to a reduction in the EOT. A reduction in EOT leads to an increased oxide electric field which leads to mobility degradation and threshold voltage instability [126, 127]. This phenomenon known as bias temperature instability (BTI) reduces the usability of the oxide for reliable operation below the industry standard duration (10 years) at the same operating voltage and high temperatures [127]. The BTI is one of the major issues concerning the reliability of the MOSFETs.

Since the BTI depends on the oxide electric field that follows different trends in JLFETs and MOSFETs due to a different conduction mechanism, the BTI is also expected to be different in JLFETs and MOSFETs. The electric field across the oxide is very high in the ON-state so as to attain inversion of carriers in MOSFETs. However, the electric field across the oxide layer is extremely low for the JLFETs in the ON-state (flat band condition) and increases only when JLFETs are operated under accumulation regime or full depletion regime (Fig. 3.12). Therefore, the BTI degradation in MOSFETs is large compared to the JLFETs due to the reduced oxide electric field as shown in Fig. 3.20.

The effect of a continuous stress on the devices over a longer period of time (say 10 years) can be mimicked by applying the bias stress and heating the devices to high temperature (~125°C) for a short time. This extended measure–stress–measure (eMSM) [85] technique can be used to analyze the BTI based on the shift in the threshold voltage with the magnitude and duration of the applied stress voltage. The eMSM measurements show that the shift in threshold voltage of both JLFETs and MOSFETs depends on the time for which stress is applied as $(t_{stress})^{1.5}$. This indicates that the BTI degradation mechanism is identical for both JLFETs and MOSFETs. However, even for a large voltage stress, the transconductance degradation in JLFETs was found to be much smaller than the MOSFETs. This is attributed to the bulk conduction in JLFETs which is less affected by the surface and traps at the oxide interface. As a result, the JLFET architecture is extremely beneficial for Ge and III-V material FETs, which suffer from high interface and border trap densities [85]. The JLFETs also exhibit the smallest negative temperature bias instability compared to the MOSFETs [85]. Therefore, JLFETs are expected to function reliably over a 10-year period unlike MOSFETs.

3.4.6 Low-Frequency Noise

Low-frequency (LF) noise is an indicator of quality and reliability of the electronic devices. A low noise is needed for radio frequency applications and for an improved signal to noise ratio in sensor applications. The LF noise generally originates due to two major factors:

a) The fluctuation of carrier concentration and carrier mobility, which is a surface phenomenon [128, 129]. The fluctuation in the carrier concentration is due to the trapping and detrapping of the carriers flowing close to the surface by the traps at the Si-gate oxide interface. The trapping of carriers modifies the oxide charges and, hence, changes the flat band voltage which in turn causes variation in the threshold voltage and the drain current. The carrier mobility fluctuation is related to the Coulomb scattering due to the trapped charges [128–133]. In a MOSFET, the carriers flow near the semiconductor surface. Therefore, the LF noise is contributed mostly by the carrier and mobility fluctuations [128–133].
b) The Hooge mobility fluctuation is a bulk phenomenon [128, 129]. In the JLFETs, since the carriers flow at the center of the silicon film in the flat band condition, i.e. the ON-state, the LF noise is expected to be dominated by the Hooge mobility fluctuation. However, the experimental measurements of LF noise in JLFETs indicate that the Hooge mobility fluctuation is not the dominant factor as explained below.

The measurement of LF noise in JLFETs and MOSFETs shows that the nature of LF noise power spectral density (S_{LF}) and the normalized noise power spectral density (S_{LF}/I_{DS}^2) is identical in both cases as shown in Fig. 3.21. While S_{LF} is proportional to $1/f$ (where f is the frequency), S_{LF}/I_{DS}^2 is proportional to g_m/I_{DS}^2 in both MOSFETs and JLFETs which is possible only when the LF noise is generated due to the carrier concentration fluctuation. This clearly indicates that the carrier concentration fluctuation is the major cause of LF noise even for the JLFETs. The carrier fluctuations in JLFETs are expected only in the accumulation mode of operation where the surface carriers are exposed to the interface traps as the current flows in the bulk along the neutral channel in other modes of operation. Therefore, the experimental measurements point toward other factors that may affect the carrier concentration and lead to LF noise as discussed below.

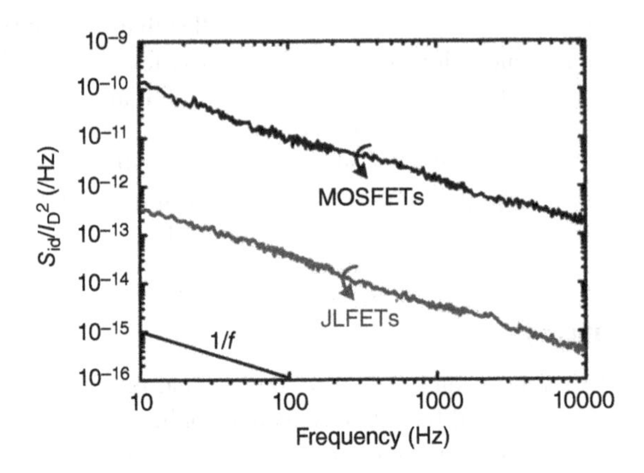

FIGURE 3.21 Normalized power spectral density for LF noise in JLFETs and MOSFETs [129].

In partial depletion mode, the carriers flow in a neutral channel separated from the oxide interface by a depletion layer. The carriers under the influence of the gate electric field may get transported through the depletion layer toward the interface where they may get trapped into the oxide trap centers. The carriers may even tunnel directly from the neutral region to the oxide through the depletion region under the influence of the gate electric field. These effects may lead to the dynamic occupancy of traps and charges in the oxide and a consequent drain current fluctuation in the partial depletion regime [128, 129].

Moreover, the presence of SRH generation–recombination centers in the changeover region between the depletion region and the neutral region may lead to generation–recombination noise [128, 129]. The trapping and subsequent detrapping of the recombination centers also causes fluctuation in the depletion layer thickness and, hence, the effective channel thickness fluctuates [128]. The LF noise behavior in the partial depletion mode of operation in JLFETs is attributed to these reasons [128, 129].

Although the behavior of LF noise in both JLFETs and MOSFETs is dominated by the carrier concentration and mobility fluctuations, the magnitude of LF noise is significantly lower in JLFETs (~2–4 orders of magnitude) compared with the MOSFETs [130, 131]. This is due to a lower interaction of the bulk carriers in JLFETs to the interface traps as compared to the MOSFETs.

A detailed analysis of LF noise for vertically stacked JLFET-based charge trap memory cell reinforces the fact that the carrier concentration fluctuation is the major contributor to the drain current fluctuations, and the carrier mobility fluctuation is negligible in JLFETs [132, 133]. However, the behavior of the normalized noise power spectral density $S_{\mathrm{LF}}/I_{\mathrm{DS}}^2$ deviates from the ideal g_m/I_{DS}^2 proportionality for carrier concentration fluctuation at high values of drain current in the JLFETs [132, 133]. The normalized noise power spectral density also increases with the reduction in the silicon film thickness [132, 133].

These two observations can be explained by including the source/drain series resistance contribution to the total noise power density along with the carrier number fluctuations [133]. The source/drain series resistance is lower in the heavily doped source/drain region of the MOSFETs. However, it is significantly high in the JLFETs with lightly doped source/drain regions ($N_{S/D} = 1 \times 10^{19}$ cm^{-3}) as compared to the MOSFETs ($N_{S/D} = 1 \times 10^{20}$ cm^{-3}). Moreover, the source/drain series resistance increases with decreasing silicon film thickness due to a reduction in the effective area for current conduction. Therefore, the deviation of the normalized noise power spectral density at high-current value and reduced silicon film thickness can be explained by the contribution from the source/drain series resistance [129, 133].

Because of the vertical scaling of the devices along with the horizontal scaling, the interaction between the carriers flowing at the surface with the traps at the Si-gate oxide interface also increases significantly. The random telegraph noise (RTN) is a measure of the electron trapping and detrapping by these defects (trap centers). The magnitude of RTN is also low in the JLFETs compared with the MOSFETs owing to the bulk conduction mechanism, which reduces the interaction of the carriers and the trapping/detrapping mechanism. However, the magnitude of RTN increases as the JLFETs are operated in the accumulation mode due to the shift in the carrier flow toward the surface where they are prone to more interactions with the surface traps.

3.4.7 Short-Channel Effects

In this section, we discuss the performance of the JLFETs when the channel lengths are scaled into the short-channel regime. Again, we would like to emphasize that since the carrier conduction mechanism is different in JLFETs and MOSFETs, the degradation in the performance of the JLFETs, owing to the short-channel effects like threshold voltage roll-off, drain-induced barrier lowering, and hot carrier effects, is expected to be different.

3.4.7.A Threshold Voltage Roll-Off The threshold voltage reduces with scaling the channel length in the MOSFETs due to the sharing of channel charge with the source/drain depletion regions [134]. The source/drain depletion region widths extend into the channel region, and the gate loses control over the channel charges at the vicinity of the source/drain depletion regions [134]. Consequently, a lower volume of channel has to be inverted by the gate which reduces the threshold voltage [134]. Furthermore, the effective channel length is also lower in the MOSFETs compared to the physical gate length due to the source/drain depletion regions extending into the channel. However, in JLFETs, due to the uniformly doped silicon film, there is an absence of source/drain depletion regions [15–20]. Therefore, the absence of source/drain metallurgical junctions not only reduces the channel charge sharing mechanism but also leads to an effective channel length which is equal to the physical gate length [15–20].

Moreover, in the OFF-state of a JLFET, the channel carriers are pushed away from the channel region into the source/drain regions to achieve volume depletion in the channel region. The movement of channel carriers into the source/channel regions

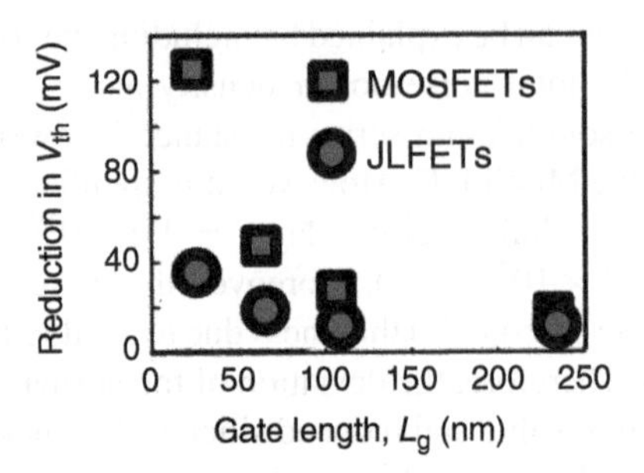

FIGURE 3.22 The threshold voltage roll-off in MOSFETs and JLFETs as a function of the gate length [16].

causes an electrostatic squeezing effect, which leads to realization of a channel length larger than even the physical gate length [15–18]. The threshold voltage roll-off is significantly reduced in JLFETs compared to the MOSFETs as shown in Fig. 3.22.

3.4.7.B Drain-Induced Barrier Lowering The drain-induced barrier lowering (DIBL) arises due to the interaction of the source/channel and channel/drain depletion regions in a short-channel MOSFET [134]. The coupling of the drain electric field to the source–channel interface leads to a reduction in the effective source-to-channel barrier height. This leads to a significant increase in the OFF-state current with increasing drain bias. However, in a JLFET, because of the absence of an inherent source/channel or channel/drain junction owing to the same doping concentration, the interaction between the source–channel interface and the drain–channel interface is limited. Although there is no metallurgical junction-induced depletion region at the channel–drain interface in JLFETs, the application of a positive drain voltage leads to the formation of an electrostatic depletion region at the channel–drain interface, which results in DIBL in JLFETs. However, the magnitude of DIBL is considerably small in JLFETs compared to the MOSFETs [15–20] as shown in Fig. 3.23.

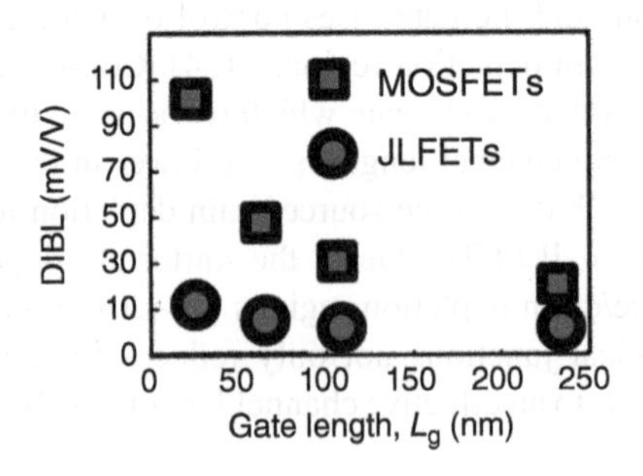

FIGURE 3.23 The drain-induced barrier lowering in MOSFETs and JLFETs as a function of the gate length [16].

3.4.7.C Hot Electron Effect The high electric field in the short-channel MOSFETs leads to a significant increase in the momentum and kinetic energy of the electrons, and such highly energetic electrons are generally referred as "hot" electrons. Since there exists a high electric field across the oxide to induce inversion layer in MOSFETs, these hot electrons are also attracted toward the gate oxide. The injection of hot electrons into the gate oxide may lead to a degradation in the quality of the oxide insulator. This may lead to a change in the threshold voltage [134].

However, the JLFETs operate close to the flat band so that the carries flow at the center (bulk) of the channel region. The vertical electric field is minimum at the flat band condition (ON-state) [15–20, 73, 85]. Even if the electrons are accelerated by the high lateral electric field, they cannot penetrate into the gate oxide owing to the low vertical electric field in the ON-state in JLFETs. In addition, the peak of the lateral electric field is lower in the JLFETs compared to the MOSFETs and located inside the drain in the JLFETs rather than at the channel–drain interface for the MOSFETs as shown in Fig. 3.24 [73]. Therefore, the JLFETs exhibit a reduced hot electron effect compared to the MOSFETs [73]. This also improves the hot carrier reliability of the JLFETs compared to the MOSFETs [85].

However, the experimental results of [135, 136] indicate that the drain voltage stress induced degradation in performance of JLFETs is more than the MOSFETs at low gate voltages. The impact ionization rate is more in JLFETs compared to the MOSFETs under low gate bias and high drain bias stress. This is attributed to the heavily doped channel region, which leads to a reduction in the band gap owing to the band gap narrowing effect [135]. A lower band gap increases the impact ionization rate in JLFETs compared to MOSFETs for a given electric field.

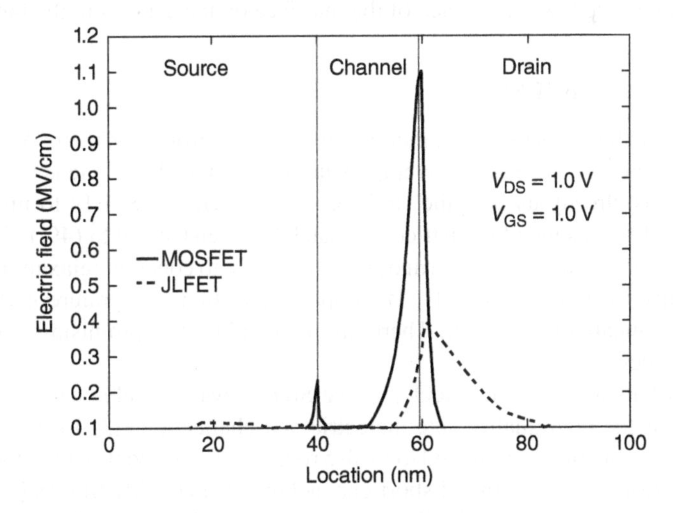

FIGURE 3.24 The lateral electric field profile in MOSFETs and JLFETs taken at a cutline 1 nm below the Si–SiO_2 interface.

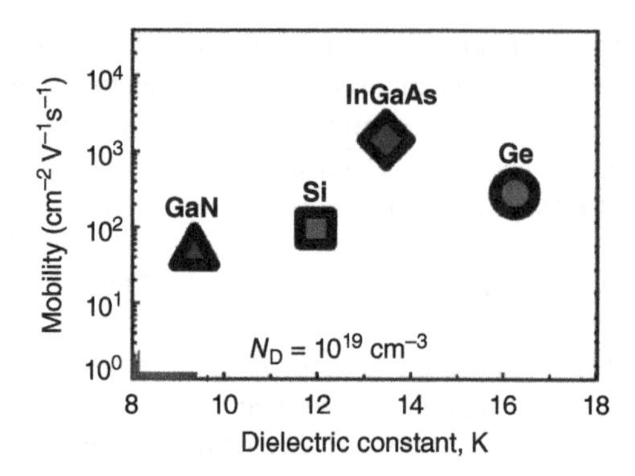

FIGURE 3.25 The electron mobility and dielectric constant for different semiconductors at a doping of $N_D = 10^{19}$ cm^{-3}.

3.5 BEYOND SILICON JLFETS: OTHER MATERIALS

In JLFETs, the interaction of the interface charges or defects with the current carrying carriers is restricted due to its bulk conduction nature. The volume conduction feature of JLFETs is quite attractive for efficient performance of not only silicon-based JLFETs but also for other materials like germanium, InGaAs, gallium nitride (GaN), etc., which do not have native oxide of their own and suffer from severe degradation issues due to the poor interface quality [85]. The use of these materials in MOSFETs is restricted due to the defects and trapped charges at the interface between the oxide and these materials. However, the junctionless architecture may pave the way for efficient realization of high mobility materials (Fig. 3.25) like InGaAs or germanium in FETs owing to the lower influence of the interface on the carriers in the bulk.

3.5.1 Germanium JLFETs

Germanium exhibits an excellent mobility for both electrons (3900 cm^2 V^{-1} s^{-1}) and holes (1900 cm^2 V^{-1} s^{-1}) at room temperature [137]. In addition, Ge is CMOS compatible and is already used in the strained silicon SiGe MOSFETs to improve the carrier mobility. In spite of the lattice mismatch between Ge and Si (4%), defect-free Ge can be grown over Si in the form of nanowires due to coherent relaxation of stress in the reduced volume [138–141]. The dopant activation temperature is also lower for Ge as compared to Si [142]. Therefore, germanium is a promising material for CMOS devices.

However, its native oxide (GeO_2) unlike SiO_2 is water soluble and volatile. The formation of the high-quality gate dielectric is a challenge [143, 144]. The higher dielectric constant of Ge compared to Si also reduces the gate control over the channel region and leads to worsening of short-channel effects in Ge MOSFETs [145].

Moreover, the Fermi level in Ge is pinned close to the valence band [146, 147]. Although this is an advantage for realizing ohmic contacts even on lightly doped Ge pMOSFETs, the pinning of the Fermi level close to the valence band does not allow the Fermi level to rise close to the conduction band and enable a strong inversion layer formation in the case of Ge n-MOSFETs [146, 147]. The Schottky barrier height is also higher for metal-n-Ge contacts making their contact resistance large [147]. Also, the smaller band gap of Ge leads to a leakage current due to a high band-to-band tunneling (BTBT) [144, 148, 149].

The junctionless architecture, owing to the volume conduction phenomenon, reduces the interaction between the dielectric interface and the carrier flow. Therefore, the Ge JLFETs are relatively immune to the interface traps and defects as compared to Ge MOSFETs [144–147, 150–153]. In addition, Fermi-level pinning near the valence band facilitates depletion of the channel region in n-Ge JLFETs [147] leading to efficient volume depletion. Single-gate [146], DG [150], TG [144, 145], and gate all around Ge JLFETs [147, 150–152] have been experimentally demonstrated on Si- or Ge-on-insulator substrates.

It was found that the use of *in situ* doping can reduce the thermal budget and result in creation of shallower junctions as compared to the ion implantation process in Ge JLFETs [151, 152]. Moreover, the ion implantation process results in a doping profile in which the peak doping concentration is located below the surface. By etching the surface, the region with the peak concentration may be exposed and a better ohmic metal germanium contact can be made for the n-Ge JLFETs [147]. A core–shell nanowire with the Ge core and Si shell was also demonstrated [142]. Such heterostructure core–shell nanowires lead to the formation of a hole gas within the channel region without any intentional doping of the channel which further boosts the performance of Ge JLFETs [141, 142].

Recently, the quantum confinement effects in ultrathin (3 nm) Ge JLFETs with ultrashort (1 nm) channel lengths were found to be helpful in achieving volume depletion and improving the performance of Ge JLFETs [153]. It was also reported that the Ge JLFETs in the quantum-confined regime exhibit an increased band gap (2.36 eV for film thickness = 1 nm and 0.86 eV for film thickness = 3 nm) and do not need to follow the conventional relation between the channel length L_g and the channel thickness t_{ch}, i.e., $L_g = t_{ch}/3$.

3.5.2 Indium Gallium Arsenide JLFETs

InGaAs has an extremely high carrier mobility even at large doping levels. For instance, the mobility of InGaAs for a film doping $N_D = 1 \times 10^{19}$ cm^{-3} is ~ 2500 cm^2 V^{-1} s^{-1}, which is nearly 25 times higher as compared to silicon (~ 100 cm^2 V^{-1} s^{-1}) [154]. However, the InGaAs MOSFETs offer a negative threshold voltage owing to the presence of traps and defects at the oxide interface [155–157]. A negative threshold voltage is undesirable for logic applications. The JLFET architecture with a lower impact of surface defects on the carriers owing to the volume conduction paves the way for realization of InGaAs JLFETs with a positive threshold voltage appropriate for logic applications [154]. The JLFET architecture also circumvents the challenges offered by the implantation process [157] or the metal organic regrowth process [158] used for realizing heavily doped (> 10^{19} cm^{-3}) source and drain regions in InGaAs MOSFETs.

TG [154] and gate-all-around InGaAs JLFETs over InGaAs on an insulator [159, 160], or a buffer layer of indium phosphide (InP) over a Si substrate [161] have been proposed in the literature. The InGaAs–InP JLFETs formed using metal organic chemical vapor deposition instead of ion implantation exhibit a high level of linearity in the transconductance [159]. A high linearity in the device characteristics eliminates the need for circuit-level implementations for achieving linear behavior which, reduces the area footprint and power dissipation. Also, to reduce the source/drain series resistance in InGaAs JLFETs, a tapered source/drain along with a dovetail (triangular) nanowire cross section was also proposed [160].

However, the InGaAs MOSFETs cannot outperform Si MOSFETs in the ballistic regime due to a lower density of states and poor short-channel effects owing to the increased dielectric constant of InGaAs [162]. However, it was shown that the InGaAs JLFETs have a lower degradation due to the impurity scattering as compared to Si JLFETs and exhibit a higher ON-state current than the Si counterpart. The Fermi-level degeneracy in the source and drain regions of InGaAs JLFETs leads to a larger source-to-channel barrier height in InGaAs JLFETs compared to the Si JLFETs. This leads to an improvement in the short-channel effects in InGaAs JLFETs and counters the impact of the high dielectric constant of InGaAs which degrades the short-channel behavior. Therefore, InGaAs JLFETs can be a promising alternative to Si JLFETs even in the ballistic regime.

3.5.3 Gallium Nitride JLFETs

GaN is a promising material for high-power and high-frequency devices owing to its high band gap, large critical electric field, and high saturation electron velocity [163]. GaN is normally employed in the high electron mobility transistors (HEMTs) as AlGaN/GaN heterostructure leading to the formation of a two-dimensional electron gas (2DEG) in GaN. Although the formation of 2DEG in the HEMT devices facilitates high-frequency operation due to the extremely high mobility (~ 1500 cm^2 V^{-1} s^{-1}), it is extremely difficult to realize normally-OFF HEMTs due to the presence of the 2DEG. However, AlGaN/GaN HEMTs can be turned OFF by reducing the GaN thickness and increasing the number of gates. The fringing fields from the side gates can easily deplete the 2DEG [164–166]. However, the ON-state performance of AlGaN/GaN HEMTs also reduces with a reduction in the width of the device.

Recently, it was demonstrated that junctionless FinFETs made of GaN as a channel material exhibit a reduced dependency on the fin width variation owing to the current flow in the heavily doped volume [167]. The GaN junctionless FinFETs also exhibit a reduced OFF-state current and an improved breakdown voltage as compared to the AlGaN/GaN FinFETs [167]. Therefore, the GaN JLFinFETs can be a lucrative alternative to AlGaN/GaN-based FinFETs [167, 168]. The large valence band offset between GaN and Si is sufficient to block tunneling and diffusion current while realizing GaN JLFETs on Si substrates [169].

3.6 CHALLENGES

3.6.1 High Source/Drain Series Resistance

As discussed in Section 3.1, the doping concentration of the channel and the silicon film thickness must be adequately designed to achieve volume depletion of the channel region in the OFF-state. However, for practically achievable silicon film thicknesses, volume depletion is possible in JLFETs only when the channel region is moderately doped ($N_D = 1 \times 10^{19}$ cm^{-3}). A higher doping concentration in the channel region not only demands an ultrathin silicon body to achieve volume depletion but also leads to a degradation in the mobility due to the increased ionized impurity scattering. A moderately doped source/drain region offers a high resistance. For instance, the sheet resistivity of n-type silicon with a phosphorus doping of 1×10^{19} cm^{-3} is 0.00543 Ω-cm and reduces by nearly seven times for a phosphorus doping of 1×10^{20} cm^{-3} (0.0008 Ω-cm) [170]. Similarly, for a boron doping of 1×10^{20} cm^{-3}, the sheet resistivity (0.0012 Ω-cm) is nearly seven times higher than for a boron doping of 1×10^{19} cm^{-3} (0.0088 Ω-cm) [170]. A higher source/drain resistance may lead to a reduced ON-state current. Furthermore, a moderate doping also leads to difficulties in ohmic contact formation. Therefore, a critical aspect while designing JLFETs is minimizing the source/drain series resistance and obtain proper ohmic contacts.

3.6.2 Random Dopant Fluctuations

As we discussed earlier in Section 3.1, the doping process, i.e. ion implantation, involves bombardment of dopant atom ions onto the silicon film which leads to discrete and random positioning of dopant atoms in the silicon film. As a result, the position of dopant atoms in the silicon film is random and the assumption of a uniform doping profile in the silicon film is vague and unrealistic. The C–V measurements of scaled MOSFETs exhibit an unusual stretch out in the C–V characteristics compared to the simulations of the Poisson equation with assumption of a uniform channel doping [171]. Earlier, this stretch out was attributed to the presence of interface traps. However, the development in the interface engineering and device processing techniques virtually eliminated the interface charge effects at the Si–SiO_2 interface [171]. Therefore, this stretch out could not be explained entirely due to the interface charges/defects.

The advent of CMOS scaling led to the requirement of a heavily doped channel as it would allow lower source–channel and channel–drain depletion region widths leading to an increased effective channel length. However, the number of dopant atoms per MOSFET increased significantly. For instance, a semiconductor film of dimensions 10 nm × 10 nm × 10 nm with a doping of 10^{20} cm^{-3} has 100 dopant atoms. However, the random positioning of the dopant atoms creates fluctuations in the surface potential by influencing the carrier flow. Each dopant atom may be treated as a source for emission of electrostatic (Coulomb attraction) field. The effective Coulomb potential at any particular position in the silicon film decays after a specific screening length,

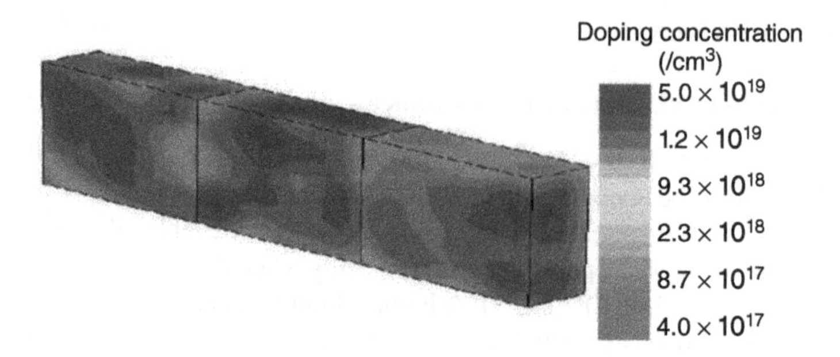

FIGURE 3.26 The random distribution of dopants in JLFETs as a result of ion implantation process [176].

which is proportional to the doping at that position. The random positioning of the dopant atoms as shown in Fig. 3.26 creates potential fluctuations in the doped silicon film [172], and these fluctuations would be larger if the silicon film is heavily doped. The potential fluctuation may lead to the formation of potential barriers, which may impede the motion of charge carriers and degrade the mobility [172]. Furthermore, the potential barriers may even lead to the formation of localized states in the band gap [172]. The measurements at low temperature that limit the impact of interface traps clearly indicate such surface potential fluctuations in the channel region of MOSFETs. The potential fluctuation translates into a random fluctuation in the threshold voltage at different locations in the MOSFETs due to the arbitrary positioning of dopant atoms. Since the potential fluctuations increase with an increase in the channel doping, the threshold voltage fluctuations would also increase with an increase in the channel doping concentration owing to the scaling process. The stretch out in the C–V characteristics also leads to degradation in the subthreshold slope of the MOSFETs [172]. Therefore, this RDF is a serious issue and hinders the scaling of MOSFETs.

The RDF leads to a significant amount of mismatch between two MOSFETs fabricated using the same process flow. Such a mismatch in the transistor characteristics renders them unfeasible for symmetric circuits such as static random access memory (SRAM) cells or differential amplifiers, which cannot tolerate any fluctuation/mismatch in the MOSFET characteristics such as threshold voltage [173]. In addition, RDF is inherent to the doping process and cannot be eliminated. Therefore, to mitigate threshold voltage fluctuations due to RDF, undoped channel regions were proposed in advanced CMOS architectures like DGFETs or FinFETs. However, the heavily doped source/drain regions still exhibit RDF and show a fluctuation in the source/drain series resistance [174] and a reduction in the effective channel length due to screening of the channel region by randomly distributed source/drain dopant atoms [172–174].

3.6.3 RDF in JLFETs

From our discussion in Section 3.1, we know that the JLFETs are heavily doped throughout the silicon film unlike MOSFETs. As a result, the impact of RDF is expected to be higher in JLFETs as compared to MOSFETs where only source and drain regions are heavily doped. A contrasting feature of the nature of RDF in JLFETs and MOSFETs is that in MOSFETs the whole transfer characteristics shift as a result of RDF, whereas in JLFETs the spread in the drain current is larger in the subthreshold (full depletion regime) and gradually reduces in the above threshold (partial depletion), flat band, and accumulation regimes [175–181]. The impact of RDF on threshold voltage and subthreshold swing variability indicates that RDF leads to a larger fluctuation in JLFETs than in MOSFETs. Therefore, JLFETs are more prone to mismatch effects than MOSFETs.

Analytically, the change in the threshold voltage (ΔV_{Th}) due to a change in the local doping (ΔN_{D}) due to RDF can be approximated as $\Delta V_{\mathrm{Th}} \approx (\partial V_{\mathrm{Th}}/\partial N_{\mathrm{D}})\Delta N_{\mathrm{D}}$. From equations (3.10) and (3.12), the change in the threshold voltage is obtained as

$$\Delta V_{\mathrm{Th}} \approx \Delta N_{\mathrm{D}} \left[-\frac{q t_{\mathrm{Si}}^2}{2\varepsilon_{\mathrm{Si}}} - \frac{q t_{\mathrm{Si}} t_{\mathrm{ox}}}{\varepsilon_{\mathrm{ox}}} \right] \quad \text{for JLFETs} \tag{3.21}$$

and

$$\Delta V_{\mathrm{Th}} \approx \Delta N_{\mathrm{A}} \left[\frac{\left\{1 + \ln\left(\frac{N_{\mathrm{A}}}{n_i}\right)\right\} \sqrt{q V_t \varepsilon_{\mathrm{Si}}}}{\varepsilon_{\mathrm{ox}} \sqrt{N_{\mathrm{A}} \ln\left(\frac{N_{\mathrm{A}}}{n_i}\right)}} t_{\mathrm{ox}} + 2\frac{V_t}{N_{\mathrm{A}}} \right] \quad \text{for MOSFETs} \tag{3.22}$$

It is not intuitive to compare the change in the threshold voltages in JLFETs and MOSFETs simply by using these expressions. Therefore, let us numerically compare the values using a SiO_2 gate oxide with a thickness of 1 nm, a silicon film thickness of 10 nm, and a silicon film doping $N_{\mathrm{D}} = 10^{19}$ cm^{-3} for JLFETs and $N_{\mathrm{A}} = 10^{16}$ cm^{-3} for MOSFETs. Substituting these values in (3.21) and (3.22), we get $\Delta V_{\mathrm{Th}} \approx 10^{-19}\, \Delta N_{\mathrm{D}}$ for JLFETs and $\Delta V_{\mathrm{Th}} \approx 5 \times 10^{-18} \Delta N_{\mathrm{A}}$ for MOSFETs. Now, using an optimistic approach, let us assume that the change in the local doping is only 5% due to RDF, i.e., $\Delta N_{\mathrm{D}} = 5 \times 10^{17}$ cm^{-3} and $\Delta N_{\mathrm{A}} = 5 \times 10^{14}$ cm^{-3}. Using these values, we obtain $\Delta V_{\mathrm{Th}} = 50$ mV for JLFETs and $\Delta V_{\mathrm{Th}} = 2.5$ mV for MOSFETs. Therefore, even using the analytical approach, we arrive at the same conclusion that JLFETs are more prone to RDF-related threshold voltage instability than MOSFETs.

The channel length scaling further leads to an enhanced threshold voltage fluctuation due to RDF as the effective channel length, which now becomes a function of the position of dopants and their screening lengths apart from the defined physical gate length, becomes very small [176–182]. For instance, in the semiconductor film example that we discussed in Section 3.6.2, a variation in the effective channel length of even 2 nm due to dopant screening may cause a significant change in the threshold voltage if the channel length is extremely small, let us say 10 nm. A reduction in the effective channel length significantly degrades the short-channel effects and leads to a considerable threshold voltage variability in JLFETs with channel length scaling.

However, scaling the channel width and the EOT has an opposite impact on the threshold voltage variation in JLFETs. Scaling the channel width and EOT enhances the gate control over the channel region and reduces the short-channel effects and the variability of JLFETs to RDF [180, 182]. Therefore, scaling the width and length simultaneously according to the technology roadmap specifications leads to compensation of the degradation caused by channel length scaling and improvement brought by the channel width scaling. The overall impact is a lower variability with technology node scaling owing to the dominance of the improvement by reducing the channel width [182].

Moreover, the impact of RDF on threshold voltage variability is more for silicon JLFETs as compared to the germanium JLFETs owing to a larger dielectric constant of germanium as compared to silicon [183]. However, the subthreshold swing variability due to RDF is larger in germanium JLFETs owing to a significant increase in the leakage current due to the lower band gap of germanium [183].

3.6.4 Sensitivity to Process Variations

3.6.4.A Meaning of Line Edge Roughness Line edge roughness (LER) refers to the imperfections in the geometry of the MOSFETs arising due to the inherent molecular structure of the photoresists [174]. The patterns are carved out on the MOSFET using a photoresist as a mask. The imperfections/roughness in the photoresists could lead to imperfect geometrical shapes, for example, a curved shape instead of a straight line along the device width or even at the source–channel and channel–drain interfaces. In scaled MOSFETs, the mean value of the dimension of the resist roughness is comparable to the device dimensions [174]. As a result, the impact of LER on the variability in performance of scaled MOSFETs is large. LER inherently leads to nonuniformity in effective channel length and the effective device thickness. Therefore, it becomes essential to improve the fabrication process and reduce the mean dimension of the roughness of the photoresists.

3.6.4.B Line Edge Roughness in JLFETs In JLFETs, the LER may lead to variation in the width of the channel region. A variation in the channel width leads to a significant change in the ON-state since the ON-current flows through the whole silicon film unlike surface conduction in MOSFETs which is insensitive to the silicon film thickness. Furthermore, the channel width also dictates the OFF-state current in JLFETs as volume depletion is achieved only when the silicon film thickness is below a particular channel thickness. Therefore, the impact of LER in inducing variations in the electrical parameters of JLFETs is considerably large compared to the MOSFETs due to the different conduction mechanism in JLFETs [184]. The variation in the threshold voltage and subthreshold swing is significantly larger in JLFETs compared to the MOSFETs.

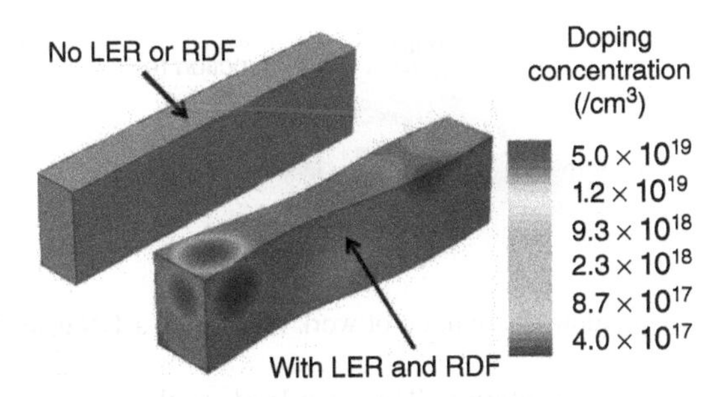

FIGURE 3.27 The impact of line edge roughness and RDF on dopant distribution and active silicon thickness in JLFETs [185].

3.6.4.C Relationship between LER and RDF Since LER and RDF are two major factors deciding the variability in JLFETs and MOSFETs, it becomes imperative to understand the interactions between the two in increasing the variability in the performance parameters, if any. We start with a discussion on the individual impact of RDF and LER on the MOSFETs: The RDF leads to a reduction in the effective channel length, whereas LER leads to a fluctuation in the channel width. A MOSFET with a reduced effective channel length would be more sensitive to the variations in the channel width due to LER. Furthermore, the overall impact of RDF depends on the magnitude of RDF and LER at the source–channel and channel–drain junctions since the junctions decide the effective channel length. Therefore, the impact of RDF is more localized in MOSFETs and there exists an interaction between LER and RDF in MOSFETs [185].

However, in JLFETs, the impact of RDF or LER is not localized and distributed over the entire channel length as shown in Fig. 3.27. The overall impact of RDF and LER get averaged out, and they are uncorrelated for the JLFETs unlike MOSFETs [185].

Moreover, as discussed in Section 2.1, TFET also builds upon the concept of utilizing a source–channel tunneling junction. Since an ultrasteep source–channel junction is required for efficient tunneling process, the impact of RDF on the source–channel junction is more important in TFETs. Therefore, the impact of RDF is more localized in TFETs (one junction) than MOSFETs (RDF localized at two junctions) and JLFETs (nonlocalized RDF). To conclude, the presence of a source–channel or channel–drain junction due to doping leads to a localization in RDF and an interaction between device variations produced by LER and RDF [185].

3.6.4.D Work Function Variability The common metals used as a gate electrode in MOSFETs do not have a uniform composition, and the work function depends on the grain orientation. For instance, in TiN which is commonly used as a gate electrode, the <200> and <111> grain orientations have work functions of 4.6 and 4.4 eV, with 60% and 40% probability of occurrence, respectively [176, 182, 183] as shown in Fig. 3.28. Therefore, different regions in the channel are subjected to a different work

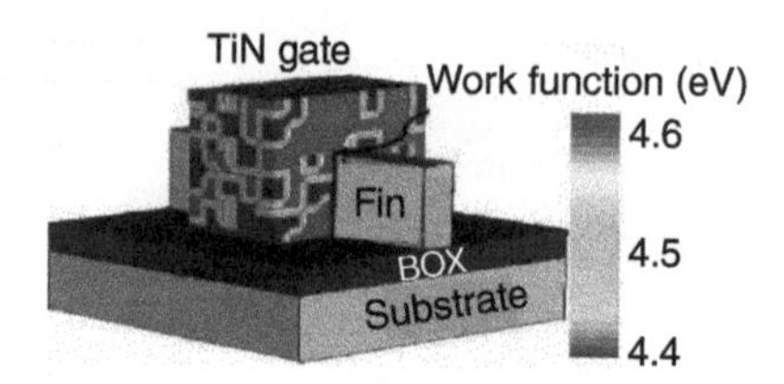

FIGURE 3.28 The random distribution of work function in a TiN gate electrode [176].

function due to the grain orientation. This also leads to threshold voltage variability in MOSFETs.

However, the impact of work function variation on the spread of ON-state current is lower in MOSFETs compared to the JLFETs owing to the surface inversion layer, which shields the channel from the impact of the work function variation. However, in JLFETs, the concentration of surface carriers is lower and no such screening effect is observed. Therefore, the ON-state current is more prone to variability due to work function variation in a JLFET compared to the MOSFET [176].

However, in a JLFET, in the partial depletion state or near threshold, the current flows at the center unlike MOSFETs. The subthreshold current is not significantly affected by the gate electrode, and the corresponding variability is lower in JLFETs compared to the MOSFETs [176].

The work function variation increases with scaling in both MOSFETs and JLFETs due to a reduction in the gate area and the consequent loss of grain orientations leading to a much more spread in the work function variation [182].

3.6.4.E Channel Width Variation As discussed in Section 3.1, an ultrathin body is required to achieve volume depletion in the OFF-state considering the highly doped channel region in JLFETs. Also, the ON-state current flows through the entire cross section in JLFETs. Therefore, JLFETs are expected to be more sensitive to the channel width variation as compared to MOSFETs where the current flows at the surface in the ON-state and the channel width only dictates the short-channel effects.

Utilizing equation (3.10), the change in the threshold voltage due to a variation in the silicon film thickness (t_{Si}) can be expressed as

$$\Delta V_{\mathrm{Th}} \approx \left(\frac{\partial V_{\mathrm{Th}}}{\partial t_{\mathrm{Si}}}\right) \Delta t_{\mathrm{Si}} = \Delta t_{\mathrm{Si}} \left[-\frac{qN_{\mathrm{D}}}{\varepsilon_{\mathrm{Si}}}\left(t_{\mathrm{Si}} - \frac{t_{\mathrm{ox}}\varepsilon_{\mathrm{Si}}}{\varepsilon_{\mathrm{ox}}}\right)\right] \tag{3.23}$$

for JLFETs. For a JLFET with a silicon film doping of 10^{19} cm^{-3}, a SiO_2 gate oxide thickness = 1 nm, a silicon thickness of 10 nm, and a change of 5% in the nominal film thickness due to process variations, $\Delta V_{\mathrm{Th}} \approx 54$ mV which is large. Ideally, the threshold voltage of the MOSFETs should not depend on the silicon film thickness according to equation (3.12).

However, in MOSFETs with lightly doped channel region, the threshold voltage decreases with an increase in the channel thickness owing to the reduced effective gate control and the consequent increase in the short-channel effects [186]. However, the

MOSFETs with a highly doped channel region exhibit a reverse trend. The threshold voltage increases with an increase in channel thickness due to a higher doping in the channel. However, a high channel doping also leads to a reduced mobility due to the impurity scattering [186]. On the other hand, in JLFETs, the threshold voltage decreases with an increase in the channel thickness since a negative voltage needs to be applied to achieve volume depletion. Moreover, the sensitivity of JLFETs to the channel width variation is significantly high compared to the MOSFETs [186].

3.6.5 Fabrication Issues

As discussed in Section 3.1, the silicon film doping in JLFETs needs to be large to avoid degradation in the ON-state current and ohmic contact quality. However, this stringent requirement puts a restriction on the silicon film thickness that can be efficiently depleted by the gate electrode. For instance, even a gate electrode with the highest possible work function of 5.93 eV corresponding to platinum can only deplete nearly 5–6 nm of n-type silicon film doped at 1×10^{19} cm^{-3} [19] as shown in Fig. 3.29. Therefore, an ultrathin body (≤ 5 nm) is indispensable to achieve volume depletion in single-gate JLFETs with platinum as a gate electrode. Moreover, it is difficult to pattern platinum as it is a noble metal. Also, utilizing platinum as the gate electrode is a costly affair. Achieving volume depletion in single-gate JLFETs is a challenge and requires ultrathin silicon film (≤ 5 nm). Realizing a uniform ultrathin silicon film is not only difficult, but it also increases the manufacturing cost and complexity. However, volume depletion may be easily realized in JLFETs with multi gate architectures due to the presence of gate electrodes on more than one side which may deplete the whole silicon film easily. Therefore, the multigate architectures provide an inherent advantage for fabricating JLFETs.

3.6.6 Band-to-Band Tunneling in OFF-State

As discussed in Section 3.2, in the OFF-state, the heavily doped channel region in a JLFET is volume depleted. This results in a significant overlap of the conduction band of the drain region and the valence band of the channel region in JLFETs as shown

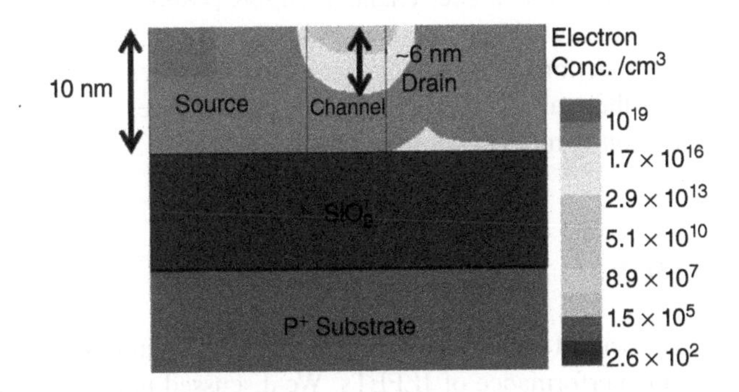

FIGURE 3.29 The extent of depletion in a SOI JLFET with an active silicon film thickness of 10 nm using a gate electrode with a work function = 5.1 eV.

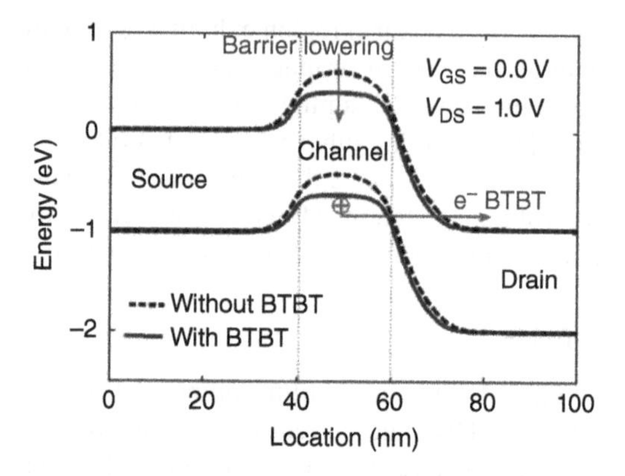

FIGURE 3.30 The energy band profiles of the SOI JLFET taken at a cutline 1 nm below the Si–SiO_2 interface with BTBT models and without BTBT models in the OFF-state.

in Fig. 3.30 [187]. This band alignment facilitates the band-to-BTBT of electrons from the channel to the drain leaving behind holes in the channel region which is similar to the lateral band-to-band tunneling (L-BTBT) gate-induced drain leakage (GIDL) in nanowire MOSFETs as discussed in Section 2.6. The high concentration of holes in the channel region leads to the formation of a parasitic bipolar junction transistor (BJT). The channel region, which is converted into p-type due to BTBT, acts as the base, and the n^+ source and the n^+ drain act as the emitter and the collector, respectively, of this parasitic BJT. Unlike the GIDL phenomenon in MOSFETs where the drain current increases significantly with a negative gate voltage, the OFF-state current in a JLFET does not have a considerable voltage dependence [187, 188]. This is attributed to the parasitic BJT which is turned-on in the OFF-state resulting in a significant increase in the OFF-state leakage current and loss of gate control over the channel region for negative gate voltages. As a result, the JLFETs suffer from a severe degradation in the ON-state to OFF-state current ratio (I_{ON}/I_{OFF}) and, therefore, this needs to be alleviated. Moreover, this BTBT-induced parasitic BJT action is more pronounced for JLFETs with smaller channel lengths posing serious challenges to the JLFET scaling in the sub-10 nm regime [188].

The parasitic BJT action is the major factor that degrades the performance of JLFETs and hinders the scaling of JLFETs to the sub-10 nm regime. We discuss this important phenomenon in more detail in Chapter 4.

3.7 CONCLUSION

In this chapter, we focused on the fundamentals of JLFETs and the various parameters that impact the performance of JLFETs. We discussed in detail the difference in the MOSFETs and JLFETs with respect to key performance parameters and how the volume conduction mechanism in JLFETs leads to an altogether different mechanism in JLFETs. We also listed the limitations of the JLFETs. In the next chapter, we give an overview of the device architectures, which have been proposed to mitigate these challenges. As we shall see, these device architectures significantly improve the performance of JLFETs.

REFERENCES

[1] J. F. Gibbons, "Ion implantation in semiconductors—Part I: Range distribution theory and experiments," *Proc. IEEE*, vol. 56, no. 3, pp. 295–319, Mar. 1968.

[2] J. F. Gibbons, "Ion implantation in semiconductors—Part II: Damage production and annealing," *Proc. IEEE*, vol. 60, no. 9, pp. 1062–1096, Sept. 1972.

[3] S. Furukawa, H. Matsumura, and H. Ishiwara, "Theoretical considerations on lateral spread of implanted ions," *Jap. J. Appl. Phys.*, vol. 11, no. 2, pp. 134, Feb. 1972.

[4] K. Boucart and A. M. Ionescu, "Double-gate tunnel FET with high-κ gate dielectric," *IEEE Trans. Electron Devices*, vol. 54, no. 7, pp. 1725–1733, July 2007.

[5] W. Y. Choi, B. G. Park, J. D. Lee, and T. J. Liu, "Tunneling field-effect transistors (TFETs) with subthreshold swing (SS) less than 60 mV/dec," *IEEE Electron Device Lett.*, vol. 28, no. 8, pp. 743–745, Aug. 2007.

[6] A. M. Ionescu and H. Riel, "Tunnel field-effect transistors as energy-efficient electronic switches," *Nature*, vol. 479, no. 7373, pp. 329–337, Nov. 2011.

[7] T. Krishnamohan, D. Kim, S. Raghunathan, and K. Saraswat, "Double-gate strained-Ge heterostructure tunneling FET (TFET) with record high drive currents and <60 mV/dec subthreshold slope," *IEDM Tech. Dig.*, pp. 1–3, Dec.2008.

[8] J. Knoch and J. Appenzeller, "Modeling of high-performance p-type III–V heterojunction tunnel FETs," *IEEE Electron Device Lett.*, vol. 31, no. 4, pp. 305–307, Apr. 2010.

[9] W. H. Brattain and B. John, "Three-electrode circuit element utilizing semiconductive materials," U.S. Patent 2524035, Oct. 1950.

[10] J. Bardeen and W. H. Brattain, "The transistor, a semiconductor triode," *Proc. IEEE*, vol. 86, no. 1, pp. 29–30, Jan. 1998.

[11] I. M. Ross, "The invention of the transistor," *Proc. IEEE*, vol. 86, no. 1, pp. 7–28, Jan. 1998.

[12] M. Riordan, L. Hoddeson, and C. Herring, "The invention of the transistor," *Rev. Mod. Phys.*, vol. 71, no. 2, S336, Mar. 1999.

[13] J. E. Lilienfield, "Method and apparatus for controlling electric currents," U.S. Patent 1745175, Jan. 1930.

[14] J. E. Lilienfield, "Device for controlling electric current," U.S. Patent 1900018, Mar. 1933.

[15] J.-P. Colinge, C.-W. Lee, A. Afzalian, N. D. Akhavan, R. Yan, I. Ferain, P. Razavi, B. O'Neill, A. Blake, M. White, A.-M. Kelleher, B. McCarthy, and R. Murphy, "Nanowire transistors without junctions," *Nature Nanotechnol.*, vol. 5, no. 3, pp. 225–229, Mar. 2010.

[16] C.-W. Lee, I. Ferain, A. Afzalian, R. Yan, N. M. Akhavan, P. Razavi, and J. P. Colinge, "Performance estimation of junctionless multigate transistors," *Solid-State Electron.*, vol. 54, no. 2, pp. 97–103, Feb. 2010.

[17] C.-W. Lee, A. Afzalian, N. D. Akhavan, R. Yan, I. Ferain, and J.-P. Colinge, "Junctionless multigate field-effect transistor," *Appl. Phys. Lett.*, vol. 94, no. 5, pp. 053511–053512, 2009.

[18] R. Rios, A. Cappellani, M. Armstrong, A. Budrevich, H. Gomez, R. Pai, N. Rahhal-Orabi, and K. Kuhn, "Comparison of junctionless and conventional trigate transistors with Lg down to 26 nm," *IEEE Electron Device Lett.*, vol. 32, no. 9, pp. 1170–1172, Sept. 2011.

[19] S. Sahay and M. J. Kumar, "Realizing efficient volume depletion in SOI junctionless FETs," *IEEE J. Electron Devices Soc.*, vol. 4, no. 3, pp. 110–115, May 2016.

[20] S. Gundapaneni, S. Ganguly, and A. Kottantharayil, "Bulk planar junctionless transistor (BPJLT): An attractive device alternative for scaling," *IEEE Electron Device Lett.*, vol. 32, no. 3, pp. 261–263, Mar. 2011.

[21] J. P. Colinge, *FinFETs and other multi-gate transistors*, vol. 73, Springer, New York, 2008.

[22] S. Sahay and M. J. Kumar, "Symmetric operation in an extended back gate JLFET for scaling to the 5 nm regime considering quantum confinement effects," *IEEE Trans. Electron Devices*, vol. 64, no. 1, pp. 21–27, Jan. 2017.

[23] V. Kumari, N. Modi, M. Saxena, and M. Gupta, "Modeling and simulation of double gate junctionless transistor considering fringing field-effects," *Solid-State Electron.*, vol. 107, pp. 20–29, May 2015.

[24] J. P. Duarte, S. J. Choi, D. I. Moon, and Y. K. Choi, "Simple analytical bulk current model for long-channel double-gate junctionless transistors," *IEEE Electron Device Lett.*, vol. 32, no. 6, pp. 704–706, June 2011.

[25] J. M. Sallese, N. Chevillon, C. Lallement, B. Iniguez, and F. Pregaldiny, "Charge-based modeling of junctionless double-gate field-effect transistors," *IEEE Trans. Electron Devices*, vol. 58, no. 8, pp. 2628–2637, Aug. 2011.

[26] J. P. Duarte, S. J. Choi, and Y. K. Choi, "A full-range drain current model for double-gate junctionless transistors," *IEEE Trans. Electron Devices*, vol. 58, no. 12, pp. 4219–4225, Dec. 2011.

[27] J. P. Duarte, M. S. Kim, S. J. Choi, and Y. K. Choi, "A compact model of quantum electron density at the subthreshold region for double-gate junctionless transistors," *IEEE Trans. Electron Devices*, vol. 59, no. 4, pp. 1008–1012, Apr. 2012.

[28] T. K. Chiang, "A quasi-two-dimensional threshold voltage model for short-channel junctionless double-gate MOSFETs," *IEEE Trans. Electron Devices*, vol. 59, no. 9, pp. 2284–2289, Sept. 2012.

[29] Z. Chen, Y. Xiao, M. Tang, Y. Xiong, J. Huang, J. Li, X. Gu, and Y. Zhou, "Surface-potential-based drain current model for long-channel junctionless double-gate MOSFETs," *IEEE Trans. Electron Devices*, vol. 59, no. 12, pp. 3292–3298, Dec. 2012.

[30] A. Gnudi, S. Reggiani, E. Gnani, and G. Baccarani, "Semianalytical model of the subthreshold current in short-channel junctionless symmetric double-gate field-effect transistors," *IEEE Trans. Electron Devices*, vol. 60, no. 4, pp. 1342–1348, Apr. 2013.

[31] J. H. Woo, J. M. Choi, and Y. K. Choi, "Analytical threshold voltage model of junctionless double-gate MOSFETs with localized charges," *IEEE Trans. Electron Devices*, vol. 60, no. 9, pp. 2951–2955, Sept. 2013.

[32] F. Jazaeri, L. Barbut, and J. M. Sallese, "Trans-capacitance modeling in junctionless symmetric double-gate MOSFETs," *IEEE Trans. Electron Devices*, vol. 60, no. 12, pp. 4034–4040, Dec. 2013.

[33] T. Holtij, M. Graef, F. M. Hain, A. Kloes, and B. Iniguez, "Compact model for short-channel junctionless accumulation mode double gate MOSFETs," *IEEE Trans. Electron Devices*, vol. 61, no. 2, pp. 288–299, Feb. 2014.

[34] F. Jazaeri, L. Barbut, and J. M. Sallese, "Generalized charge-based model of double-gate junctionless FETs, including inversion," *IEEE Trans. Electron Devices*, vol. 61, no. 10, pp. 3553–3557, Oct. 2014.

[35] F. Jazaeri, L. Barbut, and J. M. Sallese, "Modeling asymmetric operation in double-gate junctionless FETs by means of symmetric devices," *IEEE Trans. Electron Devices*, vol. 61, no. 12, pp. 3962–3970, Dec. 2014.

[36] B. W. Hwang, J. W. Yang, and S. H. Lee, "Explicit analytical current-voltage model for double-gate junctionless transistors," *IEEE Trans. Electron Devices*, vol. 62, no. 1, pp. 171–177, Jan. 2015.

[37] J. Hur, J. M. Choi, J. H. Woo, H. Jang, and Y. K. Choi, "A generalized threshold voltage model of tied and untied double-gate junctionless FETs for a symmetric and asymmetric structure," *IEEE Trans. Electron Devices*, vol. 62, no. 9, pp. 2710–2716, Sept. 2015.

[38] A. Yesayan, F. Jazaeri, and J. M. Sallese, "Charge-based modeling of double-gate and nanowire junctionless FETs including interface-trapped charges," *IEEE Trans. Electron Devices*, vol. 63, no. 3, pp. 1368–1374, Mar. 2016.

[39] X. Lin, B. Zhang, Y. Xiao, H. Lou, L. Zhang, and M. Chan, "Analytical current model for long-channel junctionless double-gate MOSFETs," *IEEE Trans. Electron Devices*, vol. 63, no. 3, pp. 959–965, Mar. 2016.

[40] B. Singh, D. Gola, K. Singh, E. Goel, S. Kumar, and S. Jit, "Analytical modeling of channel potential and threshold voltage of double-gate junctionless FETs with a vertical Gaussian-Like doping profile," *IEEE Trans. Electron Devices*, vol. 63, no. 6, pp. 2299–2305, June 2016.

[41] T. K. Chiang, "A short-channel-effect-degraded noise margin model for junctionless double-gate MOSFET working on subthreshold CMOS logic gates," *IEEE Trans. Electron Devices*, vol. 63, no. 8, pp. 3354–3359, Aug. 2016.

[42] X. Jin, X. Liu, M. Wu, R. Chuai, J. H. Lee, and J. H. Lee, "A unified analytical continuous current model applicable to accumulation mode (junctionless) and inversion mode MOSFETs with symmetric and asymmetric double-gate structures," *Solid-State Electron.*, vol. 79, pp. 206–209, Jan. 2013.

[43] Y. Xiao, X. Lin, H. Lou, B. Zhang, L. Zhang, and M. Chan, "A short channel double-gate junctionless transistor model including the dynamic channel boundary effect," *IEEE Trans. Electron Devices*, vol. 63, no. 12, pp. 4661–4667, Dec. 2016.

[44] J. P. Duarte, M. S. Kim, S. J. Choi, and Y. K. Choi, "A compact model of quantum electron density at the subthreshold region for double-gate junctionless transistors," *IEEE Trans. Electron Devices*, vol. 59, no. 4, pp. 1008–1012, Apr. 2012.

[45] A. Veloso, G. Hellings, M. J. Cho, E. Simoen, K. Devriendt, V. Paraschiv, E. Vecchio, Z. Tao, J. J. Versluijs, L. Souriau, and H. Dekkers, "Gate-all-around NWFETs vs. triple-gate FinFETs: Junctionless vs. extensionless and conventional junction devices with controlled EWF modulation for multi-VT CMOS," *VLSI Tech. Dig.*, pp. T138–T139, June 2015.

[46] T. Holtij, A. Kloes, and B. Iñíguez, "3-D compact model for nanoscale junctionless triple-gate nanowire MOSFETs, including simple treatment of quantization effects," *Solid-State Electron.*, vol. 112, pp. 85–98, Oct. 2015.

[47] F. Ávila-Herrera, B. C. Paz, A. Cerdeira, M. Estrada, and M. A. Pavanello, "Charge-based compact analytical model for triple-gate junctionless nanowire transistors," *Solid-State Electron.*, vol. 122, pp. 23–31, Aug. 2016.

[48] S. Barraud, M. Berthomé, R. Coquand, M. Cassé, T. Ernst, M. P. Samson, P. Perreau, K. K. Bourdelle, O. Faynot, and T. Poiroux, "Scaling of trigate junctionless nanowire MOSFET with gate length down to 13 nm," *IEEE Electron Device Lett.*, vol. 33, no. 9, pp. 1225–1227, Sept. 2012.

[49] R. D. Trevisoli, R. T. Doria, M. de Souza, S. Das, I. Ferain, and M. A. Pavanello, "Surface-potential-based drain current analytical model for triple-gate junctionless nanowire transistors," *IEEE Trans. Electron Devices*, vol. 59, no. 12, pp. 3510–3518, Dec. 2012.

[50] T. K. Chiang, "A new subthreshold current model for junctionless trigate MOSFETs to examine interface-trapped charge effects," *IEEE Trans. Electron Devices*, vol. 62, no. 9, pp. 2745–2750, Sept. 2015.

[51] R. Trevisoli, R. T. Doria, M. de Souza, S. Barraud, M. Vinet, and M. A. Pavanello, "Analytical model for the dynamic behavior of triple-gate junctionless nanowire transistors," *IEEE Trans. Electron Devices*, vol. 63, no. 2, pp. 856–863, Feb. 2016.

[52] Z. Guo, J. Zhang, Z. Ye, and Y. Wang, "3-D analytical model for short-channel triple-gate junctionless MOSFETs," *IEEE Trans. Electron Devices*, vol. 63, no. 10, pp. 3857–3863, Oct. 2016.

[53] T. A. Oproglidis, A. Tsormpatzoglou, D. H. Tassis, T. A. Karatsori, S. Barraud, G. Ghibaudo, and C. A. Dimitriadis, "Analytical drain current compact model in the depletion operation region of short-channel triple-gate junctionless transistors," *IEEE Trans. Electron Devices*, vol. 64, no. 1, pp. 66–72, Jan. 2017.

[54] T. Wang, L. Lou, and C. Lee, "A junctionless gate-all-around silicon nanowire FET of high linearity and its potential applications," *IEEE Electron Device Lett.*, vol. 34, no. 4, pp. 478–480, Apr. 2013.

[55] O. Moldovan, F. Lime, and B. Iniguez, "A compact explicit model for long-channel gate-all-around junctionless MOSFETs. Part II: Total charges and intrinsic capacitance characteristics," *IEEE Trans. Electron Devices*, vol. 61, no. 9, pp. 3042–3046, Sept. 2014.

[56] F. Lime, O. Moldovan, and B. Iniguez, "A compact explicit model for long-channel gate-all-around junctionless MOSFETs. Part I: DC characteristics," *IEEE Trans. Electron Devices*, vol. 61, no. 9, pp. 3036–3041.

[57] A. Martinez, M. Aldegunde, A. R. Brown, S. Roy, and A. Asenov, "NEGF simulations of a junctionless Si gate-all-around nanowire transistor with discrete dopants," *Solid-State Electron.*, vol. 71, pp. 101–105, May 2012.

[58] F. Jazaeri, L. Barbut, and J. M. Sallese, "Trans-capacitance modeling in junctionless gate-all-around nanowire FETs," *Solid-State Electron.*, vol. 96, pp. 34–37, June 2014.

[59] S. Sahay and M. J. Kumar, "Controlling the drain side tunneling width to reduce ambipolar current in tunnel FETs using heterodielectric BOX " *IEEE Trans. Electron Devices*, vol. 62, no. 11, pp. 3882–3886, Nov. 2015.

[60] A. K. Jain, S. Sahay, and M. J. Kumar, "Controlling L-BTBT in emerging nanotube FETs using dual-material gate," *IEEE J. Electron Devices Soc.*, vol. 6, pp. 611–621, June 2018.

[61] C. H. Park, M. D. Ko, K. H. Kim, C. W. Sohn, C. K. Baek, Y. H. Jeong, and J. S. Lee, "Comparative study of fabricated junctionless and inversion-mode nanowire FETs," in *IEEE Dev. Res. Conf. (DRC)*, pp. 179–180, June 2011.

[62] A. Veloso, B. Parvais, P. Matagne, E. Simoen, T. Huynh-Bao, V. Paraschiv, E. Vecchio, K. Devriendt, E. Rosseel, M. Ercken, and B. T. Chan, "Junctionless gate-all-around lateral and vertical nanowire FETs with simplified processing for advanced logic and analog/RF applications and scaled SRAM cells," *VLSI Tech. Dig.*, pp. 1–2, June 2016.

[63] S. Sahay and M. J. Kumar, "Comprehensive analysis of gate-induced drain leakage in emerging FET architectures: Nanotube FETs vs. nanowire FETs," *IEEE Access*, vol. 64, no. 7, pp. 18918–18926, Dec. 2017.

[64] S. Sahay and M. J. Kumar, "Spacer design guidelines for nanowire FETs from gate induced drain leakage perspective," *IEEE Trans. Electron Devices*, vol. 64, no. 7, pp. 3007–3015, July 2017.

[65] S. Sahay and M. J. Kumar, "Physical insights into the nature of gate induced drain leakage in ultra-short channel nanowire field-effect transistors," *IEEE Trans. Electron Devices*, vol. 64, no. 6, pp. 2604–2610, June 2017.

[66] S. Sahay and M. J. Kumar, "Nanotube junctionless field-effect transistor: Proposal, design and investigation," *IEEE Trans. Electron Devices*, vol. 64, no. 4, pp. 1851–1856, Apr. 2017.

[67] S. Sahay and M. J. Kumar, "Diameter dependency of leakage current in nanowire junctionless field-effect transistors," *IEEE Trans. Electron Devices*, vol. 64, no. 3, pp. 1330–1335, Mar. 2017.

[68] S. Sahay and M. J. Kumar, "A novel gate-stack-engineered nanowire FET for scaling to the sub-10-nm regime," *IEEE Trans. Electron Devices*, vol. 63, no. 12, pp. 5055–5059, Dec. 2016.

[69] S. Sahay and M. J. Kumar, "Insight into lateral band-to-band-tunneling in nanowire junctionless field-effect transistors," *IEEE Trans. Electron Devices*, vol. 63, no. 10, pp. 4138–4142, Oct. 2016.

[70] S. Sahay and M. J. Kumar, "Controlling L-BTBT and volume depletion in nanowire JLFETs using core-shell architecture," *IEEE Trans. Electron Devices*, vol. 63, no. 9, pp. 3790–3794, Sept. 2016.

[71] J. Fan, M. Li, X. Xu, Y. Yang, H. Xuan, and R. Huang, "Insight into gate-induced drain leakage in silicon nanowire transistors," *IEEE Trans. Electron Devices*, vol. 62, no. 1, pp. 213–219, Jan. 2015.

[72] J. Hur, B.-H. Lee, M.-H. Kang, D.-C. Ahn, T. Bang, S.-B Jeon, and Y.-K. Choi, "Comprehensive analysis of gate-induced drain leakage in vertically stacked nanowire FETs:

Inversion-mode vs. junctionless mode," *IEEE Electron Device Lett.*, vol. 37, no. 5, pp. 541–544, May 2016.

[73] J. P. Colinge, C.-W. Lee, I. Ferain, N. D. Akhavan, R. Yan, P. Razavi, R. Yu, A. N. Nazarov, and R. T. Doria, "Reduced electric field in junctionless transistors," *Appl. Phys. Lett.*, vol. 96, no. 7, 073510, Feb. 2010.

[74] J. P. Colinge, "Conduction mechanisms in thin-film accumulation-mode SOI p-channel MOSFETs," *IEEE Trans. Electron Devices*, vol. 37, no. 3, pp. 718–723, Mar. 1990.

[75] W. Cheng, A. Teramoto, M. Hirayama, S. Sugawa, and T. Ohmi, "Impact of improved high-performance Si (110)-oriented metal–oxide–semiconductor field-effect transistors using accumulation-mode fully depleted silicon-on-insulator devices," *Jap. J. Appl. Phys.*, vol. 45, no. 4S, pp. 3110, Apr. 2006.

[76] J. P. Colinge, D. Flandre and F. Van de Wiele, "Subthreshold slope of long-channel, accumulation-mode p-channel SOI MOSFETs," *Solid-State Electron.*, vol. 37, no. 2, pp. 289–294, Feb. 1994.

[77] J. W. Slotboom and H. C. De Graaff, "Measurements of bandgap narrowing in Si bipolar transistors," *Solid-State Electron.*, vol. 19, no. 10, pp. 857–862, Oct. 1976.

[78] D. B. Klaassen, J. W. Slotboom, and H. C. De Graaff, "Unified apparent bandgap narrowing in n-and p-type silicon," *Solid-State Electron.*, vol. 35, no. 2, pp. 125–129, Feb. 1992.

[79] S. C. Jain and D. J. Roulston, "A simple expression for band gap narrowing (BGN) in heavily doped Si, Ge, GaAs and Ge_xSi_{1-x} strained layers," *Solid-State Electron.*, vol. 34, no. 5, pp. 453–465, May 1991.

[80] A. Schenk, "Finite-temperature full random-phase approximation model of band gap narrowing for silicon device simulation," *J. Appl. Phys.*, vol. 84, no. 7, pp. 3684–3695, Oct. 1998.

[81] S. E. Swirhun, Y. H. Kwark, and R. M. Swanson, "Measurement of electron lifetime, electron mobility and band-gap narrowing in heavily doped p-type silicon," *IEDM Tech. Dig.*, pp. 24–27, 1986.

[82] D. J. Roulston, N. D. Arora, and S. G. Chamberlain, "Modeling and measurement of minority-carrier lifetime versus doping in diffused layers of n^+-p silicon diodes," *IEEE Trans. Electron Devices*, vol. 29, no. 2, pp. 284–291, Feb. 1982.

[83] J. G. Fossum and D. S. Lee, "A physical model for the dependence of carrier lifetime on doping density in nondegenerate silicon," *Solid-State Electron.*, vol. 25, no. 8, pp. 741–747, Aug. 1982.

[84] S. M. Wen and C. O. Chui, "CMOS junctionless field-effect transistors manufacturing cost evaluation," *IEEE Trans. Semiconductor Manufacturing*, vol. 26, no. 1, pp. 162–168, Feb. 2013.

[85] M. Toledano-Luque, P. Matagne, A. Sibaja-Hernández, T. Chiarella, L. A. Ragnarsson, B. Sorée, M. Cho, A. Mocuta, and A. Thean, "Superior reliability of junctionless pFinFETs by reduced oxide electric field," *IEEE Electron Device Lett.*, vol. 35, no. 12, pp. 1179–1181, Dec. 2014.

[86] M. Ichii, R. Ishida, H. Tsuchiya, Y. Kamakura, N. Mori, and M. Ogawa, "Computational study of effects of surface roughness and impurity scattering in Si double-gate junctionless transistors," *IEEE Trans. Electron Devices*, vol. 62, no. 4, pp. 1255–1261, Apr. 2015.

[87] A. Ueda, M. Luisier, and N. Sano, "Enhanced impurity-limited mobility in ultra-scaled Si nanowire junctionless field-effect transistors," *Appl. Phys. Lett.*, vol. 107, no. 25, 253501, Dec. 2015.

[88] N. Kadotani, T. Ohashi, T. Takahashi, S. Oda, and K. Uchida, "Experimental study on electron mobility in accumulation-mode silicon-on-insulator metal–oxide–semiconductor field-effect transistors," *Jap. J. Appl. Phys.*, vol. 50, no. 9R, 094101, Sept. 2011.

[89] K. I. Goto, T. H. Yu, J. Wu, C. H. Diaz, and J. P. Colinge, "Mobility and screening effect in heavily doped accumulation-mode metal-oxide-semiconductor field-effect transistors," *Appl. Phys. Lett.*, vol. 101, no. 7, 073503, Aug. 2012.

[90] T. Rudenko, A. Nazarov, I. Ferain, S. Das, R. Yu, S. Barraud, and P. Razavi, "Mobility enhancement effect in heavily doped junctionless nanowire silicon-on-insulator metal-oxide-semiconductor field-effect transistors," *Appl. Phys. Lett.*, vol. 101, no. 21, 213502, Nov. 2012.

[91] K. Wei, L. Zeng, J. Wang, G. Du, and X. Liu, "Physically based evaluation of electron mobility in ultrathin-body double-gate junctionless transistors," *IEEE Electron Device Lett.*, vol. 35, no. 8, pp. 817–819, Aug. 2014.

[92] B. Sorée, W. Magnus, and W. Vandenberghe, "Low-field mobility in ultrathin silicon nanowire junctionless transistors," *Appl. Phys. Lett.*, vol. 99, no. 23, 233509, Dec. 2011.

[93] x. Li, W. Han, H. Wang, L. Ma, Y. Zhang, Y. Du, and F. Yang, "Low-temperature electron mobility in heavily n-doped junctionless nanowire transistor," *Appl. Phys. Lett.*, vol. 102, no. 22, 223507, June 2013.

[94] H. Wang, W. Han, L. Ma, X. Li, and F. Yang, "Investigation of mobility enhancement of junctionless nanowire transistor at low temperatures," *J. Vac. Sci. Technol. B.*, vol. 33, no. 4, 040603, July 2015.

[95] G. Ghibaudo, "New method for the extraction of MOSFET parameters," *Electron. Lett.*, vol. 24, no. 9, pp. 543–545, Apr. 1988.

[96] F. Jazaeri and J. M. Sallese, "Carrier mobility extraction methodology in junctionless and inversion-mode FETs," *IEEE Trans. Electron Devices*, vol. 62, no. 10, pp. 3373–3378, Oct. 2015.

[97] M. K. Joo, M. Mouis, D. Y. Jeon, S. Barraud, S. J. Park, G. T. Kim, and G. Ghibaudo, "Flat-band voltage and low-field mobility analysis of junctionless transistors under low-temperature," *Semicond. Sci. Technol.*, vol. 29, no. 4, 045024, Mar. 2014.

[98] D. Y. Jeon, S. J. Park, M. Mouis, S. Barraud, G. T. Kim, and G. Ghibaudo, "New method for the extraction of bulk channel mobility and flat-band voltage in junctionless transistors," *Solid-State Electron.*, vol. 89, pp. 139–141, Nov. 2013.

[99] S. Tiwari, M. V. Fischetti, P. M. Mooney, and J. J. Welser, "Hole mobility improvement in silicon-on-insulator and bulk silicon transistors using local strain," *IEDM Tech. Dig.*, pp. 939–941, 1997.

[100] T. Ghani, M. Armstrong, C. Auth, M. Bost, P. Charvat, G. Glass, T. Hoffmann, K. Johnson, C. Kenyon, J. Klaus, and B. McIntyre, "A 90-nm high volume manufacturing logic technology featuring novel 45 nm gate length strained silicon CMOS transistors," *IEDM Tech. Digest*, pp. 11–16, 2003

[101] T. Komoda, A. Oishi, T. Sanuki, K. Kasai, H. Yoshimura, K. Ohno, A. Iwai, M. Saito, F. Matsuoka, N. Nagashima, and T. Noguchi, "Mobility improvement for 45 nm node by

combination of optimized stress and channel orientation design," *IEDM Tech. Digest*, pp. 217–220, 2004.

[102] V. Venkataraman, S. Nawal, and M. J. Kumar, "Compact analytical threshold voltage model of nanoscale fully-depleted strained-silicon on silicon-germanium-on-insulator (SGOI) MOSFETs," *IEEE Trans. Electron Devices*, vol. 54, pp. 554–562, Mar. 2007.

[103] M. J. Kumar, V. Venkataraman, and S. Nawal, "Impact of strain or Ge content on the threshold voltage of nanoscale strained-Si/SiGe bulk MOSFETs," *IEEE Trans. Device Mater. Rel.*, vol. 7, pp. 181–187, Mar. 2007.

[104] M. J. Kumar, V. Venkataraman, and S. Nawal, "A simple analytical threshold voltage model of nanoscale fully depleted single-layer strained-silicon-on-insulator (SSOI) MOSFETs," *IEEE Trans. Electron Devices*, vol. 53, pp. 2500–2506, Oct. 2006.

[105] Y. Kanda, "A graphical representation of the piezoresistance coefficients in silicon," *IEEE Trans. Electron Devices*, vol. 29, no. 1, pp. 64–70, Jan. 1982.

[106] J. P. Raskin, J. P. Colinge, I. Ferain, A. Kranti, C. W. Lee, N. D. Akhavan, R. Yan, P. Razavi, and R. Yu, "Mobility improvement in nanowire junctionless transistors by uniaxial strain," *Appl. Phys. Lett.*, vol. 97, no. 4, 042114, July 2010.

[107] C. J. Huang and S. C. Lee, "Stress effects on self-aligned silicon nanowires junctionless field-effect transistors," *IEEE-Nano*, pp. 134–138, Aug. 2011.

[108] T. K. Kang, "The piezoresistive effect in n-type junctionless silicon nanowire transistors," *Nanotechnology*, vol. 23, no. 47, 475203, Oct. 2012.

[109] M. S. Shur and L.F. Eastman, "Ballistic transport in semiconductor at low temperatures for low-power high-speed logic," *IEEE Trans. Electron Devices*, vol. 26, no. 11, pp. 1677–1683, Nov. 1979.

[110] S. Datta, *Quantum Transport: Atom to Transistor*, Cambridge University Press, Cambridge, UK, 2005.

[111] D. Sels, B. Sorée, and G. Groeseneken, "Quantum ballistic transport in the junctionless nanowire pinch-off field-effect transistor," *J. Comp. Electron.*, vol. 10, no. 1, pp. 216–221, June 2011.

[112] N. D. Akhavan, I. Ferain, P. Razavi, R. Yuand, and J. P. Colinge, "Improvement of carrier ballisticity in junctionless nanowire transistors," *Appl. Phys. Lett.*, vol. 98, no. 10, 103510, Mar. 2011.

[113] F. H. Gaensslen, V. L. Rideout, E. J. Walker, and J. J. Walker, "Very small MOSFET's for low-temperature operation," *IEEE Trans. Electron Devices*, vol. 24, no. 3, pp. 218–229, Mar. 1977.

[114] C. W. Lee, A. Borne, I. Ferain, A. Afzalian, R. Yan, N. D. Akhavan, P. Razavi, and J. P. Colinge, "High-temperature performance of silicon junctionless MOSFETs," *IEEE Trans. Electron Devices*, vol. 57, no. 3, pp. 620–625, Mar. 2010.

[115] M. de Souza, M. A. Pavanello, R. D. Trevisoli, R. T. Doria, and J. P. Colinge, "Cryogenic operation of junctionless nanowire transistors," *IEEE Electron Device Lett.*, vol. 32, no. 10, pp. 1322–1324, Oct. 2011.

[116] R. T. Doria, R. D. Trevisoli, M. de Souza, S. Das, I. Ferain, and M. A. Pavanello, "The zero temperature coefficient in junctionless nanowire transistors," *Appl. Phys. Lett.*, vol. 101, no. 6, 062101, Aug. 2012.

[117] M. H. Han, H. B. Chen, S. S. Yen, C. S. Shao, and C. Y. Chang, "Temperature-dependent characteristics of junctionless bulk transistor," *Appl. Phys. Lett.*, vol. 103, no. 13, 133503, Sept. 2013.

[118] D. Y. Jeon, S. J. Park, M. Mouis, S. Barraud, G. T. Kim, and G. Ghibaudo, "Low-temperature operation of junctionless nanowire transistors: Less surface roughness scattering effects and dominant scattering mechanisms," *Appl. Phys. Lett.*, vol. 105, no. 26, 263505, Dec. 2014.

[119] D. Y. Jeon, S. J. Park, M. Mouis, S. Barraud, G. T. Kim, and G. Ghibaudo, "Low-temperature electrical characterization of junctionless transistors," *Solid-State Electron.*, vol. 80, pp. 135–141, Feb. 2013.

[120] M. de Souza, M. A. Pavanello, R. D. Trevisoli, R. T. Doria, and J. P. Colinge, "Cryogenic operation of junctionless nanowire transistors," *IEEE Electron Device Lett.*, vol. 32, no. 10, pp. 1322–1324, Oct. 2011.

[121] H. Wang, W. Han, X. Li, Y. Zhang, and F. Yang, "Low-temperature study of array of dopant atoms on transport behaviors in silicon junctionless nanowire transistor," *J. Appl. Phys.*, vol. 116, no. 12, 124505, Sept. 2014.

[122] X. Li, W. Han, L. Ma, H. Wang, Y. Zhang, and F. Yang, "Low-temperature quantum transport characteristics in single n-channel junctionless nanowire transistors," *IEEE Electron Device Lett.*, vol. 34, no. 5, pp. 581–583, May 2013.

[123] L. Ma, W. Han, H. Wang, X. Li, and F. Yang, "Temperature dependence of electronic behaviors in n-type multiple-channel junctionless transistors," *J. Appl. Phys.*, vol. 114, no. 12, 124507, Sept. 2013.

[124] R. Trevisoli, M. de Souza, R. T. Doria, V. Kilchtyska, D. Flandre, and M. A. Pavanello, "Junctionless nanowire transistors operation at temperatures down to 4.2 K," *Semicond. Sci. Technol.*, vol. 31, no. 11, 114001, Oct. 2016.

[125] H. Wu, J. Y. Zhang, J. J. Gu, L. Dong, N. J. Conrad, and P. D. Ye, "Room-temperature quantum oscillations in Ge junctionless MOSFETs at the scaling limit," in *IEEE Dev. Res. Conf. (DRC)*, pp. 1–2, June 2013.

[126] T. Grasser, *Bias Temperature Instability for Devices and Circuits*, Springer-Verlag, New York, 2014.

[127] M. Cho, J.-D. Lee, M. Aoulaiche, B. Kaczer, P. Roussel, T. Kauerauf, R. Degraeve, J. Franco, L.-Å. Ragnarsson, and G. Groeseneken, "Insight into N/PBTI mechanisms in sub-1-nm-EOT devices," *IEEE Trans. Electron Devices*, vol. 59, no. 8, pp. 2042–2048, Aug. 2012.

[128] D. Jang, J. W. Lee, C. W. Lee, J. P. Colinge, L. Montès, J. I. Lee, G. T. Kim, and G. Ghibaudo, "Low-frequency noise in junctionless multigate transistors," *Appl. Phys. Lett.*, vol. 98, no. 13, 133502, Mar. 2011.

[129] D. Y. Jeon, S. J. Park, M. Mouis, S. Barraud, G. T. Kim, and G. Ghibaudo, "Low-frequency noise behavior of junctionless transistors compared to inversion-mode transistors," *Solid-State Electron.*, vol. 81, pp. 101–104, Mar. 2013.

[130] A. N Nazarov, I. Ferain, N. D. Akhavan, P. Razavi, R. Yu, and J. P. Colinge, "Random telegraph-signal noise in junctionless transistors," *Appl. Phys. Lett.*, vol. 98, no. 9, 092111, Feb. 2011.

[131] P. Singh, N. Singh, J. Miao, W. T. Park, and D. L. Kwong, "Gate-all-around junctionless nanowire MOSFET with improved low-frequency noise behavior," *IEEE Electron Device Lett.*, vol. 32, no. 12, pp. 1752–1754, Dec. 2011.

[132] T. Bang, B. H. Lee, C. K. Kim, D. C. Ahn, S. B. Jeon, M. H. Kang, J. S. Oh, and Y. K. Choi, "Low-frequency noise characteristics in SONOS flash memory with vertically

stacked nanowire FETs," *IEEE Electron Device Lett.*, vol. 38, no. 1, pp. 40–43, Jan. 2017.

[133] U. S. Jeong, C. K. Kim, H. Bae, D. I. Moon, T. Bang, J. M. Choi, J. Hur, and Y. K. Choi, "Investigation of low-frequency noise in nonvolatile memory composed of a gate-all-around junctionless nanowire FET," *IEEE Trans. Electron Devices*, vol. 63, no. 5, pp. 2210–2213, May 2016.

[134] A. Chaudhry and M. J. Kumar, "Controlling short-channel effects in deep-submicron SOI MOSFETs for improved reliability: A review," *IEEE Trans. Device Mater. Rel.*, vol. 4, no. 1, pp. 99–109, Mar. 2004.

[135] J. T. Park, J. Y. Kim, and J. P. Colinge, "Negative-bias-temperature-instability and hot carrier effects in nanowire junctionless p-channel multigate transistors," *Appl. Phys. Lett.*, vol. 100, no. 8, 083504, Feb. 2012.

[136] S. M. Lee, H. J. Jang, and J. T. Park, "Impact of back gate biases on hot carrier effects in multiple gate junctionless transistors," *Microelectron. Rel.*, vol. 53, no. 9, pp. 1329–1332, Nov. 2013.

[137] S. M. Sze, *Physics of Semiconductor Devices*, John Wiley & Sons, Inc., New York, pp. 122–129, 1981.

[138] M. S. Gudiksen, L. J. Lauhon, J. Wang, D. C. Smith, and C. M. Lieber, "Growth of nanowire superlattice structures for nanoscale photonics and electronics," *Nature*, vol. 415, no. 6872, pp. 617–620, Feb. 2002.

[139] M. T. Björk, B. J. Ohlsson, T. Sass, A. I. Persson, C. Thelander, M. H. Magnusson, K. Deppert, L. R. Wallenberg, and L. Samuelson, "One-dimensional steeplechase for electrons realized," *Nano Lett.*, vol. 2, no. 2, pp. 87–89, Feb. 2002.

[140] I. A. Goldthorpe, A. F. Marshall, and P. C. McIntyre, "Synthesis and strain relaxation of Ge-core/Si-shell nanowire arrays," *Nano Lett.*, vol. 8, no. 11, pp. 4081–4086, Oct. 2008.

[141] L. Chen, W. Y. Fung, and W. Lu, "Vertical nanowire heterojunction devices based on a clean Si/Ge interface," *Nano Lett.*, vol. 13, no. 11, pp. 5521–5527, Oct. 2013.

[142] L. Chen, F. Cai, U. Otuonye, and W. D. Lu, "Vertical Ge/Si core/shell nanowire junctionless transistor," *Nano Lett.*, vol. 16, no. 1, pp. 420–426, Dec. 2015.

[143] K. C. Saraswat, C. O. Chui, T. Krishnamohan, A. Nayfeh, and P. McIntyre, "Ge based high performance nanoscale MOSFETs," *Microelectron. Eng.*, vol. 80, pp. 15–21, June 2005.

[144] R. Yu, S. Das, I. Ferain, P. Razavi, M. Shayesteh, A. Kranti, R. Duffy, and J. P. Colinge, "Device design and estimated performance for p-type junctionless transistors on bulk germanium substrates," *IEEE Trans. Electron Devices*, vol. 59, no. 9, pp. 2308–2313, Sept. 2012.

[145] C. W. Chen, C. T. Chung, J. Y. Tzeng, P. S. Chang, G. L. Luo, and C. H. Chien, "Body-tied germanium tri-gate junctionless PMOSFET with in situ boron doped channel," *IEEE Electron Device Lett.*, vol. 35, no. 1, pp. 12–14, Jan. 2014.

[146] D. D. Zhao, T. Nishimura, C. H. Lee, K. Nagashio, K. Kita, and A. Toriumi, "Junctionless Ge p-channel metal–oxide–semiconductor field-effect transistors fabricated on ultrathin Ge-on-insulator substrate," *Appl. Phys. Expr.*, vol. 4, no. 3, 031302, Mar. 2011.

[147] H. Wu, M. Si, L. Dong, J. Zhang, and D. Y. Peide, "Ge CMOS: Breakthroughs of n-FETs (I_{max} = 714 mA/mm, g_{max} = 590 mS/mm) by recessed channel and S/D," *VLSI Tech. Dig.*, pp. 1–2, 2014.

[148] J. Mitard, B. De Jaeger, F. E. Leys, G. Hellings, K. Martens, G. Eneman, D. P. Brunco, R. Loo, J. C. Lin, D. Shamiryan, T. Vandeweyer, G. Winderickx, E. Vrancken, C. H. Yu, K. De Meyer, M. Caymax, L. Pantisano, M. Meuris, and M. M. Heyns, "Record I_{ON}/I_{OFF} performance for 65-nm Ge pMOSFET and novel Si passivation scheme for improved EOT scalability," *IEDM Tech. Dig.*, Dec. 2008, pp. 873–875.

[149] L. Hutin, C. Royer, J. Damlencourt, J. Hartmann, H. Grampeix, V. Mazzocchi, C. Tabone, B. Previtali, A. Pouydebasque, M. Vinet, and O. Faynot, "GeOI pMOSFETs scaled down to 30-nm gate length with record OFF-state current," *IEEE Electron Device Lett.*, vol. 31, no. 3, pp. 234–236, Mar. 2010.

[150] D. D. Zhao, C. H. Lee, T. Nishimura, K. Nagashio, G. A. Cheng, and A. Toriumi, "Experimental and analytical characterization of dual-gated germanium junctionless p-channel metal–oxide–semiconductor field-effect transistors," *Jap. J. Appl. Phys.*, vol. 51, no. 4S, 04DA03, Apr. 2012.

[151] I. H. Wong, Y. T. Chen, S. H. Huang, W. H. Tu, Y. S. Chen, T. C. Shieh, T. Y. Lin, H. S. Lan, and C. W. Liu, "In situ doped and tensile stained Ge junctionless gate-all-around n-FETs on SOI featuring I_{on} = 828 μA/μm, $I_{on}/I_{off} \sim 1 \times 10^5$, DIBL= 16 – 54 mV/V, and 1.4 X external strain enhancement," *IEDM Tech. Dig.*, pp. 6–9, Dec. 2014.

[152] I. H. Wong, Y. T. Chen, S. H. Huang, W. H. Tu, Y. S. Chen, and C. W. Liu, "Junctionless gate-all-around pFETs using in situ boron-doped Ge channel on Si," *IEEE Trans. Nanotechnol.*, vol. 14, no. 5, pp. 878–882, Sept. 2015.

[153] Y. R. Jhan, V. Thirunavukkarasu, C. P. Wang, and Y. C. Wu, "Performance evaluation of silicon and germanium ultrathin body (1 nm) junctionless field-effect transistor with ultrashort gate length (1 nm and 3 nm)," *IEEE Electron Device Lett.*, vol. 36, no. 7, pp. 654–656, July 2015.

[154] V. Djara, L. Czornomaz, V. Deshpande, N. Daix, E. Uccelli, D. Caimi, M. Sousa, and J. Fompeyrine, "Tri-gate InGaAs-OI junctionless FETs with PE-ALD Al_2O_3 gate dielectric and H_2/Ar anneal," *Solid-State Electron.*, vol. 115, pp. 103–108, Jan. 2016.

[155] M. Yokoyama, R. Iida, S. H. Kim, N. Taoka, Y. Urabe, T. Yasuda, H. Takagi, H. Yamada, N. Fukuhara, M. Hata, M. Sugiyama, Y. Nakano, M. Takenaka, and S. Takagi, "Extremely-thinbody InGaAs-on-insulator MOSFETs on Si fabricated by direct wafer bonding," *IEDM Tech. Dig.*, 2010.

[156] T. Irisawa, M. Oda, K. Ikeda, Y. Moriyama, E. Mieda, W. Jevasuwan, T. Maeda, O. Ichikawa, T. Ishihara, M. Hata, and T. Tezuka, "High mobility p–n junction-less InGaAs-OI tri-gate nMOSFETs with metal source/drain for ultra-low-power CMOS applications," in *Proc. IEEE Int. SOI Conf.*, pp. 1, 2012.

[157] V. Djara, K. Cherkaoui, S. B. Newcomb, K. Thomas, E. Pelucchi, D. O'Connell, L. Floyd, V. Dimastrodonato, L. O. Mereni, and P. K. Hurley, "On the activation of implanted silicon ions in p-In0.53Ga0.47As," *Semicond Sci Technol*, vol. 27, 082001, 2012.

[158] L. Czornomaz, M. El Kazzi, D. Caimi, P. Machler, C. Rossel, M. Bjoerk, C. Marchiori, and J. Fompeyrine, "Self-aligned S/D regions for InGaAs MOSFETs," in *Proc. ESSDERC*, vol. 219, 2011.

[159] S. Yi, Z. Chen, R. Dowdy, K. Chabak, P. K. Mohseni, W. Choi, and X. Li, "III–V junctionless gate-all-around nanowire MOSFETs for high linearity low power applications," *IEEE Electron Device Lett*, vol. 35, no. 3, pp. 324, 2014.

[160] K. H. Goh, S. Yadav, K. I. Low, G. Liang, X. Gong, and Y. C. Yeo, "Gate-all-around $In_{0.53}Ga_{0.47}As$ junctionless nanowire FET with tapered source/drain structure," *IEEE Trans. Electron Devices*, vol. 63, no. 3, pp. 1027–1033, Mar. 2016.

[161] J. H. Seo, S. Cho, and I. M. Kang, "Simulation for silicon-compatible InGaAs-based junctionless field-effect transistor using InP buffer layer," *Semicond. Sci. Technol.*, vol. 28, no. 10, 105007, Aug. 2013.

[162] A. Pan, G. Leung, and C. O. Chui, "Junctionless silicon and In 0.53 Ga 0.47 As transistors—Part I: Nominal device evaluation with quantum simulations," *IEEE Trans. Electron Devices*, vol. 62, no. 10, pp. 3199–3207, Oct. 2015.

[163] T. P. Chow and R. Tyagi, "Wide bandgap compound semiconductors for superior high-voltgae unipolar power devices," *IEEE Trans. Electron Devices*, vol. 41, no. 8, pp. 1481–1483, Aug. 1994.

[164] S. Liu, Y. Cai, G. Gu, J. Wang, C. Zeng, W. Shi, Z. Feng, H. Qin, Z. Cheng, C. Chen, and B. Zhang, "Enhancement-mode operation of nanochannel array (NCA) AlGaN/GaN HEMTs," *IEEE Electron Device Lett.*, vol. 33, no. 3, pp. 354–356, Mar. 2012.

[165] B. Lu, E. Matioli, and T. Palacios, "Tri-gate normally-off power MISFET," *IEEE Electron Device Lett.*, vol. 33, no. 3, pp. 360–362, Mar. 2012.

[166] K.-S. Im, R.-H. Kim, K.-W. Kim, D.-S. Kim, C. S. Lee, S. Cristoloveanu, and J.-H. Lee, "Normally off single nanoribbon Al_2O_3/GaN MISFET," *IEEE Electron Device Lett.*, vol. 34, no. 1, pp. 27–29, Jan. 2013.

[167] K. S. Im, C. H. Won, Y. W. Jo, J. H. Lee, M. Bawedin, S. Cristoloveanu, and J. H. Lee, "High-performance GaN-based nanochannel FinFETs with/without AlGaN/GaN heterostructure," *IEEE Trans. Electron Devices*, vol. 60, no. 10, pp. 3012–3018, Oct. 2013.

[168] K. S. Im, J. H. Seo, Y. J. Yoon, Y. I. Jang, J. S. Kim, S. Cho, J. H. Lee, S. Cristoloveanu, J. H. Lee, and I. M. Kang, "GaN junctionless trigate field-effect transistor with deep-submicron gate length: Characterization and modeling in RF regime," *Jap. J. Appl. Phys.*, vol. 53, no. 11, 118001, Oct. 2014.

[169] S. Lee, J. Lee, and S. Cho, "Design and characterization of electrically self-isolated GaN-on-Si junctionless fin-shaped-channel field-effect transistor with higher cost-effectiveness for low-power applications," *Jap. J. Appl. Phys.*, vol. 54, no. 8, 084301, July 2015.

[170] Brigham Young University Cleanroom [online]. Available: https://cleanroom.byu.edu/ResistivityCal, Accessed: Oct. 5, 2018.

[171] J. T. Watt and J. D. Plummer, "Surface potential fluctuations in MOS devices induced by the random distribution of channel dopant ions," *IEEE Trans. Electron Devices*, vol. 35, no. 12, pp. 2431, Dec. 1988.

[172] J. T. Watt and J. D. Plummer, "Dispersion of MOS capacitance-voltage characteristics resulting from the random channel dopant ion distribution," *IEEE Trans. Electron Devices*, vol. 41, no. 11, pp. 2222–2232, Nov. 1994.

[173] M. J. Pelgrom, A. C. Duinmaijer, and A. P. Welbers, "Matching properties of MOS transistors," *IEEE J. Solid-State Circuits*, vol. 24, no. 5, pp. 1433–1439, Oct. 1989.

[174] A. R. Brown, A. Asenov, and J. R. Watling, "Intrinsic fluctuations in sub 10-nm double-gate MOSFETs introduced by discreteness of charge and matter," *IEEE Trans. Nanotechnol.*, vol. 99, no. 4, pp. 195–200, Dec. 2002.

[175] A. Gnudi, S. Reggiani, E. Gnani, and G. Baccarani, "Analysis of threshold voltage variability due to random dopant fluctuations in junctionless FETs," *IEEE Electron Device Lett.*, vol. 33, no. 3, pp. 336–338, Mar. 2012.

[176] S. M. Nawaz, S. Dutta, A. Chattopadhyay, and A. Mallik, "Comparison of random dopant and gate-metal workfunction variability between junctionless and conventional FinFETs," *IEEE Electron Device Lett.*, vol. 35, no. 6, pp. 663–665, June 2014.

[177] G. Ghibaudo, "Evaluation of variability performance of junctionless and conventional Trigate transistors," *Solid-State Electron.*, vol. 75, pp. 13–15, Sept. 2012.

[178] Y. Taur, H. P. Chen, W. Wang, S. H. Lo, and C. Wann, "On–off charge–voltage characteristics and dopant number fluctuation effects in junctionless double-gate MOSFETs," *IEEE Trans. Electron Devices*, vol. 59, no. 3, pp. 863–866, Mar. 2012.

[179] M. Aldegunde, A. Martinez, and J. R. Barker, "Study of discrete doping-induced variability in junctionless nanowire MOSFETs using dissipative quantum transport simulations," *IEEE Electron Device Lett.*, vol. 33, no. 2, pp. 194–196, Feb. 2012.

[180] G. Leung and C. O. Chui, "Variability impact of random dopant fluctuation on nanoscale junctionless FinFETs," *IEEE Electron Device Lett.*, vol. 33, no. 6, pp. 767–769, June 2012.

[181] G. Giusi and A. Lucibello, "Variability of the drain current in junctionless nanotransistors induced by random dopant fluctuation," *IEEE Trans. Electron Devices*, vol. 61, no. 3, pp. 702–706, Mar. 2014.

[182] S. M. Nawaz and A. Mallik, "Effects of device scaling on the performance of junctionless FinFETs due to gate-metal work function variability and random dopant fluctuations," *IEEE Electron Device Lett.*, vol. 37, no. 8, pp. 958–961, Aug. 2016.

[183] S. M. Nawaz, S. Dutta, and A. Mallik, "A comparison of random discrete dopant induced variability between Ge and Si junctionless p-FinFETs," *Appl. Phys. Lett.*, vol. 107, no. 3, 033506, July 2015.

[184] G. Leung and C. O. Chui, "Variability of inversion-mode and junctionless FinFETs due to line edge roughness," *IEEE Electron Device Lett.*, vol. 32, no. 11, pp. 1489–1491, Nov. 2011.

[185] G. Leung and C. O. Chui, "Interactions between line edge roughness and random dopant fluctuation in nonplanar field-effect transistor variability," *IEEE Trans. Electron Devices*, vol. 60, no. 10, pp. 3277–3284, Oct. 2013.

[186] S. J. Choi, D. I. Moon, S. Kim, J. P. Duarte, and Y. K. Choi, "Sensitivity of threshold voltage to nanowire width variation in junctionless transistors," *IEEE Electron Device Lett.*, vol. 32, no. 2, pp. 125–127, Feb. 2011.

[187] S. Gundapaneni, M. Bajaj, R. K. Pandey, K. V. R. M. Murali, S. Ganguly, and A. Kottantharayil, "Effect of band-to-band tunneling on junctionless transistors," *IEEE Trans. Electron Devices*, vol. 59, no. 4, pp. 1023–1029, Apr. 2012.

[188] M. J. Kumar and S. Sahay, "Controlling BTBT induced parasitic BJT action in junctionless FETs using a hybrid channel," *IEEE Trans. Electron Devices*, vol. 63, no. 8, pp. 3350–3353, Aug. 2016.

4

DEVICE ARCHITECTURES TO MITIGATE CHALLENGES IN JUNCTIONLESS FIELD-EFFECT TRANSISTORS

In Chapter 3, we discussed the basic operating principles of junctionless FETs (JLFETs) with an emphasis on different parameters that affect the performance of JLFETs such as mobility, temperature dependence, carrier ballisticity, reliability, noise, and short-channel effects. We analyzed how the volume conduction in JLFETs leads to a higher carrier ballisticity, an improved reliability, a lower noise, and a minimized short-channel effects as compared to the metal-oxide-semiconductor field-effect transistors (MOSFETs). We also concluded that JLFETs may pave the way for utilization of other high mobility materials such as Ge, InGaAs, GaN, and so on, which was not feasible in MOSFETs due to a significant interaction of the mobile carriers with the interface traps and defects.

However, the challenges related to the high source–drain series resistance, realization of efficient volume depletion with a feasible silicon film thickness (~10 nm), enlarged sensitivity to the process variations, and the random dopant fluctuations and band-to-band tunneling (BTBT) in the OFF-state pose a severe threat to the JLFETs. These challenges must be addressed before the JLFET can be proposed as an alternative to the MOSFET for practical applications. Several device architectures have been reported to optimize the performance of JLFETs and mitigate the challenges listed in Section 3.3.6. In this chapter, we discuss these device architectures, which have the potential to enable the JLFETs to replace the conventional MOSFETs.

Junctionless Field-Effect Transistors: Design, Modeling, and Simulation, First Edition.
Shubham Sahay and Mamidala Jagadesh Kumar.

4.1 JUNCTIONLESS ACCUMULATION-MODE FIELD-EFFECT TRANSISTORS

The high source/drain series resistance in JLFETs owing to the moderately doped source/drain regions ($N_D = 1 \times 10^{19}$ cm^{-3}) reduces the ON-state current significantly. This increases the intrinsic delay of the JLFETs as compared to the MOSFETs and reduces the operating frequency of the digital circuits. Therefore, JLFETs with an additional source/drain implantation to increase the source/drain doping and reduce the series resistance were proposed. These field-effect transistors (FETs) are known as junctionless accumulation-mode field-effect transistors (JAMFETs) [1–7]. It may be noted that since the source and drain regions are heavily doped in a JAMFET high–low junctions are created at the source–channel and channel–drain interface. Therefore, JAMFETs are pseudo-junctionless.

At this juncture, we would also like to point out that the concept of accumulation-mode FETs is not new and dates back to the late 1980s and early 1990s [8–15]. Since the threshold voltage of the conventional inversion-mode MOSFETs does not scale as rapidly as the operating voltage, accumulation mode FETs with threshold voltage compatible with complementary metal-oxide-semiconductor (CMOS) circuits were introduced as a promising solution [8–10].

4.1.1 Structure

The schematic of JAMFET is shown in Fig. 4.1. The source and drain regions are heavily doped in JAMFETs as compared to JLFETs. In addition, the channel region is also somewhat lightly doped in JAMFETs ($N_D = 10^{18}$ cm^{-3}) as compared to JLFETs ($N_D = 1 \times 10^{19}$ cm^{-3}). The JAMFET may be viewed as a MOSFET with a same type (n or p) of doping impurities in source, channel, and drain regions. Unlike JLFETs, JAMFETs have a high–low junction at the source–channel and channel–drain interface owing to the difference in the doping concentration. Therefore, JAMFETs are pseudojunctionless transistors.

4.1.2 Operation

The operation of JAMFET is similar to a JLFET. The channel region is volume depleted in the OFF-state by using a gate electrode with an appropriate work function (Fig. 4.2). The lower doping concentration of the channel region in JAMFETs helps

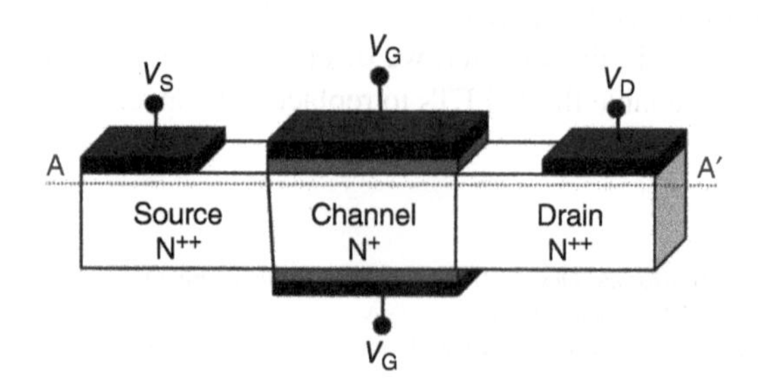

FIGURE 4.1 Three-dimensional view of a double-gate JAMFET [1].

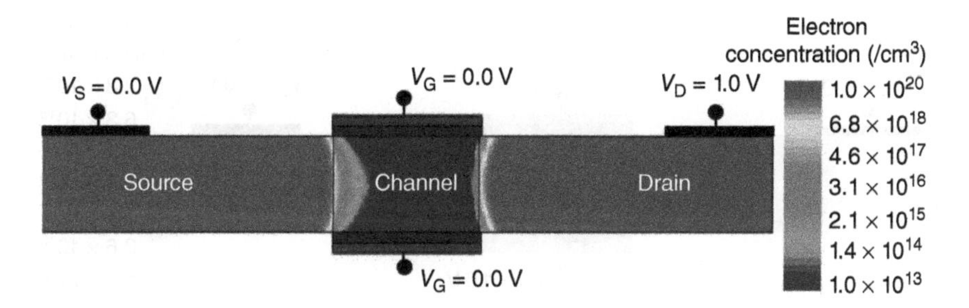

FIGURE 4.2 Electron concentration contour plot of the DGJAMFET in the OFF-state.

to achieve volume depletion for gate electrodes with even midgap work functions (4.5–4.7 eV). Therefore, the fabrication complexity is also reduced in JAMFETs as they do not require gate electrodes with extremely high work functions like JLFETs [1–3].

The application of a positive gate voltage drives the channel region from full depletion to partial depletion mode (Fig. 4.3) and further to the flat band condition in JAMFETs similar to the JLFETs. However, the distinguishing feature of the JAMFETs is the ON-state where the JAMFETs operate in the accumulation regime (Fig. 4.4), whereas the JLFETs operate in the flat band condition. The JAMFETs exhibit a higher drain current than JLFETs owing to (a) the reduced source/drain series resistance due to the heavily doped source/drain regions and (b) improved mobility due to reduced ionized impurity scattering in the lightly doped channel region [1–3]. It may also be noted that since the JAMFETs operate in the accumulation mode in the ON-state, surface scattering dominates the mobility degradation mechanism in JAMFETs [5–7].

The accumulation-mode FETs exhibit a higher mobility [11], an ultralow ON-resistance [12], a better reliability against the hot electron effect [13] and perform better than the inversion-mode MOSFETs at elevated temperatures (~300°C) [14]. Accumulation-mode FETs are also known to exhibit impact ionization induced steep subthreshold slope at relatively lower drain voltages [14]. This is attributed to the

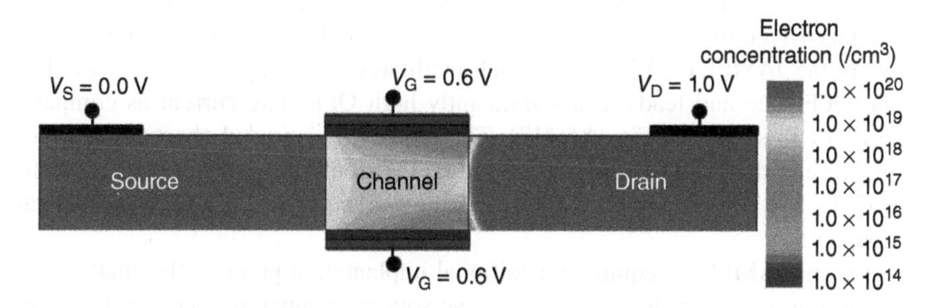

FIGURE 4.3 Electron concentration contour plot of DGJAMFETs in the partial depletion mode.

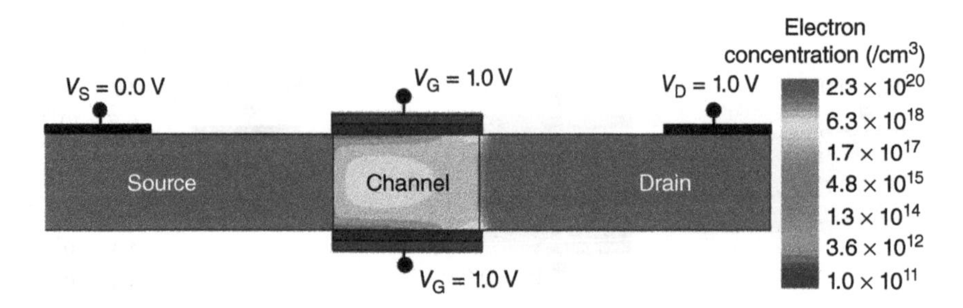

FIGURE 4.4 Electron concentration contour plot of DGJAMFET in the ON-state (accumulation mode).

increased gain of the parasitic bipolar junction transistor (BJT) in the accumulation-mode FETs [15]. In the OFF-state, due to the large electric field required for achieving volume depletion, the holes generated due to impact ionization are attracted toward the gate while the electrons are restricted at the center of the channel region. This leads to a lower interaction between the carriers and reduces the recombination rate. A reduction in the recombination rate leads to an enhanced current gain of the parasitic BJT in accumulation-mode FETs [15].

4.1.3 Challenges

Although the JAMFETs offer a significantly high drain current as compared to the JLFETs, they are more susceptible to the short-channel effects as compared to the JLFETs due to a reduction in the effective channel length owing to the high–low junction [4]. The high–low depletion region extends into the lightly doped channel region and leads to a lower effective channel length. In addition, due to a higher drain doping, the tunneling width in the OFF-state at the channel–drain interface is lower in the JAMFETs as compared to the JLFETs as shown in Fig. 4.5 [16–18]. This leads to a higher BTBT in the OFF-state of JAMFETs. Moreover, the current gain (β) of the parasitic BJT is also higher in the JAMFETs owing to the reduced base width (effective channel length) in JAMFET compared to JLFETs [17] and the reduced recombination rate [15]. The source-to-channel barrier height is also lower in the JAMFETs (Fig. 4.5), which facilitates triggering of the parasitic BJT in the OFF-state and leads to a significantly high OFF-state current as compared to JLFETs as shown in Fig. 4.6 [18]. Therefore, the degraded short-channel performance and the BTBT-induced parasitic BJT action are the major challenges for JAMFETs. We discuss these challenges in the context of JAMFETs in more detail in Chapter 5.

Since the JAMFETs require an additional implantation process, the challenge of formation of ultrasteep doping profile at the source–channel and the channel–drain junction persists in the JAMFETs unlike JLFETs. However, the JAMFETs are expected to suffer from a lower random dopant fluctuations and threshold voltage

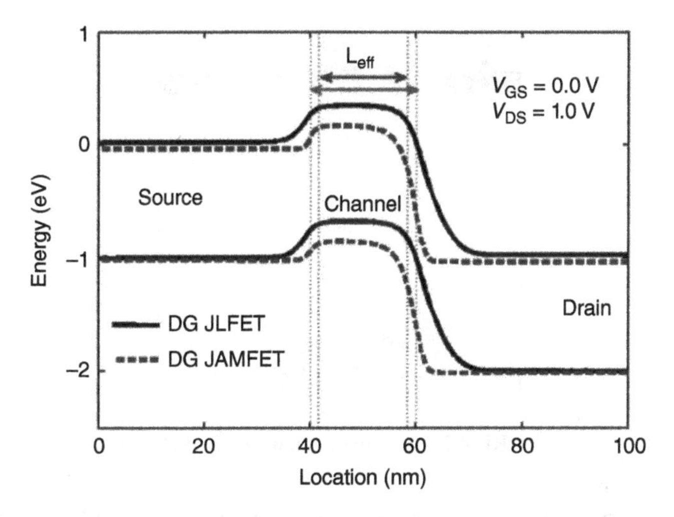

FIGURE 4.5 Energy band profiles of the DGJAMFET and DGJLFET at the cutline A–A′ which is 1 nm below the Si–SiO_2 interface.

variability as compared to the JLFETs due to a somewhat lightly doped channel region.

4.2 REALIZING EFFICIENT VOLUME DEPLETION

In JLFETs, the channel region must be volume depleted in the OFF-state to restrict the flow of carriers from the source region to the drain region. Realizing efficient volume

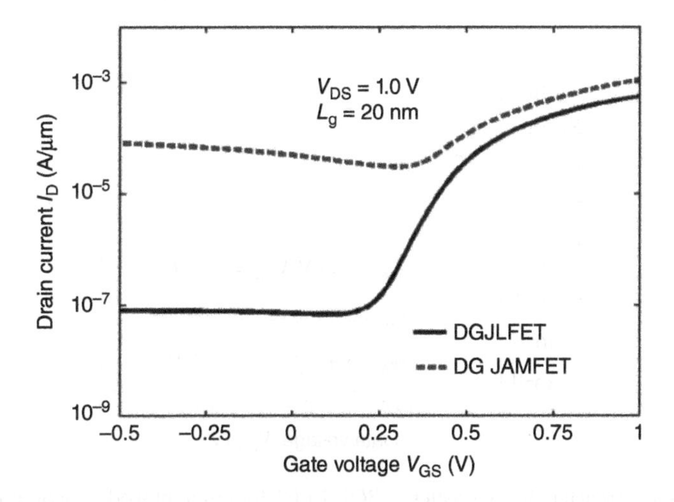

FIGURE 4.6 Transfer characteristics of the DGJAMFET and the DGJLFET.

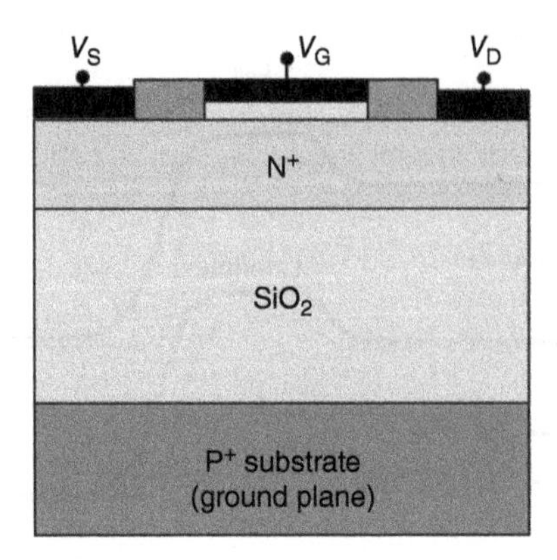

FIGURE 4.7 Schematic view of a SOI JLFET.

depletion is not possible in silicon-on-insulator (SOI) JLFETs (Fig. 4.7) unless (a) the silicon film is ultrathin (silicon film thickness ≤ 6 nm), (b) a gate electrode with a work function higher than 5.93 eV is used, or (c) a large substrate bias ($V_{sub} \leq -1.0$ V) is applied as shown in Fig. 4.8.

The use of ultrathin silicon films is costly [19, 20]. Moreover, producing ultrathin silicon films with a uniform thickness over the entire wafer is a challenge [20]. In addition, the gate electrodes with high work function such as platinum are not only costly but are also difficult to pattern as they are noble and unreactive. Also, the application of a substrate bias larger than the supply voltage requires either an additional power supply in the chip or a charge pump circuitry which could increase the voltage beyond the available power supply [19]. As a result, to attain a considerably low OFF-state current in the conventional SOI JLFETs, an additional power

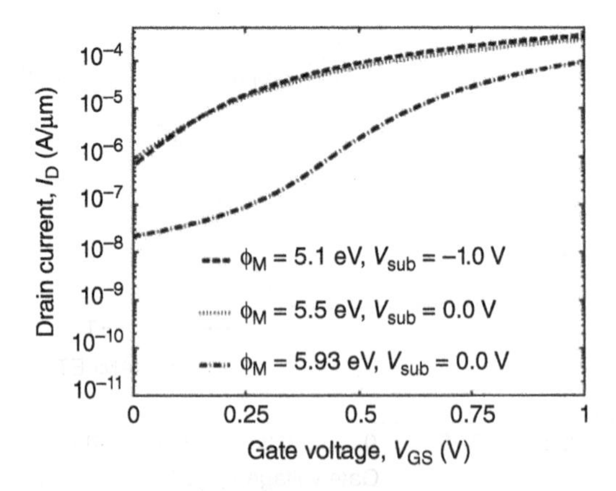

FIGURE 4.8 Transfer characteristics of SOI JLFET for different work functions of the gate metal and substrate bias (V_{sub}).

supply greater than V_{dd} or a charge pump circuit would be required for substrate biasing [19]. Both these techniques would lead to an increased area overhead and an enhanced routing complexity for the circuit designers. Therefore, achieving volume depletion in SOI JLFETs is a major challenge.

In the subsequent sections, we discuss the various architectures, which have been proposed to realize efficient volume depletion in JLFETs.

4.3 SOI JLFET WITH A HIGH-κ BOX

4.3.1 Structure

The schematic of the JLFET with a high-κ BOX (HBJLFET) is shown in Fig. 4.9. The structure of the HBJLFET is similar to the SOI JLFET (Fig. 4.7). However, a high-κ dielectric (HfO_2) is used in the buried oxide (BOX) layer and the ground plane (GP) is heavily doped (~10^{20} cm^{-3}) in the HBJLFET. The GP doping in the commercial SOI wafers is generally low (~10^{17} –10^{18} cm^{-3}). The SOI wafers with HfO_2 BOX can be realized using the smart-cut process [21], and a process flow for realizing high-κ SOI wafers is also reported in [19].

4.3.2 Transfer Characteristics

The transfer characteristics of the HBJLFET and the SOI JLFET are shown in Fig. 4.10. As can be observed from Fig. 4.10, the HBJLFET exhibits a reduced OFF-state current by six orders of magnitude as compared to the SOI JLFET. Volume depletion is not achieved in SOI JLFET with SOI layer thickness = 10 nm, and it shows a poor ON-state to OFF-state current ratio (I_{ON}/I_{OFF}) of ~6 making it unfit for applications. Although the performance of the SOI JLFET may be improved by utilizing a gate electrode with a work function in excess of 5.93 eV or a negative

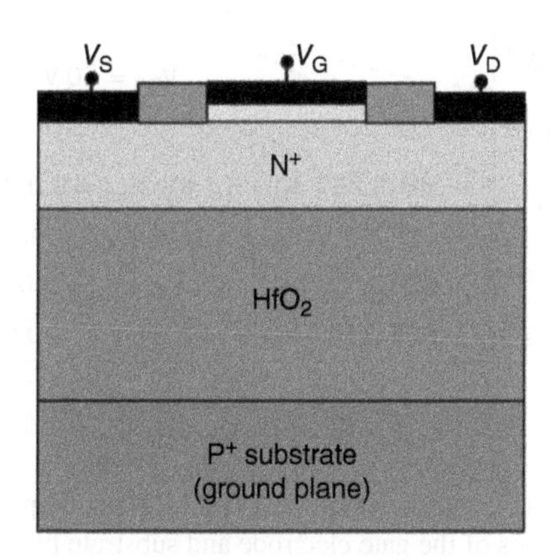

FIGURE 4.9 Schematic view of the JLFET with a high-κ BOX (HBJLFET) [19].

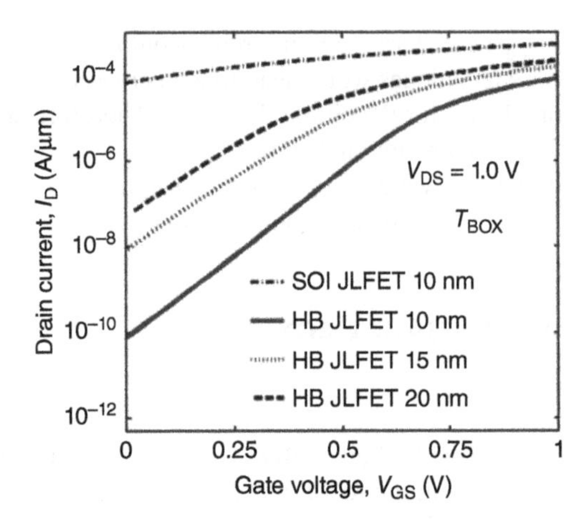

FIGURE 4.10 Transfer characteristics of the SOI JLFET and the HBJLFET for different BOX thicknesses.

substrate bias larger than V_{DD} for increasing the depletion of the channel region as shown in Fig. 4.11, realizing such a high work function is not practically feasible and inclusion of a substrate bias larger than V_{DD} would require a separate voltage source or a charge pump circuitry as explained in Section 4.2 increasing the area overhead. However, the HBJLFET outperforms the SOI JLFETs with a large work function of gate electrode or a high substrate bias and provides a simple and effective means to realize efficient volume depletion. The ON-state current of the HBJLFET is lower than the conventional SOI JLFETs.

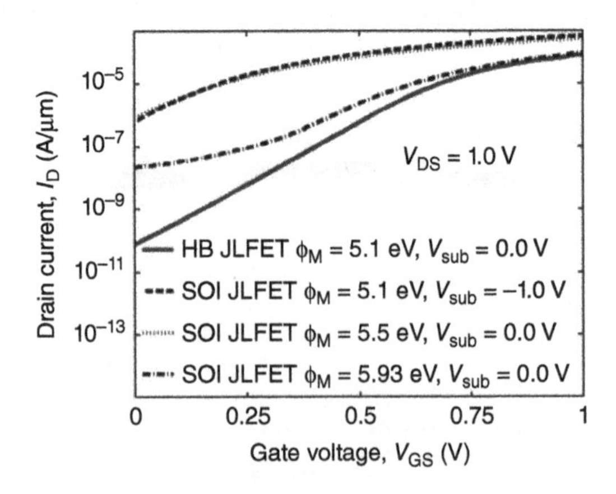

FIGURE 4.11 Transfer characteristics of the HBJLFET as compared with the SOI JLFET with different work functions of the gate electrode and substrate bias.

4.3.3 Operation

The heavily doped (p^+) ground plane acts like a metal with a high work function. This virtual metal electrode coupled with the ultrathin HfO_2 BOX layer ($t_{BOX} = 10$ nm but effective oxide thickness (EOT) ~2 nm) facilitates the depletion of the active SOI film layer from bottom. From the thermal equilibrium electron concentration contour plot of HBJLFET and conventional SOI JLFET shown in Figs. 4.12(a) and 4.12(b), we observe that the effective active layer thickness in a HBJLFET is less than the actual physical device layer thickness because of the depletion of the active device layer from the bottom. As a result, as shown in Figs. 4.12(c) and 4.12(d), the channel region in the HBJLFET is effectively volume depleted compared to the conventional SOI JLFET in the OFF-state. This significantly reduces the OFF-state leakage current in the HBJLFET compared to the conventional SOI JLFET.

However, the depletion of the source/drain regions from bottom effectively reduces the area for current flow and increases the effective source/drain series resistance leading to a reduction in the ON-state current.

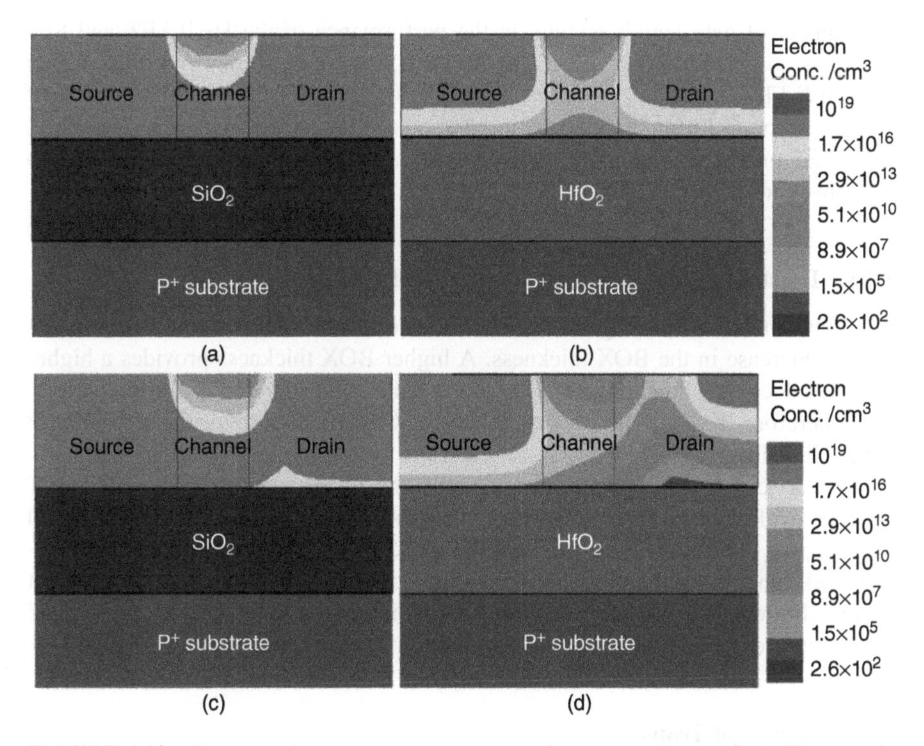

FIGURE 4.12 Electron concentration contour plot of (a) conventional SOI JLFET and (b) HBJLFET in the thermal equilibrium state ($V_{GS} = V_{DS} = 0.0$ V) and (c) SOI JLFET and (d) HBJLFET in the OFF-state ($V_{GS} = 0.0$ V, $V_{DS} = 1.0$ V).

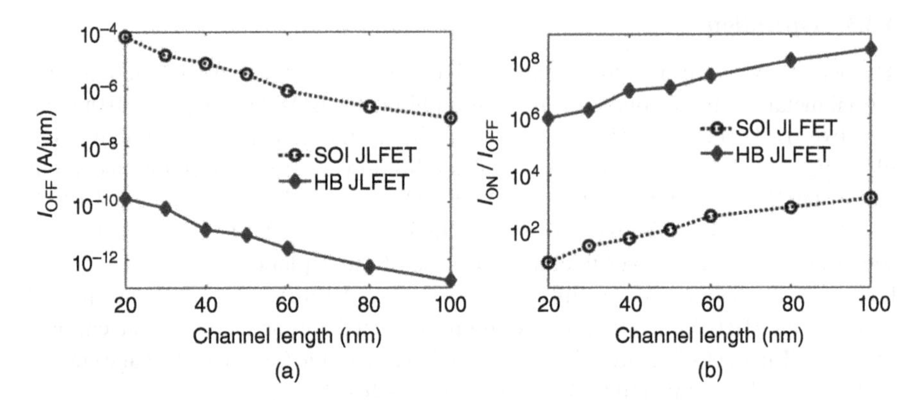

FIGURE 4.13 Variation of (a) OFF-state current and (b) ON-state to OFF-state current ratio of the SOI JLFET and the HBJLFET with gate length scaling.

4.3.4 Impact of Gate Length Scaling

The impact of gate length scaling on the performance of the HBJLFET and SOI JLFET is compared in Fig. 4.13. The HBJLFET performs significantly better than the SOI JLFET, exhibiting a significantly high ON-state to OFF-state current ratio of $\sim 10^6$ even for a gate length of 20 nm as compared to the SOI JLFET ($\sim$6). Therefore, HBJLFET provides an inherent scaling benefit and can be a lucrative alternative to the conventional SOI JLFETs.

4.3.5 Impact of BOX Thickness and Ground Plane Doping

As can be observed from Fig. 4.10, the OFF-state current of the HBJLFET increases with an increase in the BOX thickness. A higher BOX thickness provides a higher EOT and does not allow the active SOI layer to be efficiently depleted from the bottom. Therefore, the BOX thickness should be $\leq$ 10 nm for optimizing the performance of the HBJLFET.

Also, as can be observed from Fig. 4.14, the OFF-state current of HBJLFET increases with a reduction in the doping of the ground plane. A reduction in the ground plane doping reduces the effective work function of the ground plane. Since the effective work function of the virtual bottom gate is reduced, the active SOI layer is not effectively depleted, resulting in a high OFF-state current. Therefore, the ground plane must be heavily doped to achieve optimum performance of the HBJLFET.

4.3.6 Impact of Traps

The fabrication process for growing SiO_2 over silicon is mature enough to eliminate the possibility of interface traps at the Si–SiO_2 interface. However, the interface between silicon and HfO_2 is not defect-free and interface traps are bound to be present

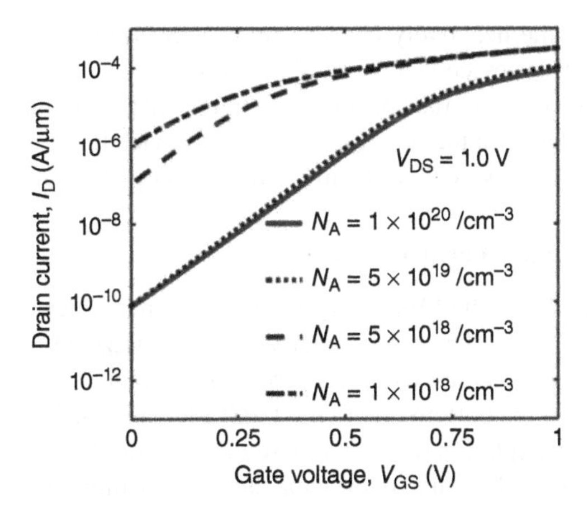

FIGURE 4.14 Impact of ground plane doping on the transfer characteristics of the HBJLFET.

at the HfO_2–silicon interface due to the immature process technology for growing high-κ dielectrics over silicon [19, 22]. As a result, it becomes necessary to investigate the impact of the traps at the Si–HfO_2 BOX interface on the performance of the HBJLFET as shown in Fig. 4.15. For both acceptor and donor traps, a trap density of 1.5×10^{12} cm^{-2} located at the midgap [23] with a capture cross section of 2×10^{-13} cm^2 for the acceptor traps and 7×10^{-16} cm^2 for the donor traps [24] was used in [19].

We observe from Fig. 4.15 that the OFF-state current decreases in the presence of the acceptor-type traps while it increases when donor-type traps are present. The

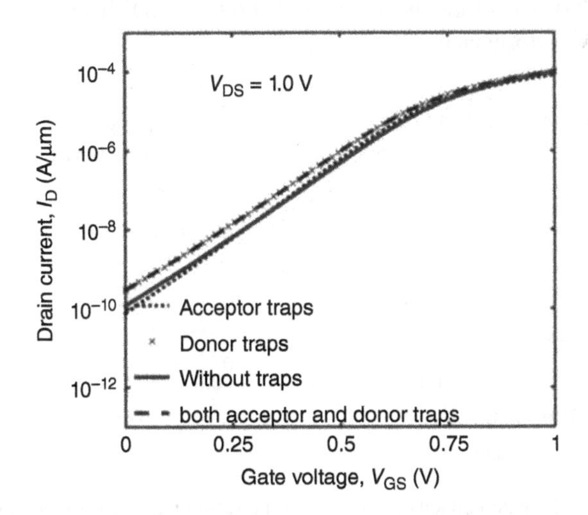

FIGURE 4.15 Impact of traps at the Si–HfO_2 BOX interface on the performance of HBJLFET.

acceptor traps become negatively charged upon ionization and hence facilitate the depletion of the active device layer from the bottom [22].

However, the donor-type traps become positively charged upon ionization leading to electron accumulation at the bottom of the active device layer and hence screen the depletion of the active device layer from the heavily doped ground plane [22].

4.3.7 High-Frequency Performance

The gate capacitance and the Miller capacitance govern the high-frequency (analog) performance of the FETs. The gate capacitance of the HBJLFETs is lower than the SOI JLFETs due to the increased depletion region width in the channel region owing to the depletion from the bottom due to the HBJLFET and the heavily doped ground plane. The increased depletion region width reduces the depletion capacitance, which appears in series with the gate oxide capacitance and minimizes the overall gate capacitance as shown in Fig. 4.16. A lower gate capacitance implies a smaller delay and is beneficial for digital circuit applications.

The drain region at the channel–drain interface is also depleted in the HBJLFET as shown in Figs. 4.12(c) and 4.12(d). This results in a depletion capacitance appearing in series with the effective Miller capacitance (gate to drain feedback capacitance, C_{gd}). Consequently, the Miller capacitance is also significantly reduced in the HBJLFET as compared to the SOI JLFET. A smaller Miller capacitance implies a low overshoot in the voltage transfer characteristics of the CMOS inverter and also improves the analog performance. Also, the ratio of gate to source capacitance C_{gs} and the Miller capacitance C_{gd}, which is an important analog performance metric, is significantly larger in the HBJLFET compared to the conventional SOI JLFET. A high C_{gs}/C_{gd} yields a better control of channel charge by the gate and an improved analog performance [25]. Therefore, the use of HBJLFET in a SOI JLFET also improves the high-frequency performance.

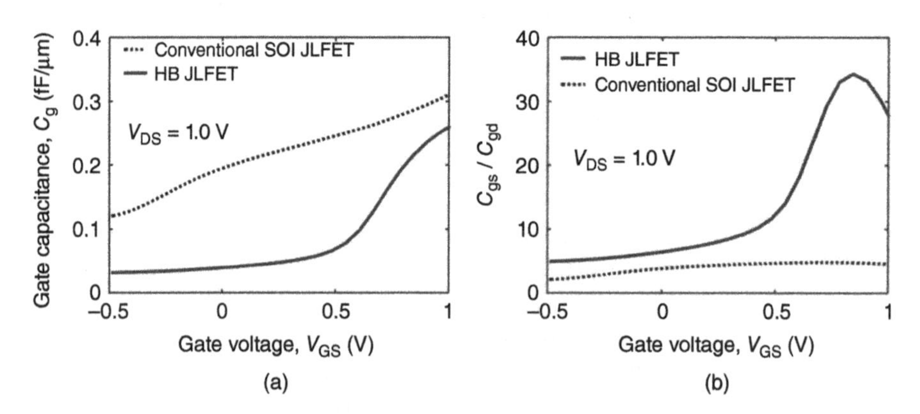

FIGURE 4.16 The gate capacitance (C_g) and the ratio of gate to source capacitance (C_{gs}) and the gate to drain capacitance (C_{gd}) as a function of the gate voltage (V_{GS}) for the HBJLFET and the SOI JLFET.

4.3.8 Challenges

Although HBJLFET appears to be a lucrative alternative to the conventional SOI JLFET, the realization of a good quality defect-free HfO_2–silicon interface in HBJLFET is a challenge. Moreover, the cost of using HfO_2 BOX should also be kept in mind while evaluating its potential for commercial usage. However, the significantly enhanced performance of the HBJLFET may provide incentive for its experimental realization.

4.4 BULK PLANAR JLFET

4.4.1 Structure

Figure 4.17 shows the schematic view of the bulk planar JLFET (BPJLFET). The BPJLFET consists of a uniformly doped active device layer, which does not have any source/channel or channel/drain metallurgical junction. However, the n^+ active device layer is placed over a p^+ substrate for providing a p–n junction electrical isolation [20]. Therefore, the BPJLFET is essentially junction-free along the channel direction, i.e. from source to the drain, but has a vertical metallurgical junction due to the presence of the complementary-doped substrate. The BPJLFET is a planar device and is compatible with the conventional CMOS twin well process. Also, the presence of bulk substrate reduces the processing cost of BPJLFET as compared to the costly SOI wafers [20, 26].

4.4.2 Transfer Characteristics

The transfer characteristics of the BPJLFET and the SOI JLFET are compared in Fig. 4.18. As discussed in Section 4.2, volume depletion is not achieved in SOI JLFT for a SOI thickness of 10 nm resulting in a significantly high OFF-state current and a poor ON-state to OFF-state current ratio. However, the BPJLFET exhibits a considerably low OFF-state current and a significantly high ON-state to OFF-state current ratio of $\sim 10^8$ as compared to the SOI JLFETs ($\sim$6). However, the ON-state current of the BPJLFET is also lower than the conventional SOI JLFET.

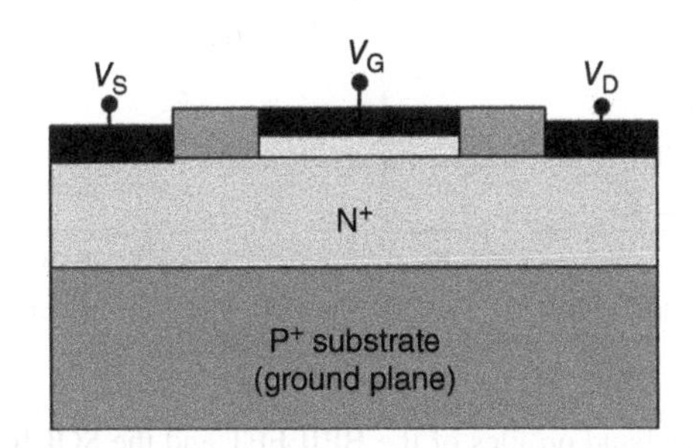

FIGURE 4.17 Schematic view of a BPJLFET.

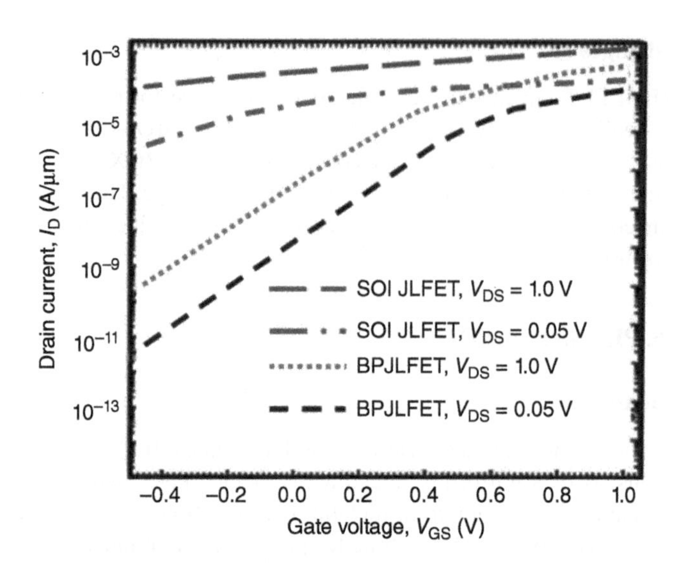

FIGURE 4.18 Transfer characteristics of the BPJLFET and SOI JLFET [20].

4.4.3 Operation

To understand the operation of the BPJLFET, let us look at Fig. 4.19(a) and first focus on the energy band profiles of both devices in the OFF-state. In the SOI JLFET, only a portion of the active device layer is depleted due to the influence of the gate electrode leaving a significant portion of the active device layer near the BOX interface in the flat band mode. This facilitates the leakage current flow through

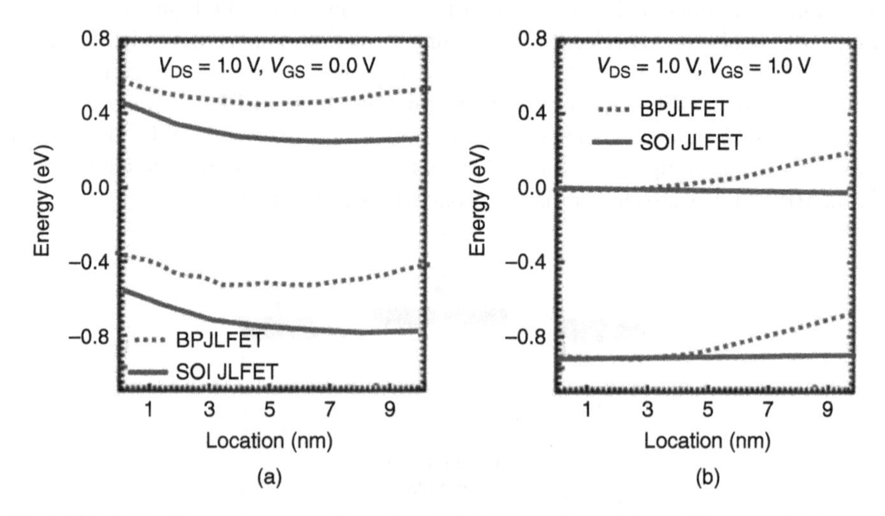

FIGURE 4.19 Energy band profiles of the BPJLFET and the SOI JLFET in (a) OFF-state and (b) ON-state [20].

the undepleted region resulting in a large OFF-state current as observed in Fig. 4.18. However, in the BPJLFET, the active device layer is depleted from top owing to the gate electrode and from bottom due to the inherent p–n junction depletion owing to the complementary-doped substrate. Therefore, in a BPJLFET, efficient volume depletion is achieved leading to a significant reduction in the OFF-state current as shown in Fig. 4.18. It may also be noted that there is no BTBT leakage at the vertical n^+–p^+ junction in the BPJLFET.

The depletion of the active device layer from bottom due to the inherent p–n junction depletion leads to a lower effective active layer thickness than the physical layer thickness. As a result, the gate electrode has to deplete only the effective active layer thickness rather than the physical layer thickness in the BPJLFET to achieve volume depletion. However, the p–n junction depletion region also extends under the source and drain regions resulting in an increased source/drain series resistance. Also, as can be observed from Fig. 4.19(b), only the effective active layer thickness remains in the flat band condition during the ON-state unlike SOI JLFET where the entire physical thickness of the active layer is in flat band and conducts current. This reduction in the effective area for current flow leads to a lower ON-state current in BPJLFET compared to the SOI JLFET as observed in Fig. 4.18.

4.4.4 Impact of Gate Length Scaling

The impact of gate length scaling on the BPJLFET and the SOI JLFET is compared in Fig. 4.20. The lower effective active layer thickness in the BPJLFET as compared to the physical layer thickness enhances the electrostatic integrity of the BPJLFET and leads to a reduced drain influence on the channel electrostatics. As a result, the short-channel effects such as drain-induced barrier lowering are also reduced in BPJLFET as compared to the SOI JLFET. A reduction in the short-channel effects and the provision for realizing efficient volume depletion allows BPJLFET to exhibit an inherent scaling benefit. Therefore, the BPJLFETs exhibit an ON-state to OFF-state current ratio of $\sim 10^3$ even at a gate length of 10 nm as compared to the SOI JLFET (~ 3).

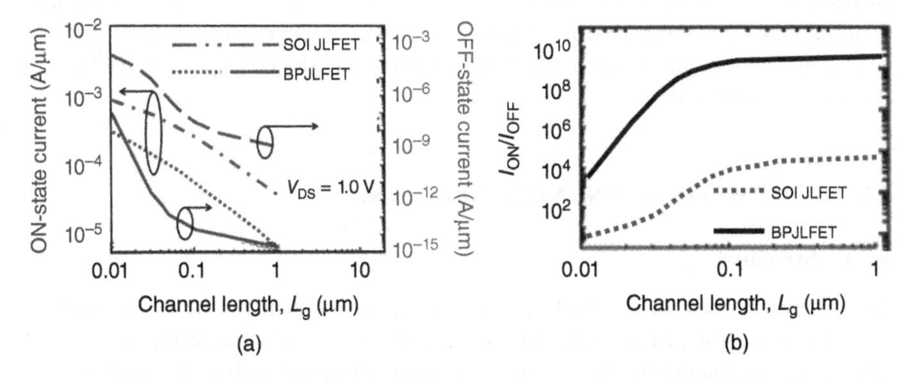

FIGURE 4.20 Impact of gate length scaling on the BPJLFET and SOI JLFET [20].

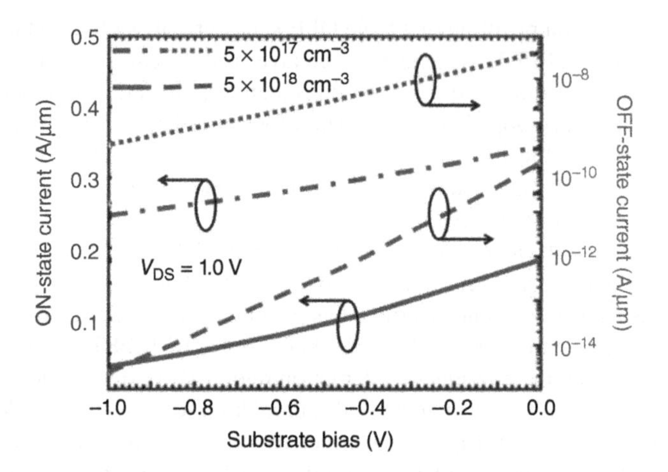

FIGURE 4.21 Impact of substrate doping and substrate bias on the BPJLFET [20].

4.4.5 Impact of Substrate Doping and Substrate Bias

Since the depletion width at the bottom of the active device layer due to the p–n junction isolation can be modulated by changing the substrate doping and substrate bias, the impact of these additional design parameters for tuning the performance of the BPJLFET is also analyzed as shown in Fig. 4.21. A larger substrate doping and a more negative substrate bias leads to a larger depletion of the active device layer from the bottom yielding efficient volume depletion and a lower OFF-state current.

4.4.6 Challenges

The BPJLFET exhibits a significantly low leakage current as compared to the conventional SOI JLFETs. However, the reduced ON-state current is detrimental from delay perspective and needs to be improved. Although the fabrication process for BPJLFETs is expected to be simple and cost-effective [20], it suffers from the fabrication challenges inherent to the JLFET architecture. For instance, it is difficult to realize source/drain ohmic contacts on a moderately doped ($N_D = 1 \times 10^{19}$ cm^{-3}) active layer in BPJLFET.

4.5 JLFET WITH A NONUNIFORM DOPING

4.5.1 Structure

The schematic view of the SOI JLFET with a nonuniform profile in the vertical direction is shown in Fig. 4.22. The doping profile is uniform in the active SOI layer along the lateral direction (source to drain) along the cutline A–A′. However, the doping profile is Gaussian along the thickness of the active SOI layer (vertical

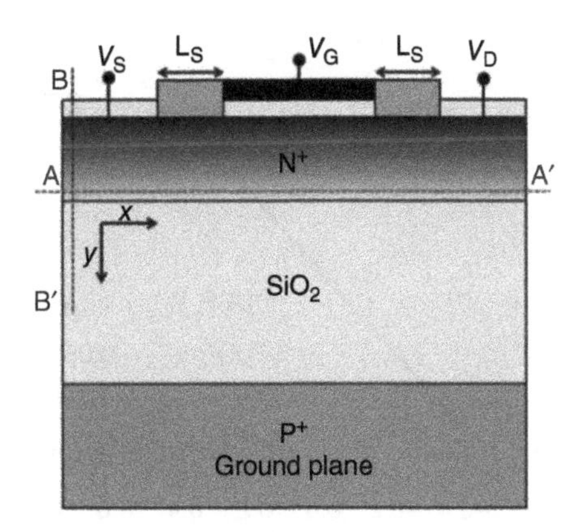

FIGURE 4.22 JLFET with a nonuniform vertical Gaussian doping profile.

direction cutline B–B′). The doping profile obtained from the ion-implantation process follows a Gaussian distribution given as [27]

$$N_D(x, y) = N_{peak} \exp\left(-\frac{y^2}{2\sigma^2}\right) \tag{4.1}$$

where N_D is the effective doping at any position, N_{peak} is the doping concentration at the top surface of the SOI film, and σ is the standard deviation denoting the spread around the peak doping concentration. The doping gradient can be changed by modulating the value of σ. Smaller the σ, steeper the doping gradient. The doping concentration at the surface is taken as 10^{19} cm^{-3}, and a standard deviation of 4 nm is taken so that the doping concentration at the bottom of the active SOI layer (SOI thickness = 10 nm) is ~10^{17} cm^{-3} [27]. The performance of the JLFET with the Gaussian doping profile (GDP) in the vertical direction has been compared against a JLFET with a uniform doping profile (UDP).

4.5.2 Transfer Characteristics

The transfer characteristics of the JLFET with GDP and UDP are compared in Fig. 4.23. As discussed in Section 4.2, volume depletion is not achieved in SOI JLFET with a UDP and the leakage current flows through the undepleted (neutral) portion near the bottom of the SOI layer in the OFF-state leading to a poor ON-state to OFF-state current ratio of ~6. However, the OFF-state current reduces by nearly four orders of magnitude in the JLFET with a GDP leading to a high ON-state to OFF-state current ratio (I_{ON}/I_{OFF}) of ~10^4.

4.5.3 Operation

As discussed in Section 4.2 and observed from Fig. 4.24(a), the extent of depletion owing to the gate electrode reduces along the thickness of the SOI layer and an

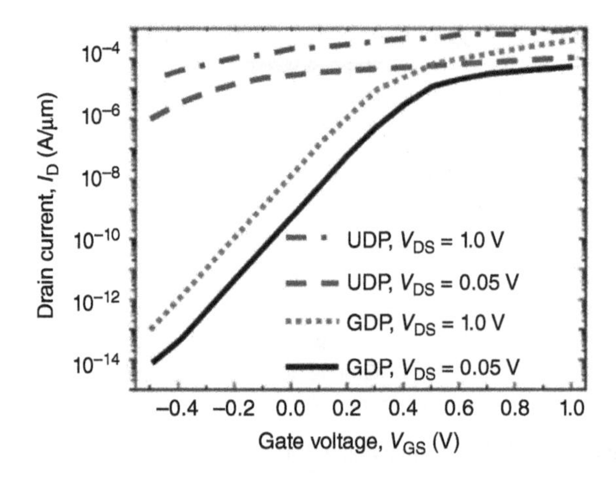

FIGURE 4.23 Transfer characteristics of the JLFET with GDP and UDP [27].

undepleted neutral region (in flat band) exists at the bottom of the active SOI layer, which conducts the leakage current. However, in the case of a JLFET with GDP, the non-UDP with a high doping concentration at the top surface and a low doping concentration at the bottom serves two purposes. First, it further extends the depletion region in the active SOI layer owing to the formation of a high–low junction with a continuous doping gradient in the SOI layer as shown in Fig. 4.24(a). This facilitates the realization of efficient volume depletion. Second, it reduces the available mobile carriers for carrying the leakage current at the bottom of the SOI layer due to the reduced doping concentration.

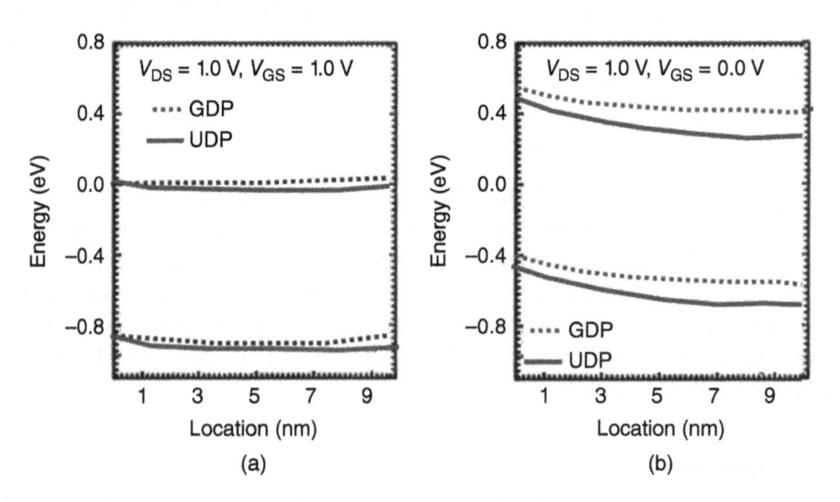

FIGURE 4.24 Energy band profiles of the JLFET with GDP and UDP in (a) ON-state and (b) OFF-state [27].

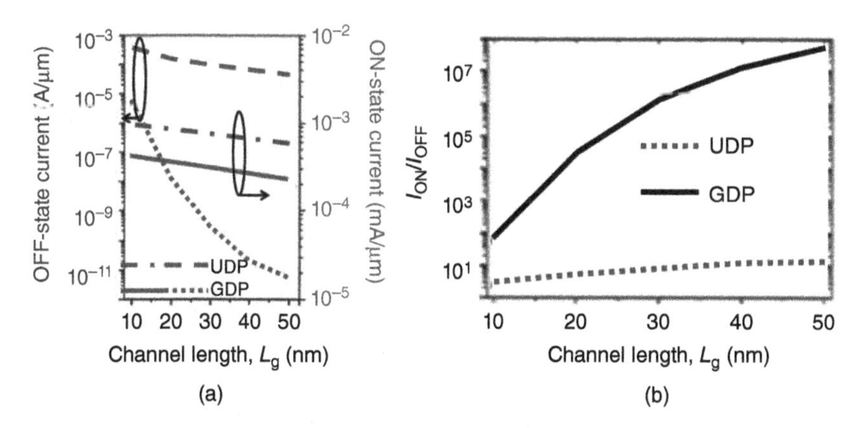

FIGURE 4.25 Impact of gate length scaling on the JLFET with GDP and UDP [27].

The presence of the GDP allows achieving efficient volume depletion and a low leakage current in JLFETs. However, in the ON-state, due to the inherent depletion region owing to the high–low junction (Fig. 4.24(b)) and the lower doping concentration at the bottom of the SOI film, the bottom of the active SOI layer offers a large resistance and leads to a smaller ON-state current.

4.5.4 Impact of Gate Length Scaling

The impact of gate length scaling on the JLFETs with UDP and GDP is analyzed in Fig. 4.25. The JLFET with a GDP offers a significantly low OFF-state current and a considerably high I_{ON}/I_{OFF} ratio even for a channel length of 10 nm. However, the HBJLFET [19] (Section 4.2.1) and the BPJLFET [20] (Section 4.2.2) perform significantly better than the JLFETs with GDP in terms of scalability.

4.5.5 Impact of the Steepness of Doping Profile

The impact of changing the doping gradient on the JLFETs with GDP is analyzed in Fig. 4.26. Increasing the steepness of the doping gradient by reducing the standard deviation (sigma, σ) of the GDP results in a larger extension of the depletion region due to the high–low junction at the bottom of the active SOI layer. This also lowers the doping concentration at the bottom of the SOI layer and reduces the number of mobile carriers available for conducting the leakage current in the GDP JLFET. Therefore, the JLFETs with a GDP perform significantly better when a doping profile with a smaller σ is used. The standard deviation of the GDP can be used as an effective parameter for optimizing the performance of the JLFETs with a GDP.

4.5.6 Challenges

Although the JLFETs with a GDP offer a significantly reduced OFF-state current, realizing such a doping profile over the entire 300-mm wafer with precise control on

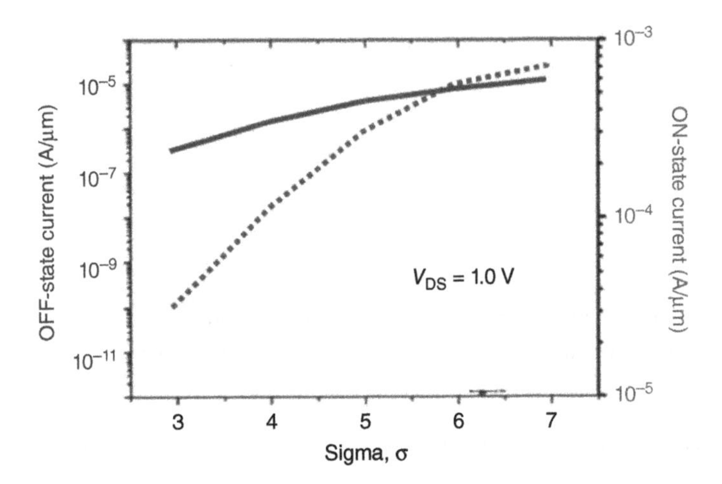

FIGURE 4.26 Impact of the standard deviation (sigma, σ) of the GDP on the GDP JLFETs [27].

the doping depth (dictated by sigma, σ) is a technological challenge. Also, the performance of the JLFETs with a GDP degrades significantly with gate length scaling making them unfit for future technology nodes.

4.6 JLFET WITH A STEP DOPING PROFILE

To achieve volume depletion in a JLFET, an ultrathin active silicon film is required and the film doping should be high enough to achieve a decent source/drain series resistance while realizing efficient volume depletion. However, the high doping of the channel region offers a low mobility and the ultrathin silicon film decreases the effective area through which the carriers flow reducing the ON-state current. Also, an increase in the doping of the silicon film to enhance the ON-state current by reducing series resistance of the source/drain region facilitates a larger coupling of the drain region to the source region increasing the short-channel effects. Therefore, the JLFETs with a step doping profile was proposed to achieve a higher ON-state current and a lower OFF-state current at the same time.

4.6.1 Structure

The three-dimensional (3D) structure and the schematic view of the JLFET with step doping are shown in Fig. 4.27. It is essentially a trigate (TG) structure. The active device layer consists of two regions with distinct doping: (a) the region near the Si–SiO_2 surface (shell region) close to the gate is doped with a doping concentration N_1 and (b) the region far from the surface (bulk or core region) is doped with a doping

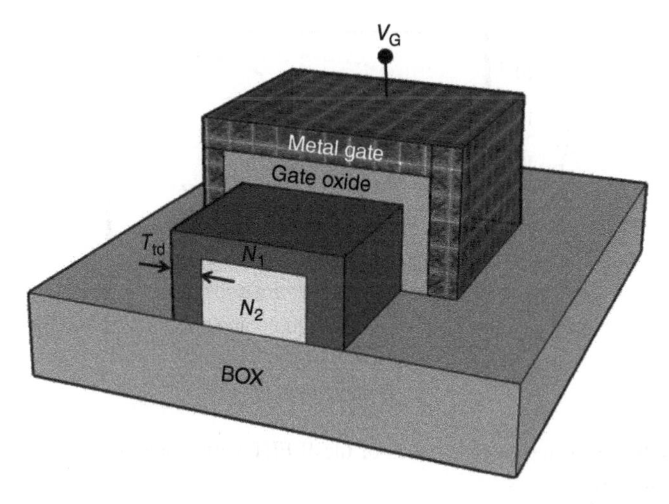

FIGURE 4.27 Three-dimensional view of the JLFET with a step doping profile.

concentration N_2 as shown in Fig. 4.27. If $N_1 > N_2$, the doping profile is called step doping, if $N_1 = N_2$ the JLFET is uniformly doped and if $N_2 > N_1$, the doping profile is termed as retrograde doping [28]. Therefore, the JLFET with a step doping is free of a metallurgical junction at the source/channel and channel/drain interface but consists of a high–low junction throughout the active device layer close to the doping transition length (T_{td}). The values of N_1 and N_2 are selected as 5×10^{19} cm^{-3} and 5×10^{18} cm^{-3}, respectively, for the step doping, 5×10^{18} cm^{-3} and 5×10^{19} cm^{-3} for the retrograde doping. The performance of these devices is compared against a control device with uniformly doped active layer having doping concentration of $(N_1 + N_2)/2$ so that the total number of dopants remains the same in all three structures.

4.6.2 Transfer Characteristics

The transfer characteristics of the JLFET with different doping profiles are compared in Fig. 4.28. The threshold voltage is defined according to the constant current method as the gate voltage for which the current is $10^{-7}\ W/L$ A. As can be observed from Fig. 4.28, the JLFET with retrograde doping exhibits the most negative threshold voltage followed by the JLFET with a UDP. The JLFET with a step doping profile exhibits a threshold voltage useful for logic applications. As a result, the OFF-state current (drain current at $V_{GS} = 0.0$ V) is smallest in the JLFET with step doping. The subthreshold slope of the JLFET with step doping is significantly improved to 88 mV/decade as compared to 98 mV/decade for uniformly doped JLFET and 107 mV/decade for the JLFET with retrograde doping.

The ON-state current, defined as the drain current at $V_{GS} = V_{th} + \frac{2}{3}V_{DS}$, where V_{th} is the threshold voltage and V_{DD} is the supply voltage, is also highest for the JLFET with step doping.

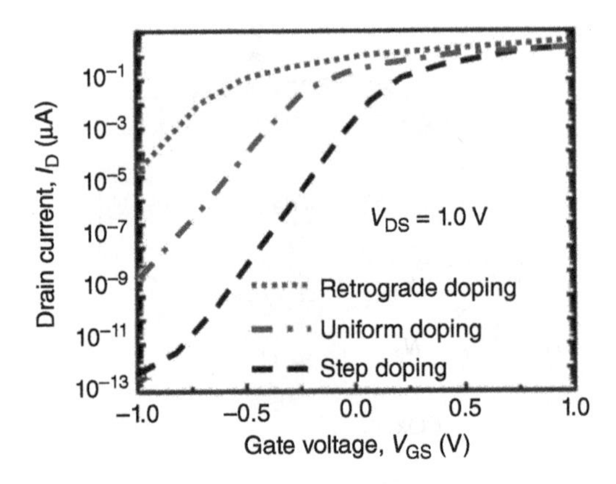

FIGURE 4.28 Transfer characteristics of the JLFET with a step, uniform, and retrograde doping profile [28].

4.6.3 Operating Principle

A large fin width is required to obtain a large current drivability for the JLFETs. However, the maximum fin width is limited by the maximum depletion region width, W_{DM}, induced by the gate electrode. By solving the Poisson equation along the direction normal to the Si–SiO_2 interface, the maximum depletion width for any arbitrary two-step doping scheme can be obtained as [27]

$$W_{DM} = \sqrt{T_{td}^2 + \frac{2\varepsilon_{Si}\varphi_s}{qN_2} - T_{td}^2 \frac{N_1}{N_2}} \tag{4.2}$$

where ϕ_s is the surface potential and T_{td} is the thickness of the region doped to N_1 (shell region).

The maximum depletion width of the JLFET with step doping was found to be higher than the JLFET with uniform doping (obtained by substituting $N_1 = N_2 = \left(\frac{N_1+N_2}{2}\right)$) and retrograde doping (obtained by substituting N_1 in place of N_2 and vice versa). Therefore, the fin width can be increased for the JLFET with step doping yielding a higher effective drive current while achieving efficient volume depletion and a lower OFF-state current at the same time.

The current driving capability depends not only on the number of carriers (which is kept same for the three cases owing to same effective doping concentration) but also on the mobility of the individual carriers. In JLFETs, the current is mostly conducted by the carriers flowing in the bulk of the device. Therefore, it is the doping of the subsurface or the bulk (core) region (N_2), which dictates the mobility of the carriers as they mostly flow in the bulk region. The mobility values for doping of 5×10^{19} cm^{-3} (N_2 for retrograde doping) or 2.25×10^{19} cm^{-3} (N_2 for uniform doping) are close to ~ 100 cm^2 $V\text{-}s^{-1}$. However, the mobility value for a doping of 5×10^{18} cm^{-3} (N_2 for step doping) increases significantly to ~ 160 cm^2 $V\text{-}s^{-1}$ [29–31] and increases

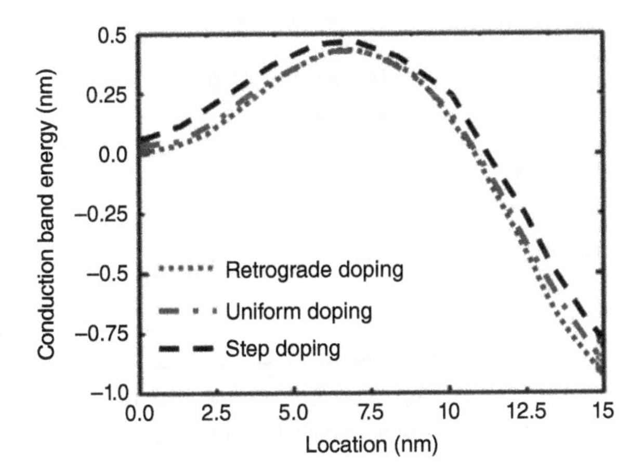

FIGURE 4.29 Conduction band energy profile of the JLFET with a step, uniform, and retrograde doping profile in the OFF-state (V_{GS} = 0.0 V) [28].

the effective current driving capability of the JLFET with a step doping profile. As a result, the JLFET with step doping exhibit an increased driving strength for the same overdrive voltage ($V_{GS} - V_{th}$) and a larger ON-state current compared to the other two configurations as observed from Fig. 4.28.

Figure 4.29 compares the energy band diagram of the JLFETs with the three doping configurations. The source-to-channel barrier height is larger in the JLFET with step doping. The effective channel length is also larger in the JLFET with a step doping. The lower OFF-state current of the JLFET with step doping is attributed to these factors. In addition, these attributes also lead to an increased immunity of JLFET with step doping against the short-channel effects as observed from Figs. 4.30 and 4.31.

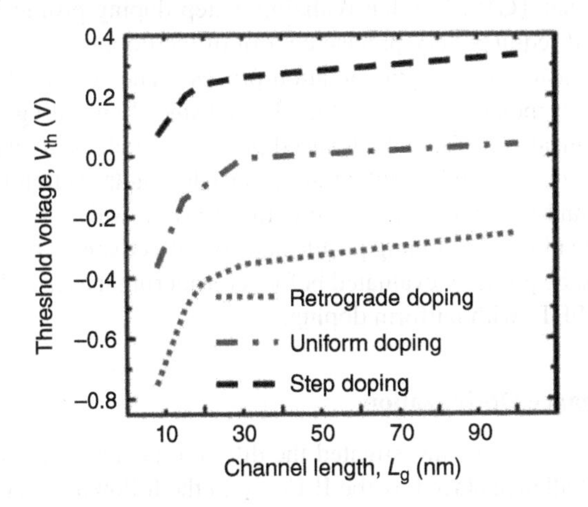

FIGURE 4.30 Variation of threshold voltage with gate length scaling on JLFETs with step, uniform, and retrograde doping profile [28].

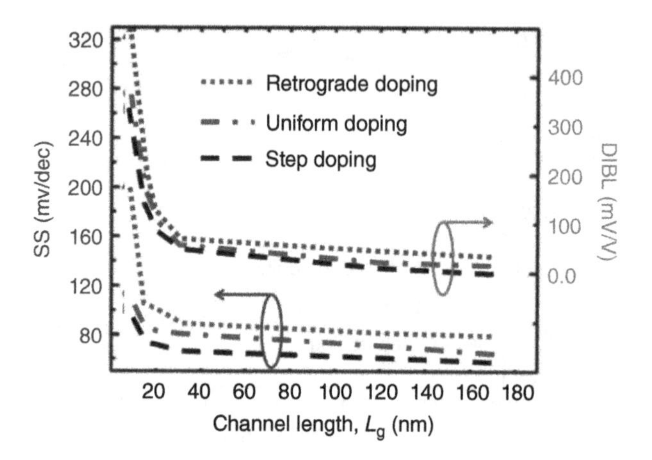

FIGURE 4.31 Variation of subthreshold slope and DIBL with gate length scaling on JLFETs with step, uniform, and retrograde doping profile [28].

Although the threshold voltage decreases, the drain-induced barrier lowering (DIBL) increases and the subthreshold slope degrades as the gate length is reduced in all the three doping configurations, the rate of (a) decay of threshold voltage, (b) degradation in the subthreshold slope, and (c) an increase in the DIBL is least for the JLFET with step doping [28]. Therefore, the JLFET with step doping provides an inherent scaling benefit to the JLFETs.

4.6.4 Challenges

The JLFET with a step doping appears to be a promising alternative to the uniformly doped conventional TGJLFETs but realizing a step doping profile is a technological challenge and requires very precise control of the doping process, which is difficult to achieve using the ion implantation process. The advanced doping process such as monolayer molecular doping [32–34], advanced annealing techniques such as microwave annealing [35], and advanced processes such as *in situ* doping during epitaxial growth are required to obtain ultrasharp doping transition between regions of high doping and low doping over the entire 300 mm wafer. Therefore, the challenge of realizing ultrasteep doping profile and cost effectiveness of the JLFETs with step doping must be properly evaluated before considering it as a replacement for the conventional JLFETs with uniform doping.

4.6.5 Performance Optimization

In Sections 4.1 and 4.2, we investigated the different device architectures proposed to mitigate the challenges faced by the JLFETs. In the following sections, we look at the different ways in which the performance of the JLFETs can be further optimized. As discussed in Section 3.2.1, utilizing multigate architectures [36] like double gate

(DG) [25], [37–75], TG [7, 76–91], or gate-all around [92–116] in JLFETs can lead to an improved performance. Apart from using these multigate architectures, we see how the gate sidewall spacer acts like an effective design parameter to further enhance the performance of JLFETs.

4.7 MULTIGATE JLFET

4.7.1 Structure

The multigate architecture essentially consists of more than one gate electrode modulating the potential of the channel region. The structure of the multigate JLFET consisting of two gate electrodes is called DG architecture as shown in Fig. 4.32(a), whereas the structure in which the gate electrode surrounds the fin-channel region from three sides is called the TG architecture (Fig. 4.32(b)). The gate-all-around nanowire (GAANW) structure, shown in Fig. 4.32(c), is considered to be the most promising architecture for ultimate scaling of FETs and the gate electrode wraps around the entire channel region providing the best possible gate control [116].

4.7.2 Transfer Characteristics

The transfer characteristics of the SOI JLFET with a single gate are compared against the multigate JLFETs without considering the BTBT in Fig. 4.33. As discussed in Section 4.2, efficient volume depletion is not achieved in SOI JLFETs leading to significantly high leakage current. However, the multigate JLFET architectures perform significantly better than the SOI JLFET. Also, the OFF-state leakage current reduces as the number of gates in the multigate architecture increases as shown in Fig. 4.33. Therefore, multigate architectures can be a lucrative alternative to the conventional SOI JLFET.

4.7.3 Operation

In JLFETs, the OFF-state leakage current is mainly determined by the gate-induced depletion of the channel region. In multigate architectures, the channel region is under the influence of the gate electrode from more than one side. Therefore, the total channel volume to be depleted gets shared by the multiple gates. This implies that the gate electrode on the one side has to deplete an effective volume which is lower than the total volume. For instance, in SOI JLFET with a single gate, length $= L_g$, width $=$ thickness $= t_{Si}$, the gate electrode has to deplete a volume of $L_g t_{Si}^2$. However, for the DGJLFET with a similar dimension, a single-gate electrode has to deplete only half of the total volume, i.e. $(L_g t_{Si}^2)/2$. Similarly, for a GAANW with a square cross section and similar dimension, a single-gate electrode has to deplete only one fourth of the total volume, i.e. $\left(L_g t_{Si}^2\right)/4$. As a result, due to enhanced gate control, DG, TG, and GAANW configurations would be more efficient in achieving volume depletion compared to SOI JLFETs with a top gate [19] as shown in Fig. 4.34.

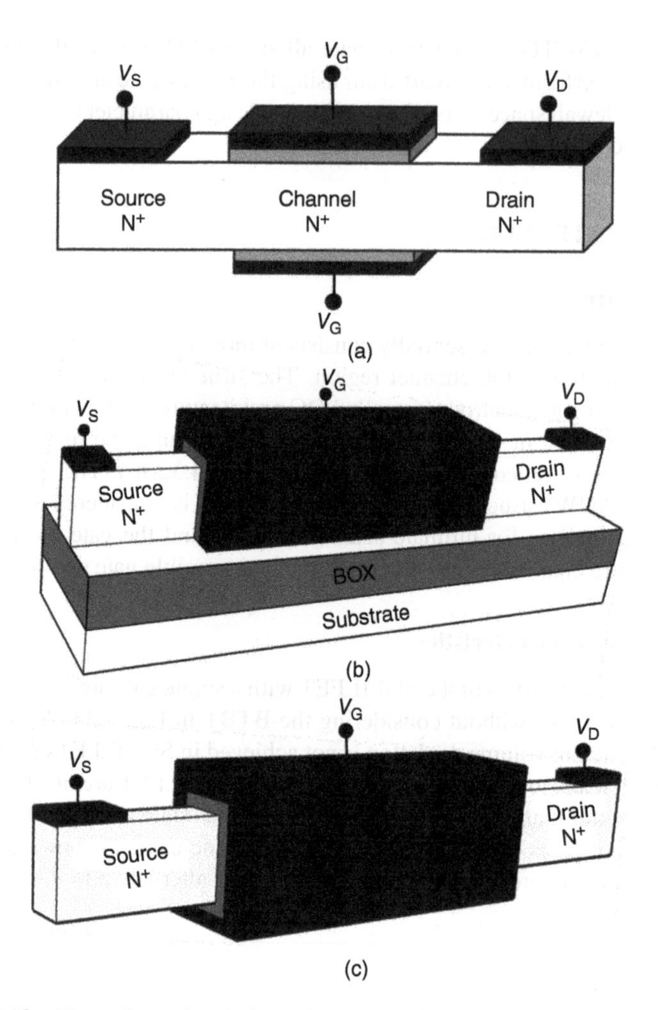

FIGURE 4.32 Three-dimensional view of (a) DGJLFET, (b) TGJLFET. and (c) GAANW JLFET.

The threshold voltage also increases as the number of gates increases. This is evident from the analytical expression derived for the threshold voltage of single-gate JLFET:

$$V_{\mathrm{Th}} = V_{\mathrm{FB}} - \frac{qN_{\mathrm{D}}t_{\mathrm{Si}}^2}{2\varepsilon_{\mathrm{Si}}} - \frac{qN_{\mathrm{D}}t_{\mathrm{Si}}t_{\mathrm{ox}}}{\varepsilon_{\mathrm{ox}}} \tag{4.3}$$

and the DGJLFETs:

$$V_{\mathrm{Th}} = V_{\mathrm{FB}} - \frac{qN_{\mathrm{D}}t_{\mathrm{Si}}^2}{8\varepsilon_{\mathrm{Si}}} - \frac{qN_{\mathrm{D}}t_{\mathrm{Si}}t_{\mathrm{ox}}}{2\varepsilon_{\mathrm{ox}}} \tag{4.4}$$

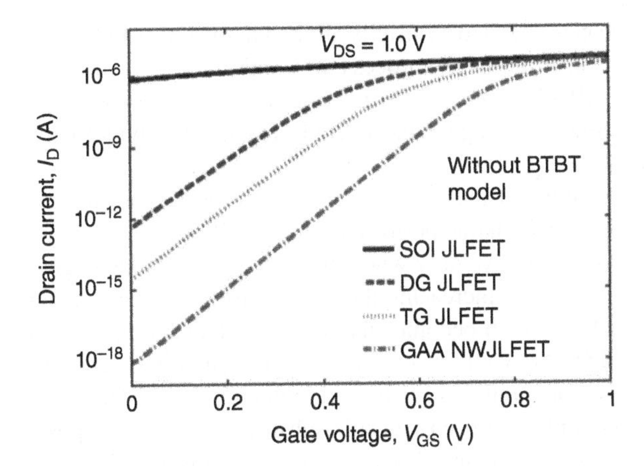

FIGURE 4.33 Transfer characteristics of the SOI JLFET and the multigate JLFETs such as DG, TG, and GAANW JLFET.

in Chapter 3. Clearly, the negative term is lower for the DGJLFET as compared to the single JLFET which leads to an increased threshold voltage as observed in Fig. 4.33.

However, an increase in the gate control leads to a significant overlap of the valence band of the channel region with the conduction band of the drain region leading to a lateral band-to-band tunneling (L-BTBT), which manifests itself as the dominant leakage mechanism in multigate architectures such as GAA NWFETs [105–115]. This new component of gate-induced drain leakage (GIDL) is unique and severely degrades the performance of GAA NWFETs and JLFETs. Therefore, we have dedicated the entire Chapter 5 to understanding the impact of L-BTBT on JLFETs.

4.7.4 Challenges

Although multigate JLFET architectures perform significantly better than the conventional SOI JLFETs, their fabrication process is quite complex and expensive. For

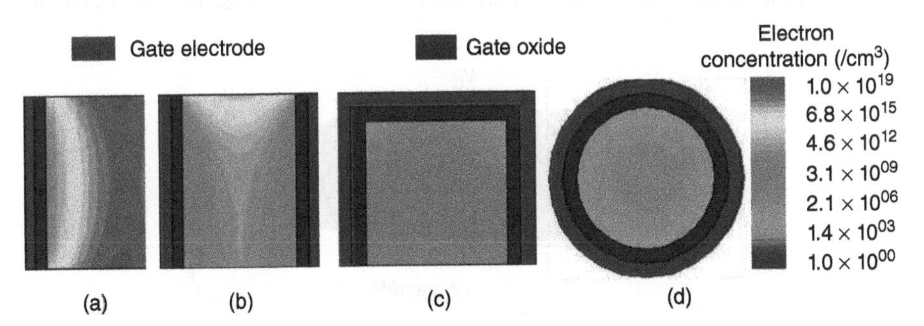

FIGURE 4.34 Electron concentration contour plot of the channel region (thickness = 10 nm and doping concentration = 10^{19} cm^{-3}) in single and multigate JLFETs in the OFF-state (V_{GS} = 0.0 V): (a) SOI JLFET, (b) DGJLFET, (c) TGJLFET, and (d) GAANW JLFET.

instance, the two gates need to be strictly aligned in the DGJLFET architecture, which puts a stringent constraint on the process flow and makes it complex [38]. Moreover, fabrication of TGJLFETs requires formation of ultrathin fins which is difficult. The formation of nanowires using a commercially feasible top-bottom approach is still a technological challenge [116].

The use of the multigate architectures also increases the parasitic capacitance components degrading the dynamic performance [116]. Therefore, the multigate JLFETs can be a lucrative alternative only when the technological challenges for fabricating them are overcome. Also, increasing the gate control over the channel region results in a L-BTBT GIDL, which becomes the dominant leakage mechanism in multigate JLFETs [105–115]. We would discuss this phenomenon in detail in Chapter 5.

In addition to the multigate JLFET architectures on SOI wafer as shown in Fig. 4.32(b), junctionless FinFETs on bulk substrate were also proposed for improved gate controllability and enhanced performance. We discuss this architecture in the next section.

4.7.5 Junctionless bulk FinFET

4.7.5.A Structure The 3D view of a junctionless (JL) bulk FinFET is shown in Fig. 4.35. It consists of a uniformly doped n-type active silicon fin on a p-type bulk substrate. This architecture is junctionless in the sense that there are no source and drain junctions, but a p–n junction is needed underneath the active silicon fin to isolate the source, drain, and channel from the substrate. It may be viewed as the 3D version of the BPJLFET discussed in Section 4.4. Similar to the BPJLFET, the fabrication cost of the JL bulk FinFET is low owing to the absence of SOI wafer. Moreover, it is compatible with the industry standard bulk FinFET CMOS technology [74].

4.7.5.B Operation The underlying p-substrate depletes the active n-type silicon fin channel from bottom. As a result, the effective channel thickness is lower than the physical channel thickness. This leads to a significant improvement in the gate control of the channel region as compared to the SOI JLFinFETs. The short-channel

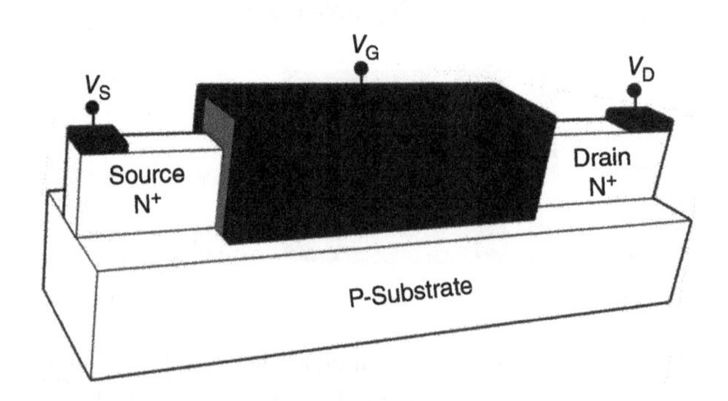

FIGURE 4.35 Three-dimensional view of a junctionless FinFET on a bulk substrate.

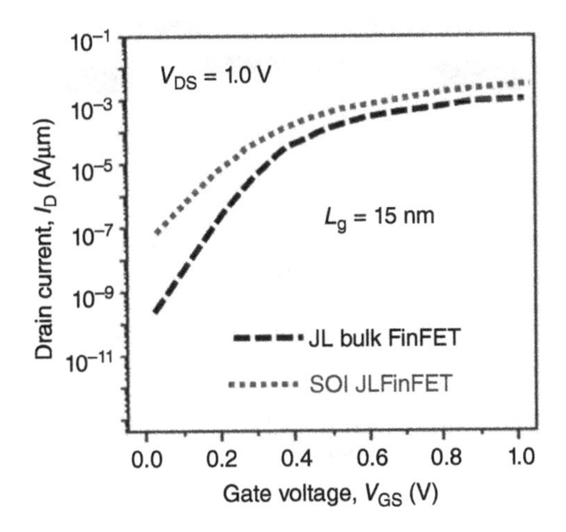

FIGURE 4.36 Transfer characteristics of the JL bulk FinFET and the SOI JLFinFET [74].

effects are significantly suppressed in JL bulk FinFETs as compared to the SOI JLFinFETs. The lower OFF-state current and the improved subthreshold swing in the JL bulk FinFETs (Fig. 4.36) is attributed to the reduced short-channel effects. Moreover, the channel–substrate depletion region width may be tuned by varying the substrate doping. Therefore, the substrate doping could be used as an additional parameter to optimize the performance of the JL bulk FinFETs.

A comparative analysis of germanium and silicon bulk inversion-mode FinFET, accumulation-mode FinFET, and JL bulk FinFET with a gate length of 3 nm confirms the superior immunity of the JLFETs against the short-channel effects [72].

4.7.5.C Challenges The ON-state current is somewhat reduced in the JL bulk FinFET as compared to the SOI JLFinFET. This is attributed to (a) the reduced effective channel thickness, which decreases the effective area for current flow, and (b) the reduced effective thickness of the source/drain regions, which increases the source/drain series resistance. The lower ON-state current of JL bulk FinFETs is detrimental from a delay perspective and needs to be improved for practical applications.

4.8 JLFET WITH A HIGH-κ SPACER

4.8.1 Structure

The schematic view of the SOI JLFET with gate sidewall spacer is shown in Fig. 4.37. The use of a material with a high dielectric constant (high-κ) as gate sidewall spacer was proposed in [117]. Since the source/drain extension regions are not present in JLFETs owing to the inherent uniform doping of the source, channel, and drain regions, a gate sidewall spacer is not required from this perspective. However, the gate

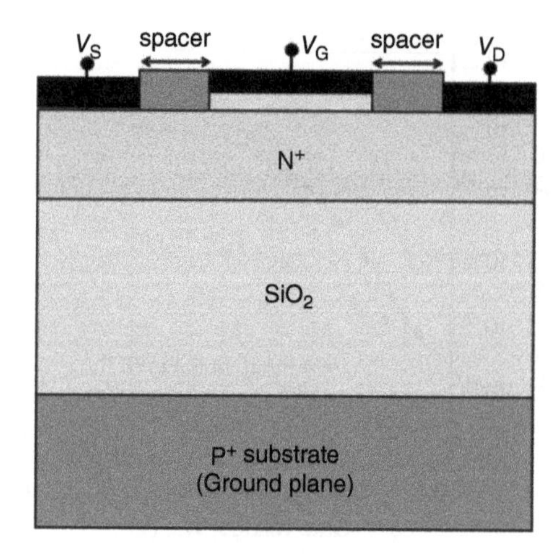

FIGURE 4.37 Schematic view of the SOI JLFET with a spacer.

sidewall spacer becomes inevitable to isolate the gate electrode and the source/drain silicide contacts. In the recent past, the application of (a) high-κ gate sidewall spacer [118–123] and (b) asymmetric and dual-κ spacers [123–128] have been investigated in nanoscale devices. Therefore, it would be interesting to study the impact of gate sidewall spacers on the performance of JLFETs.

4.8.2 Transfer Characteristics

The transfer characteristics of the SOI JLFET with different materials in the gate sidewall spacer are shown in Fig. 4.38. The SOI JLFET with a HfO_2 spacer exhibits the least OFF-state current and a significantly high ON-state to OFF-state current ratio of $\sim 10^7$ compared to the SOI JLFET with an air spacer ($\sim 10^3$). It may be noted

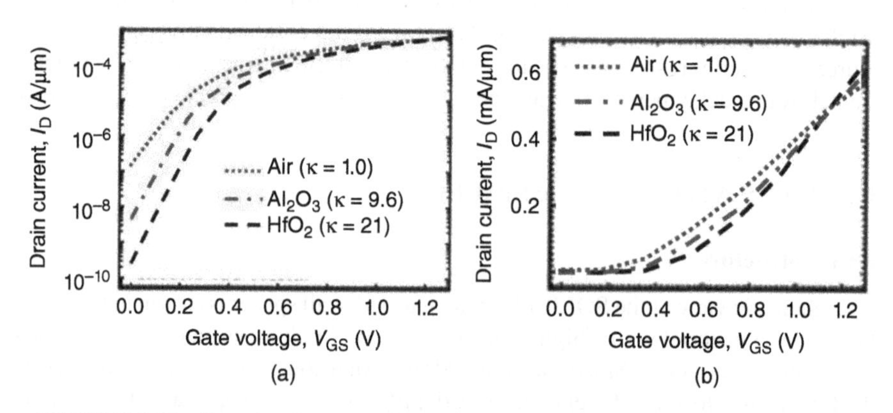

FIGURE 4.38 Transfer characteristics of the SOIJLFET with different spacers [117].

that the active SOI layer thickness has been taken as 6 nm to achieve volume depletion using the gate electrode with work function of 5.1 eV.

The ON-state current and the subthreshold swing are also improved in the SOI JLFETs with a high-κ spacer.

4.8.3 Operating Principle

The improvement in the performance of the JLFETs with the high-κ spacer can be understood from Figs. 4.33 and 4.34. The inclusion of a high-κ spacer adjacent to the gate increases the fringing field through the spacer. As a result, the gate effectively modulates even the active SOI layer under the spacer region via the fringing fields. This leads to a larger accumulation of electrons under the spacer region for JLFETs with high-κ spacer resulting in a lower parasitic series resistance and a higher current driving capability.

The gate modulation of the SOI layer below the spacer through the fringing fields facilitates the depletion of the active SOI layer beyond the physical gate length as shown in Figs. 4.39(a) and 4.39(b). Therefore, the effective channel length is increased in the SOI JLFETs with a high-κ spacer as shown in Fig. 4.40. A larger effective channel length results in a significant suppression of the short-channel effects. The source-to-channel barrier height is also increased in the SOI JLFETs with a high-κ spacer, indicating a reduced DIBL as shown in Fig. 4.41. The significantly low OFF-state current and the higher ON-state to OFF-state current ratio of SOI JLFETs with high-κ spacer are attributed to these factors.

Also, the improvement obtained from utilizing a high-κ dielectric is evident in SOI JLFETs even when different materials are used as the gate dielectric. However, the performance improvement is maximum when SiO_2 is used as the gate dielectric and HfO_2 is used as the spacer as shown in Fig. 4.41.

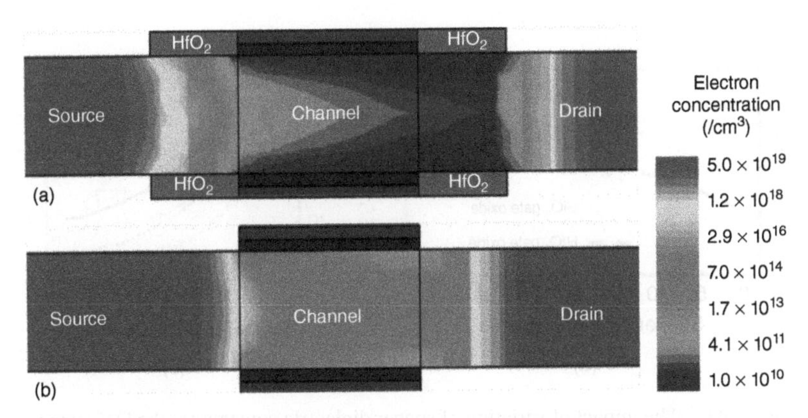

FIGURE 4.39 Electron concentration contour plot of SOIJLFET in the OFF-state (V_{GS} = 0.0 V) with (a) HfO_2 spacer and (b) air spacer.

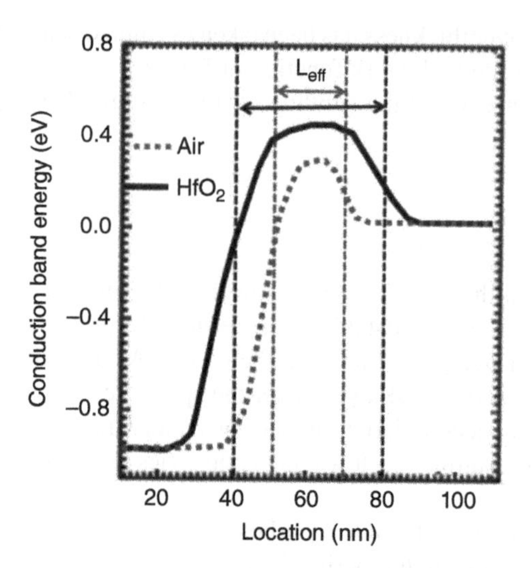

FIGURE 4.40 Energy band profiles of the SOI JLFET with different spacers [117].

4.8.4 Challenges

The high-κ spacers may appear to be a promising design parameter as they allow tuning the I_{ON}/I_{OFF} by more than four orders of magnitude. However, the fringing fields through the high-κ spacer increase the outer fringe capacitance component which leads to an enlarged gate capacitance [106]. The increase in the gate capacitance results in a higher intrinsic delay of the JLFETs with the high-κ spacer [106]. The intrinsic delay of the inverter is an important aspect while designing JLFETs. Therefore, there is a trade-off between the intrinsic delay of the inverter and the ON-state to OFF-state current ratio exhibited by the JLFETs with high-κ spacer. The spacer

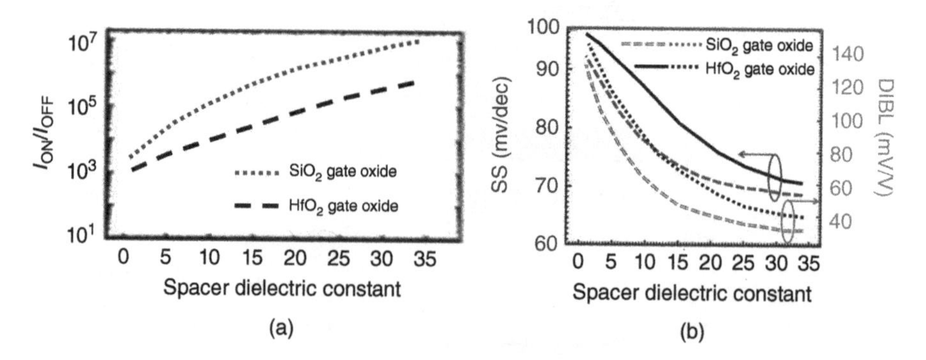

FIGURE 4.41 The impact of variation of spacer dielectric constant (a) the ON-state to OFF-state current ratio of the SOI JLFET and (b) the subthreshold slope (SS) and DIBL of SOI-JLFET with different gate dielectric [117].

dielectric constant should be carefully chosen to obtain a lower intrinsic delay while achieving a high I_{ON}/I_{OFF}.

4.9 JLFET WITH A DUAL MATERIAL GATE

4.9.1 Structure

The 3D view and the schematic view of a dual material gate (DMG) JLFET is shown in Fig. 4.42. The gate in a DMG JLFET consists of two materials with different work functions placed adjacent to each other: (a) material M_1 with work function ϕ_{M1} is called the control gate and (b) material M_2 with a work function ϕ_{M2} is called the screen gate. When $\phi_{M1} > \phi_{M2}$, the DMG architecture is known to escalate the carrier transport efficiency and enhance the transconductance and current driving capability while simultaneously reducing the short-channel effects [130–139]. Therefore, it is quite interesting to analyze the efficacy of the DMG architecture in improving the performance of the JLFETs. The DMG architecture can be realized using tilt angle evaporation and lithography, metal interdiffusion technique, metal wet etch process, or other advanced fabrication processes reported in [130–139]. The conventional single material gate (SMG) JLFET utilizes a material with a single work function for the gate electrode.

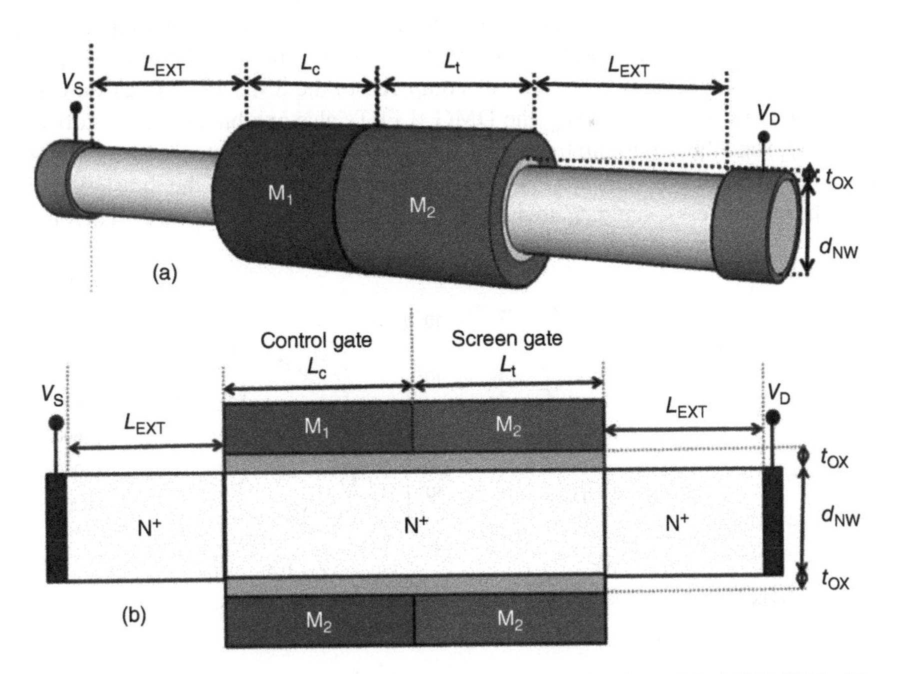

FIGURE 4.42 (a) Three-dimensional view and (b) schematic view of the NWJLFET with a dual material gate.

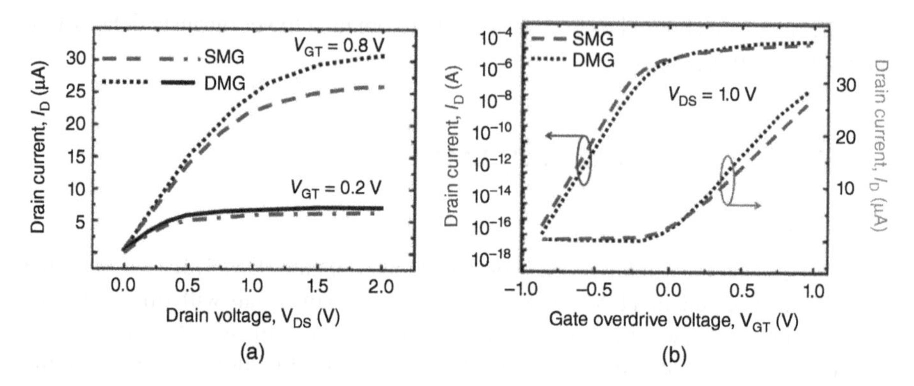

FIGURE 4.43 (a) The output characteristics and (b) the transfer characteristics of the SMG and DMG JLFETs [129].

4.9.2 Characteristics

The output characteristics and the transfer characteristics of the DMG and SMG JLFETs are shown in Fig. 4.43. The DMG JLFETs exhibit a higher ON-state current as well as a lower OFF-state current as compared to the SMG JLFETs. As a result, the DMG JLFETs offer a higher ON-state to OFF-state current ratio. Also, the DMG JLFETs exhibit a higher transconductance (G_m) in the saturation region leading to a better analog performance and a lower G_m in the subthreshold region resulting in a better turnoff property as compared to the SMG JLFETs as shown in Fig. 4.44. Owing to a larger G_m, the DMG JLFETs also exhibit a higher cutoff frequency (f_T) and the maximum frequency of operation (f_{max}) for gate overdrive voltage $V_{GT} = V_{GS} - V_{th} < 0.5$ V, where V_{GS} is the gate voltage and V_{th} is the threshold voltage. The parameters f_T and f_{max} are important analog performance metrics. Therefore,

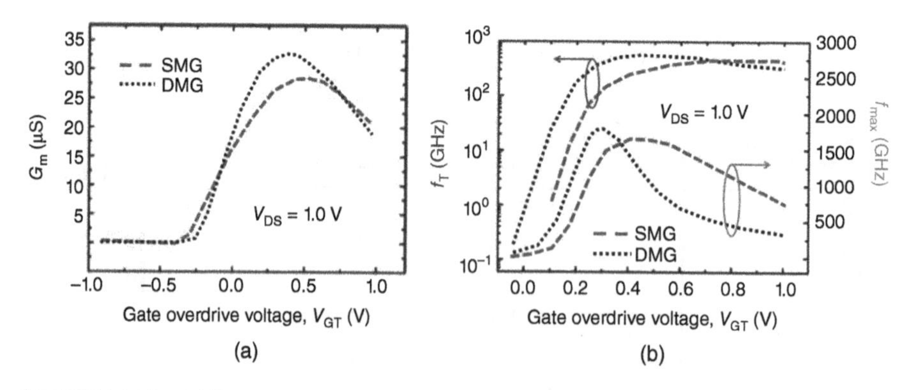

FIGURE 4.44 (a) Transconductance (G_m) and (b) cutoff frequency (f_T), and maximum operating frequency (f_{max}) of the SMG and DMG JLFETs [129].

the DMG JLFETs perform better than the SMG JLFETs for both analog and digital applications.

4.9.3 Operation

The improvement in the performance of the DMG JLFETs can be understood from Fig. 4.45. The presence of the DMG induces an abrupt potential step in the channel region of DMG JLFET as shown in Fig. 4.45(a) due to the difference in the work functions of materials M_1 and M_2 [129–147]. This abrupt potential change translates into an increased electric field at the source–channel interface and within the channel region in case of DMG JLFET as compared to the SMG JLFETs (Fig. 4.45(b)). The enhanced electric field accelerates the carriers and improves the carrier transport efficiency leading to a higher ON-state current in DMG JLFETs.

In addition, the peak electric field at the channel–drain interface is significantly reduced in the DMG JLFETs as shown in Fig. 4.45(b) [129–147]. A reduction in the electric field at the channel–drain interface suppresses the short-channel effects considerably. In addition, the increased field within the channel region screens the source–channel interface from the drain electric field. Therefore, the OFF-state current is also reduced in the DMG JLFETs compared to the SMG JLFETs.

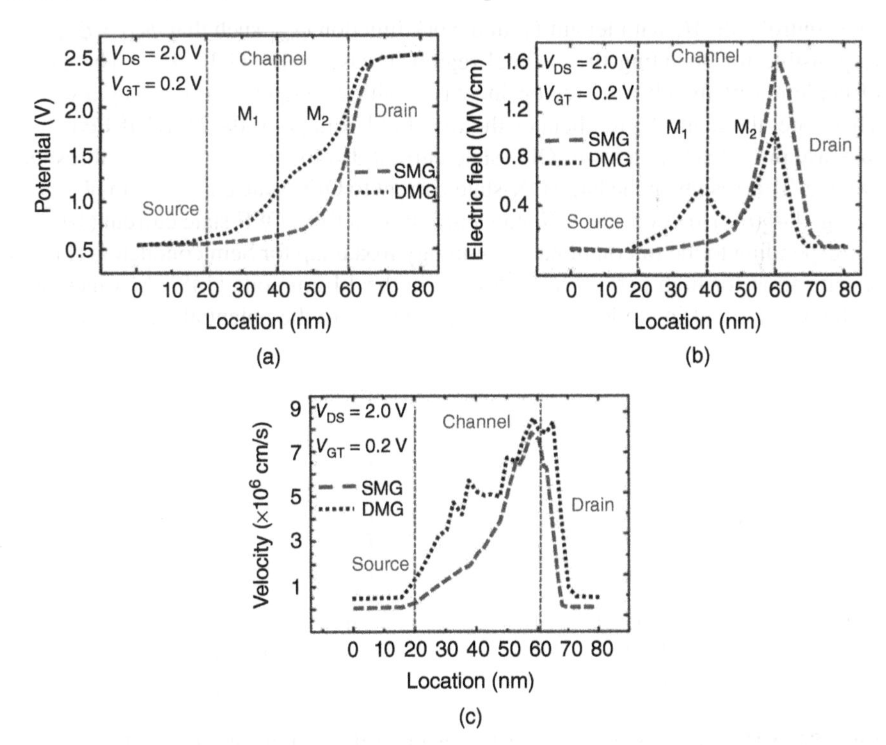

FIGURE 4.45 (a) Potential profile, (b) electric field profile, and (c) electron velocity of the SMG and DMG JLFETs [129].

The electron velocity is also enhanced in the DMG JLFETs by nearly three times as compared to the SMG JLFETs as shown in Fig. 4.45(c) owing to the abrupt potential change and the high electric field at the source–channel interface and within the channel region [129–147]. This also contributes to the increased ON-state current in the DMG JLFETs. Also, the abrupt potential change in the channel region owing to the DMG results in a larger potential change in response to the same applied voltage as compared with the DMG JLFETs. This enlarges the transconductance of the DMG JLFETs as compared to the SMG JLFETs [129–147].

Although the large G_m of the DMG JLFETs leads to a higher f_T and f_{max}, the abrupt change of potential due to the different work functions of M_1 and M_2 leads to an inherent electric field at the M_1/M_2 interface. This electric field increases the field lines through the gate oxide increasing the parasitic capacitances such as inner fringe capacitance. At high gate overdrive voltages, the increase in the capacitance surpasses the increment in the G_m leading to a lower f_T and f_{max} in DMG JLFETs as compared to the SMG JLFETs.

4.9.4 Impact of Length of Control Gate and Work Function Difference between Control and Screen Gates

The control gate M_1 with length L_C and work function ϕ_{M1} such that $\phi_{M1} > \phi_{M2}$ are responsible for depleting the channel region in case of DMG JLFETs. The extent of depletion of the channel region increases with an increase in the length of L_C. Therefore, the control gate dictates the threshold voltage of the DMG JLFETs. An increase in L_C leads to a lower OFF-state current due to an increase in the threshold voltage. This results in the large ON-state current to OFF-state current ratio observed in Fig. 4.46(a). However, the ON-state current for a fixed OFF-state current (10^{-7} A, corresponding to the International Technology Roadmap for Semiconductors (ITRS) specification for high-performance devices in the 14-nm node [148]) first increases with an increment in the length of control gate due to the potential step induced by

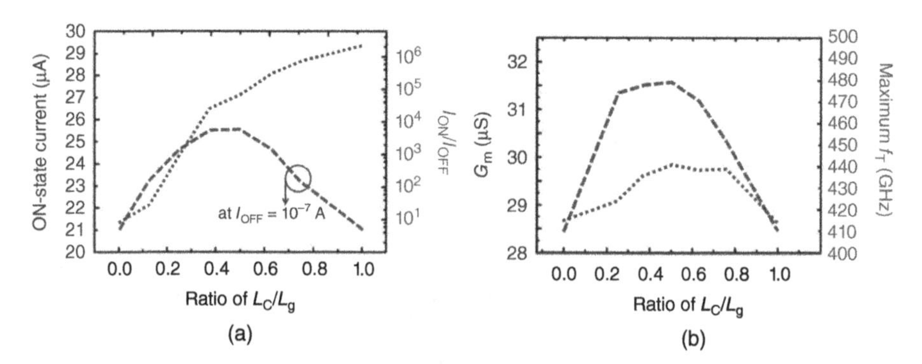

FIGURE 4.46 The impact of ratio of the length of the control gate (L_C) to the total gate length ($L_g = L_C + L_t$) on (a) ON-state current and ON-state to OFF-state current ratio and (b) transconductance and maximum f_T of DMG JLFET [129].

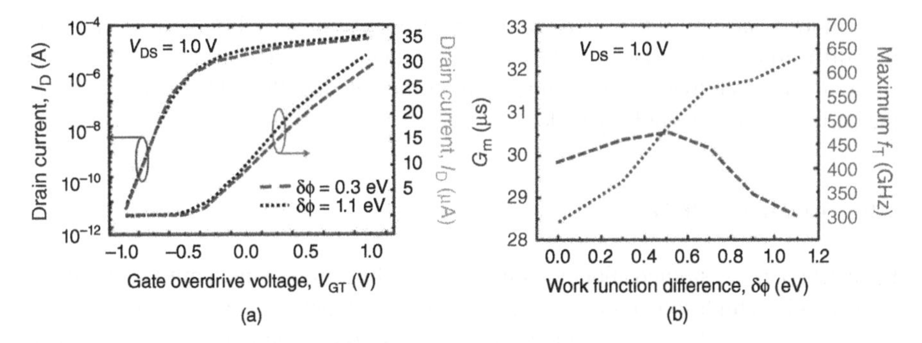

FIGURE 4.47 Impact of increasing the work function difference between ϕ_{M1} and ϕ_{M2} on (a) transfer characteristics of DMG, (b) transconductance and maximum f_T of the DMG JLFETs [129].

DMG, reaches a peak value and then reduces due to the shift in the threshold voltage (Fig. 4.46(b)). Moreover, the transconductance and the maximum f_T also follow the same trend as the ON-state current. Therefore, the length of the control gate should be chosen as half of the gate length for optimizing performance of DMG JLFETs. Such a choice is also favorable from a fabrication perspective.

To investigate the impact of increasing the work function difference ($\delta\phi = \phi_{M1} - \phi_{M2}$), Fig. 4.47(a) shows the transfer characteristics of the DMG JLFETs for different values of $\delta\phi$. The ON-state current increases with an increase in the $\delta\phi$ owing to the enlarged potential step and the consequent electric field enhancement at the source–channel interface and within the channel. The influence of the drain field also reduces owing to the enhanced screening from the electric field within the channel region. A larger $\delta\phi$ also results in a reduction in the electric field at the channel–drain interface owing to the field enhancement in the channel region. Therefore, a larger $\delta\phi$ is beneficial for attaining a higher ON-state to OFF-state current ratio. The enlarger potential step also results in a larger transconductance for larger $\delta\phi$ as shown in Fig. 4.47(b).

However, an increase in $\delta\phi$ leads to a larger inherent electric field originating due to the work function difference at the interface of M_1/M_2. This results in larger field lines through the gate oxide increasing the parasitic inner fringe capacitance significantly. As a result, the increment in the transconductance is dominated by the increased parasitic capacitance due to a larger $\delta\phi$. This results in the trend observed for the maximum f_T which first increases due to an enhancement in the transconductance but reaches a peak and reduces because of the dominance of the increased parasitic capacitance.

4.9.5 Challenges

Although the DMG leads to a higher ON-state current along with a lower OFF-state current, the application of DMG in JLFETs is limited due to the fabrication difficulties. The processes such as tilt angle deposition and lithography or metal interdiffusion used to realize DMG are not expected to yield an abrupt transition from one material to another within a span of the gate length (~22 nm) specified for the current 14 nm technology node.

4.10 CONCLUSION

In this chapter, we studied about various device architectures, which mitigate the challenges such as high source/drain series resistance and inefficient volume depletion in JLFETs. These architectures could pave the way for realization of extremely efficient JLFETs, which could potentially replace the conventional MOSFETs. However, as you might have observed, each of these architectures have some associated technological challenges, which needs to be addressed for fully exploiting the performance benefits offered by them.

As discussed in Section 3.6.5, one important problem degrading the performance of JLFETs and hindering their scaling is the BTBT-induced parasitic BJT action in the OFF-state. Although the architectures proposed in this chapter efficiently improve the source/drain series resistance and provide efficient volume depletion, their performance has been analyzed without including the BTBT model, which accounts for tunneling and the consequent parasitic BJT action. Since BTBT-induced parasitic BJT action has been identified as the major problem hindering the scaling of JLFETs, the next chapter is dedicated to explain this important phenomenon in detail and discuss about the various device architectures that may help to diminish the BTBT-induced parasitic BJT action.

REFERENCES

[1] T. K. Kim, D. H. Kim, Y. G. Yoon, J. M. Moon, B. W. Hwang, D. I. Moon, G. S. Lee, D. W. Lee, D. E. Yoo, H. C. Hwang, and J. S. Kim, "First demonstration of junctionless accumulation-mode bulk FinFETs with robust junction isolation," *IEEE Electron Device Lett.*, vol. 34, no. 12, pp. 1479–1481, Dec. 2013.

[2] D. H. Kim, T. K. Kim, Y. G. Yoon, B. W. Hwang, Y. K. Choi, B. J. Cho, and S. H. Lee, "First demonstration of ultra-thin SiGe-channel junctionless accumulation-mode (JAM) bulk FinFETs on Si substrate with PN junction-isolation scheme," *IEEE J. Electron Devices Soc.*, vol. 2, no. 5, pp. 123–127, Sept. 2014.

[3] J. H. Choi, T. K. Kim, J. M. Moon, Y. G. Yoon, B. W. Hwang, D. H. Kim, and S. H. Lee, "Origin of device performance enhancement of junctionless accumulation-mode (JAM) bulk FinFETs with high-κ gate spacers," *IEEE Electron Device Lett.*, vol. 35, no. 12, pp. 1182–1184, Dec. 2014.

[4] R. Rios, A. Cappellani, M. Armstrong, A. Budrevich, H. Gomez, R. Pai, N. Rahhal-Orabi, and K. Kuhn, "Comparison of junctionless and conventional trigate transistors with L_g down to 26 nm," *IEEE Electron Device Lett.*, vol. 32, no. 9, pp. 1170–1172, Sept. 2011.

[5] N. Kadotani, T. Ohashi, T. Takahashi, S. Oda, and K. Uchida, "Experimental study on electron mobility in accumulation-mode silicon-on-insulator metal-oxide-semiconductor field-effect transistors," *Jap. J. Appl. Phys.*, vol. 50, no. 9R, 094101, Sept. 2011.

[6] K. I. Goto, T. H. Yu, J. Wu, C. H. Diaz, and J. P. Colinge, "Mobility and screening effect in heavily doped accumulation-mode metal-oxide-semiconductor field-effect transistors," *Appl. Phys. Lett.*, vol. 101, no. 7, 073503, Aug. 2012.

[7] C. W. Lee, I. Ferain, A. Afzalian, R. Yan, N. D. Akhavan, P. Razavi, and J. P. Colinge, "Performance estimation of junctionless multigate transistors," *Solid-State Electron.*, vol. 54, no. 2, pp. 97–103, Feb. 2010.

[8] J. P. Colinge, "Thin-film accumulation-mode p-channel SOI MOSFETs," *Electron. Lett.*, vol. 24, no. 5, pp. 257–258, 1988.

[9] J. Colinge, "Conduction mechanisms in thin-film accumulation-mode SOI p-channel MOSFET's," *IEEE Trans. Electron Devices*, vol. 37, pp. 718–723, Mar. 1990.

[10] O. Faynot, S. Cristoloveanu, A. J. Auberton-Herve, and C. Raynaud, "Performance and potential of ultrathin accumulation-mode SIMOX MOSFET's," *IEEE Trans. Electron Devices*, vol. 42, pp. 713–719, Apr. 1995.

[11] L. K. Wang, J. Seliskar, T. Bucelot, A. Edenfeld, and N. Haddad, "Enhanced performance of accumulation mode 0.5-μm CMOS/SOI operated at 300 K and 85 K," *IEDM Tech. Dig.*, pp. 679–682, 1991.

[12] B. J. Baliga, T. Syau, and P. Venkatraman, "The accumulation-mode field-effect transistor: A new ultralow on-resistance MOSFET," *IEEE Electron Device Lett.*, vol. 13, no. 8, pp. 427–429, Aug. 1992.

[13] D. Flandre and S. Cristoloveanu, "Latch and hot-electron gate current in accumulation-mode SOI p-MOSFETs," *IEEE Electron Device Lett.*, vol. 15, no. 5, pp. 157–159, May 1994.

[14] D. Flandre, A. Terao, P. Francis, B. Gentinnea, and J. P. Colinge, "Demonstration of the potential of accumulation-mode MOS transistors on SOI substrates for high-temperature operation (150–300°C)," *IEEE Electron Device Lett.*, vol. 14, no. 1, pp. 10–12, Jan. 1993.

[15] S. Madan, "Enhanced leakage current due to a high-gain stray bipolar transistor in accumulation-mode SOI MOSFET's," *IEEE Trans. Electron Devices*, vol. 32, no. 11, pp. 2549–2550, Nov. 1985.

[16] S. Sahay and M. J. Kumar, "Insight into lateral band-to-band-tunneling in nanowire junctionless field-effect transistors," *IEEE Trans. Electron Devices*, vol. 63, no. 10, pp. 4138–4142, Oct. 2016.

[17] S. Sahay and M. J. Kumar, "Symmetric operation in an extended back gate JLFET for scaling to the 5 nm regime considering quantum confinement effects," *IEEE Trans. Electron Devices*, vol. 64, no. 1, pp. 21–27, Jan. 2017.

[18] S. Sahay and M. J. Kumar, "Physical insights into the nature of gate induced drain leakage in ultra-short channel nanowire field-effect transistors," *IEEE Trans. Electron Devices*, vol. 64, no. 6, pp. 2604–2610, June 2017.

[19] S. Sahay and M. J. Kumar, "Realizing efficient volume depletion in SOI junctionless FETs," *IEEE J. Electron Devices Soc.*, vol. 4, no. 3, pp. 110–115, May 2016.

[20] S. Gundapaneni, S. Ganguly, and A. Kottantharayil, "Bulk planar junctionless transistor (BPJLT): An attractive device alternative for scaling," *IEEE Electron Device Lett.*, vol. 32, no. 3, pp. 261–263, Mar. 2011.

[21] G. K. Celler and S. Cristoloveanu, "Frontiers of silicon-on-insulator," *J. Appl. Phys.*, vol. 93, no. 9, pp. 4955–4978, May 2003.

[22] S. Sahay and M. J. Kumar, "Controlling the drain side tunneling width to reduce ambipolar current in tunnel FETs using heterodielectric BOX," *IEEE Trans. Electron Devices*, Vol. 62, no. 11, pp. 3882–3886, Nov. 2015.

[23] Y. Wang, P. H. K. Wei, L. Zeng, X. Liu, G. Du, X. Zhang, and J. Kang, "Impact of random interface traps and random dopants in high-κ/metal gate junctionless FETs," *IEEE Trans. Nanotechnol.*, vol. 13, no. 3, pp. 584–588, May 2014.

[24] Y. Wang, K. Wei, X. Liu, G. Du, and J. Kang, "Random interface trap induced fluctuation in 22 nm high-k/metal gate junctionless and inversion mode FETs," in *Proc. IEEE Int. Symp. VLSI Tech., Syst., Appl.*, 2013, pp. 177–178.

[25] D. Ghosh, M. S. Parihar, G. A. Armstrong, and A. Kranti, "High-performance junctionless MOSFETs for ultralow-power analog/RF applications," *IEEE Electron Device Lett.*, vol. 33, no. 10, pp. 1477–1479, Oct. 2012.

[26] S. Gundapaneni, M. Bajaj, R. K. Pandey, K. V. R. M. Murali, S. Ganguly, and A. Kottantharayil, "Effect of band-to-band tunneling on junctionless transistors," *IEEE Trans. Electron Devices*, vol. 59, no. 4, pp. 1023–1029, Apr. 2012.

[27] P. Mondal, B. Ghosh, and P. Bal, "Planar junctionless transistor with non-uniform channel doping," *Appl. Phys. Lett.*, vol. 102, no. 13, 133505, Apr. 2013.

[28] Y. Song and X. Li, "Scaling junctionless multigate field-effect transistors by step-doping," *Appl. Phys. Lett.*, vol. 105. no. 22, 223506, Dec. 2014.

[29] Brigham Young University Cleanroom [online]. Available: https://cleanroom.byu.edu/ResistivityCal, Accessed: Oct. 5, 2018.

[30] S. C. Sun and J. D. Plummer, "Electron mobility in inversion and accumulation layers on thermally oxidized silicon surfaces," *IEEE J. Solid-State Circuits*, vol. 15, no. 4, pp. 562–273, Aug. 1980.

[31] G. Masetti, M. Severi, and S. Solmi, "Modeling of carrier mobility against carrier concentration in arsenic-, phosphorus-, and boron-doped silicon," *IEEE Trans. Electron Devices*, vol. 30, no. 7, pp. 764–769, July 1983.

[32] J. C. Ho, R. Yerushalmi, Z. A. Jacobson, Z. Fan, R. L. Alley, and A. Javey, "Controlled nanoscale doping of semiconductors via molecular monolayers," *Nature Mater.*, vol. 7, no. 1, pp. 62–67, Jan. 2008.

[33] J. C. Ho, R. Yerushalmi, G. Smith, P. Majhi, J. Bennett, J. Halim, V. N. Faifer, and A. Javey, "Wafer-scale, sub-5 nm junction formation by monolayer doping and conventional spike annealing," *Nano Lett.*, vol. 9, no. 2, pp. 725–730, 2009.

[34] K.-W. Ang, J. Barnett, W.-Y. Loh, J. Huang, B.-G. Min, P. Y. Hung, I. Ok, J. H. Yum, G. Bersuker, M. Rodgers, V. Kaushik, S. Gausepohl, C. Hobbs, P. D. Kirsch, and R. Jammy, "300 mm FinFET results utilizing conformal, damage free, ultra-shallow junctions ($X_j \sim 5$ nm) formed with molecular monolayer doping technique," *IEDM Tech. Dig.*, pp. 35.5.1–35.5.4, Dec. 2011.

[35] Y.-J. Lee, T.-C. Cho, S.-S. Chuang, F.-K. Hsueh, Y.-L. Lu, P.-J. Sung, H.-C. Chen, M. I. Current, T.-Y. Tseng, T.-S. Chao, C. Hu, and F.-L. Yang, "Low-temperature

microwave annealing processes for future IC fabrication—A review," *IEEE Trans. Electron Devices*, vol. 61, no. 3, pp. 651–665, Mar. 2014.

[36] J. P. Colinge, *FinFETs and Other Multi-Gate Transistors*, vol. 73, Springer, New York, 2008.

[37] J.-M. Sallese, F. Jazaeri, L. Barbut, N. Chevillon, and C. Lallement, "A common core-Model for junctionless nanowires and symmetric double gate FETs," *IEEE Trans. Electron Devices*, vol. 60, pp. 4277–4280, Dec. 2013.

[38] S. I. Amin and R. K. Sarin, "The impact of gate misalignment on the analog performance of a dual-material double gate junctionless transistor," *J. Semicond.*, vol. 36, no. 9, 094001, Sept. 2015.

[39] X. Jin, X. Liu, M. Wu, R. Chuai, J. H. Lee and J. H. Lee, "Modelling of the nanoscale channel length effect on the subthreshold characteristics of junctionless field-effect transistors with a symmetric double-gate structure," *J. Appl. Phys. D*, vol. 45, no. 37, 375102, Aug. 2012.

[40] M. S. Parihar and A. Kranti, "Enhanced sensitivity of double gate junctionless transistor architecture for biosensing applications," *Nanotechnology*, vol. 26, no. 14, 145201, Mar. 2015.

[41] C. Jiang, R. Liang, J. Wang, and J. Xu, "A two-dimensional analytical model for short channel junctionless double-gate MOSFETs," *AIP Adv.*, vol. 5, no. 5, 057122, May 2015.

[42] T. Holtij, M. Graef, A. Kloes, and B. Iñíguez, "Modeling and performance study of nanoscale double gate junctionless and inversion mode MOSFETs including carrier quantization effects," *Microelectron. J.*, vol. 45, no. 9, pp. 1220–1225, Sept. 2014.

[43] E. Chebaki, F. Djeffal, H. Ferhati, and T. Bentrcia, "Improved analog/RF performance of double gate junctionless MOSFET using both gate material engineering and drain/source extensions," *Superlattices Microstruct.*, vol. 92, pp. 80–91, Apr. 2016.

[44] J. P. Duarte, M. S. Kim, S. J. Choi, and Y. K. Choi, "A compact model of quantum electron density at the subthreshold region for double-gate junctionless transistors," *IEEE Trans. Electron Devices*, vol. 59, no. 4, pp. 1008–1012, Apr. 2012.

[45] R. Narang, M. Saxena, and M. Gupta, "Investigation of dielectric modulated (DM) double gate (DG) junctionless MOSFETs for application as a biosensors," *Superlattices Microstruct.*, vol. 85, pp. 557–572, Sept. 2015.

[46] D. Munteanu and J. L. Autran, "Radiation sensitivity of junctionless double-gate 6T SRAM cells investigated by 3-D numerical simulation," *Microelectron. Rel.*, vol. 54, no. 9, pp. 2284–2288, Oct. 2014.

[47] J. Alvarado, P. Flores, S. Romero, F. Ávila-Herrera, V. González, B. S. Soto-Cruz, and A. Cerdeira, "Verilog-A implementation of a double-gate junctionless compact model for DC circuit simulations," *Semicond. Sci. Technol.*, vol. 31, 075002, July 2016.

[48] B. C. Paz, F. Ávila-Herrera, A. Cerdeira, and M. A. Pavanello, "Double-gate junctionless transistor model including short-channel effects," *Semicond. Sci. Technol.*, vol. 30, no. 5, 055011, Apr. 2015.

[49] E. Y. Jeong, M. J. Deen, C. H. Chen, R. H. Baek, J. S. Lee, and Y. H. Jeong, "Physical DC and thermal noise models of 18 nm double-gate junctionless p-type MOSFETs for low noise RF applications," *Jap. J. Appl. Phys.*, vol. 54, no. 4S, 04DC08, Feb. 2015.

[50] Y. Taur, H. P. Chen, W. Wang, S. H. Lo, and C. Wann, "On–off charge–voltage characteristics and dopant number fluctuation effects in junctionless double-gate MOSFETs," *IEEE Trans. Electron Devices*, vol. 59, no. 3, pp. 863–866, Mar. 2012.

[51] J. P. Duarte, S. J. Choi, D. I. Moon, and Y. K. Choi, "Simple analytical bulk current model for long-channel double-gate junctionless transistors," *IEEE Electron Device Lett.*, vol. 32, no. 6, pp. 704–706, June 2011.

[52] J. M. Sallese, N. Chevillon, C. Lallement, B. Iniguez, and F. Pregaldiny, "Charge-based modeling of junctionless double-gate field-effect transistors," *IEEE Trans. Electron Devices*, vol. 58, no. 8, pp. 2628–2637, Aug. 2011.

[53] J. P. Duarte, S. J. Choi, and Y. K. Choi, "A full-range drain current model for double-gate junctionless transistors," *IEEE Trans. Electron Devices*, vol. 58, no. 12, pp. 4219–4225, Dec. 2011.

[54] J. P. Duarte, M. S. Kim, S. J. Choi, and Y. K. Choi, "A compact model of quantum electron density at the subthreshold region for double-gate junctionless transistors," *IEEE Trans. Electron Devices*, vol. 59, no. 4, pp. 1008–1012, Apr. 2012.

[55] T. K. Chiang, "A quasi-two-dimensional threshold voltage model for short-channel junctionless double-gate MOSFETs," *IEEE Trans. Electron Devices*, vol. 59, no. 9, pp. 2284–2289, Sept. 2012.

[56] Z. Chen, Y. Xiao, M. Tang, Y. Xiong, J. Huang, J. Li, X. Gu, and Y. Zhou, "Surface-potential-based drain current model for long-channel junctionless double-gate MOSFETs," *IEEE Trans. Electron Devices*, vol. 59, no. 12, pp. 3292–3298, Dec. 2012.

[57] A. Gnudi, S. Reggiani, E. Gnani, and G. Baccarani, "Semianalytical model of the subthreshold current in short-channel junctionless symmetric double-gate field-effect transistors," *IEEE Trans. Electron Devices*, vol. 60, no. 4, pp. 1342–1348, Apr. 2013.

[58] J. H. Woo, J. M. Choi, and Y. K. Choi, "Analytical threshold voltage model of junctionless double-gate MOSFETs with localized charges," *IEEE Trans. Electron Devices*, vol. 60, no. 9, pp. 2951–2955, Sept. 2013.

[59] F. Jazaeri, L. Barbut, and J. M. Sallese, "Trans-capacitance modeling in junctionless symmetric double-gate MOSFETs," *IEEE Trans. Electron Devices*, vol. 60, no. 12, pp. 4034–4040, Dec. 2013.

[60] T. Holtij, M. Graef, F. M. Hain, A. Kloes, and B. Iniguez, "Compact model for short-channel junctionless accumulation mode double gate MOSFETs," *IEEE Trans. Electron Devices*, vol. 61, no. 2, pp. 288–299, Feb. 2014.

[61] F. Jazaeri, L. Barbut L, and J. M. Sallese, "Generalized charge-based model of double-gate junctionless FETs, including inversion," *IEEE Trans. Electron Devices*, vol. 61, no. 10, pp. 3553–3557, Oct. 2014.

[62] F. Jazaeri, L. Barbut, and J. M. Sallese, "Modeling asymmetric operation in double-gate junctionless FETs by means of symmetric devices," *IEEE Trans. Electron Devices*, vol. 61, no. 12, pp. 3962–3970, Dec. 2014.

[63] B. W. Hwang, J. W. Yang, and S. H. Lee, "Explicit analytical current-voltage model for double-gate junctionless transistors," *IEEE Trans. Electron Devices*, vol. 62, no. 1, pp. 171–177, Jan. 2015.

[64] M. Ichii, R. Ishida, H. Tsuchiya, Y. Kamakura, N. Mori, and M. Ogawa, "Computational study of effects of surface roughness and impurity scattering in Si double-gate

junctionless transistors," *IEEE Trans. Electron Devices*, vol. 62, no. 4, pp. 1255–1261, Apr. 2015.

[65] J. Hur, J. M. Choi, J. H. Woo, H. Jang, and Y. K. Choi, "A generalized threshold voltage model of tied and untied double-gate junctionless FETs for a symmetric and asymmetric structure," *IEEE Trans. Electron Devices*, vol. 62, no. 9, pp. 2710–2716, Sept. 2015.

[66] A. Yesayan, F. Jazaeri, and J. M. Sallese, "Charge-based modeling of double-gate and nanowire junctionless FETs including interface-trapped charges," *IEEE Trans. Electron Devices*, vol. 63, no. 3, pp. 1368–1374, Mar. 2016.

[67] X. Lin, B. Zhang, Y. Xiao, H. Lou, L. Zhang, and M. Chan, "Analytical current model for long-channel junctionless double-gate MOSFETs," *IEEE Trans. Electron Devices*, vol. 63, no. 3, pp. 959–965, Mar. 2016.

[68] B. Singh, D. Gola, K. Singh, E. Goel, S. Kumar, and S. Jit, "Analytical modeling of channel potential and threshold voltage of double-gate junctionless FETs with a vertical Gaussian-like doping profile," *IEEE Trans. Electron Devices*, vol. 63, no. 6, pp. 2299–2305, June 2016.

[69] T. K. Chiang, "A short-channel-effect-degraded noise margin model for junctionless double-gate MOSFET working on subthreshold CMOS logic gates," *IEEE Trans. Electron Devices*, vol. 63, no. 8, pp. 3354–3359, Aug. 2016.

[70] F. Jazaeri and J. -M. Sallese, *Modeling Nanowire and Double-Gate Junctionless Field-Effect Transistors*, Cambridge University Press, New York, 2018.

[71] Y. Xiao, X. Lin, H. Lou, B. Zhang, L. Zhang, and M. Chan, "A short channel double-gate junctionless transistor model including the dynamic channel boundary effect," *IEEE Trans. Electron Devices*, vol. 63, no. 12, pp. 4661–4667, Dec. 2016.

[72] V. Thirunavukkarasu, J. L. Jaehyun, T. T. Sadi, V. P. Georgiev, F. A. Lema, K. P. Soundarapandian, Y. R. Jhan, S. Y. Yang, Y. R. Lin, E. D. Kurniawan, Y. C. Wu, and A. Asenov, "Investigation of inversion, accumulation and junctionless mode bulk germanium FinFETs," *SuperlatticesMicrostruct.*, vol. 111, pp. 649–655, 2017

[73] A. Cerdeira, M. Estrada, B. Iniguez, R. D. Trevisoli, R. T. Doria, M. de Souza, and M. A. Pavanello, "Charge-based continuous model for long-channel symmetric double-gate junctionless transistors," *Solid-State Electron.*, vol. 85, pp. 59–63, July 2013.

[74] M. H. Han, C. Y. Chang, H. B. Chen, J. J. Wu, Y. C. Cheng, and Y. C. Wu, "Performance comparison between bulk and SOI junctionless transistors," *IEEE Electron Device Lett.*, vol. 34, no. 2, pp. 169–171, 2013.

[75] B. Singh, D. Gola, E. Goel, S. Kumar, K. Singh, and S. Jit, "Dielectric pocket double gate junctionless FET: A new MOS structure with improved subthreshold characteristics for low power VLSI applications," *J. Comp. Electron.*, vol. 15, no. 2, pp. 502–507, June 2016.

[76] A. Veloso, G. Hellings, M. J. Cho, E. Simoen, K. Devriendt, V. Paraschiv, E. Vecchio, Z. Tao, J. J. Versluijs, L. Souriau, and H. Dekkers, "Gate-all-around NWFETs vs. triple-gate FinFETs: Junctionless vs. extensionless and conventional junction devices with controlled EWF modulation for multi-VT CMOS," *VLSI Tech. Dig.*, pp. T138–T139, June 2015.

[77] C. W. Chen, C. T. Chung, J. Y. Tzeng, P. S. Chang, G. L. Luo, and C. H. Chien, "Body-tied germanium tri-gate junctionless PMOSFET with in situ boron doped channel," *IEEE Electron Device Lett.*, vol. 35, no. 1, pp. 12–14, Jan. 2014.

[78] G. Ghibaudo, "Evaluation of variability performance of junctionless and conventional Trigate transistors," *Solid-State Electron.*, vol. 75, pp. 13–15, Sept. 2012.

[79] D. Y. Jeon, S. J. Park, M. Mouis, S. Barraud, G. T. Kim, and G. Ghibaudo, "Effects of channel width variation on electrical characteristics of tri-gate Junctionless transistors," *Solid-State Electron.*, vol. 81, pp. 58–62, Mar. 2013.

[80] S. Das, R. Yu, K. Cherkaoui, P. Razavi, and S. Barraud, "Performance of 22 nm tri-gate junctionless nanowire transistors at elevated temperatures," *ECS Solid-State Lett.*, vol. 2, no. 8, pp. Q62–Q65, Jan. 2013.

[81] S. Barraud, M. Berthomé, R. Coquand, M. Cassé, T. Ernst, M. P. Samson, P. Perreau, K. K. Bourdelle, O. Faynot, and T. Poiroux, "Scaling of trigate junctionless nanowire MOSFET with gate length down to 13 nm," *IEEE Electron Device Lett.*, vol. 33, no. 9, pp. 1225–1227, Sept. 2012.

[82] R. D. Trevisoli, R. T. Doria, M. de Souza, S. Das, I. Ferain, and M. A. Pavanello, "Surface-potential-based drain current analytical model for triple-gate junctionless nanowire transistors," *IEEE Trans. Electron Devices*, vol. 59, no. 12, pp. 3510–3518, Dec. 2012.

[83] T. K. Chiang, "A new subthreshold current model for junctionless trigate MOSFETs to examine interface-trapped charge effects," *IEEE Trans. Electron Devices*, vol. 62, no. 9, pp. 2745–2750, Sept. 2015.

[84] R. Trevisoli, R. T. Doria, M. de Souza, S. Barraud, M. Vinet, and M. A. Pavanello, "Analytical model for the dynamic behavior of triple-gate junctionless nanowire transistors," *IEEE Trans. Electron Devices*, vol. 63, no. 2, pp. 856–863, Feb. 2016.

[85] Z. Guo, J. Zhang, Z. Ye, and Y. Wang, "3-D analytical model for short-channel triple-gate junctionless MOSFETs," *IEEE Trans. Electron Devices*, vol. 63, no. 10, pp. 3857–3863, Oct. 2016.

[86] T. A. Oproglidis, A. Tsormpatzoglou, D. H. Tassis, T. A. Karatsori, S. Barraud, G. Ghibaudo, and C. A. Dimitriadis, "Analytical drain current compact model in the depletion operation region of short-channel triple-gate junctionless transistors," *IEEE Trans. Electron Devices*, vol. 64, no. 1, pp. 66–72, Jan. 2017.

[87] D. Y. Jeon, S. J. Park, M. Mouis, S. Barraud, G. T. Kim, and G. Ghibaudo, "A new method for the extraction of flat-band voltage and doping concentration in tri-gate junctionless transistors," *Solid-State Electron.*, vol. 81, pp. 113–118, Mar. 2013.

[88] S. J. Park, D. Y. Jeon, L. Montès, S. Barraud, G. T. Kim, and G. Ghibaudo, "Back biasing effects in tri-gate junctionless transistors," *Solid-State Electron.*, vol. 87, pp. 74–79, Sept. 2013.

[89] T. Holtij, A. Kloes, and B. Iñíguez, "3-D compact model for nanoscale junctionless triple-gate nanowire MOSFETs, including simple treatment of quantization effects," *Solid-State Electron.*, vol. 112, pp. 85–98, Oct. 2015.

[90] F. Ávila-Herrera, B. C. Paz, A. Cerdeira, M. Estrada, and M. A. Pavanello, "Charge-based compact analytical model for triple-gate junctionless nanowire transistors," *Solid-State Electron.*, vol. 122, pp. 23–31, Aug. 2016.

[91] G. Saini and S. Choudhahry, "Investigation of trigate JLT with dual-k sidewall spacers for enhanced analog/RF FOMs," *J. Comp. Electron.*, vol. 15, no. 3, pp. 856–873, Sept. 2016.

[92] M. Najmzadeh, J. M. Sallese, M. Berthomé, W. Grabinski, and A. M. Ionescu, "Local volume depletion/accumulation in GAA Si nanowire junctionless nMOSFETs," *IEEE Trans. Electron Devices*, vol. 59, no. 12, pp. 3519–3526, Dec. 2012.

[93] M. Cho, G. Hellings, A. Veloso, E. Simoen, P. Roussel, B. Kaczer, H. Arimura, W. Fang, J. Franco, P. Matagne, and N. Collaert, "On and off state hot carrier reliability in junctionless high-κ MG gate-all-around nanowires," *IEDM Tech. Dig.*, pp. 14–15, Dec. 2015.

[94] P. Singh, N. Singh, J. Miao, W. T. Park, and D. L. Kwong, "Gate-all-around junctionless nanowire MOSFET with improved low-frequency noise behaviour," *IEEE Electron Device Lett.*, vol. 32, no. 12, pp. 1752–1754, Dec. 2011.

[95] M. H. Han, C. Y. Chang, Y. R. Jhan, J. J. Wu, H. B. Chen, Y. C. Cheng, and Y. C. Wu, "Characteristic of p-type junctionless gate-all-around nanowire transistor and sensitivity analysis," *IEEE Electron Device Lett.*, vol. 34, no. 2, pp. 157–159, Feb. 2013.

[96] T. Wang, L. Lou, and C. Lee, "A junctionless gate-all-around silicon nanowire FET of high linearity and its potential applications," *IEEE Electron Device Lett.*, vol. 34, no. 4, pp. 478–480, Apr. 2013.

[97] O. Moldovan, F. Lime, and B. Iniguez, "A compact explicit model for long-channel gate-all-around junctionless MOSFETs. Part II: Total charges and intrinsic capacitance characteristics," *IEEE Trans. Electron Devices*, vol. 61, no. 9, pp. 3042–3046, Sept. 2014.

[98] F. Lime, O. Moldovan, and B. Iniguez, "A compact explicit model for long-channel gate-all-around junctionless MOSFETs. Part I: DC characteristics," *IEEE Trans. Electron Devices*, vol. 61, no. 9, pp. 3036–3041.

[99] A. Martinez, M. Aldegunde, A. R. Brown, S. Roy, and A. Asenov, "NEGF simulations of a junctionless Si gate-all-around nanowire transistor with discrete dopants," *Solid-State Electron.*, vol. 71, pp. 101–105, May 2012.

[100] F. Jazaeri, L. Barbut, and J. M. Sallese, "Trans-capacitance modeling in junctionless gate-all-around nanowire FETs," *Solid-State Electron.*, vol. 96, pp. 34–37, June 2014.

[101] F. Djeffal, H. Ferhati, and T. Bentrcia, "Improved analog and RF performances of gate-all-around junctionless MOSFET with drain and source extensions," *Superlattices Microstruct.*, vol. 90, pp. 132–140, Feb. 2016.

[102] O. Moldovan, F. Lime, and B. Iñiguez, "A complete and verilog-A compatible gate-all-around long-channel junctionless MOSFET model implemented in CMOS inverters," *Microelectron. J.*, vol. 46, no. 11, pp. 1069–1072, Nov. 2015.

[103] S. J. Choi, D. I. Moon, J. P. Duarte, S. Kim, and Y. K. Choi, "A novel junctionless all-around-gate SONOS device with a quantum nanowire on a bulk substrate for 3D stack NAND flash memory," *VLSI Tech. Dig.*, pp. 74–75, June 2011.

[104] A. Veloso, B. Parvais, P. Matagne, E. Simoen, T. Huynh-Bao, V. Paraschiv, E. Vecchio, K. Devriendt, E. Rosseel, M. Ercken, and B. T. Chan, "Junctionless gate-all-around lateral and vertical nanowire FETs with simplified processing for advanced logic and analog/RF applications and scaled SRAM cells," *VLSI Tech. Dig.*, pp. 1–2, June 2016.

[105] S. Sahay and M. J. Kumar, "Comprehensive analysis of gate-induced drain leakage in emerging FET architectures: Nanotube FETs vs. nanowire FETs," *IEEE Access*, vol. 5, pp. 18918–18926, Dec. 2017.

[106] S. Sahay and M. J. Kumar, "Spacer design guidelines for nanowire FETs from gate induced drain leakage perspective," *IEEE Trans. Electron Devices*, vol. 64, no. 7, pp. 3007–3015, July 2017.

[107] E. Gnani, A. Gnudi, S. Reggiani and G. Baccarani, "Theory of the junctionless nanowire FET," *IEEE Trans. Electron Devices*, vol. 58, no. 9, pp. 2903–2910, Sep 2011.

[108] S. Sahay and M. J. Kumar, "Nanotube junctionless field-effect transistor: Proposal, design and investigation," *IEEE Trans. Electron Devices*, vol. 64, no. 4, pp. 1851–1856, Apr. 2017.

[109] S. Sahay and M. J. Kumar, "Diameter Dependency of Leakage Current in Nanowire Junctionless Field-Effect Transistors," *IEEE Trans. Electron Devices*, vol. 64, no. 3, pp. 1330–1335, Mar. 2017.

[110] S. Sahay and M. J. Kumar, "A novel gate-stack-engineered nanowire FET for scaling to the sub-10-nm regime," *IEEE Trans. Electron Devices*, vol. 63, no. 12, pp. 5055–5059, Dec. 2016.

[111] M. J. Kumar and S. Sahay, "Controlling BTBT induced parasitic BJT action in junctionless FETs using a hybrid channel," *IEEE Trans. Electron Devices*, vol. 63, no. 8, pp. 3350–3353, Aug. 2016.

[112] S. Sahay and M. J. Kumar, "Controlling L-BTBT and volume depletion in nanowire JLFETs using core-shell architecture," *IEEE Trans. Electron Devices*, vol. 63, no. 9, pp. 3790–3794, Sept. 2016.

[113] J. Fan, M. Li, X. Xu, Y. Yang, H. Xuan, and R. Huang, "Insight into gate-induced drain leakage in silicon nanowire transistors," *IEEE Trans. Electron Devices*, vol. 62, no. 1, pp. 213–219, Jan. 2015.

[114] J. Hur, B.-H. Lee, M.-H. Kang, D.-C. Ahn, T. Bang, S.-B Jeon, and Y.-K. Choi, "Comprehensive analysis of gate-induced drain leakage in vertically stacked nanowire FETs: Inversion-mode vs. junctionless mode," *IEEE Electron Device Lett.*, vol. 37, no. 5, pp. 541–544, May 2016.

[115] J.-P. Colinge, C.-W. Lee, A. Afzalian, N. D. Akhavan, R. Yan, I. Ferain, P. Razavi, B. O'Neill, A. Blake, M. White, A.-M. Kelleher, B. McCarthy, and R. Murphy, "Nanowire transistors without junctions," *Nature Nanotechnol.*, vol. 5, no. 3, pp. 225–229, Mar. 2010.

[116] K. J. Kuhn, "Considerations for ultimate CMOS scaling," *IEEE Trans. Electron Devices*, vol. 59, no. 7, pp. 1813–1828, July 2012.

[117] S. Gundapaneni, S. Ganguly, and A. Kottantharayil, "Enhanced electrostatic integrity of short-channel junctionless transistors with high-κ spacers," *IEEE Electron Dev. Lett.*, vol. 32, no. 10, pp. 1325–1327, Oct. 2011.

[118] R. K. Baruah and R. P. Paily, "Impact of high-k spacer on device performance of a junctionless transistor," *J. Comput. Electron*, vol. 12, pp. 14–19, 2013.

[119] A. B. Sachid, R. Francis, M. S. Baghini, D. K. Sharma, K.-H. Bach, R. Mahnkopf, and V. R. Rao, "Sub-20 nm gate length FinFET design: Can high-κ spacers make a difference?" *IEDM Tech. Dig.*, Dec. 2008, pp. 697–700.

[120] A. B. Sachid, M.-C. Chen, and C. Hu, "FinFET with high-κ spacers for improved drive current," *IEEE Electron Device Lett.*, vol. 37, no. 7, pp. 835–838, July 2016.

[121] J.-H. Hong, S.-H. Lee, Y.-R. Kim, E.-Y. Jeong, J.-S. Yoon, J.-S. Lee, R.-H. Baek, and Y.-H. Jeong, "Impact of the spacer dielectric constant on parasitic RC and design

guidelines to optimize DC/AC performance in 10-nm-node Si-nanowire FETs," *Jpn. J. Appl. Phys.*, vol. 54, 04DN05, Feb. 2015.

[122] K. Koley, A. Dutta, S. K. Saha, and C. K. Sarkar, "Analysis of high-κ spacer asymmetric underlap DGMOSFET for SOC application," *IEEE Trans. Electron Devices*, vol. 62, no. 6, pp. 1733–1738, June 2015.

[123] S. Chakraborty, A. Dasgupta, R. Das, A. Kundu, and C. K. Sarkar, "Impact of asymmetric dual-κ spacer in the underlap regions of sub 20 nm NMOSFET with gate stack," *Superlattices Microstruct.*, vol. 98, pp. 448–457, 2016.

[124] P. K. Pal, B. K. Kaushik, and S. Dasgupta, "Asymmetric dual-spacer trigate FinFET device-circuit co-design and its variability analysis," *IEEE Trans. Electron Devices*, vol. 62, no. 4, pp. 1105–1112, Apr. 2015.

[125] A. Nandi, A. K. Saxena, and S Dasgupta, "Impact of dual-k spacer on analog performance of underlap FinFET," *Microelectron. J.* vol. 43, pp. 883–887, 2012.

[126] H. G. Virani, R. B. R. Adari, and A. Kottantharayil, "Dual-k spacer device architecture for the improvement of performance of silicon n-channel tunnel FETs," *IEEE Trans. Electron Devices*, vol. 57, no. 10, pp. 2410–2417, Oct. 2010.

[127] H. G. Virani, S. Gundapaneni, and A. Kottantharayil, "Double dielectric spacer for the enhancement of silicon p-channel tunnel field-effect transistor performance," *Jpn. J. Appl. Phys.*, vol. 50, 04DC04, Apr. 2011.

[128] P. K. Pal, B. K. Kaushik, and S. Dasgupta, "Investigation of symmetric dual-k spacer trigate FinFETs from delay perspective," *IEEE Trans. Electron Devices*, vol. 61, no. 11, pp. 3579–3585, Nov. 2014.

[129] H. Lou, L. Zhang, Y. Zhu, X. Lin, S. Yang, J. He, and M. Chan, "A junctionless nanowire transistor with a dual-material gate," *IEEE Trans. Electron Devices*, vol. 59, no. 7, pp. 1829–1836, July 2012.

[130] W. Long, H. Ou, J. M. Kuo, and K. K. Chin, "Dual-material gate (DMG) field-effect transistor," *IEEE Trans. Electron Devices*, vol. 46, no. 5, pp. 865–870, May 1999.

[131] A. Chaudhary and M. J. Kumar, "Investigation of the novel attributes of a fully depleted dual-material gate SOI MOSFET," *IEEE Trans. Electron Devices*, vol. 51, no. 9, pp. 1463–1467, Sept. 2004.

[132] K. Na and Y. Kim, "Silicon complementary metal-oxide-semiconductor field-effect transistors with dual work function gate," *Jpn. J. Appl. Phys.*, vol. 45, no. 12, pp. 9033–9036, Dec. 2006.

[133] X. Zhou and W. Long, "A novel hetero-material gate (HMG) MOSFET for deep-submicron ULSI technology," *IEEE Trans. Electron Devices*, vol. 45, no. 12, pp. 2546–2548, Dec. 1998.

[134] S. Saurabh and M. J. Kumar, "Novel attributes of a dual material gate nanoscale tunnel field-effect transistor," *IEEE Trans. Electron Devices*, vol. 58, no. 2, pp. 404–410, Feb. 2011.

[135] H. Lou, L. Zhang, Y. Zhu, X. Lin, S. Yang, J. He, and M. Chan, "A junctionless nanowire transistor with a dual-material gate," *IEEE Trans. Electron Devices*, vol. 59, no. 7, pp. 1829–1836, July 2012.

[136] H. Lou, D. Li, Y. Dong, X. Lin, J. He, S. Yang, and M. Chan, "Suppression of tunneling leakage current in junctionless nanowire transistors," *Semicond. Sci. Technol.*, vol. 28, pp. 125016–125022, Nov. 2013.

[137] C.-Y. Hsui, C.-Y. Changi, E. Y. Changi, and C. Hu, "Suppressing non-uniform tunneling in InAs/GaSb TFET with dual-metal gate," *IEEE J. Electron Devices Soc.*, vol. 4, no. 2, pp. 60–65, Mar. 2016.

[138] I. Polishchuk, P. Ranade, T. King, and C. Hu, "Dual work function metal gate CMOS technology using metal interdiffusion," *IEEE Electron Device Lett.*, vol. 22, no. 9, pp. 444–446, Sept. 2001.

[139] S. Song, Z. Zhang, M. Hussain, C. Huffman, J. Barnett, S. Bae, H. Li, P. Majhi, C. Park, B. Ju, H. K. Park, C. Y. Kang, R. Choi, P. Zeitzoff, H. H. Tseng, B. H. Lee, and R. Jammy, "Highly manufacturable 45 nm LSTP CMOSFETs using novel dual high-k and dual metal gate CMOS integration," in *VLSI Symp. Tech. Dig.*, 2006, pp. 13–14.

[140] R. K. Baruah and R. P. Paily, "A dual-material gate junctionless transistor with high-κ spacer for enhanced analog performance," *IEEE Trans. Electron Devices*, vol. 61, no. 1, pp. 123–128, Jan. 2014.

[141] V. Kumari, N. Modi, M. Saxena, and M. Gupta, "Theoretical investigation of dual material junctionless double gate transistor for analog and digital performance," *IEEE Trans. Electron Devices*, vol. 62, no. 7, pp. 2098–2105, July 2015.

[142] J. Singh, V. Gadi, and M. J. Kumar, "Modeling a dual-material-gate junctionless FET under full and partial depletion conditions using finite-differentiation method," *IEEE Trans. Electron Devices*, vol. 63, no. 6, pp. 2282–2287, June 2016.

[143] J. C. Pravin, D. Nirmal, P. Prajoon, and M. A. Menokey, "A new drain current model for a dual metal junctionless transistor for enhanced digital circuit performance," *IEEE Trans. Electron Devices*, vol. 63, no. 9, pp. 3782–3789, Sept. 2016.

[144] A. K. Agrawal, P. N. Koutilyan and M. J. Kumar, "A pseudo 2-D surface potential model of a dual material double gate junctionless field-effect transistor," *J. Comp. Electron.*, vol. 14, no. 3, pp. 686–693, Sept. 2015.

[145] P. Wang, Y. Zhuang, C. Li, Y. Li and Z. Jiang, "Subthreshold behavior models for nanoscale junctionless double-gate MOSFETs with dual-material gate stack," *Jap. J. Appl. Phys.*, vol. 53, no. 8, pp. 084201, July 2014.

[146] S. M. Biswal, B. Baral, D. De, and A. Sarkar, "Analytical subthreshold modeling of dual material gate engineered nano-scale junctionless surrounding gate MOSFET considering ECPE," *Superlattices Microstruct.*, vol. 82, pp. 103–112, June 2015.

[147] A. K. Jain, S. Sahay, and M. J. Kumar, "Controlling L-BTBT in emerging nanotube FETs using dual-material gate," *IEEE J. Electron Dev. Soc.*, vol. 6, pp. 611–621, June 2018.

[148] International Technology Roadmap for Semiconductors [online]. Available: http://www.itrs2.net/2012-itrs.html, Accessed: Oct. 5, 2018.

5

GATE-INDUCED DRAIN LEAKAGE IN JUNCTIONLESS FIELD-EFFECT TRANSISTORS

As discussed in Section 4.2, achieving volume depletion in junctionless FETs (JLFETs) requires an efficient gate control of the channel region, which may be achieved by utilizing gate electrodes with a very high work function (~5.1 eV) for n-JLFETs and extremely low work function (~3.9 eV) for p-JLFETs [1–5]. In addition, for efficient gate control, the silicon film thickness needs to be ultrathin (< 6 nm) so that the gate electric field penetrates the entire volume of the channel region to deplete it in the OFF-state [4, 5]. Although such an efficient gate control leads to a high source-to-channel barrier height which reduces short-channel effects and hinders the transport of electrons from source to drain, it leads to a considerable overlap between the valence band of the channel region with the conduction band of the drain in n-JLFETs as shown in Fig. 5.1. The significant proximity between the valence band of the channel region and the conduction band of the drain region facilitates the band-to-band tunneling (BTBT) of electrons from the channel region to the drain region [6]. This is similar to the lateral band-to-band tunneling (L-BTBT) gate-induced drain leakage (GIDL) phenomenon in the nanowire metal-oxide-semiconductor field-effect transistors (MOSFETs) [7–12], which also originates due to an efficient gate control as discussed in Section 2.6. It may be noted that we refer the BTBT from the channel to the drain as L-BTBT in nanowire metal-oxide-semiconductor field-effect transistor (NWMOSFETs) just to differentiate it from the transverse BTBT, which occurs in the gate-on drain overlap region when the gate voltage is negative ($V_{GS} \ll 0.0$ V) [7–12] as shown in Fig. 5.2.

Junctionless Field-Effect Transistors: Design, Modeling, and Simulation, First Edition.
Shubham Sahay and Mamidala Jagadesh Kumar.

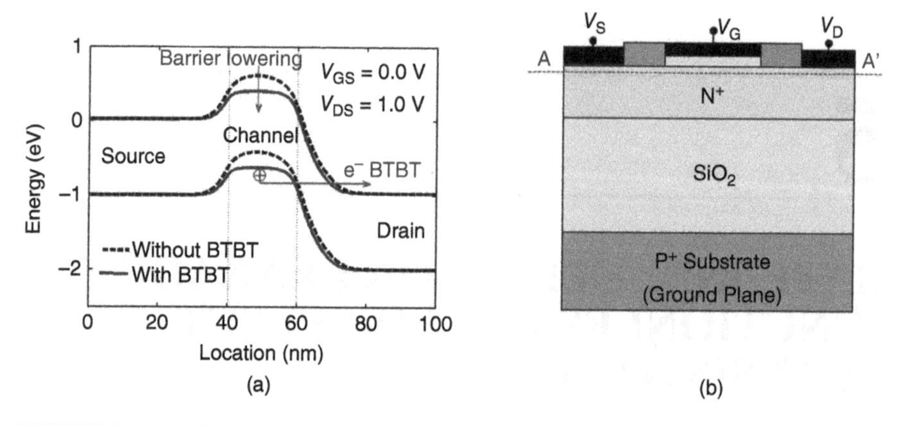

FIGURE 5.1 (a) Energy band profiles of the silicon-on-insulator (SOI) JLFET taken at a cutline 1 nm below the Si–SiO_2 interface along A–A′ as shown in (b).

5.1 HOLE ACCUMULATION

The electrons that tunnel from the channel to the drain region leave behind holes in the channel region. The significantly high hole generation rate due to BTBT as shown in Fig. 5.3 corroborates this phenomenon. Now, these BTBT-generated holes encounter

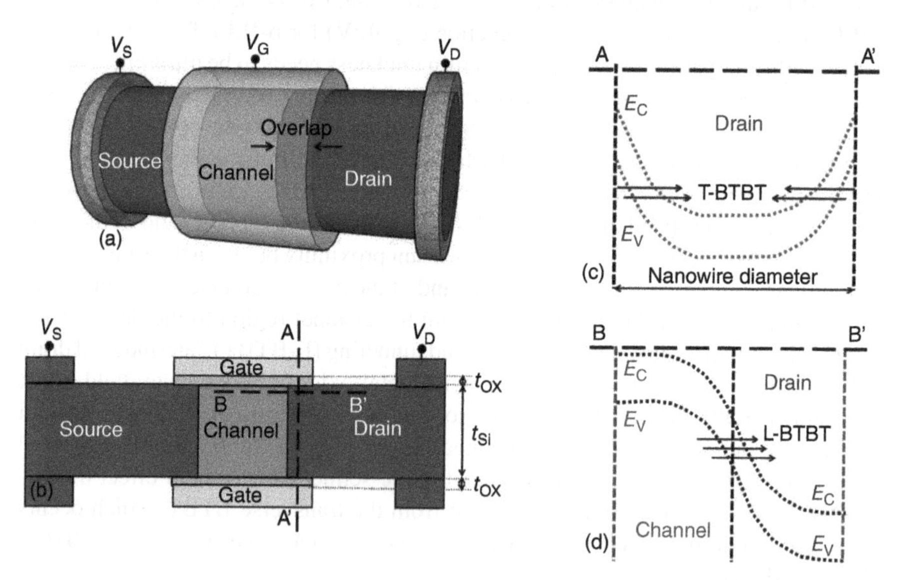

FIGURE 5.2 (a) Three-dimensional view, (b) two-dimensional schematic of a nanowire MOSFET (NWMOSFET) and energy band profile in the OFF-state of NWMOSFET along (c) the longitudinal direction (cutline A–A′), indicating transverse band-to-band tunneling (T-BTBT) and (d) the lateral direction (cutline B–B′) indicating L-BTBT.

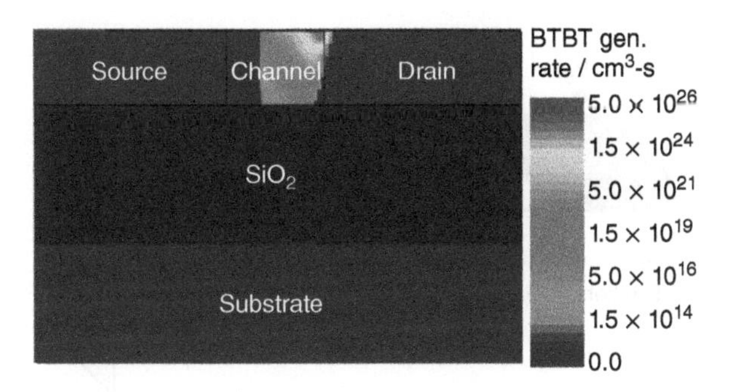

FIGURE 5.3 Contour plot showing the hole generation rate due to BTBT in SOI JLFETs in the OFF-state ($V_{GS} = 0.0$ V, $V_{DS} = 1.0$ V).

the source-to-channel potential barrier and start accumulating in the channel region. The hole concentration contour plot shown in Fig. 5.4 clearly indicates that the holes begin to pile up in the channel region in the OFF-state of JLFETs converting the channel into a p-type region.

As the deposition of positive charge induces a positive potential, the accumulation of the holes increases the potential of the channel region. The amount by which the potential of the channel region is raised depends upon the number of accumulated holes. An increase in the channel potential lowers the barrier height between the source and channel region and leads to forward biasing of the source–channel junction. Consequently, the holes flow into the source contact due to the minority carrier injection mechanism just as in the case of a forward biased p–n junction. Although we started with an n$^+$ source, n$^+$ channel, and n$^+$ drain in an n-JLFET, due to the BTBT in the OFF-state, the channel region gets converted into p-type. This leads to the formation of a parasitic bipolar junction transistor (BJT) with the n$^+$ source as the emitter, p-type channel as the base, and the n$^+$ drain as the collector as shown in Fig. 5.5.

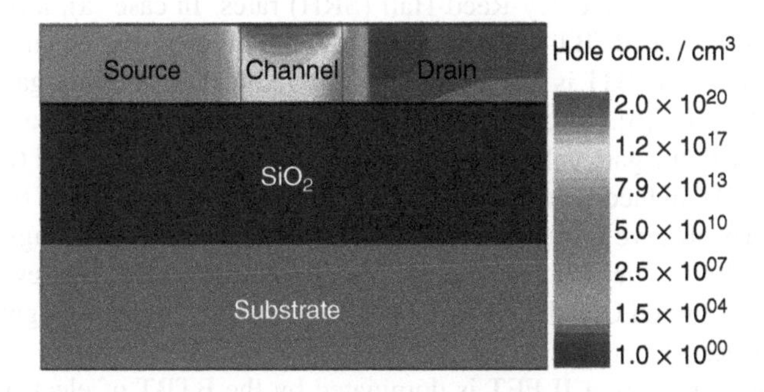

FIGURE 5.4 Hole concentration contour plot of the SOI JLFET (channel thickness = 10 nm and doping concentration = 10^{19} cm^{-3}) in the OFF-state ($V_{GS} = 0.0$ V, $V_{DS} = 1.0$ V).

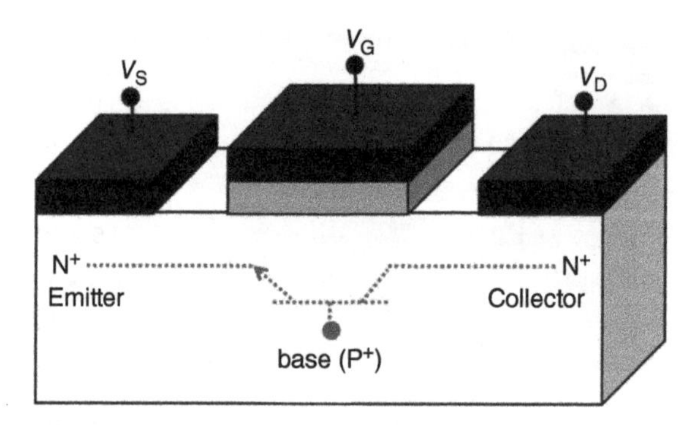

FIGURE 5.5 Schematic view of the parasitic BJT formed in the JLFETs due to BTBT in the OFF-state ($V_{GS} = 0.0$ V, $V_{DS} = 1.0$ V).

5.2 PARASITIC BJT ACTION

The base current of the parasitic BJT is the BTBT-generated current, which gets amplified by the current gain (β) of the parasitic BJT, and results in a large collector current, which is essentially the drain current of the n-JLFET. Therefore, due to the parasitic BJT action, the OFF-state current increases drastically in a JLFET [6]. Moreover, as the gate voltage is reduced below 0 V, the hole accumulation increases and further lowers down the source-to-channel barrier height. This increases the magnitude of forward bias of the n^+ source-p-channel junction leading to saturation in the source-to-channel barrier height. As a result, the drain current can no longer be controlled by the gate in JLFETs for negative gate voltages and becomes saturated and constant as the parasitic BJT is triggered in JLFETs [6].

The drain current due to GIDL increases in a MOSFET as the gate voltage becomes more negative [7]. However, in a JLFET, the situation is different due to the presence of the BTBT-induced parasitic BJT. To understand this, let us look at Fig. 5.6. We need to focus on the two curves: (a) with the BTBT model and (b) with the BTBT model and increased Shockley-Reed-Hall (SRH) rates. In case (a), a parasitic BJT is present due to the BTBT-induced holes accumulating in the channel region. In case (b), the parasitic BJT is made redundant by reducing its current gain due to the increased SRH rates [6]. We can clearly distinguish between the variation of the OFF-state current with the gate voltage when the parasitic BJT is turned ON (case (a)) and when only BTBT-induced carriers are present without the parasitic BJT (case (b)). We can now clearly observe from Fig. 5.6 that the OFF-state current is strongly dependent on the gate voltage when the parasitic BJT action is suppressed. However, when the parasitic BJT is turned ON, the OFF-state current not only increases significantly but also becomes less dependent on the gate voltage.

The GIDL current in a JLFET is dominated by the BTBT of electrons from the channel to drain region. Moreover, in the JLFETs, the channel is depleted in the OFF-state and the channel carriers are pushed into the source–drain regions. This leads to

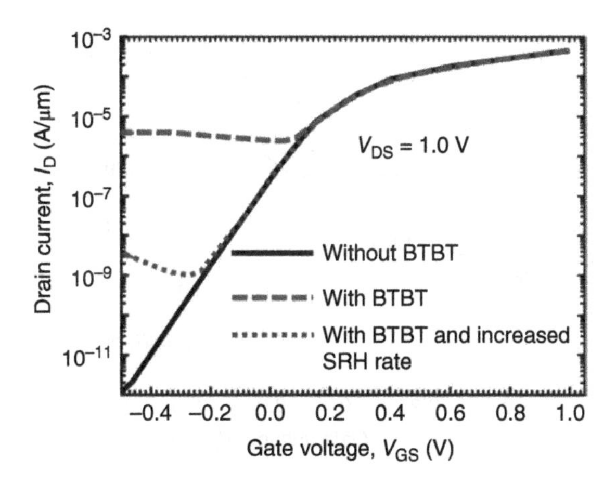

FIGURE 5.6 Transfer characteristics of SOI JLFET without the BTBT model and with the BTBT model and increased SRH rates to diminish the parasitic BJT action [6].

an electrostatic squeezing of the carriers and results in an unintentional gate–drain underlap in JLFETs [1–3]. As a result, the possibility of GIDL due to transverse tunneling in the depletion region caused due to the gate-on-drain overlap region is also eliminated in JLFETs [12–18]. On the other hand, in a MOSFET, the GIDL is contributed both by the transverse tunneling (T-BTBT) in the gate-on-drain overlap region and the lateral tunneling from the channel to drain region [19, 20]. Therefore, the nature of GIDL in JLFETs is significantly different from MOSFETs [8]. We discuss more about the difference in nature of GIDL in different nanowire FET architectures in Section 5.5.2.

5.3 IMPACT OF BTBT-INDUCED PARASITIC BJT ACTION ON SCALING

The BTBT-induced parasitic BJT action is the major problem concerning the gate length scaling of JLFETs [17]. A smaller gate length leads to a higher short-channel effect such as DIBL, which leads to a lower source-to-channel barrier height in JLFETs. A lower source-to-channel barrier height not only facilitates the triggering of the parasitic BJT but also leads to an increased OFF-state current even if BTBT-induced parasitic BJT action was not present [17]. This is due to the ability of the source electrons to easily jump over the small source-to-channel barrier height.

A smaller channel length also implies a reduced base width for the parasitic BJT. As the gain of the BJT increases with a reduction in the base width, the short-channel JLFETs would be more susceptible to the parasitic BJT action induced increase in the leakage current. As shown in Fig. 5.7, the reduction in channel length in ultrashort-channel JLFETs results in a poor ON-state to OFF-state current ratio (I_{ON}/I_{OFF}).

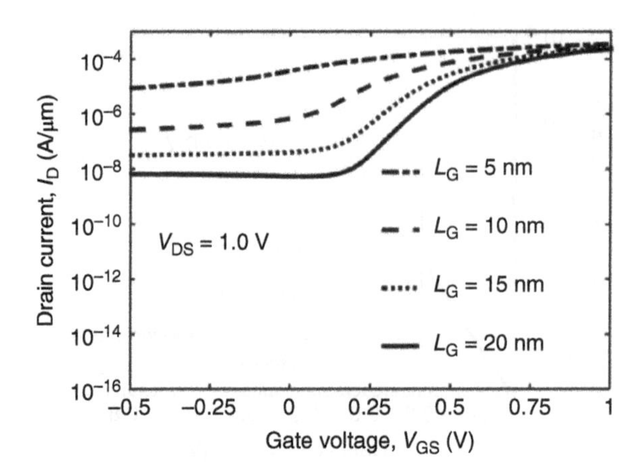

FIGURE 5.7 Impact of gate length scaling on the performance of SOI JLFETs.

Therefore, the parasitic BJT action is detrimental for scaling JLFETs and needs to be mitigated [6, 17].

Although the BTBT-induced parasitic BJT action can be inferred from the experimental results in [21–24], the BTBT has been overlooked in these studies. Even in [25] which analyses the OFF-state leakage current of nanowire junctionless field-effect transistor (NWJLFETs), the increase in the OFF-state leakage current has been attributed to the gate leakage current via the direct tunneling and trap-assisted tunneling mechanisms. Since a long channel (~50 nm–1 μm) JLFET has been used in [25], the parasitic BJT action could not have been recognized due to a longer base width. However, the results of [24] as shown in Fig. 5.8 clearly indicate that the

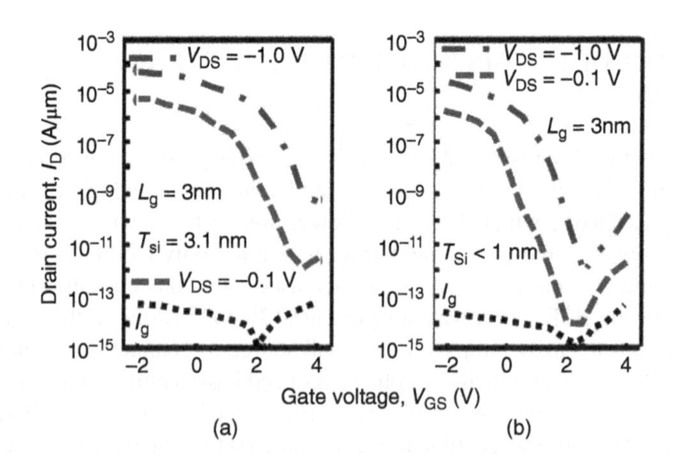

FIGURE 5.8 Transfer characteristics and the gate leakage current of ultrashort-channel (L_g = 3 nm) JLFETs with silicon film thickness (T_{Si}) of (a) 3.1 nm and (b) sub-1 nm [24].

magnitude of gate leakage current is significantly lower than the OFF-state current in ultra-short-channel JLFETs.

The magnitude of gate leakage current decreases with gate length scaling due to a reduction in the effective gate area over which tunneling occurs. If gate leakage would have been the dominant mechanism for leakage current in JLFETs, the OFF-state current in ultra-short-channel JLFETs would have been much lower than the long channel JLFETs. This is contrary to the experimental results of [21–24]. Therefore, the BTBT-induced parasitic BJT action is the dominant mechanism for leakage current in JLFETs and hinders the scaling of JLFETs [6].

So far, we discussed the impact of BTBT-induced parasitic BJT action on scaling the channel length. In the following section, we focus on the impact of the silicon film thickness on the BTBT-induced parasitic BJT action, which is the dominant GIDL mechanism in JLFETs. The silicon film thickness translates into the diameter in the nanowire architecture, which has been proposed as the most promising architecture for scaling the device dimension in the sub-10 nm length [26]. As we shall see, the dominant GIDL mechanism depends significantly on the nanowire diameter or the channel thickness in JLFETs.

5.4 IMPACT OF SILICON FILM THICKNESS ON GIDL

The experimental results of [21], as shown in Fig. 5.9, clearly indicate that the OFF-state current in JLFETs depends significantly on the silicon film thickness (nanowire diameter in case of NWJLFETs) [15]. An increase in the silicon film thickness leads to a reduction in the threshold voltage of a JLFET as the negative term increases in the threshold voltage expression given below:

$$V_{\mathrm{Th}} = V_{\mathrm{FB}} - \frac{qN_{\mathrm{D}}t_{\mathrm{Si}}^2}{2\varepsilon_{\mathrm{Si}}} - \frac{qN_{\mathrm{D}}t_{\mathrm{Si}}t_{\mathrm{ox}}}{\varepsilon_{\mathrm{ox}}} \tag{5.1}$$

This is also evident from Figs. 5.9 and 5.10. However, this diameter dependency of the OFF-state current in NWJLFETs cannot be explained entirely due to a shift in the threshold voltage with changing diameter [21].

A comprehensive analysis of the diameter-dependent dominant leakage mechanism was performed in [15]. As can be observed from Figs. 5.9 and 5.10, the OFF-state current of NWJLFETs reduces significantly as the nanowire diameter (silicon film thickness) decreases. The threshold voltage also increases as the nanowire diameter is reduced. The drain current would continue to decrease for negative gate voltages in the absence of BTBT in NWJLFETs [15] as shown in Fig. 5.10. However, the drain current becomes independent of the gate voltage in NWJLFETs for negative gate voltages when BTBT is taken into account. This behavior corroborates with the experimental characteristics [21–24] shown in Fig. 5.9 and clearly indicates the presence of BTBT-induced parasitic BJT action in NWJLFETs.

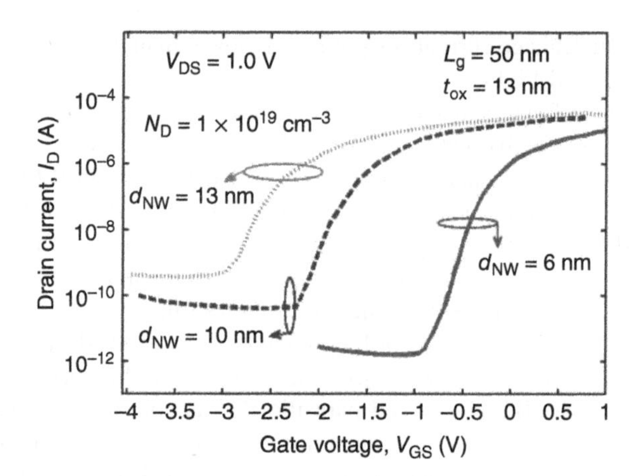

FIGURE 5.9 Transfer characteristics of NWJLFETs with different nanowire diameters (d_{NW}) [21].

As observed from Fig. 5.11, there exists a significant spatial proximity between the valence band of the channel region and the conduction band of the drain region in the OFF-state of NWJLFETs with the nanowire diameter ranging from 5 to 20 nm. This band alignment clearly indicates that BTBT is inevitable in NWJLFETs [15].

As discussed in Section 4.2, a reduction in the nanowire diameter (silicon film thickness) increases the gate control of the channel region. As a result, at the first instance, you may think that a reduction in the nanowire diameter should lead to a larger BTBT and increase the OFF-state current in NWJLFETs [7] since L-BTBT arises due to the efficient gate control as discussed in Section 2.6.4. However, the experimental results of [21] shown in Fig. 5.9 and the simulated transfer characteristics shown in Fig. 5.10 indicate that the OFF-state current decreases with a reduction in the nanowire diameter. This can be understood from Figs. 5.11–5.13.

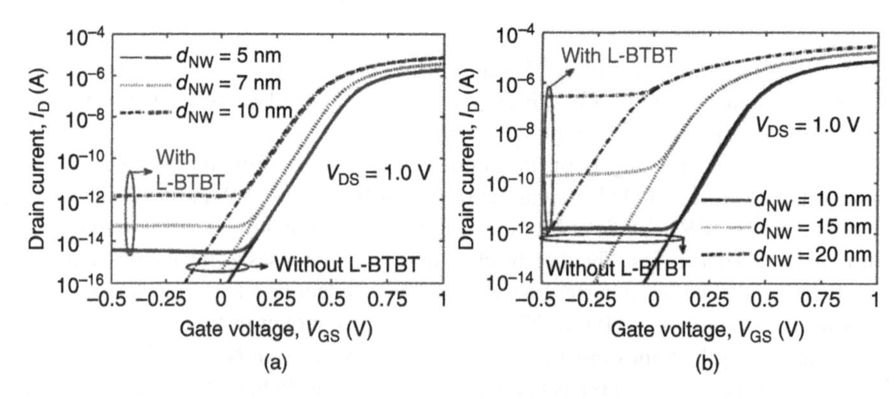

FIGURE 5.10 The transfer characteristics of NWJLFETs with (a) $d_{NW} \leq 10$ nm and (b) $d_{NW} \geq 10$ nm.

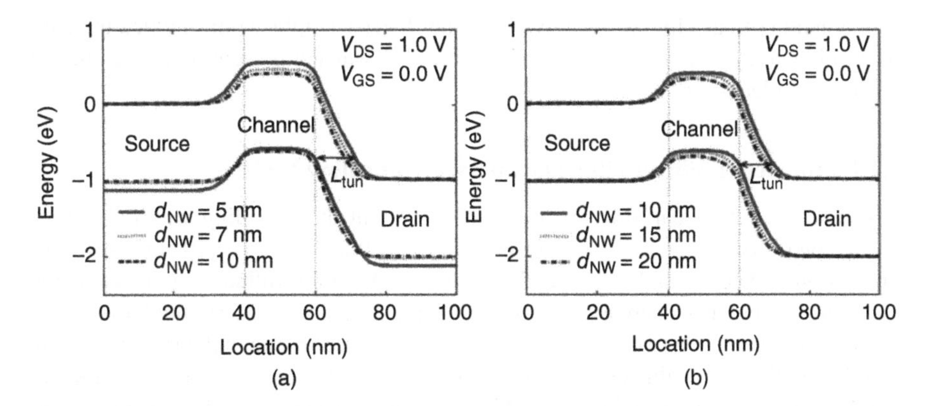

FIGURE 5.11 The energy-band profiles of the NWJLFET with (a) $d_{NW} \leq 10\,\text{nm}$ and (b) $d_{NW} \geq 10\,\text{nm}$ at 1 nm below the Si–SiO_2 interface.

Undoubtedly, a reduction in the nanowire diameter leads to an enhanced gate control of the channel region and increases the electrostatic integrity as explained in Section 4.2. However, the improved gate control suppresses the influence of the drain electric fields on the channel region. Therefore, the drain-induced barrier lowering (DIBL) is reduced in the NWJLFETs as the nanowire diameter is decreased as shown in Fig. 5.12(a). Consequently, the source-to-channel barrier height is also increased in the NWJLFETs (Fig. 5.12(b)). It may be noted that the DIBL has been calculated without including the BTBT model [15] in Fig. 5.12 to decouple these two phenomena. A reduction in DIBL and the consequent increase in the source-to-channel barrier height can be observed even in the presence of BTBT from Fig. 5.11. A higher source-to-channel barrier height inhibits the triggering of the parasitic BJT. You may

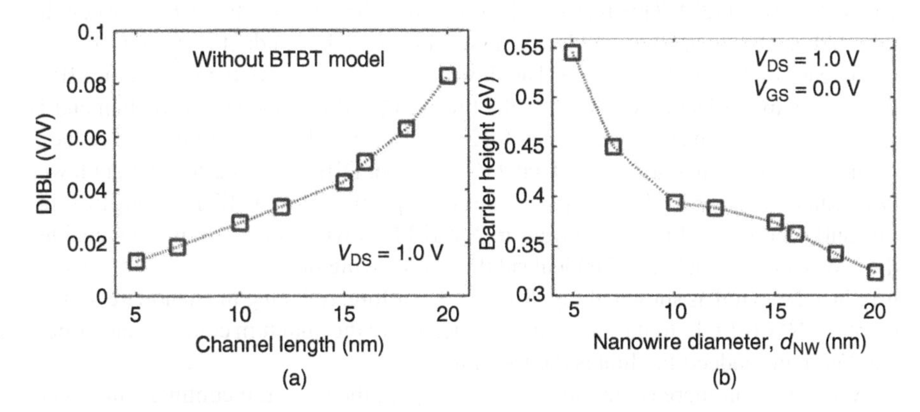

FIGURE 5.12 The variation of (a) DIBL without including the BTBT model and (b) the source-to-channel barrier height at 1 nm below the Si–SiO_2 interface in the OFF-state for different nanowire diameters.

also conclude that a higher source-to-channel barrier height would reduce the subthreshold leakage current even when parasitic BJT was not present by increasing the barrier for injection of source electrons.

5.4.1 Impact of Silicon Film Thickness on Tunneling Width

If you see Fig. 5.11 carefully, you may observe that the tunneling width (L_{tun}) for the nanowire diameter, $d_{NW} = 5$ nm is more than that for $d_{NW} = 20$ nm. From our discussion in Section 5.2, we already know that the drain influence reduces as the nanowire diameter is decreased from 20 to 5 nm. The drain electric field affects the device performance in two ways: (a) It lowers the source-to-channel barrier height by coupling to the source through the channel region and (b) it also reduces the tunneling width at the channel–drain interface. A high drain electric field leads to a larger band bending and results in a lower tunneling width at the channel–drain interface. This phenomenon is known as the drain-induced barrier thinning (DIBT) [27, 28].

The DIBT also reduces as the influence of the drain electric field is decreased with reducing d_{NW}. This reduction in DIBT leads to a larger tunneling width in NWJLFETs as the nanowire diameter is reduced from 20 to 5 nm as shown in Fig. 5.12(a). It may also be noted that the minimum tunneling width at a particular nanowire diameter depends explicitly on the efficiency of the gate control since L-BTBT originates due to the enhanced gate control. Therefore, the minimum tunneling width reduces as the nanowire diameter is decreased from 15 to 10 nm as observed in Fig. 5.12(a).

5.4.2 Impact of Quantum Confinement Effects

As the nanowire diameter is reduced below 7 nm, the quantum confinement effects start dominating and lead to a significant discretization of the valence band and the conduction band [29]. This restricts the tunneling phenomenon to occur between the first subband of the valence band to the first subband of the conduction band leading to an increase in the band gap [30–32] as shown in Fig. 5.13(a). An increase in the effective band gap masks the reduction in the tunneling width as the nanowire diameter is reduced below 7 nm as shown in Fig. 5.12(a). The tunneling probability also reduces significantly with increasing band gap [15]. A lower DIBL, and the consequent lower DIBT, an increase in the band gap, and the consequent reduced L-BTBT leads to a significantly reduced OFF-state current in NWJLFETs with $d_{NW} \leq 7$ nm. This enables the NWJLFETs to achieve the ideal subthreshold swing of 60 mV/decade as the d_{NW} is reduced to 5 nm as shown in Fig. 5.12(b). Therefore, reducing the nanowire diameter of the NWJLFETs to the sub-7 nm regime, where the quantum confinement effects are dominant, indeed facilitates their scaling.

Apart from an increase in the energy band gap, the quantum confinement effects also lead to a modification in the carrier concentration in the ON-state and a shift in the threshold voltage [31]. The energy subband discretization of conduction band due to

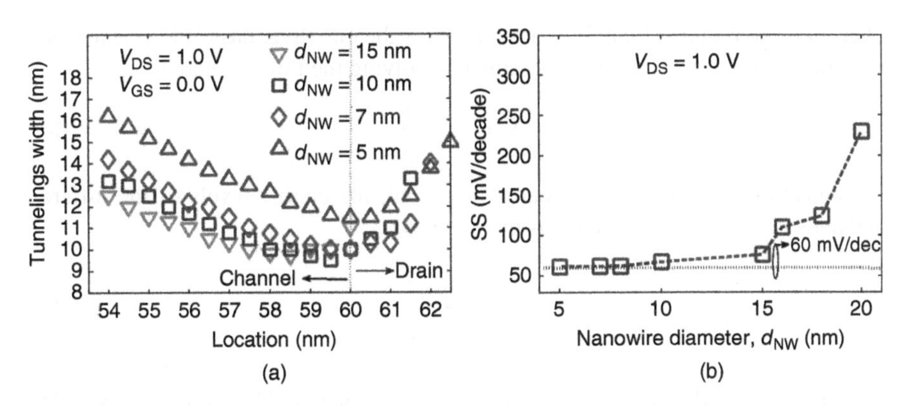

FIGURE 5.13 The variation of (a) the tunneling width at 1 nm below the Si–SiO_2 interface and (b) the average subthreshold swing of the NWJLFETs with diameter.

quantization effects is shown in Fig. 5.14(b). We can observe that the subbands in the conduction band due to quantization are confined within the potential barrier offered by the classical conduction band energy. This system resembles a quantum well confined within two large potential barriers with a perturbation potential at the bottom. We may calculate the energy of the different subbands by solving the Schrödinger equation using the effective mass approximation. The solution of the Schrödinger equation for the ith wave function (ψ_i) and the corresponding subband energy level E_i can be given as

$$\left[-\frac{\hbar^2}{2m^*}\nabla^2 - q\varphi\right]\psi_i = E_i\psi_i \tag{5.2}$$

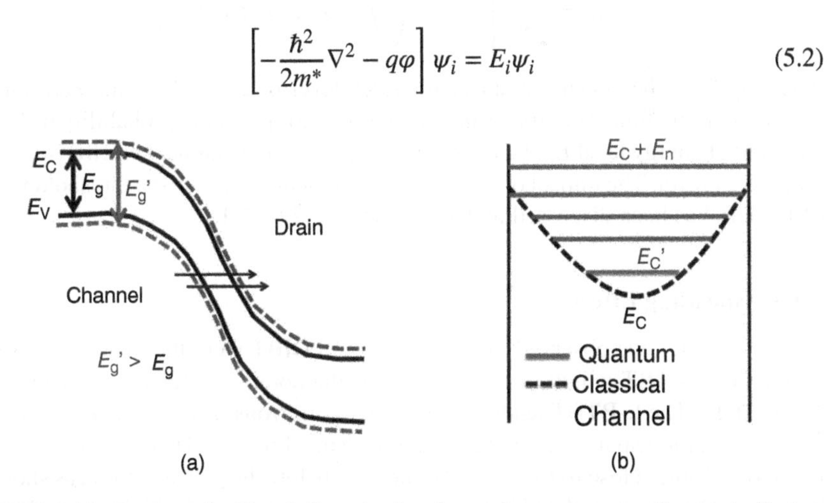

FIGURE 5.14 Impact of subband discretization due to the quantum confinement effect on the NWJLFETs: (a) lateral energy band profile (along cutline B–B′ in Fig. 5.2(a)). E_c' is the first subband of E_C and E_v' is the first subband of E_V and (b) transverse energy band profile (along cutline A–A′ in Fig. 5.2(a)).

where $\hbar$ is the reduced Planck's constant, m^* is the effective mass, and φ is the electrostatic potential, which can be obtained by solving the Poisson equation:

$$\nabla^2 \varphi = -\frac{\rho}{\varepsilon_{\text{Si}}} \tag{5.3}$$

Neglecting the bottom perturbation potential, the energy levels of the different subbands in the conduction band is obtained as

$$E_n = \frac{\hbar^2 n^2}{8m^* t_{\text{Si}}^2}, \tag{5.4}$$

for $n = 1,2,3\ldots$ and so on. The energy of the first subband (E_c') from where the electrons tunnel into the first subband of the valence band due to the quantum confinement effects can be obtained by putting $n = 1$ and is given by

$$E_c' = E_c + \frac{\hbar^2}{8m^* t_{\text{Si}}^2}. \tag{5.5}$$

The self-consistent solution of the Schrödinger–Poisson equation (equations 5.2 and 5.3) also yields the quantum electron density, which is different from the electron density obtained using classical mechanics [31]. The quantum electron density is given as

$$n_{\text{QM}} = \sum_i \left[(\psi_i \psi_i^*) \left(\int_{E_i}^{\infty} g_i(E) f(E)\, dE \right) \right] \tag{5.6}$$

where $g_i(E)$ is the density of states at energy level E_i and $f(E)$ is the Fermi–Dirac distribution function. The first term in equation (5.6) gives the probability of finding electrons at energy level E_i and the second term gives the classical electron density at that energy level. The impact of quantum confinement effects on the threshold voltage of JLFETs has been discussed in detail in Section 7.6 [87].

5.4.3 Shielding Effect

In Section 5.4.1, we observed that the DIBL and DIBT were the major factors governing the L-BTBT which manifests itself as the dominant leakage mechanism for NWJLFETs. The L-BTBT leads to tunneling of electrons from the channel region to the drain region leaving behind holes in the channel region. This results in an accumulation of holes close to the Si–SiO_2 interface below the gate electrode as shown in Fig. 5.15. These accumulated holes, originating from L-BTBT, shield the underlying channel from getting depleted due to the gate electrode.

As can be observed from Fig. 5.16, the extent of depletion in the presence of L-BTBT is smaller as compared to that without L-BTBT. As a result, volume depletion is not achieved as the nanowire diameter (silicon film thickness) is increased above 15 nm and an undepleted neutral region exists at the center of the nanowire which

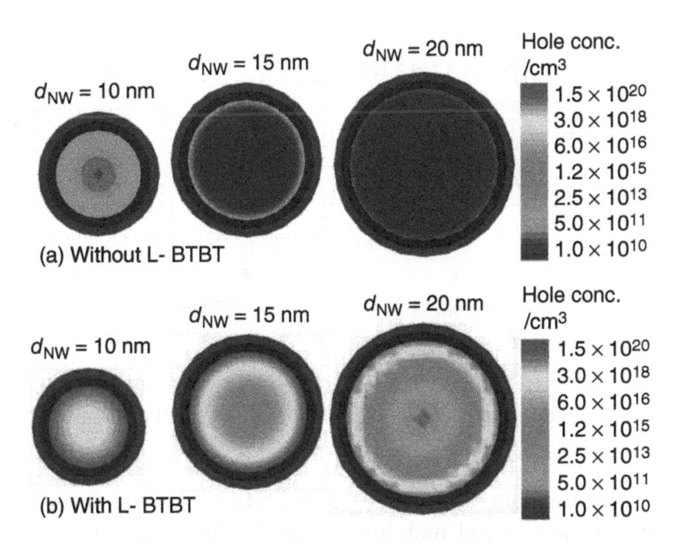

FIGURE 5.15 The hole concentration contour plot of the NWJLFET at middle of the channel for various d_{NW} (a) without including the BTBT model and (b) with the BTBT model in the OFF-state ($V_{GS} = 0.0$ V and $V_{DS} = 1.0$ V).

conducts the leakage current [15, 33]. At this juncture, we like to point out that the large leakage current observed for NWJLFETs with $d_{NW} \geq 15$ nm is attributed to the insufficient volume depletion due to the shielding effect induced by L-BTBT rather than the tunneling current and the parasitic BJT action due to L-BTBT.

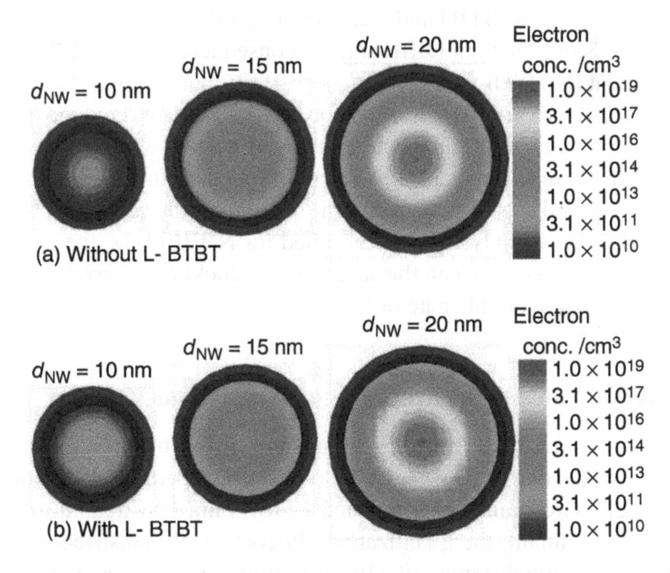

FIGURE 5.16 The electron concentration contour plot of the NWJLFET at middle of the channel for various d_{NW} (a) without including the BTBT model and (b) with the BTBT model in the OFF-state ($V_{GS} = 0.0$ V and $V_{DS} = 1.0$ V).

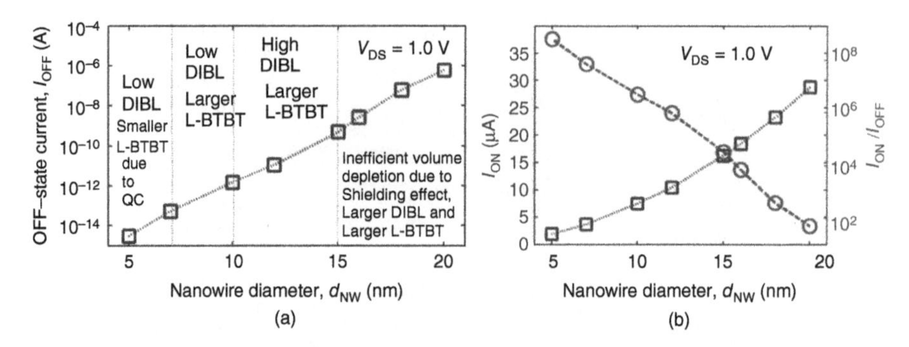

FIGURE 5.17 The variation in (a) the OFF-state current and (b) ON-state current and ON-state to OFF-state current ratio (I_{ON}/I_{OFF}) of the NWJLFET with the nanowire diameter.

We already discussed in Section 4.1 that JLFETs suffer from a smaller ON-state current owing to the reduced mobility due to the larger impurity scattering and a high source/drain series resistance. The ON-state current could be improved by using a NWJLFET with a larger diameter. However, the large leakage current in the NWJLFETs with a large diameter (> 15 nm) degrades the ON-state to OFF-state current ratio (I_{ON}/I_{OFF}) significantly. Therefore, increasing the nanowire diameter to increase the effective area for current flow and enhance the current driving capability of NWJLFETs is not a viable option considering the poor I_{ON}/I_{OFF}.

Figure 5.17 summarizes the dominant mechanisms that govern the leakage current of NWJLFETs with a different nanowire diameter. While the inefficient volume depletion owing to the L-BTBT-induced shielding effect increases the leakage current for d_{NW} > 15 nm, the high DIBL and the consequent higher DIBT increases the OFF-state current for NWJLFETs with d_{NW} > 10 nm (but less than 15 nm). However, the DIBL and DIBT reduce significantly for $d_{NW} \leq 10$ nm and reduce the OFF-state current significantly. In addition, the quantum confinement effects become dominant for $d_{NW} \leq 7$ nm and result in an increased band gap which further minimizes the tunneling probability and the OFF-state current facilitating the scaling of NWJLFETs.

Although the above analysis was performed for NWJLFETs, the physical mechanisms would remain same when the same silicon thickness is used in JLFETs with other architectures like double gate or triple gate.

5.4.4 Impact of Silicon Film Thickness on the ON-State Current

In general, a reduction in the nanowire diameter (silicon film thickness) lowers the effective area for the current flow and results in a decreased ON-state current. However, as the nanowire diameter is reduced below 7 nm and quantum confinement effects become dominant, the quantization effects tend to redistribute the electrons in different valleys with different effective transport masses owing to the valley edge shifts [34] as given by equation (5.6). Such an analysis is limited in the commercial device simulators due to the simultaneous implementation of the Schrödinger–Poisson model (which accurately describes the Eigen values of the wave functions and hence the population of electrons in the different valleys along with effective transport mass) with the nonlocal BTBT model due to incompatibilities of the numerical solvers used for them [30, 31]. Therefore, a full quantum simulation based analysis is desirable to gain further new insights into the electron redistribution in the ON-state of JLFETs due to quantum confinement [15].

5.5 IMPACT OF DOPING ON GIDL

The transfer characteristics of the NWJLFET with different nanowire dopings is shown in Fig. 5.18. The OFF-state current increases significantly as the doping is increased in NWJLFETs. This can be understood from Fig. 5.19. The tunneling width increases as the doping is reduced. This leads to a lower L-BTBT in NWJLFETs with smaller doping. Also, a larger doping increases the coupling of the drain electric field through the channel to the source–channel interface [1–3]. This results in larger short-channel effects like DIBL and a consequent lower source-to-channel barrier height in the NWJLFETs with large doping as shown in Fig. 5.19. The reduced source-to-channel barrier height in the NWJLFETs with large doping allows the source electrons to easily surmount the barrier and flow to the drain resulting in a large leakage current. On the other hand, a reduction in the silicon film doping also leads to an increase in the threshold voltage of a JLFET as the negative term in equation (5.1) diminishes which is evident from Fig. 5.19.

As can be observed from Fig. 5.20 and discussed in Section 4.2, volume depletion is not achieved in NWJLFETs when a large silicon film doping is used and an electron channel exists at the center of the nanowire, which conducts the leakage current. This

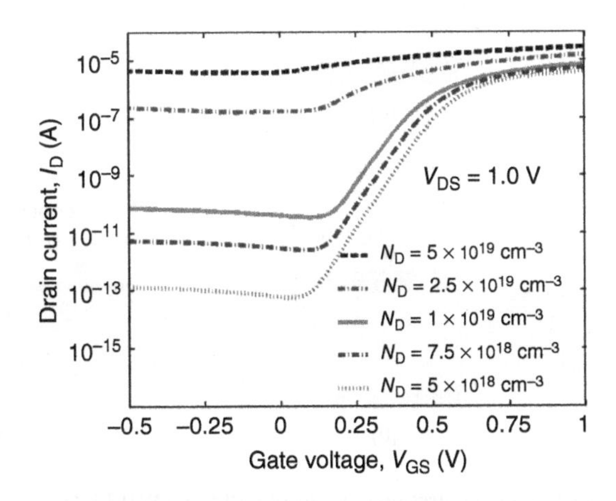

FIGURE 5.18 Transfer characteristics of NWJLFETs with different dopings.

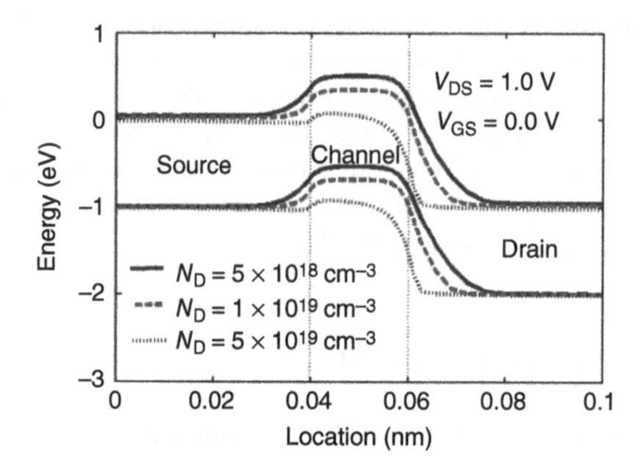

FIGURE 5.19 Energy band profiles of the NWJLFET with different silicon film doping taken at a cutline 1 nm below the Si–SiO_2 interface.

insufficient volume depletion is attributed to the L-BTBT-induced shielding effect due to hole accumulation discussed in Section 5.2.3. The loss of gate control for NWJLFETs with large silicon film doping is due to the holes accumulated at the Si–SiO_2 interface adjacent to the gate electrode, which shield the underlying channel region from influence of the gate electric field.

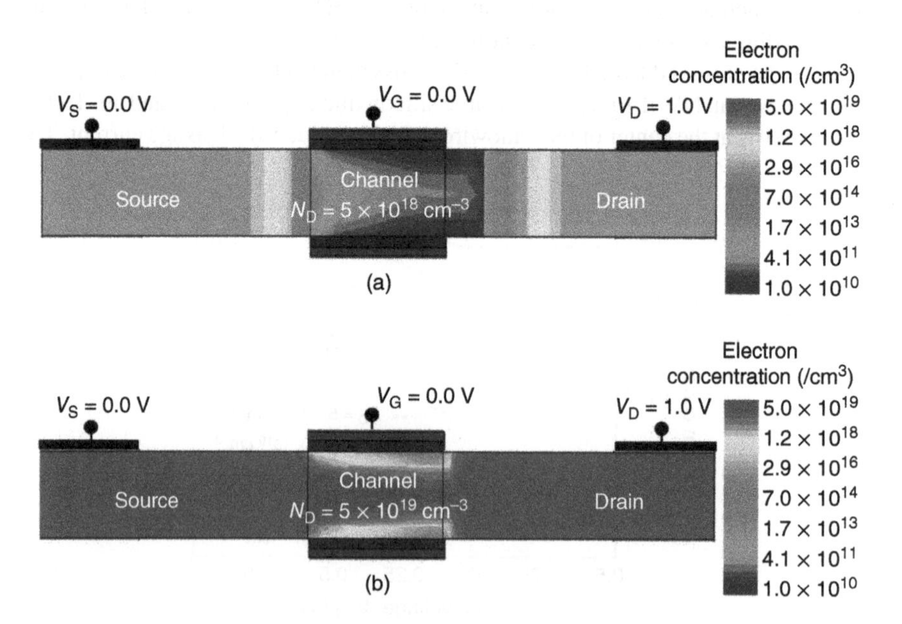

FIGURE 5.20 Electron concentration contour plot of the NWJLFETs with different silicon film doping concentrations.

5.6 IMPACT OF SPACER DESIGN ON GIDL

As discussed in Section 4.8, a gate sidewall spacer is inevitable to separate the gate contact from the source/drain silicide contacts in a JLFET. As you may recall from our discussions in Section 4.8.3, the use of a high-κ gate sidewall spacer facilitates the gate modulation of the silicon film region under the spacer through the fringing fields [35]. This results in a larger effective channel length as compared to the physical gate length. A larger effective length compared to the physical length reduces the short-channel effects in JLFETs with high-κ spacer and improves the electrostatic integrity. However, the L-BTBT GIDL was not included in the analysis presented in Section 4.8 [35]. As you may have noted by now that L-BTBT is the most important factor degrading the performance of the NWJLFETs. Therefore, the impact of gate sidewall spacer design on the NWJLFETs needs to be studied from a GIDL perspective.

Figure 5.21(a) shows the transfer characteristics of the NWJLFETs with different gate sidewall spacers. The OFF-state current reduces by ~4 orders of magnitude with the inclusion of a high-κ spacer even when L-BTBT is considered. This considerable improvement in the performance of the NWJLFETs with high-κ spacer can be understood from Fig. 5.21(b). We explain this improvement through the L-BTBT-induced parasitic BJT action theory: The inclusion of the high-κ spacer increases the fringing fields in the drain region under the spacer and reduces the peak electric field at the channel–drain interface. This results in a lower band bending and attenuates the transition of the energy bands. As a result, the tunneling width increases significantly and leads to a smaller L-BTBT in NWJLFETs with a high-κ spacer. In addition, the increase in the source-to-channel barrier height in the NWJLFETs with a high-κ spacer inhibits the triggering of the L-BTBT-induced parasitic BJT. Furthermore, the larger effective channel length due to the inclusion of the high-κ spacer leads to an increased effective base width of the parasitic BJT as shown in Fig. 5.21(b) and reduces its current gain. These factors diminish the L-BTBT-induced parasitic BJT action in NWJLFETs with a high-κ spacer [15].

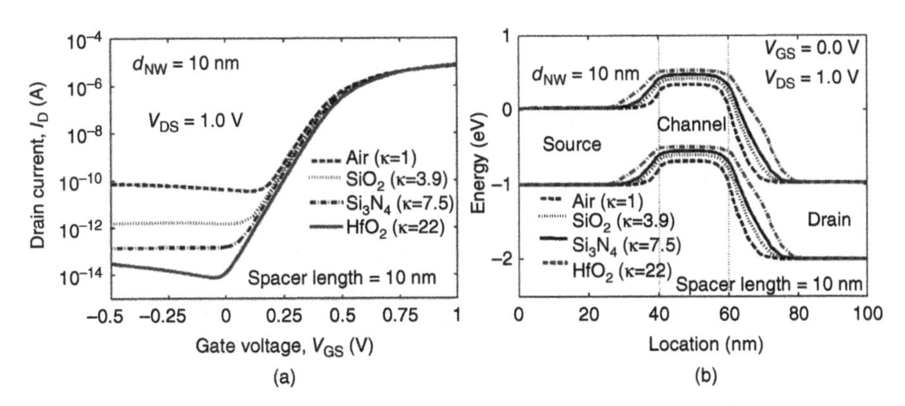

FIGURE 5.21 (a) The transfer characteristics and (b) the energy band profiles at 1 nm below the Si–SiO_2 interface of the NWJLFET with different gate sidewall spacers.

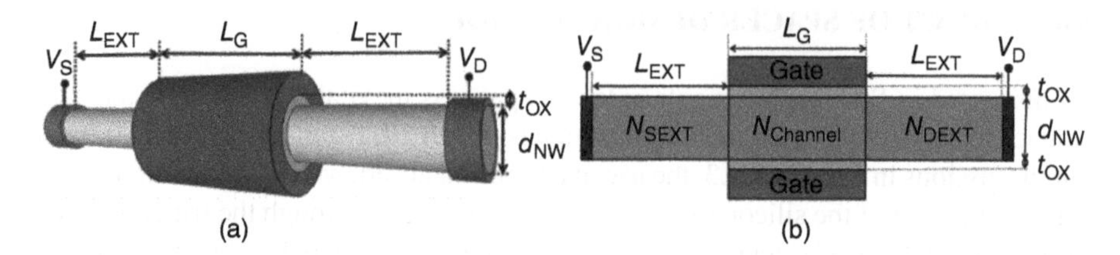

FIGURE 5.22 (a) Three-dimensional view and (b) the cross-sectional view of a nanowire FET.

You may appreciate now that there is so much of physics involved in dictating the OFF-state current of JLFETs, which we may have missed if we did not consider BTBT.

5.7 NATURE OF GIDL IN DIFFERENT NWFET CONFIGURATIONS

At this juncture, we would also like to discuss the nature of GIDL in different FET configurations. We focus on nanowire architecture (Fig. 5.22) as it is considered the most promising architecture for ultimate scaling of FETs [26]. The experimental results of [8], as shown in Fig. 5.23, clearly indicate that the OFF-state current in NWMOSFETs is larger than the OFF-state current in NWJLFETs. This was attributed in [8] to the larger L-BTBT in NWMOSFETs due to a higher drain extension doping which leads to a larger band bending (Fig. 5.24(b)). Also, the ON-state current is larger in the NWMOSFETs as compared to the NWJLFET owing to the reduced series resistance offered by the highly doped source/drain extension regions [9–18].

From Fig. 5.23, you may clearly observe that while the drain current is independent of gate voltage (V_{GS}) for negative values of V_{GS} in NWJLFETs, the drain current increases with negative gate voltages for MOSFETs. If the tunneling current due to

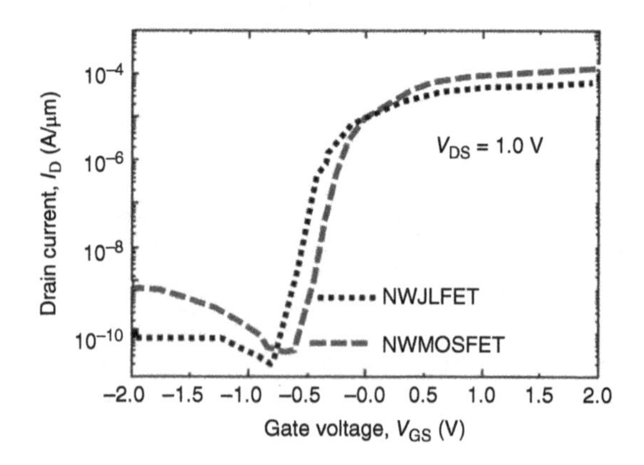

FIGURE 5.23 Experimental transfer characteristics of NWMOSFET and NWJLFET extracted from [8].

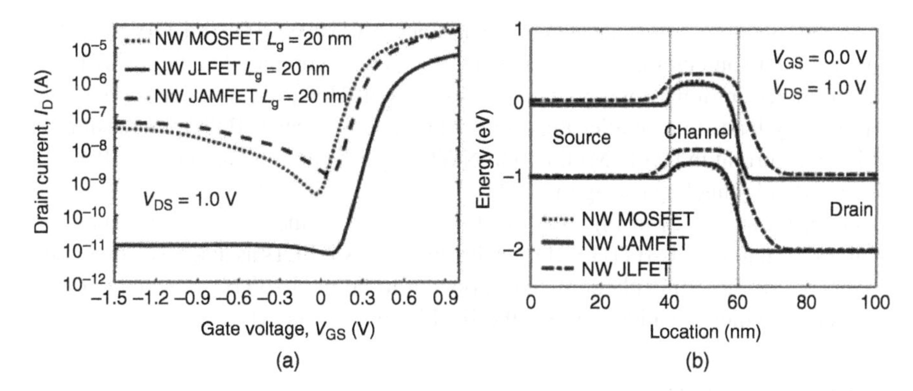

FIGURE 5.24 (a) Transfer characteristics of different NWFETs and (b) energy band profiles taken at a cutline 1 nm below the Si–SiO_2 interface of different NWFETs.

L-BTBT were the only dominating leakage mechanism in NWFETs, as explained in [7, 8], for all the NWFET configurations, the drain current should have increased with negative gate voltages. Therefore, the difference in the nature of GIDL behavior of NWMOSFETs and NWJLFETs cannot be explained entirely by the L-BTBT GIDL mechanism [8]. In the subsequent sections, we try to explain this difference in the nature of GIDL between different configurations of NWFETs, i.e., NWMOSFET, nanowire junctionless accumulation-mode field-effect transistor (NWJAMFET), and NWJLFET.

5.7.1 Transfer Characteristics

The transfer characteristics of different NWFET configurations, namely NWMOSFET, NWJAMFET, and NWJLFET, are presented in Fig. 5.24(a). The NWJLFET exhibits the least OFF-state current, whereas the NWJAMFET exhibits the highest OFF-state leakage current. However, the NWMOSFET and NWJAMFET offer a high ON-state current as compared to the NWJLFET owing to the reduced source/drain series resistance due to the heavily doped source/drain regions ($N_D = 1 \times 10^{20}$ cm^{-3}) as compared to NWJLFETs ($N_D = 1 \times 10^{19}$ cm^{-3}). However, due to the significant reduction in the OFF-state leakage current, the NWJLFET exhibits the highest ON-state to OFF-state current ratio (I_{ON}/I_{OFF}) of $\sim 10^6$as compared to the NWMOSFET ($\sim 10^4$) and NWJAMFET ($\sim 10^3$). As a result, NWJLFETs tend to appear as the most promising nanowire (NW) configuration in terms of I_{ON}/I_{OFF}. However, the low ON-state current should lead to an increased delay of the NWJLFETs as compared to the NWJAMFETs and NWMOSFETs. Also, as can be clearly seen from Fig. 5.24(a), the drain current becomes nearly constant for negative gate voltages in case of NWJLFETs whereas it increases with negative gate voltages for NWJAMFETs and NWMOSFETs. Therefore, the drain current follows a different behavior for NWJLFETs as compared to NWMOSFETs and NWJAMFETs for negative gate voltages [12].

At this juncture, you may wonder that since NWMOSFETs and NWJAMFETs have a similar source/drain doping which is different as compared to the NWJLFETs, the difference in drain doping causes this behavior. However, a larger source/drain doping only leads to a smaller tunneling width and a larger L-BTBT increasing the OFF-state current in NWJAMFETs and NWMOSFETs as compared to NWJLFETs. Since vertical tunneling owing to T-BTBT was not considered while obtaining transfer characteristics [12] in Fig. 5.24(a), there must be some other phenomenon that changes the nature of L-BTBT GIDL in the different NWFETs as discussed in Section 5.5. As we discuss, this phenomenon turns out to be the L-BTBT-induced parasitic BJT action, which we addressed for the JLFETs in Section 5.1.2.

5.7.2 Parasitic BJT Action

To understand the difference in the nature of GIDL in different NWFET configurations, let us focus our attention toward Fig. 5.24(b). It can be observed that there exists a significant spatial proximity (overlap) between the valence band of the channel region and the conduction band of the drain region. This band alignment facilitates the tunneling of electrons from the channel region to the drain region. As discussed in Section 5.1.1, this L-BTBT leads to an accumulation of electrons in all the NWFET configurations as shown in Fig. 5.25. The accumulation of holes increases the potential of the channel region and lowers the source-to-channel barrier height. This phenomenon is similar to the kink effect observed in partially depleted (PD) SOI MOSFETs [36–39].

The channel region is converted into p-type due to the accumulation of holes in all the NWFET configurations in the OFF-state irrespective of the initial doping configuration of the channel region (lightly doped p-type for MOSFET, n-type for JAMFET, or heavily doped n-type for JLFET) as shown in Fig. 5.25. A parasitic BJT is formed in all the NWFET configurations in the OFF-state with the source acting as the emitter, the drain acting as the collector and the p-type converted channel acting as the base region [12]. The tunneling current acts as the base current and gets amplified by the current gain (β) of the parasitic BJT to yield the large OFF-state current observed for the NWFET configurations in Fig. 5.24(a). Therefore, similar to the JLFETs, a parasitic BJT is formed in all the NWFET configurations due to L-BTBT.

We would like to underline another important point from Fig. 5.24(b). Although the hole accumulation leads to lowering of source-to-channel barrier height, the extent of barrier lowering depends on the tunneling probability. A smaller tunneling width in the NWJAMFET and NWMOSFET (Fig. 5.24(b)) leads to a larger tunneling probability. Also, the formation of depletion regions at the source/channel and channel/drain interface in NWJAMFET (high–low junction) and NWMOSFET (p–n junction) reduces the effective channel length as compared to the gate length. A lower effective channel length makes NWJAMFET and NWMOSFET more susceptible to the short-channel effects such as DIBL [12]. As a result, the source-to-channel barrier height is lower in the NWJAMFET and NWMOSFET as compared to the

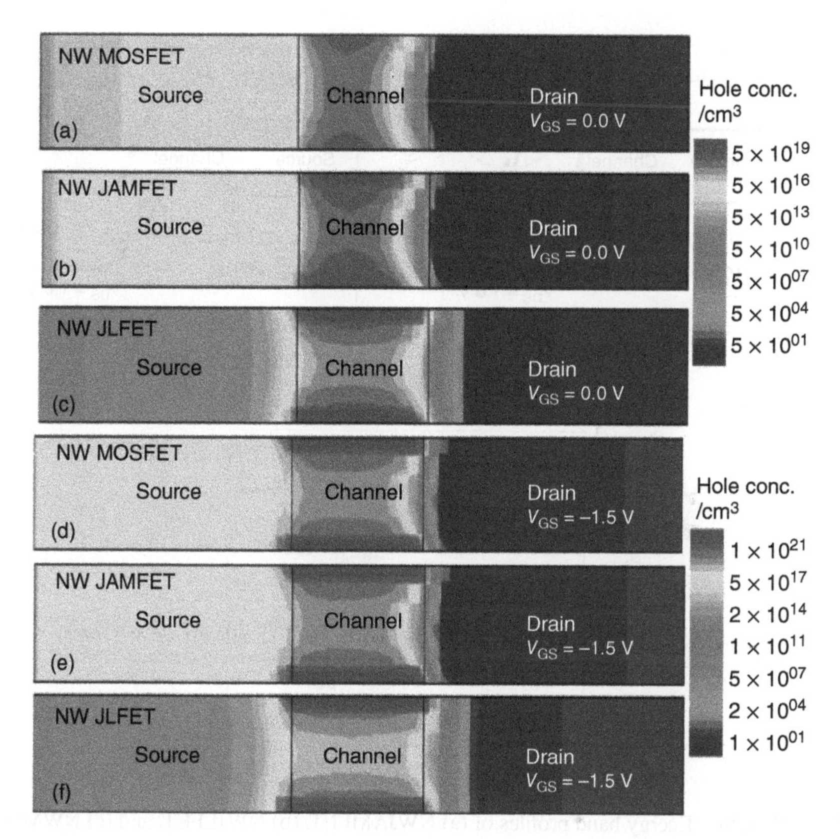

FIGURE 5.25 Hole concentration contour plot of (a) NWMOSFET, (b) NWJAMFET, (c) NWJLFET in the OFF-state ($V_{GS} = 0.0\,V$) and (d) NWOSFET, (e) NWJAMFET, and (f) NWJLFET for $V_{GS} = -1.5\,V$.

NWJLFETs. Also, since the effective channel length is larger in the NWJLFET as compared to the NWJAMFET or NWMOSFET for the same physical gate length, the effective base width of the parasitic BJT is also larger in NWJLFETs. The current gain of the BJT increases with a reduction in the base width. Therefore, the amplification of the tunneling current is also larger in NWJAMFETs and NWMOSFETs as compared to NWJLFETs. A larger tunneling width, a higher source-to-channel barrier height, and a lower gain of the parasitic BJT lead to a significantly improved I_{ON}/I_{OFF} observed for the NWJLFETs.

5.7.3 Origin of Difference in Nature of GIDL

Now, ideally, turning ON of the parasitic BJT should lead to a loss of gate control on the drain current in NWFETs as observed for the JLFET due to the saturation in the source-to-channel barrier height as explained in Section 5.1. However, such a loss of gate control is observed only for the NWJLFETs and there is an increase in

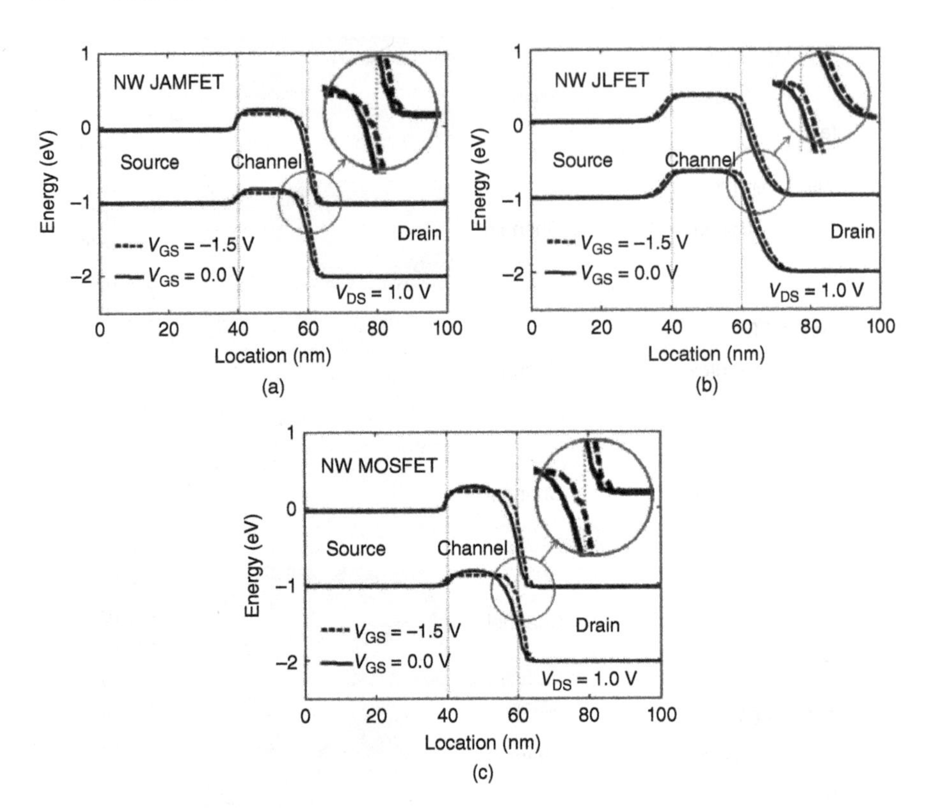

FIGURE 5.26 Energy band profiles of (a) NWJAMFET, (b) NWJLFET, and (c) NWMOSFET for different gate voltages.

the drain current with increasing negative gate voltages for the NWJAMFET and NWMOSFET unlike NWJLFET [8]. This clearly indicates that the nature of the GIDL in NWJLFETs is different than that of NWJAMFETs and NWMOSFETs for $V_{GS} < 0.0\,V$. This can also be explained using the parasitic BJT theory proposed in [12].

As can be observed from Fig. 5.26, the tunneling width remains almost the same in NWJLFETs even at large and negative gate voltages. However, the tunneling width reduces significantly in the NWJAMFETs and NWMOSFETs when a large and negative gate voltage is applied. A smaller tunneling width leads to a larger tunneling probability and a consequent larger accumulation of holes in the channel region for NWJAMFETs and NWMOSFETs as shown in Fig. 5.25. This further increases the potential of the channel region in the NWMOSFETs and NWJAMFETs reducing the source-to-channel barrier height significantly. The source-to-channel barrier height remains almost same for the NWJLFETs even when the gate voltage is large and negative and restricts the accumulation of holes to the channel region. However, the source-to-channel barrier height reduces significantly in the NWJAMFETs and NWMOSFETs as shown in Fig. 5.26. This allows the source electrons to surmount the small energy barrier and flow into the drain through the channel. Therefore, the hole concentration increases even in the source region adjacent to the source–channel interface as shown in Fig. 5.25.

5.7.4 Gain of Parasitic BJT in NWFETs

The accumulation of holes in the source region adjacent to the source–channel interface increases the effective base width of the parasitic BJT when the gate voltage is large and negative. An increase in the effective base width reduces the current gain of the BJT significantly, which can be observed from Fig. 5.27. The spreading of holes in the source region facilitates the collection of holes by the source contact dissuading the parasitic BJT action. Therefore, the parasitic BJT becomes redundant when the gate voltage becomes large and negative.

The ratio of the electron current to the hole current $I_{\mathrm{n}}/I_{\mathrm{p}}$ in the base region effectively determines the gain of the parasitic BJT [40]. Although the electron current remains higher than the hole current in the base (channel region) of the NWJLFET yielding an appreciable gain of the parasitic BJT, the hole current in the base (channel region) of the NWMOSFET is comparable to the electron current while the hole current is more than the electron current in the case of the NWJAMFET. This clearly indicates that the parasitic BJT action is diminished in the NWJAMFET and NWMOSFET when a large and negative gate voltage is applied while the parasitic BJT action continues to dominate the drain current in NWJLFETs.

Due to the suppressed parasitic BJT action, the source-to-channel barrier height is not saturated in the NWMOSFET and NWJAMFET and the gate is able to modulate the tunneling width. Therefore, for NWMOSFETs and NWJAMFETs, it is the gate-modulated tunneling current that dictates the drain current for large and negative gate voltages whereas the parasitic BJT action dictates the drain current for NWJLFETs [12]. Also, there is an inherent unintentional gate-on drain overlap in the NWJAMFETs and NWMOSFETs owing to the source/drain dopant diffusion in the channel region and the consequent doping gradient [7, 8]. This unintentional overlap facilitates the T-BTBT when the gate voltage is large and negative further increasing the drain current in NWJAMFETs and NWMOSFETs. However, due to the electrostatic squeezing of carriers in NWJLFETs and the consequent underlap architecture [1–3], T-BTBT is not possible in NWJLFETs. Therefore, the nature of GIDL in NWFETs remains the same even in the presence of T-BTBT.

It is clear now that the gate-modulated tunneling current can increase with the applied negative voltage whereas the current-dominated by parasitic BJT action may not be controlled by the gate voltage once the parasitic BJT is triggered. The parasitic BJT theory explains this difference in the nature of GIDL in different configurations of NWFETs [12]. At this juncture, we would also like to discuss the anomalous behavior shown by the conventional carbon nanotube FETs (CNFETs). The symmetrically doped (n–i–n or p–i–p) conventional CNFETs exhibit a very high leakage current as compared to the CNFETs with a Schottky source and drain region or the p–i–n tunneling CNFETs [41–43]. The parasitic BJT theory can explain

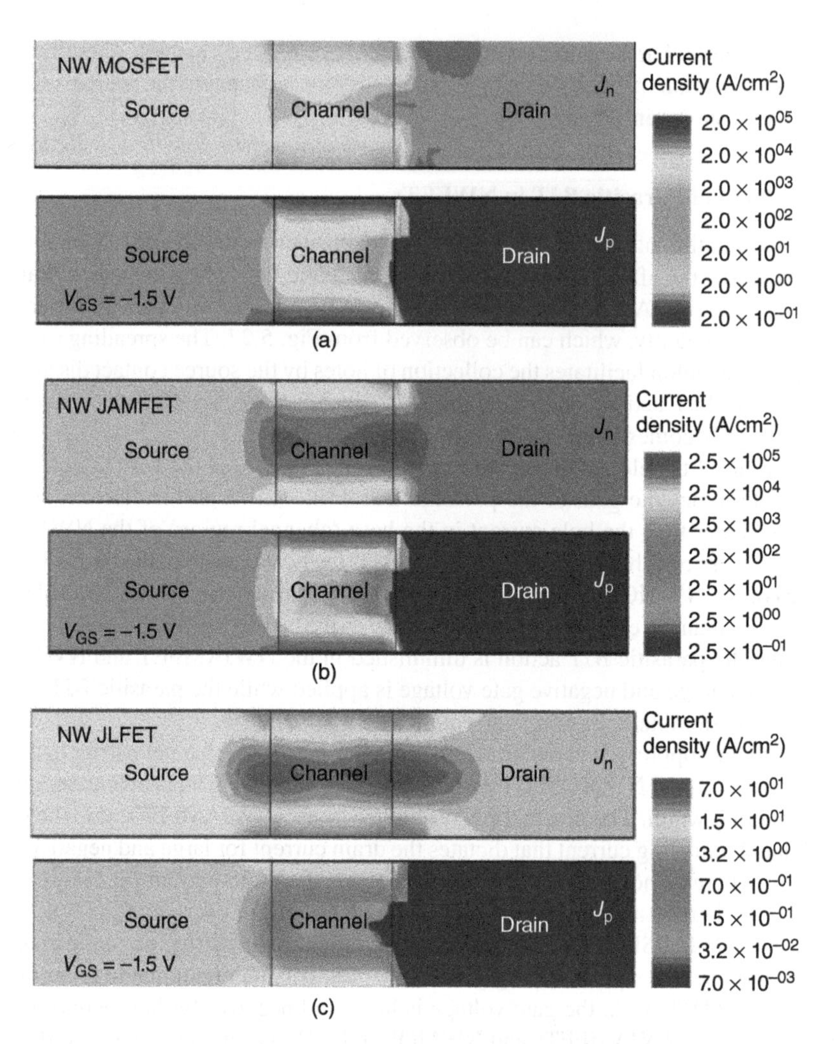

FIGURE 5.27 The electron current density and hole current density contour plots of (a) NWMOSFET, (b) NWJAMFET, and (c) NWJLFET for $V_{GS} = -1.5$ V.

this anomalous behavior: At high drain bias, L-BTBT takes place in conventional CNFETs which leads to accumulation of holes in the channel region due to the BTBT [12]. Consequently, the source-to-channel barrier height is lowered and the parasitic BJT action is triggered, which leads to a significantly high OFF-state leakage current in conventional CNFETs. The drain current also becomes less dependent on the gate voltage in the case of conventional CNFETs owing to the parasitic BJT action [41–43].

The high OFF-state leakage current even in the case of quantum well InGaAs MOSFETs could not be explained entirely due to BTBT or the gate leakage current

[40]. However, the results obtained from the parasitic BJT theory could easily corroborate that the experimentally observed OFF-state current for the quantum well MOSFETs are amplified versions of the tunneling current. The parasitic BJT theory could also explain the gate length dependence of the leakage current for the quantum well InGaAs MOSFETs [40].

In the future, if the readers encounter a tunneling component in any new device, they could relate it to the existence of a parasitic BJT and its implications on the device characteristics. From our discussions in Sections 5.2.3 and 5.5.2, it becomes clear that a loss of gate control on the drain current for negative gate voltages should be attributed either to BTBT-induced parasitic BJT action or the shielding effect due to the accumulation of charge carriers owing to BTBT or some other phenomenon such as impact ionization.

5.7.5 Impact of Gate Length Scaling on Nature of GIDL

Figure 5.28 shows the impact of scaling the gate length down to 10 nm (with a nanowire diameter = 10 nm) on the different NWFET configurations. Gate length scaling in NWFETs is equivalent to reducing the effective base width of the

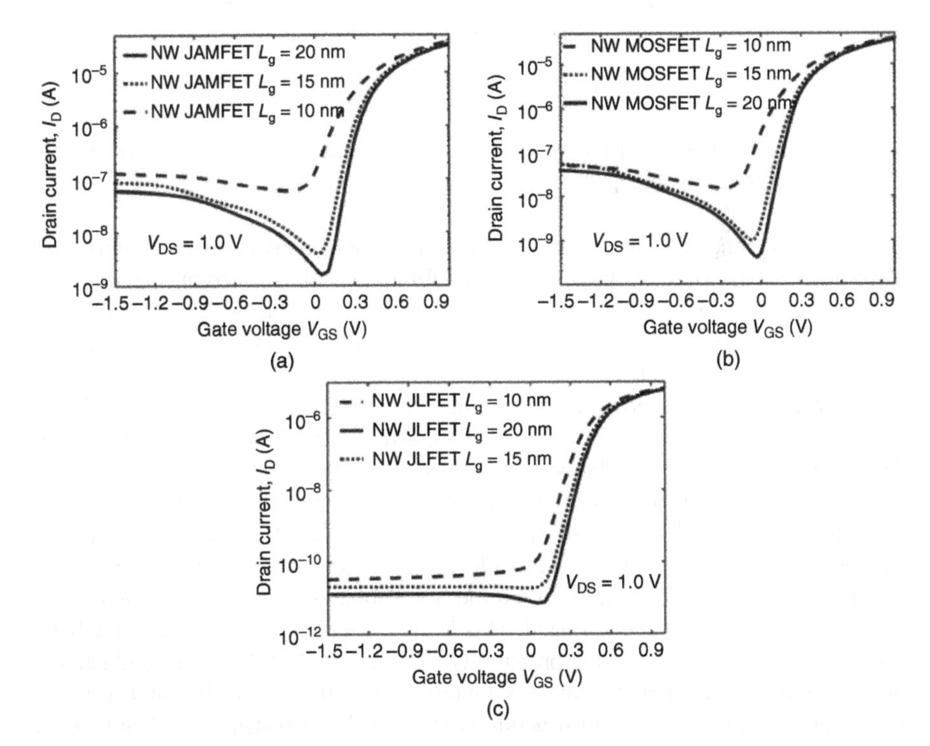

FIGURE 5.28 Transfer characteristics of (a) NWJAMFET, (b) NWMOSFET, and (c) NWJLFET for different gate lengths.

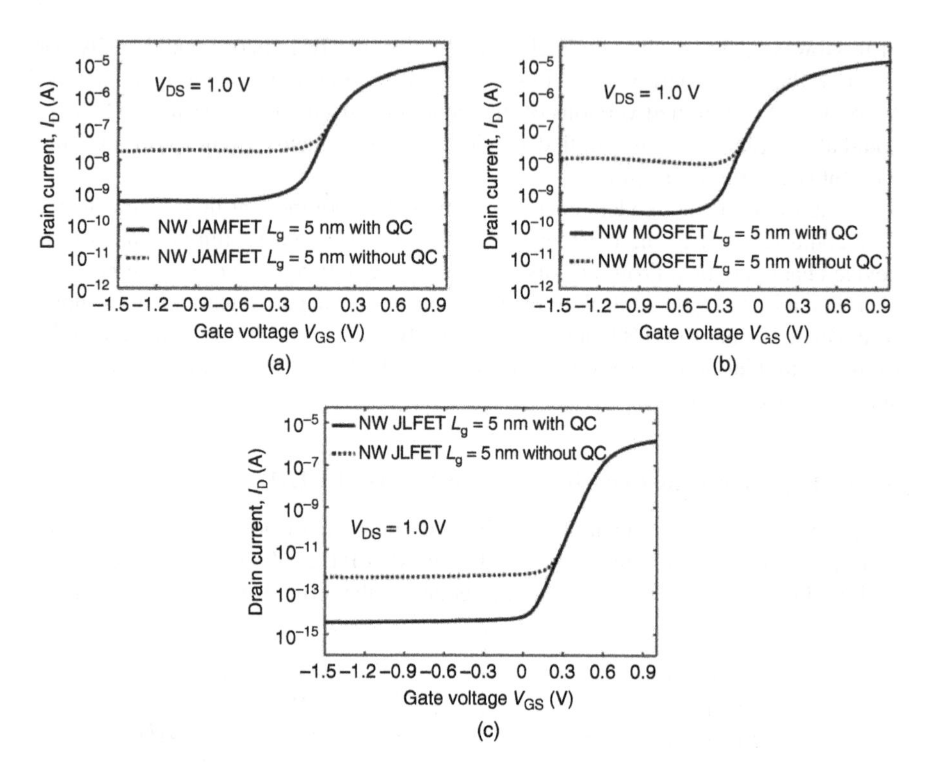

FIGURE 5.29 Transfer characteristics of (a) NWJAMFET, (b) NWMOSFET, and (c) NWJLFET for $L_g = 5\,\text{nm}$, $d_{NW} = 5\,\text{nm}$.

parasitic BJT. Since the current gain of the parasitic BJT increases significantly with a reduction in the effective base width, the OFF-state leakage current increases significantly with the gate length scaling. Therefore, the L-BTBT-induced parasitic BJT action increases the OFF-state current significantly for the NWFETs when they are scaled to the 10-nm regime.

However, as can be observed from Fig. 5.29, the OFF-state current in all the NWFET configurations: NWMOSFET, NWJAMFET, and NWJLFET, reduces significantly as the nanowire diameter is scaled to 5 nm despite a reduction in the gate length to 5 nm. As discussed in Section 5.2.2, the quantum confinement effects become significant as the silicon film thickness is reduced below 7 nm [29]. The quantization effects lead to a considerable discretization of the energy bands, which restricts the L-BTBT to occur between the first subband of the valence band to the first subband of the conduction band. This leads to an effective increase in the band gap which reduces the tunneling probability significantly [29–32]. The band gaps of silicon for different silicon thicknesses are reported in [6, 32], and the band structure can be analyzed using the band structure lab tool [44]. For instance, the band gap of silicon increases to $E_g = 1.241\,\text{eV}$ for $d_{NW} = 5\,\text{nm}$ and $E_g = 1.137\,\text{eV}$ for $d_{NW} = 7\,\text{nm}$ as compared to $E_g = 1.12\,\text{eV}$ for $d_{NW} = 10\,\text{nm}$ at 300 K. In scaled devices with dimensions less than 7 nm, it is advisable to use this increased value of band gap in the simulations along with other quantum correction models such as the density gradient model or modified local density approximation to correctly analyze the impact of quantum confinement effects [11].

This increase in the band gap masks the impact of increased current gain of the parasitic BJT due to a reduction in gate length leading to a significantly low OFF-state current and a diminished L-BTBT. Therefore, quantum confinement effects facilitate the scaling of the NWFETs to the sub-5 nm regime where the quantization effects are significant. The significantly reduced L-BTBT GIDL when the FET dimensions are reduced below 5 nm has been recently observed in the experimental results of [45] and corroborate with the results presented in [12].

5.8 DEVICE ARCHITECTURES TO MITIGATE GIDL

So far, we have discussed the physics of the BTBT-induced parasitic BJT in JLFETs and NWFETs. We saw how scaling the gate length increases the current gain of the parasitic BJT and increases the leakage current by several orders of magnitude. We also concluded that it is indeed the parasitic BJT action, which is the dominant mechanism hindering the scaling of JLFETs. In subsequent sections, we discuss the various device architectures that have been proposed to mitigate the parasitic BJT action in JLFETs. Diminishing the BTBT-induced parasitic BJT action in JLFETs requires architectures that either (a) increase the tunneling width at the channel–drain interface, which reduces BTBT, or (b) increase the source-to-channel barrier height, which hinders the triggering of the parasitic BJT, or (c) increase the effective length of the channel region (base width), which reduces the current gain of the parasitic BJT, or (d) provide a path to the holes accumulated in the channel region to the ground.

5.8.1 JLFETs with a Hybrid Channel

5.8.1.A Structure The three-dimensional (3D) view of the JLFET with a hole sink (HS) is shown in Fig. 5.30. The HSJLFET consists of a heavily doped n^+ active device layer ($N_D = 1 \times 10^{19}$ cm^{-3}) over a p^+ layer ($N_D = 5 \times 10^{18}$ cm^{-3}) which we call the hole sink. The HSJLFET contains no metallurgical junction in the lateral direction, i.e. along the channel. However, there is an inherent p–n junction in the vertical direction transverse to the channel. Starting with a p^+ active SOI layer, the HSJLFET requires the formation of an abrupt p^+–n^+ vertical junction, which can be achieved using the recently reported molecular monolayer doping technique [46–48] and the microwave annealing technique [49, 50] over the entire 300-mm wafer. The thickness of the hole sink layer is denoted by t_{HS}. The SOI JLFET can be considered as a special case of HSJLFET with $t_{HS} = 0$ nm. The hole sink layer is kept floating.

5.8.1.B Operation The p^+ layer underlying the active n^+ device layer facilitates the depletion of the drain region at the channel–drain interface in the HSJLFET

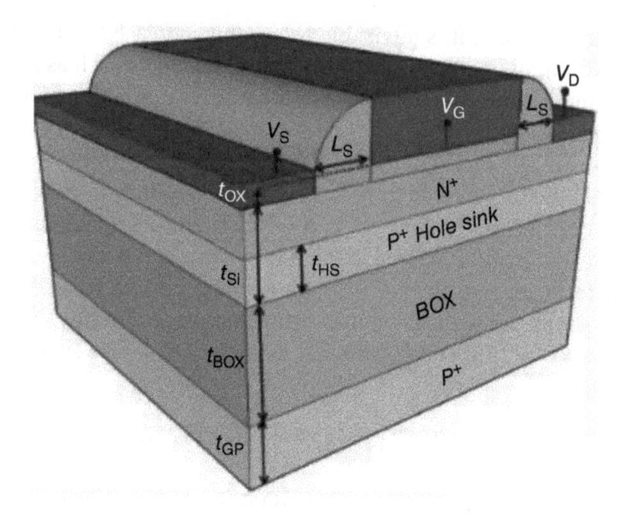

FIGURE 5.30 Three-dimensional view of a JLFET with a hole sink.

due to the p–n junction depletion. This leads to an increased tunneling width at the channel–drain interface in the HSJLFET as compared to the SOI JLFET as shown in Fig. 5.31. An increased tunneling width reduces the tunneling probability and results in a decreased BTBT generation rate as shown in Fig. 5.32. A reduced BTBT is reflected as the diminished hole concentration in the channel region of the HSJLFET as compared to the SOI JLFET as shown in Fig. 5.33.

The underlying p^+ layer also facilitates the depletion of the source region at the source–channel interface. This leads to a larger source-to-channel barrier height in the HSJLFET as compared to the SOI JLFET as shown in Fig. 5.31. A higher source-to-channel barrier height inhibits the triggering of the parasitic BJT in the HSJLFET

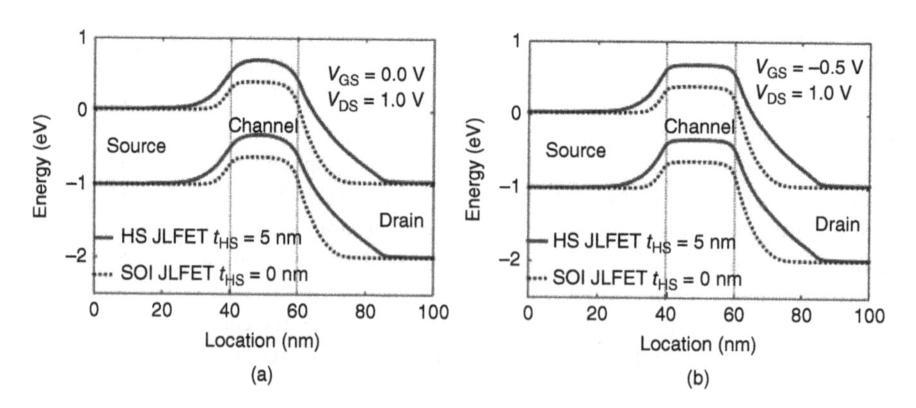

FIGURE 5.31 Energy band profiles of the SOI JLFET and a HSJLFET in (a) the OFF-state ($V_{GS} = 0.0$ V) and (b) $V_{GS} = -0.5$ V at $V_{DS} = 1.0$ V.

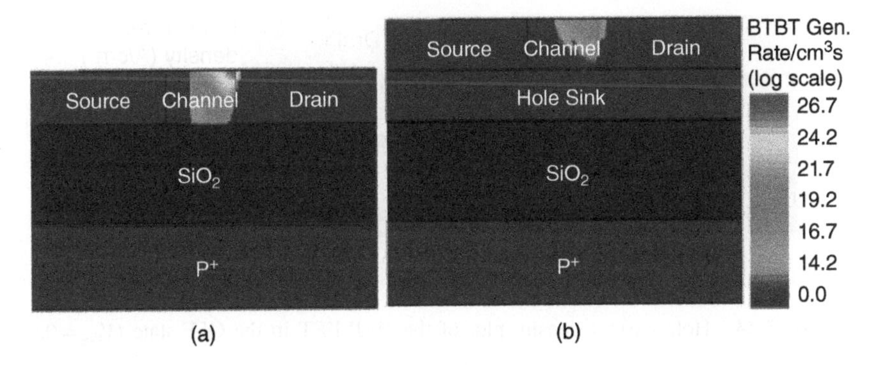

FIGURE 5.32 BTBT generation rate contour plot of (a) SOI JLFET and (b) HSJLFET in the OFF-state ($V_{GS} = 0.0$ V, $V_{DS} = 1.0$ V).

as compared to the SOI JLFET. You may comprehend that a higher source-to-channel barrier height would not allow the electrons to surmount the barrier and flow into the drain region. Therefore, the higher source-to-channel barrier height would lead to a lower OFF-state current even when parasitic BJT action is not present. It may also be noted that there is no BTBT leakage at the vertical n^+–p^+ junction in the HSJLFET.

5.8.1.C Hole Sink Mechanism You might wonder why we are continuously referring the underlying p^+ layer as "hole sink." Is it really sinking holes? To answer this question, let us look at Fig. 5.34. The hole density plot clearly indicates that the BTBT-generated holes flow to the source electrode through the hole sink. This observation can be explained as follows: The BTBT leads to the accumulation of holes in the channel region. Owing to the inherent p–n junction electric field at the n^+ active layer–p^+ hole sink layer, which is directed from an active layer to the hole sink layer, the accumulated holes are carried into the hole sink layer. Now, the hole sink layer starts acting like the base region of the parasitic BJT. Since the length of

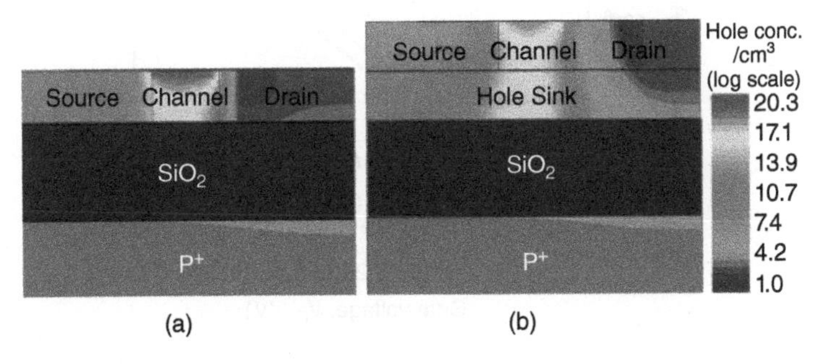

FIGURE 5.33 Hole concentration contour plot of (a) SOI JLFET and (b) HSJLFET in the OFF-state ($V_{GS} = 0.0$ V, $V_{DS} = 1.0$ V).

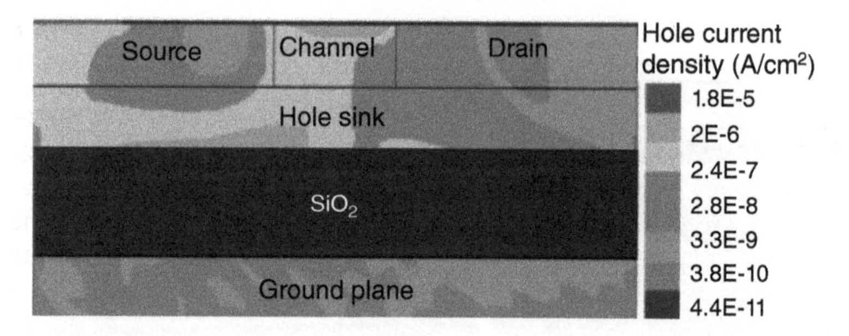

FIGURE 5.34 Hole current density plot of the HSJLFET in the OFF-state ($V_{GS} = 0.0$ V, $V_{DS} = 1.0$ V).

the base region (which now spans the entire device length) is increased, the current gain of the parasitic BJT is reduced [17].

The accumulation of the holes in the hole sink layer increases its potential. This inherently forward biases the p$^+$ hole sink–n$^+$ active device layer junction below the source contact and provides a path for the holes to flow to the ground. This is similar to the kink effect observed in the PD SOI MOSFETs [36–39]. In other words, the floating underlying p$^+$ layer first attracts the holes preventing the hole accumulation in the channel region of HSJLFET and provides them a path to flow to the ground via the source contact. Therefore, we call it a hole sink as it sinks the BTBT-generated holes to the ground via the source contact.

5.8.1.D Transfer Characteristics The transfer characteristics of the SOI JLFET and the HSJLFET for different hole sink thicknesses are shown in Fig. 5.35. A larger tunneling width, a higher source-to-channel barrier height, and a provision for collecting the BTBT-generated holes lead to a diminished parasitic BJT action in HSJLFET.

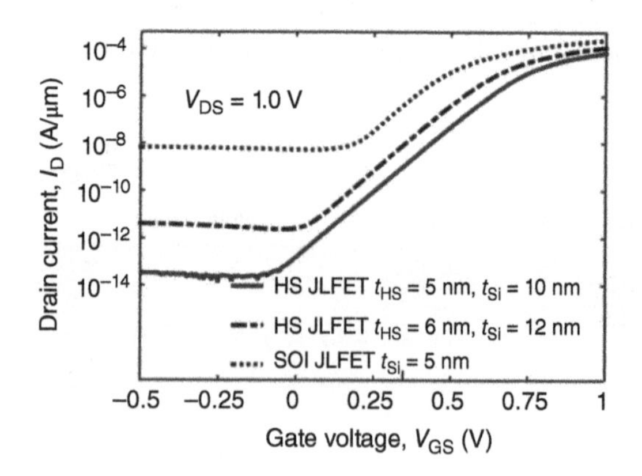

FIGURE 5.35 Transfer characteristics of the SOI JLFET and the HSJLFET for different hole sink thicknesses (t_{HS}).

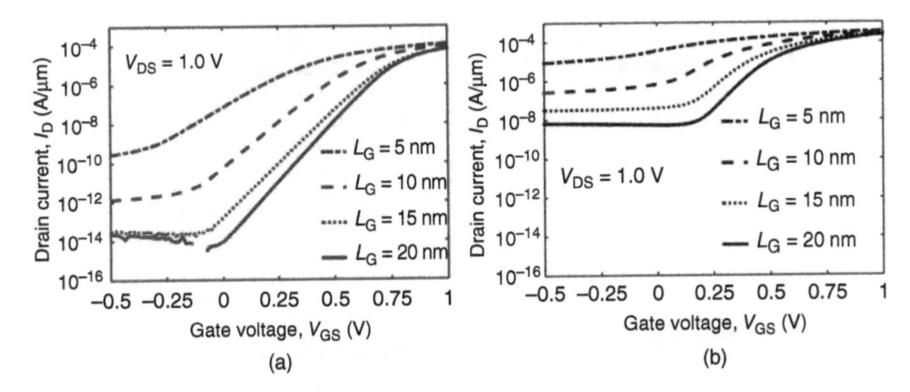

FIGURE 5.36 Impact of gate length scaling on the transfer characteristics of (a) HSJLFET and (b) SOI JLFET.

This results in a significantly low OFF-state current in the HSJLFET as compared to the SOI JLFET as shown in Fig. 5.35.

5.8.1.E Impact of Scaling We discussed in Section 5.3 that the BTBT-induced parasitic BJT action is more pronounced in ultrascaled JLFETs. It would be interesting to analyze the efficacy of HSJLFETs in mitigating the BTBT-induced parasitic BJT action for ultra-short channel lengths. The impact of gate length scaling is shown in Fig. 5.36 from which we can infer that BTBT-induced parasitic BJT action significantly increases as the gate length (base width) reduces degrading the performance of both HSJLFET and SOI JLFET. However, the HSJLFET performs significantly better than the SOI JLFET, exhibiting an ON-state to OFF-state current ratio (I_{ON}/I_{OFF}) of ~10^4 even for a gate length of 5 nm as compared to SOI JLFET (I_{ON}/I_{OFF} ~6). Therefore, the presence of hole sink in a JLFET provides inherent scaling benefit [17].

5.8.1.F Challenges Although the HSJLFET appears to be a lucrative alternative to the SOI JLFETs, there are several hurdles in its pathway to replace the SOI JLFET. First, creation of such an abrupt junction is not possible using the industry standard ion implantation process and requires advanced techniques like molecular monolayer doping and microwave annealing. Second, the ON-state current of HSJLFET is low (~3.3 times) as compared to the SOI JLFET and enhancement in the ON-state current by increasing the n^+ silicon film thickness leads to a significant increase in the OFF-state current. Therefore, there is a trade-off between the dynamic performance and static performance in HSJLFET. Third, the creation of source/drain contacts over the ultrathin film (~5 nm) is also a technological challenge. Since the HSJLFET consists of a doped p–n junction, it is also expected to be susceptible to performance parameter variations due to the random dopant fluctuation as discussed in Section 3.6.3.

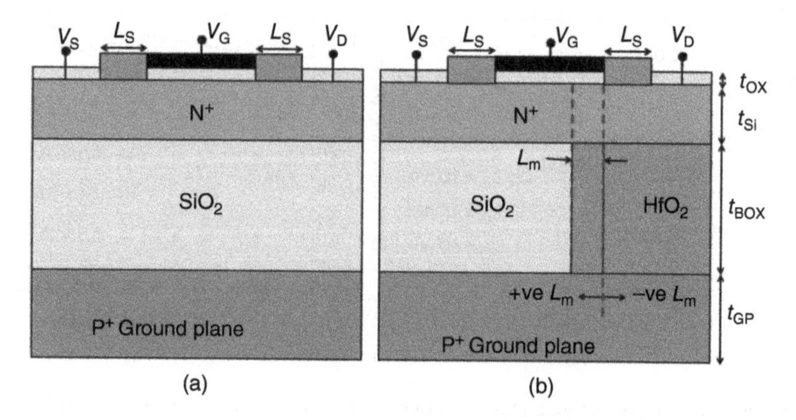

FIGURE 5.37 Schematic view of (a) conventional SOI JLFET and (b) HDB JLFET.

5.8.2 JLFETs with a Heterodielectric Buried Oxide

5.8.2.A Structure The JLFET with a heterodielectric (HDB) buried oxide (BOX) is essentially a SOI JLFET where a high-κ dielectric replaces the SiO_2 in the BOX just below the drain region as shown in Fig. 5.37 . In addition, the HDB JLFET uses a heavily P^+ doped substrate (ground plane) unlike conventional SOI JLFETs, which utilize a moderately doped substrate (ground plane). The parameter L_m indicates the length by which the placement of high-κ BOX with respect to the channel–drain interface is misaligned. If $L_m = 0$ nm, the high-κ BOX is exactly aligned with the channel–drain interface. A positive value of L_m means that the high-κ BOX is extended under channel region, whereas a negative value of L_m signifies the placement of high-κ BOX edge is under the drain region.

5.8.2.B Operation As discussed earlier, to mitigate GIDL in JLFETs, it is essential to increase the depletion region width on the drain side in the OFF-state. In HDB JLFET, as can be observed from Fig. 5.38, the heterodielectric BOX facilitates the

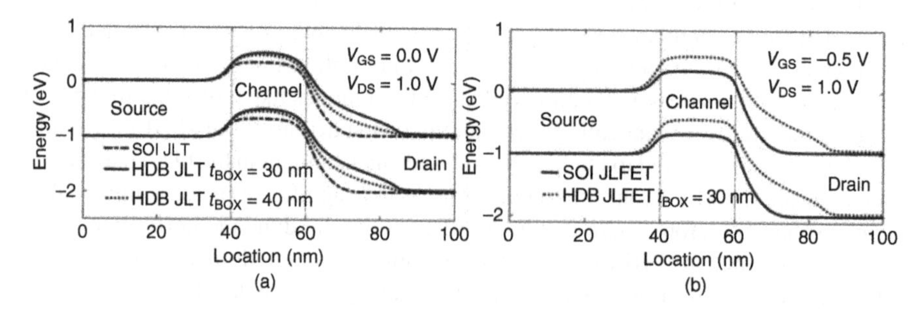

FIGURE 5.38 Energy-band profiles of the conventional SOI JLFET and the HDB JLFET at 1 nm below Si–SiO_2 interface for different V_{GS}.

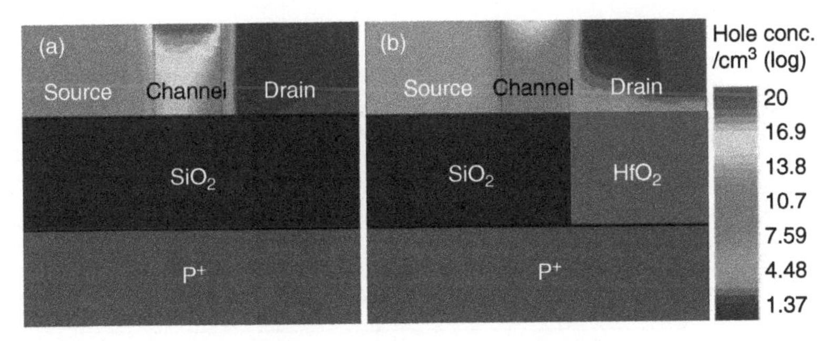

FIGURE 5.39 Hole concentration contour plot of (a) the conventional SOI JLFET and (b) the HDB JLFET in the OFF-state ($V_{GS} = 0.0$ V and $V_{DS} = 1.0$ V).

depletion of the drain region at the channel–drain interface. The heavily doped ground plane acts like a metal with a high work function (~5.1 eV). This ground plane coupled with the small effective BOX thickness just below the drain region due to the presence of high-κ BOX leads to efficient depletion of the drain region at the channel drain interface [51]. This leads to an increase in the tunneling width resulting in a significant reduction of the BTBT of electrons from the channel to the drain. In the OFF-state, as can be observed from Fig. 5.39, a reduction in tunneling leads to a reduced hole accumulation in the channel region of the HDB JLFET as compared to the SOI JLFET. The concentration of holes in the channel region in a HDB JLFET is significantly low (~10^9 times less) compared to the conventional JLFET. This low concentration of the holes in the OFF-state is due to the absence of channel–drain tunneling as the valence band in the channel is not aligned with the conduction band in the drain region in a HDB JLFET. Because of the presence of a low hole concentration, the parasitic BJT action is diminished leading to a reduction in the OFF-state leakage current in a HDB JLFET. Further, as can be observed from Fig. 5.38, the source-to-channel barrier height is also larger in the HDB JLFET compared to the SOI JLFET. The triggering of the parasitic BJT would, therefore, be difficult in the HDB JLFET than the SOI JLFET.

5.8.2.C Transfer Characteristics The transfer characteristics of the conventional SOI JLFET, the proposed HDB JLFET for different BOX thicknesses (t_{BOX}), and the JLFET with high-κ (HfO_2) spacers (spacer length = 15 nm, which shows the least I_{OFF}) [35] are shown in Fig. 5.40. We observe that the leakage current is drastically reduced in a HDB JLFET by three orders of magnitude when a BOX thickness of 30 nm is used increasing the I_{ON}/I_{OFF} ratio by a factor of 10^3. Moreover, the HDB JLFET also shows a lower leakage current than the JLFET with high-κ spacers [35]. It may also be pointed out that the OFF-state current (~10^{-9} A/μm at $V_{GS} = 0$ V) in the proposed HDB JLFET is significantly less than the OFF-state current (~10^{-7} A/μm at $V_{GS} = 0$ V) of a bulk planar junctionless transistor [5] (discussed in Section 5.8.3), which has a direct path for the holes to the ground. Therefore, it is clear that the

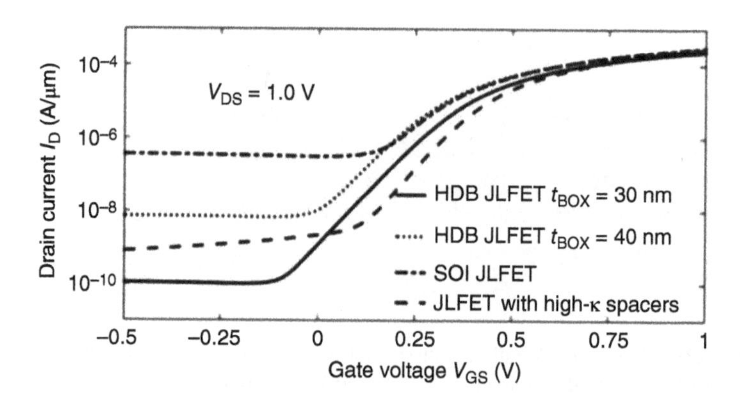

FIGURE 5.40 Transfer characteristics of the conventional SOI JLFET, the HDB JLFET for different BOX thicknesses, and JLFET with high-κ spacers [35] at $V_{DS} = 1.0$ V.

HDB JLFET would perform better than a grounded body contact approach in terms of reducing the parasitic BJT effect. However, it may be noted that the ON-state current of the HDB JLFET is ~1.3 times less than that of the SOI JLFET because of the increased depletion of the drain region which adds to the series resistance.

5.8.2.D Challenges In HDB JLFET, the interface of the high-κ dielectric and SiO_2 in the BOX region has to be appropriately aligned with the channel–drain interface in the SOI layer in HDB JLFET [51]. This alignment requires a stringent control over the wafer fabrication process. The HDB JLFETs may be realized only when the alignment challenges are overcome. In addition, the interface between high-κ dielectric and SOI layer is bound to have defects and traps. A buffer layer of SiO_2 must be used above the high-κ dielectric to ensure a defect-free interface at the SOI layer–BOX interface below the drain region. In the subsequent sections, we discuss the impact of misalignment due to process variation and the traps on the performance of HDB JLFET.

5.8.2.E Impact of Misalignment The impact of process variations in the placement of high-κ BOX with respect to the drain was investigated as shown in Fig. 5.41(a). The transfer characteristics of the HDB JLFET do not change significantly when the edge of the high-κ BOX is misaligned by 10% of the channel length ($L_m = \pm 2$ nm). In addition, the leakage current reduces when the high-κ BOX is misaligned by 50% of the channel length ($L_m = 10$ nm). This reduction in the leakage current with positive values of L_m is attributed to the increased depletion of the channel region (Fig. 5.41(b)), which further increases the tunneling width and reduces the BTBT. Also, as shown in Fig. 5.42, the source-to-channel barrier height is increased and hence the parasitic BJT action is diminished further. However, this also leads to a reduction in the ON-state current when $L_m = 10$ nm as the depletion of the channel region increases the channel resistance. Further, the leakage current increases for negative values of L_m as

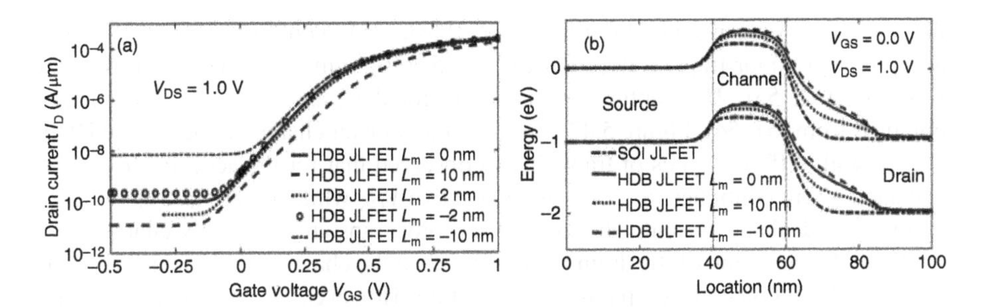

FIGURE 5.41 (a) The impact of heterodielectric BOX misalignment (L_m) on the transfer characteristics of the HDB JLFET with t_{BOX} = 30 nm and (b) the energy band profile of the SOI JLFET and the HDB JLFET for different values of heterodielectric BOX misalignment (L_m).

the high-κ BOX is not able to effectively deplete the drain region at the channel–drain interface as can be observed from Fig. 5.41(b). As a result, the tunneling width and the source-to-channel barrier height reduces slightly increasing the leakage current when the high-κ BOX is misaligned by 50% of the channel length (L_m = − 10 nm). In spite of this small reduction in the tunneling width and source-to-channel barrier height and the consequent leakage current enhancement for L_m = −10 nm, the HDB JLFET exhibits a reduction in the leakage current by more than two orders of magnitude compared to the SOI JLFET.

5.8.2.F Impact of Traps The HDB JLFET was also analyzed in the presence of (a) only acceptor traps, (b) only donor traps, and (c) both acceptor donor traps at the Si/HfO_2 interface [51]. Generally, an ultrathin (0.7–1 nm) buffer layer of SiO_2 is used over HfO_2 to reduce the trap density and the stress [51]. However, to consider

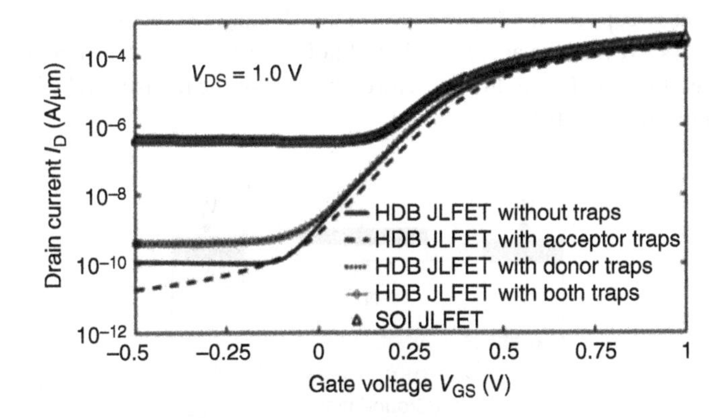

FIGURE 5.42 The impact of traps at the Si/HfO_2 interface on the transfer characteristics of the HDB JLFET with t_{BOX} = 30 nm.

the worst-case scenario, no buffer layer was considered in our analysis. For both the acceptor and donor traps, a trap density of 1.5×10^{12} cm^{-2} located at the midgap was used [52, 53]. The capture cross section for the traps (2×10^{-13} cm^{-2}) was obtained from [51, 54]. Figure 5.42 compares the transfer characteristics of the HDB JLFET with different types of traps. It was observed that for the acceptor traps, the leakage current continues to reduce with negative V_{GS} while the donor traps further increase the leakage current. The ionized acceptor traps are negatively charged and facilitate the depletion of the drain region further at the channel–drain interface [51], reducing the BTBT-induced parasitic BJT action. However, the ionized donor traps are positively charged and lead to accumulation of electrons at the bottom of the drain and screen the depletion of the drain region at the channel–drain interface by the high-κ BOX [51]. However, when both types of traps are used, since the donor-type traps are ionized near the channel–drain interface, they dominate the transfer characteristics [4].

5.8.3 Bulk Planar JLFET

5.8.3.A Structure The structure of the bulk planar junctionless FET (BPJLFET) was already discussed in Section 4.4. It consists of a thin n$^+$ active device layer over a p-silicon substrate as shown in Fig. 5.43. It can be simply viewed as the JLFET in which the isolation is provided using a vertical p–n junction unlike an oxide layer in SOI JLFET. The effective active layer thickness is defined as the thickness of the silicon film, which is depleted only by the gate electrode in the OFF-state. The depletion of the active n$^+$ layer by the p-substrate from bottom helps to attain an effective active layer thickness, which is lower than the physical silicon film thickness and helps to improve the electrostatic integrity of the device.

5.8.3.B Operation The BPJLFET has a vertical p–n junction, which consists of the n$^+$ active device layer and the p-substrate. The in-built electric field at this metallurgical junction sweeps the holes generated in the channel region due to BTBT into the p-substrate. As a result, the accumulation of holes is reduced in the channel region of BPJLFET as shown in Fig. 5.44. Therefore, the gain of the parasitic BJT in BPJLFET is significantly reduced [6].

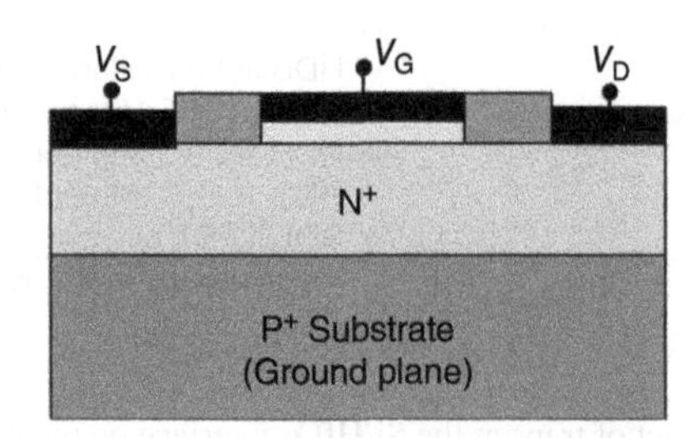

FIGURE 5.43 Schematic view of the BPJLFET.

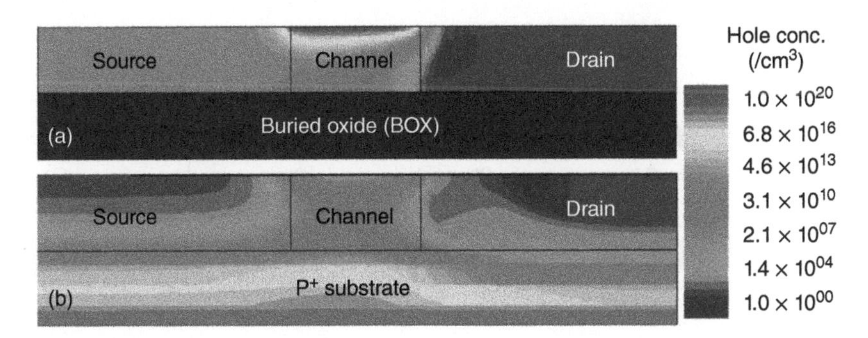

FIGURE 5.44 Hole concentration contour plot of (a) SOI JLFET and (b) BP JLFET in the OFF-state ($V_{GS} = 0.0$ V, $V_{DS} = 1.0$ V).

5.8.3.C Transfer Characteristics Figure 5.45 compares the transfer characteristics of BPJLFET and the SOI JLFET with a silicon film thickness of 6 nm. It can be observed that the BPJLFET exhibits a significantly improved performance compared to the SOI JLFET owing to the diminished parasitic BJT action in BPJLFETs as compared to the SOI JLFETs. Also, the leakage current is dominated by the BTBT-induced parasitic BJT current at a more negative gate voltage ($V_{GS} \sim -0.3$ V) in BPJLFET as compared to the SOI JLFET ($V_{GS} \sim 0.0$ V). This clearly indicates the effectiveness of BPJLFET in suppressing the OFF-state current. The effective active layer thickness is lower in the BPJLFET even in the ON-state due to the vertical depletion region induced by the p-substrate. Therefore, the ON-state current in a BPJLFET is less than SOI JLFET as discussed in Section 4.4.

5.8.3.D Challenges Although the fabrication process for BPJLFETs is expected to be simple and cost-effective [5], it suffers from the fabrication challenges inherent to the JLFET architecture as discussed in Section 2.6.5. For instance, it is difficult to realize source/drain ohmic contacts on a lightly doped ($N_D = 1 \times 10^{19}$ cm^{-3}) active

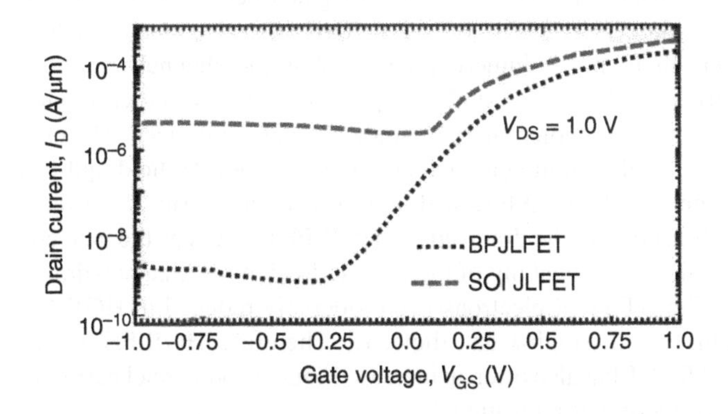

FIGURE 5.45 Transfer characteristics of the BPJLFET and the SOI JLFET [6].

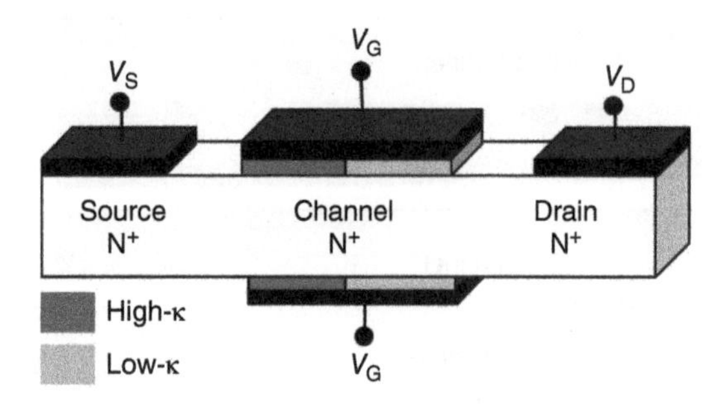

FIGURE 5.46 Three-dimensional view of a JLFET with a heterogate dielectric (HGJLFET).

layer in BPJLFET. Also, realizing a high ON-state current in BPJLFET is difficult due to the lower effective active layer thickness even in the ON-state.

5.8.4 JLFETs with Heterogate Dielectric

5.8.4.A Structure The schematic view of a JLFET with a heterogate dielectric (HGJLFET) is shown in Fig. 5.46, where the gate dielectric is different on the source side and the drain side. While a high-κ gate dielectric (HfO_2) is used as the gate dielectric from the source–channel interface to the middle of the channel region, a low-κ gate dielectric (SiO_2) is employed as the gate dielectric from the middle of the channel region to the channel–drain interface. This heterogate dielectric may be realized using selective etching of SiO_2 followed by deposition of high-κ dielectric using the atomic layer deposition process [55, 56].

5.8.4.B Operation The application of a high-κ gate dielectric leads to a stronger gate to channel coupling and a consequent higher gate–induced electric field as compared to a low-κ gate dielectric. As a result, the JLFETs with high-κ gate dielectric suffer from a large electric field at both the source–channel and channel–drain interface as shown in Fig. 5.47. The stronger gate to channel coupling on the source side near the source–channel interface increases the gate electric field as shown in Fig. 5.47 and facilitates volume depletion below the channel region under high-κ gate dielectric [55]. Moreover, it also helps to increase the source-to-channel barrier height due to efficient depletion at the source–channel interface. The use of a high-κ gate dielectric on the drain side increases the gate electric field and leads to a significantly high BTBT in JLFETs with high-κ gate dielectric. However, the use of a low-κ gate dielectric on the drain side in HGJLFETs reduces the gate coupling leading to a significant diminishing of the electric field at the channel–drain interface in HGJLFET. The BTBT of electrons is considerably reduced in HGJLFETs as compared to JLFETs with a high-κ gate dielectric. Therefore, HGJLFETs not only exhibit a diminished BTBT but also offer an increased source-to-channel barrier height which hinders the triggering of parasitic BJT [55].

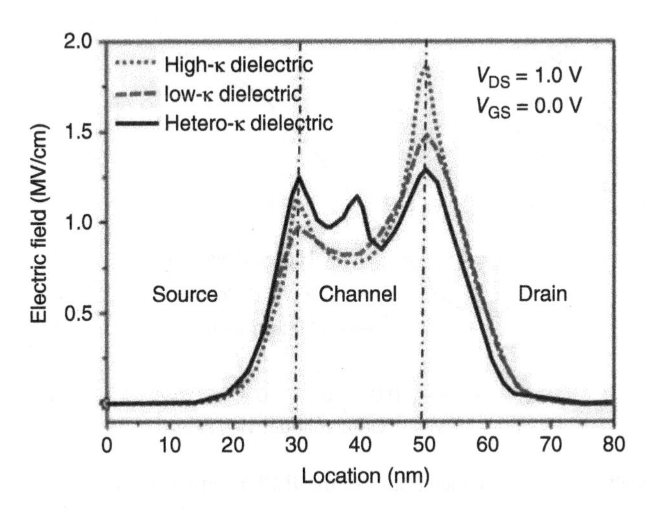

FIGURE 5.47 Electric field profile in the JLFETs with different gate dielectrics in the OFF-state ($V_{GS} = 0.0$ V, $V_{DS} = 1.0$ V) [55].

Also, the use of a heterogate dielectric leads to a gradual transition in the energy bands from the channel region to the drain region in HGJLFETs as compared to JLFETs with a high-κ or low-κ gate dielectric as shown in Fig. 5.48. This results in a larger tunneling width in the HGJLFET compared to even the JLFETs with low-κ gate dielectric. Therefore, BTBT-induced parasitic BJT action is significantly reduced in HGJLFET compared to JLFETs with both high-κ and low-κ gate dielectric.

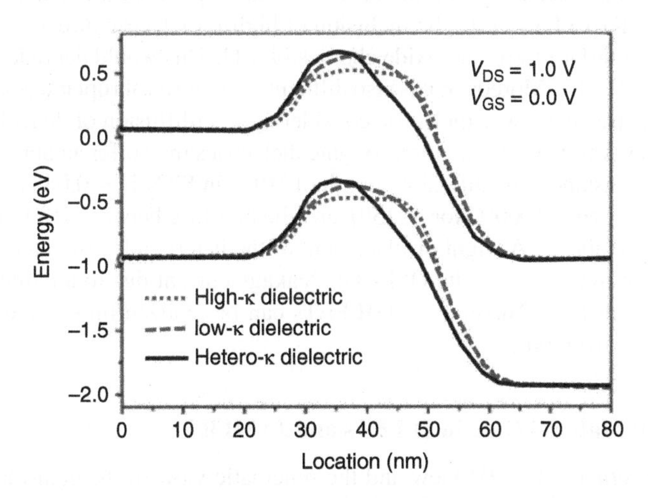

FIGURE 5.48 Energy band profiles of the JLFETs with different gate dielectrics in the OFF-state ($V_{GS} = 0.0$ V, $V_{DS} = 1.0$ V) [55].

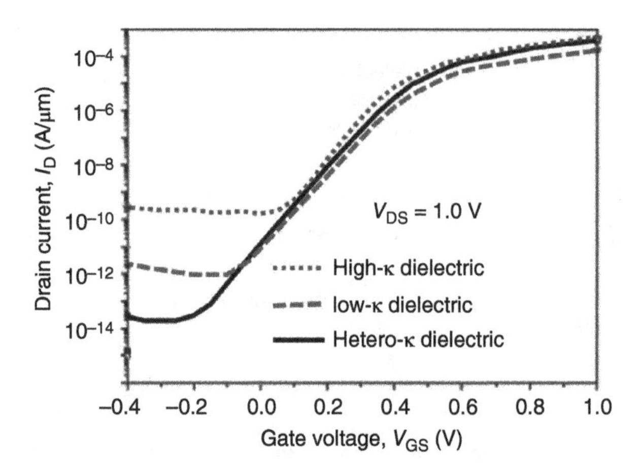

FIGURE 5.49 Transfer characteristics of the JLFET with different gate dielectrics [55].

5.8.4.C Transfer Characteristics The transfer characteristics of the HGJLFET are compared with JLFETs with only high-κ and low-κ gate dielectric of same gate oxide thickness in Fig. 5.49. The HGJLFETs exhibit a lower OFF-state leakage current by orders of magnitude as compared to the JLFETs with only high-κ or low-κ gate dielectric. Therefore, HGJLFETs perform significantly better as far as the GIDL is concerned.

5.8.4.D Challenges Fabricating the heterogate dielectric and alignment of the heterodielectric interface at the middle of the channel region are the major issues concerning the HGJLFETs [55]. The inclusion of high-κ dielectric directly over silicon may lead to interface traps and oxide charges [4, 51]. This would degrade the performance of HGJLFET. Moreover, it is also difficult to create an abrupt interface between high-κ dielectric and low-κ dielectric considering interdiffusion of dielectric materials [55]. However, few combinations of gate dielectrics may offer an abrupt interface [57, 58]. For instance, the diffusion length of HfO_2 in SiO_2 is $\sim$ 0.035 nm when the device is annealed at 1000°C for 5 s [59], and the interface between HfO_2 and SiO_2 is expected to be abrupt. A slight displacement of the heterodielectric interface toward the drain side may increase the OFF-state leakage current due to a reduction in the tunneling width [51]. Therefore, HGJLFETs can be realized only if the alignment challenges are overcome.

5.8.5 Dual Material Gate in JLFETs and JAMFETs

5.8.5.A Structure The 3D view and the schematic view of the dual-material gate (DMG) NWJLFETs is shown in Fig. 5.50. As discussed in Section 4.10, the gate electrode in the DMG JLFETs consists of two materials placed adjacent to each other. The length and work function of the gate material close to the source–channel interface is

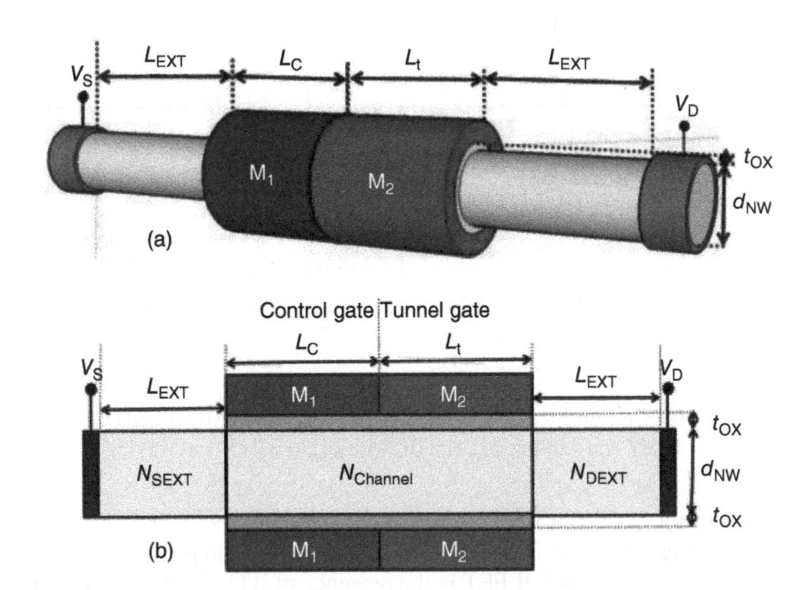

FIGURE 5.50 (a) Three-dimensional view and (b) the cross-sectional view of the DMG NWJLFET [13].

L_c and $\phi_{con,}$ respectively, while that of the material close to the channel–drain interface is L_t and ϕ_{tun}, respectively. The NWJLFETs with a single-material gate (SMG) on the other hand utilizes a single material for the gate electrode. Two configurations of NWJLFETs are considered: (a) the entire nanowire, i.e. source, channel, and drain, regions are uniformly doped to 10^{19} cm^{-3} and (b) the source and drain regions are heavily doped to 10^{20} cm^{-3} to reduce the source/drain series resistance while the channel doping is reduced to 10^{18} cm^{-3} to improve the mobility. As you may also conclude, the second configuration is NWJAMFET, which is pseudojunctionless as discussed in Section 4.1 [60–62].

5.8.5.B Transfer Characteristics As can be observed from Fig. 5.24(a) and discussed in Section 5.7.3, the OFF-state current exhibited by NWJAMFETs is significantly higher as compared to the NWJLFET. Although NWJAMFETs exhibit a high ON-state current as compared to the NWJLFETs which improves the dynamic performance, the large OFF-state current reduces its I_{ON}/I_{OFF} ratio significantly. If somehow the OFF-state current is reduced in NWJAMFETs without affecting the ON-state current, the utility of NWJAMFETs for digital applications would increase significantly.

As explained in Section 4.10, the application of DMG in NWJLFETs leads to an increased ON-state current owing to the abrupt potential step offered by the DMG architecture and a lower OFF-state current due to a reduction in the short-channel effects owing to the reduced electric field at the channel–drain interface [63, 64]. It would be interesting to analyze the efficacy of the DMG architecture in reducing the

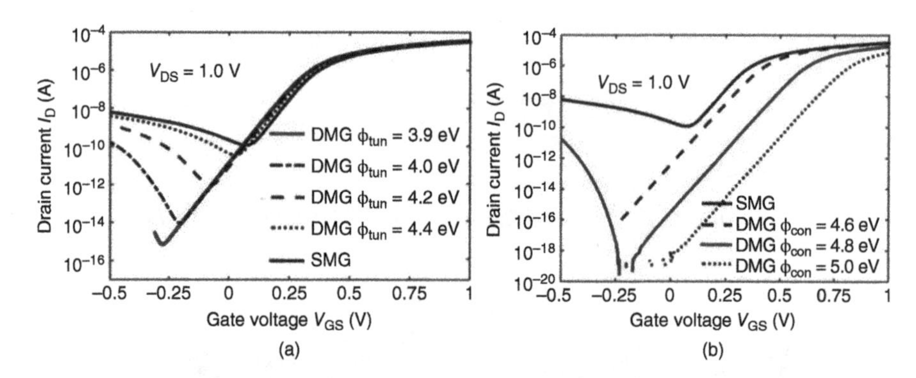

FIGURE 5.51 (a) The transfer characteristics of the SMG NWJAMFET and DMG NWJAMFET with $\phi_{con} = 4.5$ eV and varying ϕ_{tun} and (b) with $\phi_{tun} = 3.9$ eV and varying ϕ_{con}.

OFF-state current of NWJAMFETs. Moreover, the analysis in [63] (Section 4.10) was performed ignoring the BTBT. Therefore, it becomes imperative to also study the performance of DMG NWJLFET in the presence of BTBT-induced parasitic BJT action. In this section, we analyze the impact of the tunnel gate and the control gate on the transfer characteristics of DMG NWJAMFETs first and then compare the performance of the DMG and SMG NWJLFET and NWJAMFET considering parasitic BJT action [13].

To understand the impact of variation of the work function of the tunnel gate on DMG NWJAMFETs, let us observe the transfer characteristics of the SMG and DMG NWJAMFET ($L_c/L_g = 0.5$) with $\phi_{con} = 4.5$ eV and varying ϕ_{tun} is compared in Fig. 5.51(a). As you may observe, the drain current first reduces with decreasing gate voltage, reaches a minimum value, and then starts increasing because of GIDL. The L-BTBT GIDL is triggered at a gate voltage for which the minimum drain current is observed. Reducing ϕ_{tun} while keeping the ϕ_{con} constant decreases the value of the gate voltage at which the drain current is minimum. The onset of L-BTBT GIDL is delayed (shifted to more negative voltages) when a low ϕ_{tun} is used. Therefore, ϕ_{tun} dictates the tunneling phenomenon and hence we call it the tunnel gate [13].

5.8.5.C Impact of Length of Control Gate To understand the impact of variation of the work function of the control gate on DMG NWJAMFETs, Fig. 5.51(b) shows the transfer characteristics of the SMG and DMG NWJAMFET ($L_c/L_g = 0.5$) with $\phi_{tun} = 3.9$ eV and varying ϕ_{con}. As you may observe, the transfer characteristics shift toward right when the work function of the control gate is increased. The control gate only shifts the threshold voltage of the DMG NWJAMFET. Since ϕ_{con} controls the threshold voltage of the device, we call it the control gate.

5.8.5.D Operation A reduction in ϕ_{tun} leads to a lower peak electric field at the channel–drain interface in the DMG NWJAMFET as shown in Fig. 5.52(a). This leads to a lower band bending and a consequent larger tunneling width as shown in

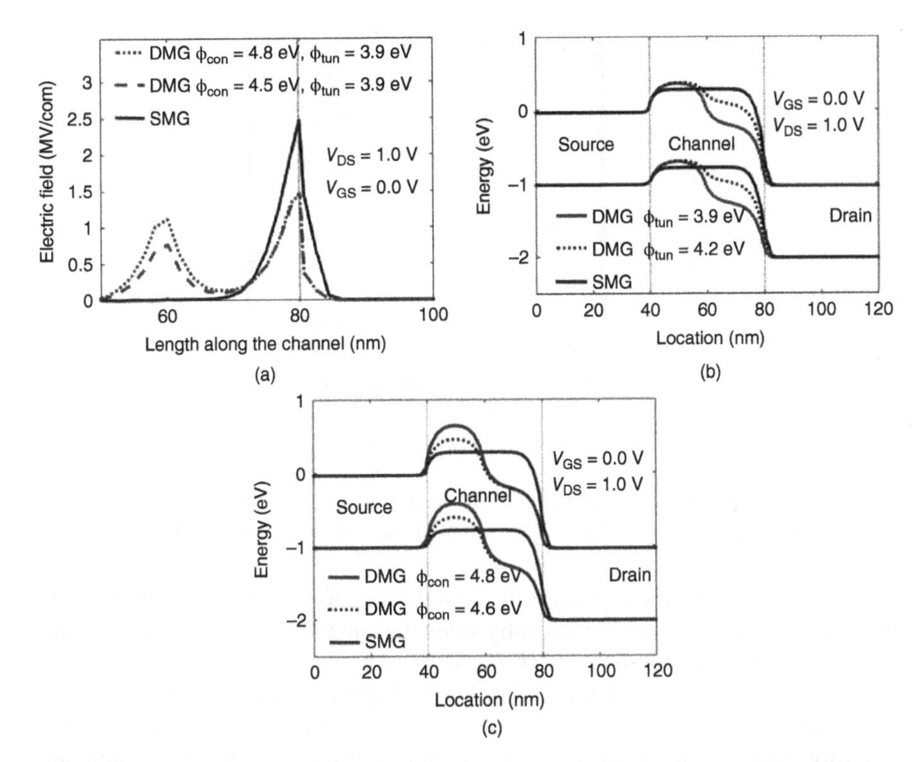

FIGURE 5.52 (a) The lateral electric field distribution of the SMG and DMG NWJAMFET and (b) the energy band profiles of the SMG NWJAMFET and DMG NWJAMFET with $\phi_{con} = 4.5$ eV and varying ϕ_{tun} and (c) with $\phi_{tun} = 3.9$ eV and varying ϕ_{con} taken at a cutline 1 nm below the Si–SiO_2 interface.

Fig. 5.52(b). Therefore, the L-BTBT GIDL is suppressed in the DMG NWJAMFET with a lower ϕ_{tun}. The source-to-channel barrier height is also increased in the DMG NWJAMFET with lower ϕ_{tun} leading to a diminished parasitic BJT action.

A higher ϕ_{con} leads to a shift in the threshold voltage of the DMG NWJAMFETs. Also, an increase in the ϕ_{con} leads to a higher electric field at the source–channel interface as shown in Fig. 5.52(c). This results in a larger source-to-channel barrier height, which suppresses the triggering of the parasitic BJT and lowers the short-channel effects. Similarly, even for the DMG NWJLFETs, the OFF-state current reduces as the work function difference between the control gate and the tunnel gate is increased as shown in Fig. 5.53. However, a larger shift in the threshold voltage, for instance, $\phi_{con} = 5.0$ eV leads to a reduction in the ON-state current. Therefore, the ϕ_{con} must be chosen between 4.6 and 4.8 eV for realizing optimum performance of the DMG NWJAMFETs.

5.8.5.E Impact of Length of Control Gate and Tunnel Gate Figure 5.54(a) compares the performance of the DMG NWJAMFET with different lengths of the control

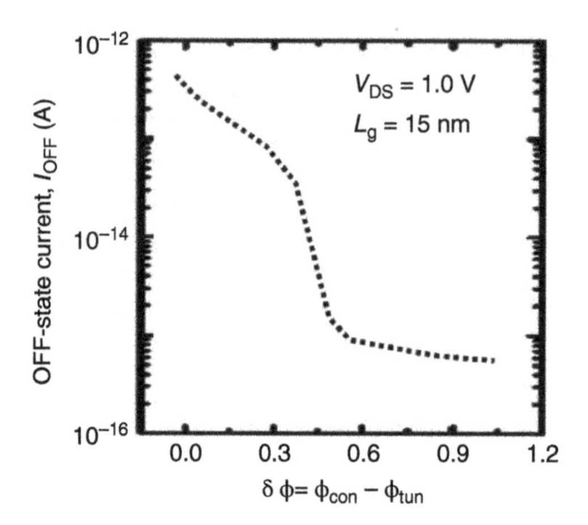

FIGURE 5.53 The impact of variation of the work function difference between the tunnel gate and the control gate on the OFF-state current of the DMG NWJLFET [64].

gate expressed as the ratio of the length of control gate to the gate length (L_c/L_g). The OFF-state current increases significantly when the length of the control gate is either too small ($L_c/L_g = 0.2$) or too large ($L_c/L_g = 0.8$) [13].

Since the control gate dictates the threshold voltage, a reduction in the L_c results in a lower effective channel length and a consequent higher short-channel effects leading to a lower source-to-drain barrier height as shown in Fig. 5.54(b). This results in a higher OFF-state current when the control gate is too small.

When the control gate length is too large compared to the gate length, the tunnel gate length reduces significantly. As a result, the efficacy of the tunnel gate in increasing the tunneling width reduces considerably and the OFF-state current

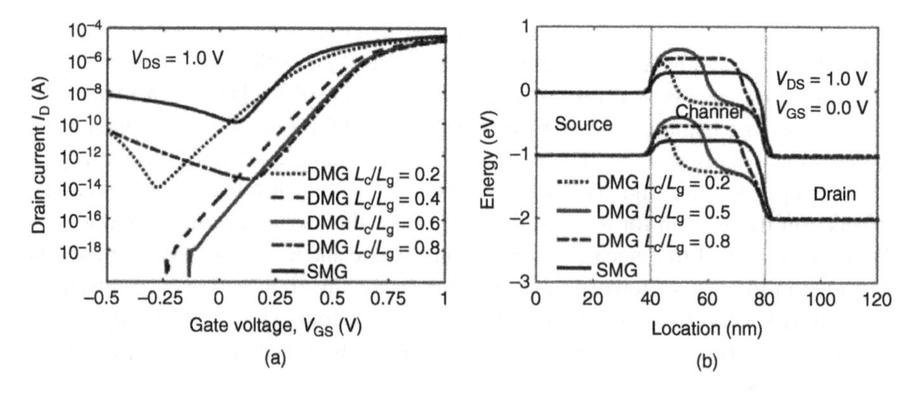

FIGURE 5.54 (a) The transfer characteristics and (b) the energy band profiles taken at a cutline 1 nm below the Si–SiO_2 interface of the SMG NWJAMFET and DMG NWJAMFET with varying L_c/L_g ratio.

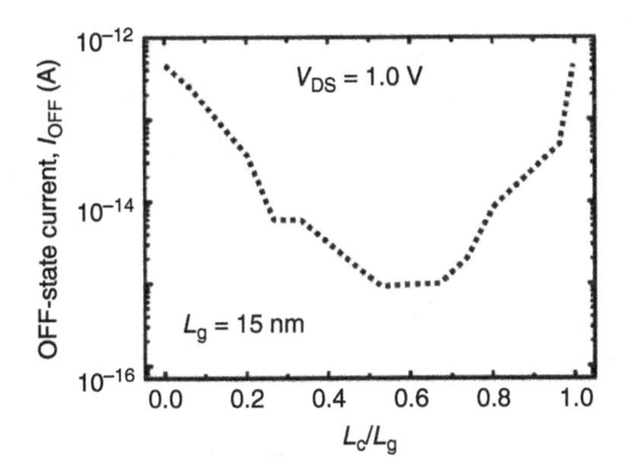

FIGURE 5.55 The impact of variation in the length of the control gate on the OFF-state current of the DMG NWJLFET [64].

increases due to L-BTBT. This corroborates with Fig. 5.54(b), which shows that the tunneling width reduces as the length of the control gate is increased. Therefore, the DMG NWJAMFET must be designed such that L_c/L_g is between 0.4 and 0.6 [13].

Similarly, the impact of L_c/L_g on the OFF-state current of DMG NWJLFETs is shown in Fig. 5.55. Even for NWJLFETs, due to the reasons mentioned earlier, the optimum performance and the lowest OFF-state current are observed when the L_c/L_g is ~0.5. As discussed in Section 4.10, the choice of $L_c/L_g = 0.5$ is suitable even from fabrication perspective.

5.8.5.F Comparison of SMG and DMG NWJLFET and NWJAMFET The transfer characteristics of the SMG and DMG NWJLFET and NWJAMFET are compared in Fig. 5.56(a). The DMG NWJAMFET performs significantly better than

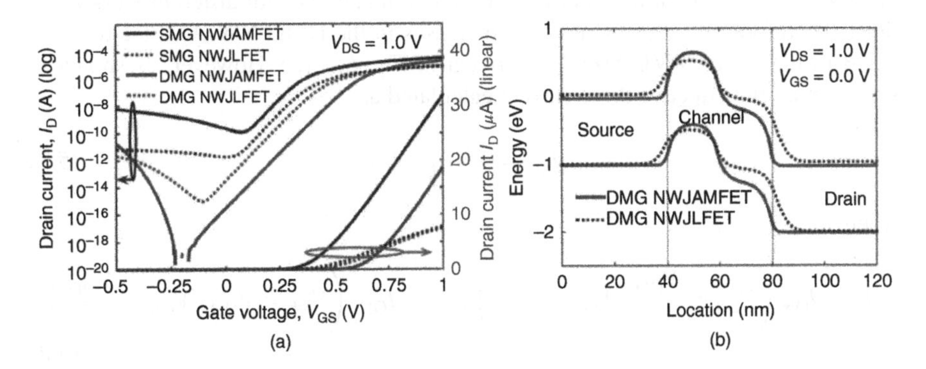

FIGURE 5.56 (a) The transfer characteristics and (b) the energy band profiles taken at a cutline 1 nm below the Si–SiO_2 interface of the SMG NWJAMFET, SMG NWJLFET, DMG NWJAMFET, and DMG NWJLFET ($\phi_{tun} = 4.2$ eV and $\phi_{con} = 4.7$ and $L_c/L_g = 0.5$).

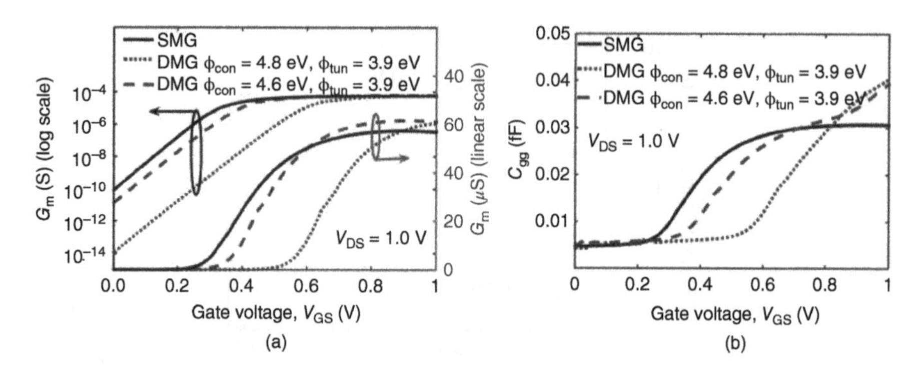

FIGURE 5.57 The variation of (a) the transconductance (G_m) and (b) the total gate capacitance (C_{gg}) with the gate voltage for SMG and DMG NWJAMFET.

SMG NWJAMFET and the SMG and DMG NWJLFET. The DMG NWJAMFET exhibits both a higher ON-state current improving the dynamic performance and a lower OFF-state current enhancing the static performance. This is attributed to a higher source-to-channel barrier height and a larger tunneling width in the DMG NWJAMFET as compared to the DMG NWJLFET as shown in Fig 5.56(b). Therefore, the DMG NWJAMFET could be a lucrative alternative to the NWJLFETs [13].

5.8.5.G Transconductance, Dynamic Performance, and Cutoff Frequency The transconductance (G_m) of the DMG NWJAMFET has been compared against the SMG NWJAMFET for different ϕ_{con} in Fig. 5.57(a). The DMG NWJAMFET exhibits a higher G_m in the saturation regime, which is beneficial for analog applications, and a lower G_m in the subthreshold regime, which facilitates a better turn-off characteristic for the DMG NWJAMFET [63].

As discussed in Section 4.10, the introduction of the DMG leads to an increased gate capacitance (C_{gg}) owing to the increased fringe capacitance arising due to the inherent electric field at the interface of the two materials with different work functions. Therefore, the gate capacitance increases in the DMG NWJAMFET as compared to the SMG NWJAMFET as shown in Fig. 5.57(b). However, as shown in Fig. 5.58(a), the effective drive current calculated as [65]

$$I_{eff} = 0.5(I_H + I_L) \tag{5.7}$$

where

$$I_H = I_{ON}\left(V_{DS} = \frac{V_{DD}}{2},\ V_{GS} = V_{DD}\right);\ I_L = I_{ON}\left(V_{DS} = V_{DD},\ V_{GS} = \frac{V_{DD}}{2}\right) \tag{5.8}$$

is larger for the DMG NWJAMFET as compared to the SMG NWJAMFET owing to the abrupt potential step offered by the DMG, which not only accelerates the electrons

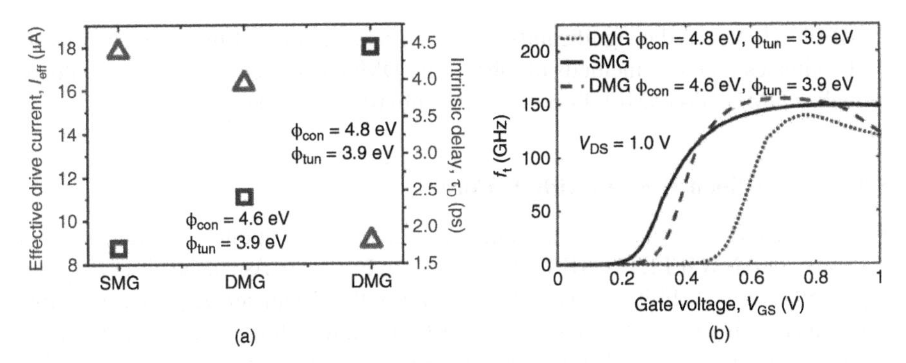

FIGURE 5.58 (a) The effective drive current (I_{eff}) and the intrinsic delay (τ_D) and (b) the cutoff frequency (f_t) of the SMG and the DMG NWJAMFET.

but also increases the electron velocity as discussed in Section 4.10. However, the increment in the effective drive current in the DMG NWJAMFET is lower than the increase in the gate capacitance. The intrinsic delay (τ_D) of the inverter calculated using the effective drive current method [65] given by

$$\tau_D = (C_{gg} \times V_{DD})I_{eff}^{-1} \tag{5.9}$$

increases for the DMG NWJAMFET as compared to the SMG NWJAMFET (Fig. 5.58(a)). Also, there is a trade-off between the dynamic and static performance of the DMG NWJAMFET based on the ϕ_{con}. A higher ϕ_{con} leads to a lower OFF-state current as shown in Fig. 5.58(a). However, a higher ϕ_{con} also leads to a larger delay and a degraded dynamic performance. Therefore, ϕ_{con} must be appropriately chosen [13].

The cutoff frequency (f_t) [13] defined as

$$f_t = \frac{G_m}{2\pi C_{gg}} \tag{5.10}$$

is higher for the DMG NWJAMFETs in the subthreshold regime owing to the larger G_m (Fig. 5.58(b)). However, the cutoff frequency reduces at higher gate voltages owing to the increased gate capacitance, which masks the improvement brought by the increased G_m.

5.8.5.J Challenges Although DMG JAMFET exhibits the best performance among SMG or DMG NWJLFET and NWJAMFET, the fabrication of DMG in the sub-20 nm regime is quite difficult as discussed in Section 4.10. Achieving distinct gate materials within a span of 20 nm itself is a technological challenge. Therefore, the fabrication of DMG in the sub-10 nm regime is the major hurdle for scaling of DMG

JAMFET. However, the significantly improved performance of the DMG NWJAMFET indicates that new methods of fabricating DMG architectures must be explored to realize DMG NWJAMFETs even for the sub-10 nm regime.

5.8.6 Core–Shell JLFETs with P⁺ Core

5.8.6.A Structure The cross-sectional view of the conventional n-NWJLFET and the core–shell NWJLFET (CSJLFET) is shown in Fig. 5.59. As can be observed from Fig. 5.59, the CSJLFET consists of a p^+ core with a diameter d_{core} and n^+ active shell layer with a thickness t_{shell}. The CSJLFET requires formation of an abrupt p–n junction at the core–shell interface which may be realized experimentally using the recently reported in-situ doping approach or molecular monolayer doping and microwave annealing technique [46–50].

It may also be noted that in the CSJLFET, the carriers flow in the shell layer while the core is responsible for improving electrostatic integrity and reducing L-BTBT. Therefore, the active device layer thickness (twice shell thickness) is taken same as the thickness (nanowire diameter) of the conventional n-NWJLFET to have a fair comparison between the two [14]. However, the physical thickness of the CSJLFET

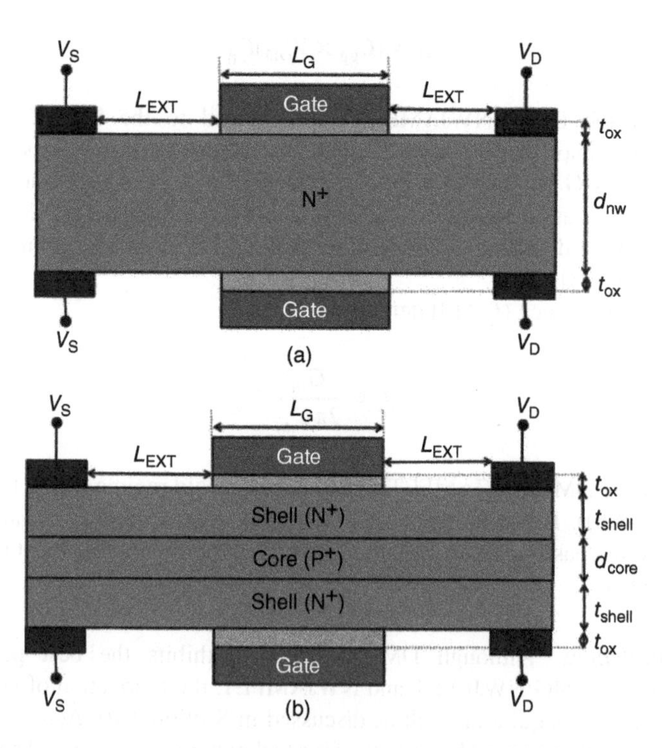

FIGURE 5.59 Cross-sectional view of the (a) conventional and (b) core–shell nanowire JLFET [14].

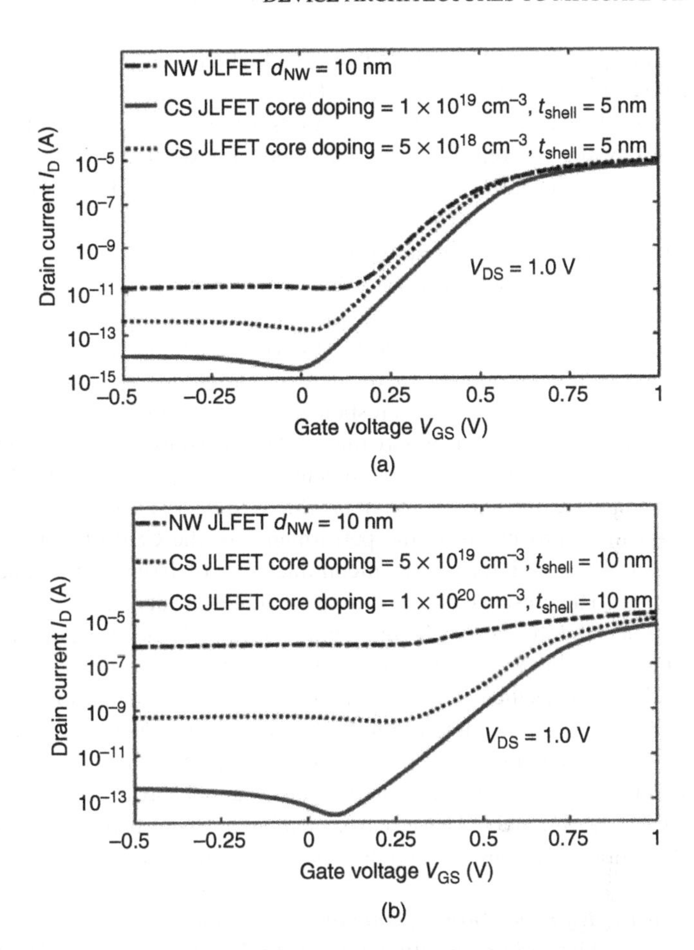

FIGURE 5.60 (a) The transfer characteristics of the conventional NWJLFET (d_{NW} = 10 nm) and the CSJLFET (d_{NW} = 15 nm) with different core dopings [active silicon layer thickness = 10 nm] and (b) conventional NWJLFET (d_{NW} = 20 nm) and the CSJLFET (d_{NW} = 25 nm) with different core dopings [active silicon layer thickness = 20 nm.

exceeds the conventional NWJLFET by the core diameter. As you may observe, this does not restrict the scaling of the CSJLFETs since these are essentially lateral nanowires which may even be stacked to extract higher drive current.

5.8.6.B Transfer Characteristics The transfer characteristics of the conventional NWJLFETs and the CSJLFETs have been compared in Fig. 5.60(a) for different active layer thicknesses (nanowire diameter in case of NWJLFETs and twice the shell thickness in case of CSJLFETs). For an active silicon layer thickness of 10 nm, the CSJLFET exhibits a significantly reduced OFF-state current by four orders of magnitude rendering a considerably high ON-state to OFF-state current ratio of 10^{10}. Also, the OFF-state current increases as the doping of the p^+ core is reduced.

Although the conventional NWJLFET offers a high ON-state current by ~2 times when the active silicon thickness is increased to 20 nm, it exhibits an extremely poor I_{ON}/I_{OFF} of ~10. However, the CSJLFET exhibits a significantly low OFF-state current, a higher ON-state current by ~1.5 times, and a consequent high I_{ON}/I_{OFF} ratio of ~ 10^9 even when the active silicon layer thickness is 20 nm (Fig. 5.60(b)). Also, the OFF-state leakage current increases as the p^+ core doping reduces. As you may observe, a higher core doping is required for achieving low OFF-state current as the active silicon thickness is increased.

5.8.6.C Mitigating Parasitic BJT Action As discussed in Section 5.4, the leakage current depends significantly on the active silicon film thickness [15]. While the L-BTBT induced parasitic BJT action is responsible for the OFF-state leakage current when silicon film thickness (or nanowire diameter) is ≤ 10 nm, the hole accumulation induced shielding effect results in an inefficient volume depletion when silicon film thickness (or nanowire diameter) ≥ 15 nm and leads to a large leakage current [15].

The significant improvement in the performance of the CSJLFET as compared to conventional NWJLFET for active silicon thickness of 10 nm can be understood from Fig. 5.61(a). The presence of the p^+ core facilitates the depletion of the n^+ drain extension region in the shell at the channel–drain interface. This results in a larger tunneling width observed in Fig. 5.61(a) as compared to the conventional NWJLFETs. The p^+ core also depletes the source extension region adjacent to the source–channel interface in the shell region leading to a higher source-to-channel barrier height in the CSJLFET as compared to the NWJLFET as shown in Fig. 5.61(a). A larger tunneling width reduces the L-BTBT and a higher source-to-channel barrier height inhibits the triggering of the parasitic BJT. These reasons lead to a diminished L-BTBT induced parasitic BJT action and a consequent reduction in the OFF-state leakage current [14].

5.8.6.D Realizing Efficient Volume Depletion As discussed in Section 5.4, volume depletion is not achieved in conventional NWJLFETs when the nanowire diameter is more than 15 nm due to the hole accumulation induced shielding effect and an electron channel exists at the center of the channel region which conducts leakage current [15] as shown in Fig. 5.62. However, the p^+ core facilitates the depletion of the active shell region and results in an effective shell thickness which is less than the physical shell thickness. Consequently, the gate electrode has to deplete a lower effective active shell thickness to achieve volume depletion. Therefore, the electrostatic integrity of the NWJLFETs is also improved due to the presence of p^+ core. In addition, the depletion of the drain region at the channel–drain interface leads to a larger tunneling width as shown in Fig. 5.61(b). A lower L-BTBT due to a larger tunneling width leads to a reduced hole accumulation, which diminishes the shielding effect. These reasons help in achieving efficient volume depletion and reducing the leakage current significantly in the CSJLFET even when the active silicon film thickness is 20 nm [14].

The presence of the p^+ core (a) leads to a reduced L-BTBT-induced parasitic BJT action in NWJLFETs with the nanowire diameter ≤ 10 nm and (b) helps in

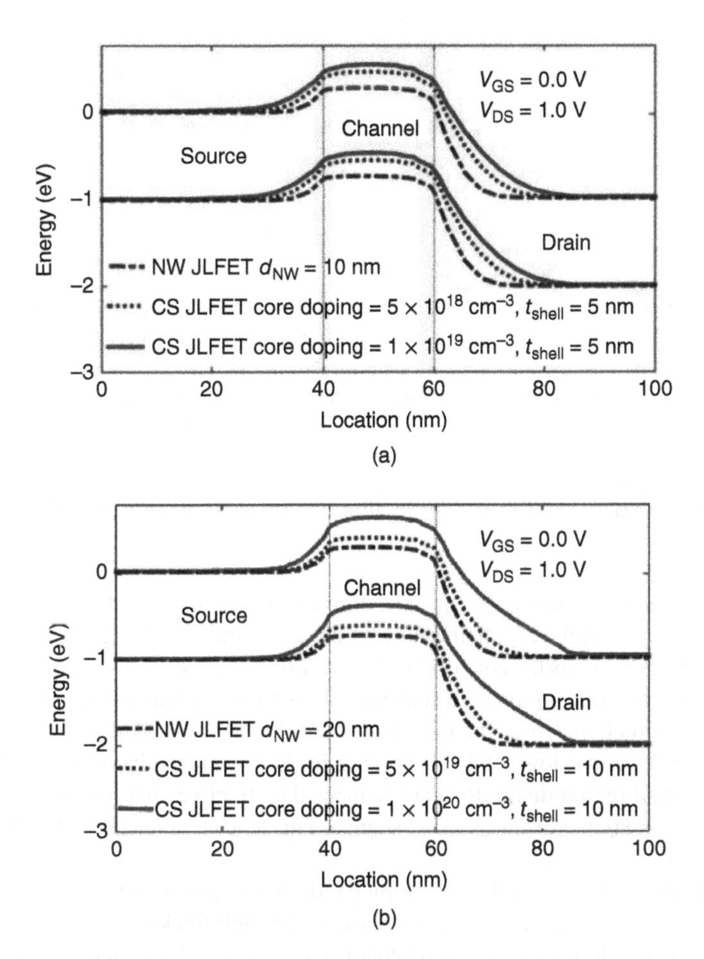

FIGURE 5.61 (a) The energy-band profiles in the OFF-state at a lateral cutline taken 1 nm below Si–SiO_2 interface of the conventional NWJLFET ($d_{NW} = 10$ nm) and the CSJLFET ($d_{NW} = 15$ nm) and (b) of the conventional NWJLFET ($d_{NW} = 20$ nm) and the CSJLFET ($d_{NW} = 25$ nm).

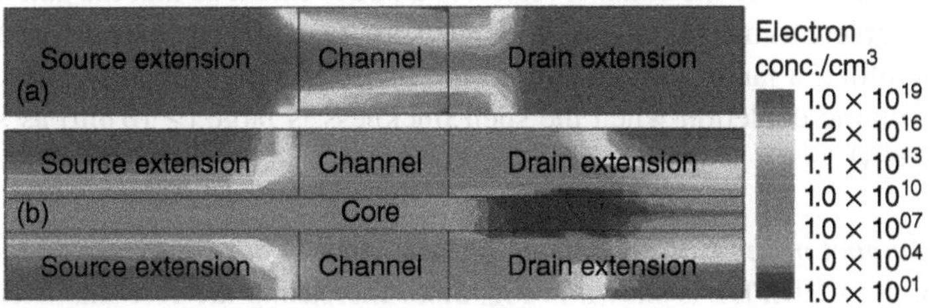

FIGURE 5.62 The electron concentration contour plot of (a) the conventional NWJLFET ($d_{NW} = 20$ nm) and (b) CSJLFET ($d_{NW} = 25$ nm, $t_{shell} = 10$ nm) with a core doping of $N_A = 1 \times 10^{20}$ cm^{-3} at $V_{GS} = 0.0$ V and $V_{DS} = 1.0$ V.

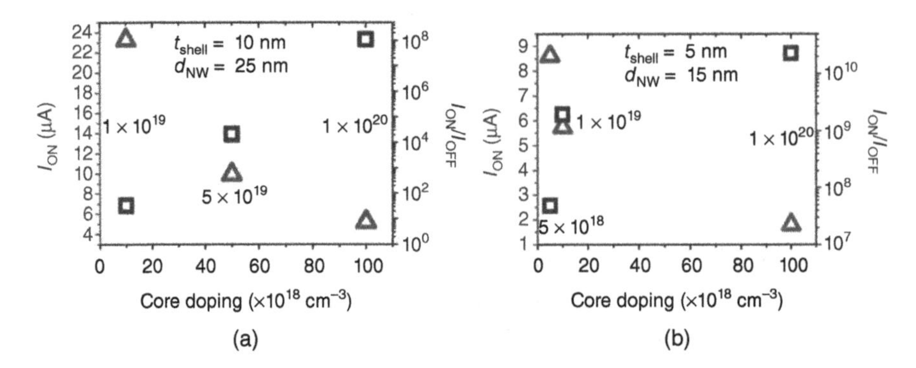

FIGURE 5.63 The impact of variation in the core doping on I_{ON} and I_{ON}/I_{OFF} of the CSJLFET with (a) $d_{NW} = 15$ nm, $t_{shell} = 5$ nm and (b) $d_{NW} = 25$ nm, $t_{shell} = 10$ nm.

achieving efficient volume depletion in NWJLFETs with the nanowire diameter ≥ 10 nm [14].

5.8.6.E Impact of Core Doping The impact of core doping on the performance of CSJLFET with different shell thickness has been analyzed in Fig. 5.63. The OFF-state current reduces as the core doping is increased. This is attributed to the increased depletion of the drain region at the channel–drain interface owing to the highly doped core region which increases the tunneling width further [14]. This results in a diminished L-BTBT and a lower OFF-state current. However, the increased depletion of the entire shell region due to the core reduces the effective area for current flow and leads to a large source/drain series resistance. This leads to a smaller ON-state current observed in Fig. 5.60(a).

It may also be noted that the core doping required to achieve optimum performance of the CSJLFET increases with an increase in the shell thickness (Fig. 5.63). This is attributed to the different leakage mechanisms when the shell thickness is increased from 10 to 20 nm [15]. For a shell thickness of 10 nm, since the core has to deplete only the drain region at the channel–drain interface to increase the tunneling width, a core doping of 5×10^{18} cm^{-3} is sufficient. However, for a shell thickness of 20 nm, the core has to deplete the channel region to a larger extent so that the gate electrode is able to deplete a channel thickness which is lower than the physical channel thickness. Therefore, a larger core doping in the range of 1×10^{20} cm^{-3} is required to achieve volume depletion when the shell thickness is large (≥ 10 nm) as shown in Fig. 5.64 [14].

5.8.6.F Impact of the Core Diameter The impact of core diameter on the performance of the CSJLFET is shown in Fig. 5.65. Increasing the core diameter without increasing the diameter of the CSJLFET leads to a reduction in the effective active shell thickness. This improves the electrostatic integrity of the CSJLFET since the

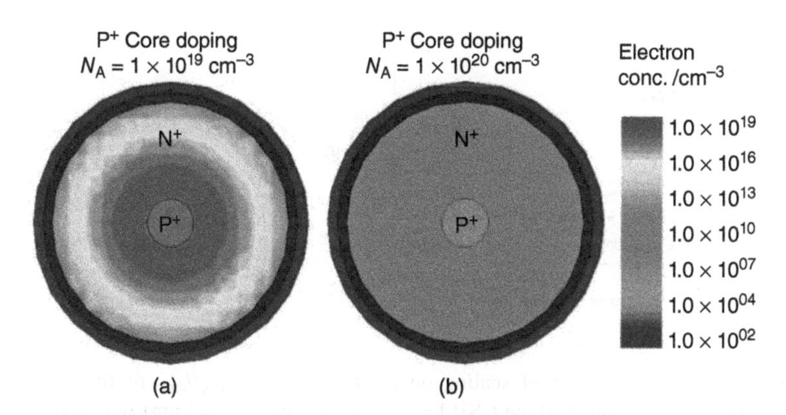

FIGURE 5.64 The electron concentration contour plot of the cross section of CSJLFET ($d_{NW} = 25$ nm, $t_{shell} = 10$ nm) at the center of the channel region for a core doping of (a) $N_A = 1 \times 10^{19}$ cm^{-3} and (b) $N_A = 1 \times 10^{20}$ cm^{-3} at $V_{GS} = 0.0$ V and $V_{DS} = 1.0$ V.

gate has to deplete a lower effective thickness [4]. A reduction in the active shell thickness leads to a lower DIBL and consequent DIBT and the L-BTBT-induced parasitic BJT action as explained in Section 5.4.1. Therefore, an increase in the core diameter and the consequent reduction in the active shell thickness lead to a reduced OFF-state current [14].

However, a reduction in the effective active layer thickness translates into an increased source/drain series resistance and a lower area for the current to flow. Consequently, the ON-state current reduces as the core diameter increases. However, the significant reduction in the OFF-state current leads to the large I_{ON}/I_{OFF} ratio observed for the CSJLFET when the core diameter is increased.

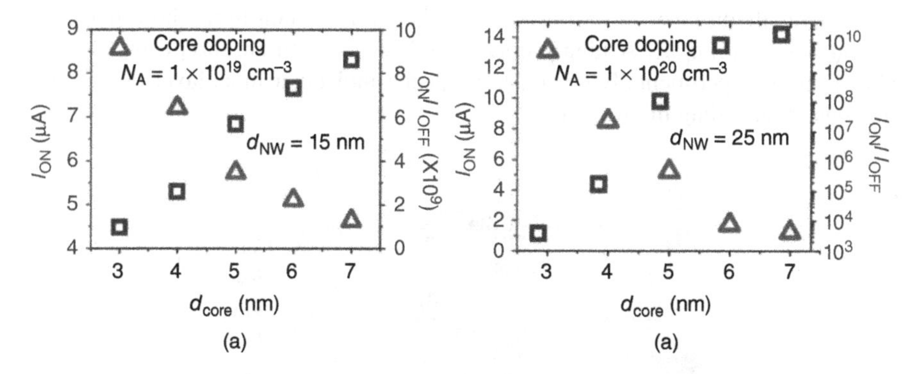

FIGURE 5.65 The impact of variation in the core diameter on I_{ON} and I_{ON}/I_{OFF} of the CSJLFET with (a) $d_{NW} = 15$ nm, $t_{shell} = 5$ nm and (b) $d_{NW} = 25$ nm, $t_{shell} = 10$ nm.

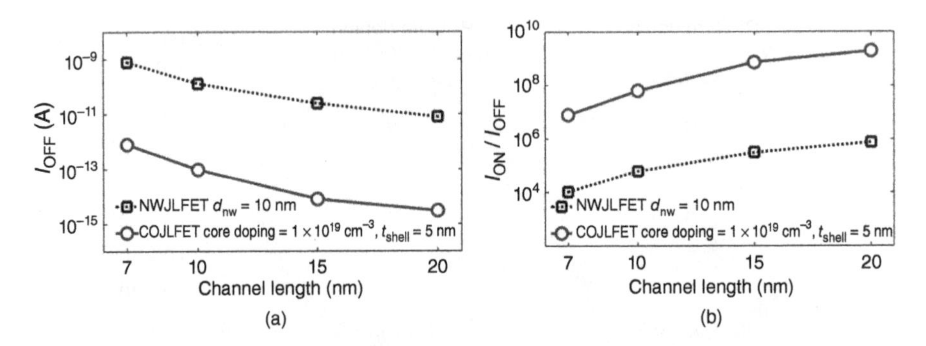

FIGURE 5.66 The impact of scaling on (a) I_{OFF} and (b) I_{ON}/I_{OFF} of the conventional NWJLFET (d_{NW} = 10 nm) and the CSJLFET (d_{NW} = 15 nm, t_{shell} = 5 nm) with a core doping of $N_A = 1 \times 10^{19}$ cm^{-3}.

5.8.6.G Impact of Scaling The impact of scaling the gate length on the performance of the NWJLFETs and CSJLFETs is shown in Fig. 5.66. The OFF-state current increases significantly for both the devices with channel length scaling due to the increased gain of the parasitic BJT action owing to the reduced base width (channel length). However, the CSJLFET performs significantly better than the conventional NWJLFET and exhibits a considerably high I_{ON}/I_{OFF} of ~10^7 even for a channel length of 7 nm. Therefore, the presence of the p^+ core facilitates the scaling of the conventional NWJLFETs to the sub-10 nm regime and provides an inherent scaling advantage.

5.8.7 CSJLFETs with Intrinsic Core

5.8.7.A Structure The structure of the CSJLFET with an intrinsic core is shown in Fig. 5.67. It consists of an undoped core and a heavily doped shell region. The active silicon layer in this structure is the entire silicon film whose thickness is $d_{NW} = 2 \bullet t_{shell} + d_{core}$. A highly doped shell region with an abrupt junction has already been experimentally demonstrated using molecular monolayer doping and microwave annealing in [46–50].

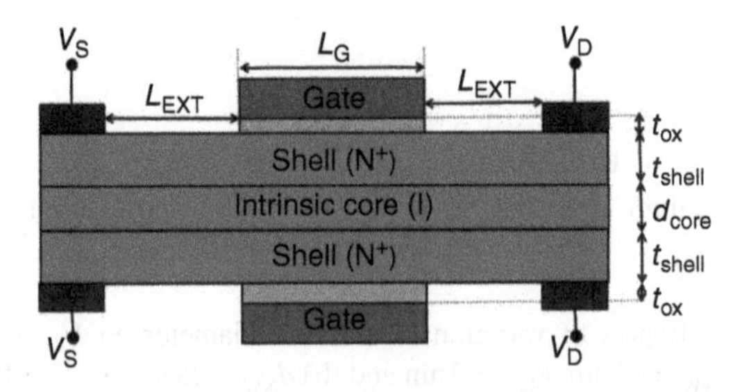

FIGURE 5.67 Schematic view of the CSJLFET with an intrinsic core region.

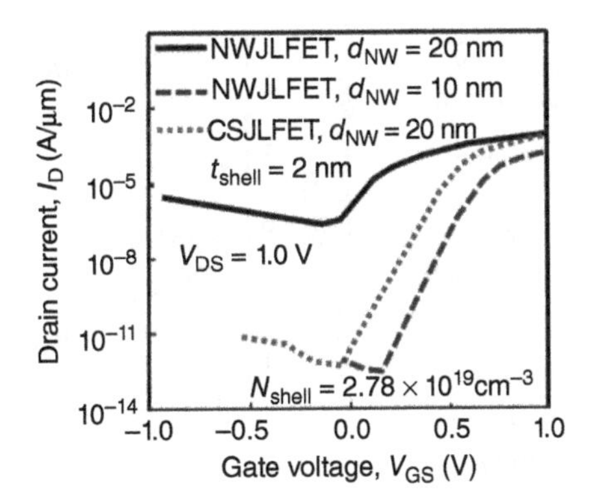

FIGURE 5.68 Transfer characteristics of the NWJLFET for different nanowire diameters and the CSJLFET with an intrinsic core [66].

5.8.7.B Transfer Characteristics The transfer characteristics of the CSJLFET with a shell thickness of 2 nm have been compared against the NWJLFET with a diameter of 10 and 20 nm in Fig. 5.68. To have a fair comparison, the total number of dopants has been kept same in the CSJLFET as in NWJLFET with a nanowire diameter (d_{NW}) = 20 nm by adjusting the doping of the shell region as 2.78×10^{19} cm^{-3}. The NWJLFET with d_{NW} = 20 nm exhibits a high OFF-state current due to the insufficient volume depletion as discussed in Section 5.4. However, the CSJLFET with an intrinsic core with a d_{NW} = 20 nm exhibits a significantly reduced leakage current by ~5 orders of magnitude leading to considerably high I_{ON}/I_{OFF}. Although the OFF-state leakage current of the CSJLFET with an intrinsic core is comparable to the NWJLFET with d_{NW} = 10 nm, it exhibits a higher ON-state current. Therefore, the CSJLFET with an intrinsic core leads to both a higher ON-state current and a significantly low OFF-state current.

5.8.7.C Operation The CSJLFET offers a higher electrostatic integrity as compared to the NWJLFETs owing to the restricted volume (shell region) of the active dopant concentration and their proximity to the gate electrode which facilitates efficient depletion of the shell region even when the doping concentration is high [66]. As a result, the source/drain series resistance can also be improved and better ohmic contacts may be realized with the source/drain electrodes in CSJLFETs when a large shell doping is utilized while achieving efficient volume depletion.

The L-BTBT occurs only in the shell region at the channel–drain interface in a CSJLFET. The lower volume of the shell provides a lesser area for the occurrence of BTBT. Therefore, the hole concentration in reduced in the CSJLFET with an intrinsic core. This diminishes the BTBT-induced parasitic BJT action.

In addition, due to the symmetric nature of the CSJLFET, there is a potential mirroring effect, i.e. with respect to the center of the nanowire, the potential profiles are symmetrical [66]. This effect results in a negligible potential barrier for the electrons

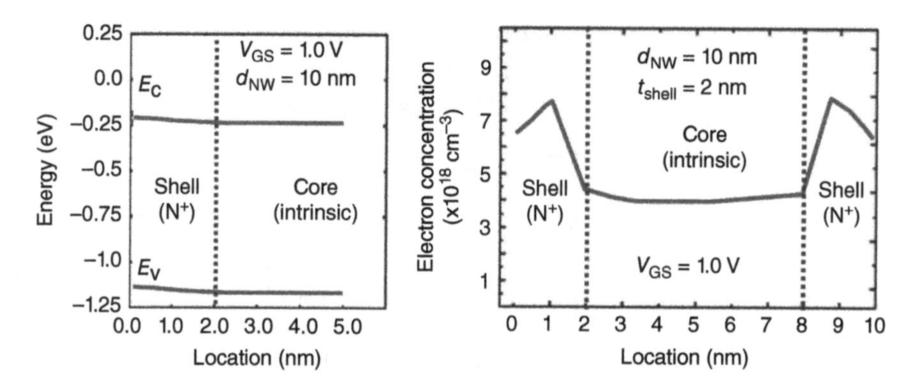

FIGURE 5.69 (a) Energy band profiles of the CSJLFET with an intrinsic core taken at a cutline 1 nm below the Si–SiO_2 interface and (b) electron concentration contour plot of the CSJLFET with an intrinsic core in the ON-state [66].

in the shell region to flow to the intrinsic core as shown in Fig. 5.69(a). The electron concentration plot shown in Fig. 5.69(b) clearly indicates that the electrons also flow into the intrinsic core region although it is initially undoped. The transport of the electrons in the intrinsic region, which exhibits a higher mobility due to a reduction in the impurity scattering mechanism, enhances the ON-state current. A lower source/drain series resistance also contributes to the higher ON-state current.

5.8.7.D Impact of Shell Thickness The transfer characteristics of the CSJLFET with an intrinsic core for different shell thicknesses are shown in Fig. 5.70. The OFF-state current decreases with a reduction in the shell thickness due to an increased electrostatic integrity resulting in an improved gate control of the channel region reducing the short-channel effects. Also, the effective area over which the tunneling takes place is reduced. These factors lead to a significant reduction in the OFF-state current in CSJLFETs with a reduced shell thickness [66]. The subthreshold swing also improves as the shell thickness is reduced.

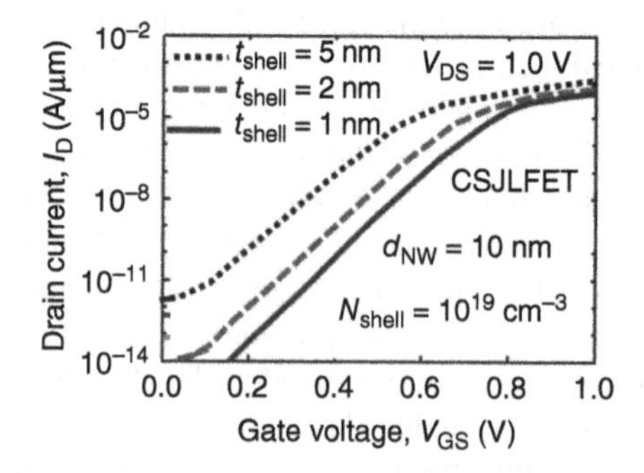

FIGURE 5.70 Transfer characteristics of the CSJLFET with an intrinsic core for different shell thickness [66].

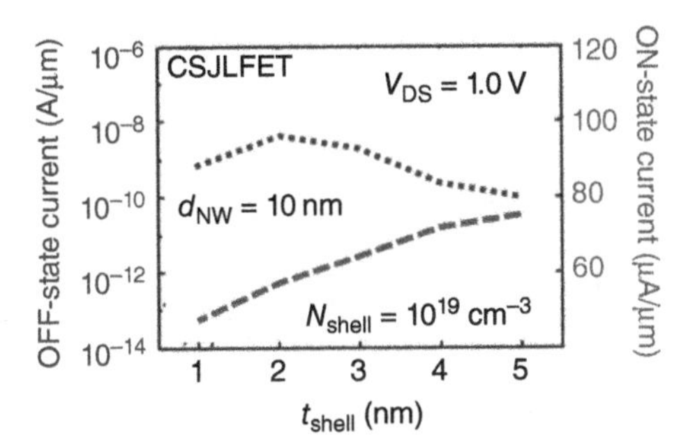

FIGURE 5.71 The variation in the ON-state current and the OFF-state current with the shell thickness [66].

Also, as can be observed from Fig. 5.71, the ON-state current also improves as the shell thickness is reduced. This may seem contradicting from source/drain series resistance perspective. You may wonder that a reduction in the shell thickness in the source/drain region should lead to an increased source/drain series resistance. You are indeed correct, and the increase in the source/drain series resistance is inevitable when the shell thickness is reduced. However, a lower shell thickness also means a larger intrinsic core area where the majority of electrons flow with an enhanced mobility [66]. Also, the improved subthreshold swing with reduced shell thickness results in a larger current for the same gate overdrive voltage. The increase in the ON-state current is attributed to these factors.

However, the increased source/drain series resistance opposes the improvement in the ON-state current and its impact becomes severe when the shell thickness becomes less than 2 nm. The reduction in the ON-state current for a shell thickness of 1 nm is attributed to the increased source/drain series resistance [66].

5.8.7.E Comparison with Conventional NWJLFET and NWMOSFET The performance of the NWJLFET and NWMOSFET has been compared against the CSJLFET with an intrinsic core in Fig. 5.72. The source/drain shell regions in the CSJLFETs and extension regions in NWJLFET and NWMOSFET have been additionally doped to 1×10^{20} cm^{-3} in all the three structures to reduce the source/drain series resistance. The CSJLFET exhibits the lowest OFF-state current and the highest I_{ON}/I_{OFF} ratio. This is attributed to the lower area over which tunneling occurs in the CSJLFET (shell region) as compared to the NWJLFET and NWMOSFET [66].

5.8.7.F Challenges Although CSJLFET performs significantly better than the conventional NWJLFETs even for a channel length of 7 nm [14], the fabrication of CSJLFET is a challenge. Creation of an abrupt junction at the core–shell interface using the industry standard ion implantation process is not feasible, and advanced techniques like molecular monolayer doping and microwave annealing are required. Since CSJLFETs are essentially grown as nanowires, they are expected to suffer from

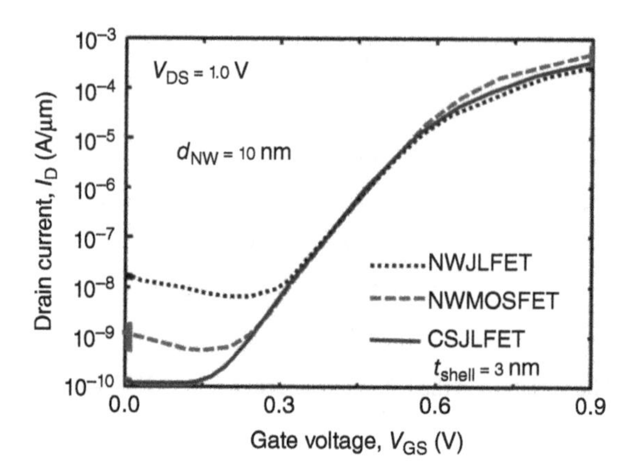

FIGURE 5.72 Transfer characteristics of the NWMOSFET, NWJLFET, and the CSJLFET with an intrinsic core [66].

the same technological challenges faced while fabricating nanowires such as lack of control over uniformity and diameter of nanowire which are discussed in Section 2.6.4. Also, the lack of a top-bottom approach for fabricating nanowires restricts the mass production of CSJLFETs.

5.8.8 Extended Back Gate Double-Gate JLFET

5.8.8.A Structure The 3D view of the conventional double-gate junctionless field-effect transistor (DGJLFET) and the DGJLFET with an extended back gate (EBG DGJLFET) is shown in Fig. 5.73. The EBG DGJLFET consists of a back gate, which spans from the source end to the drain end as compared to the DGJLFET in which the

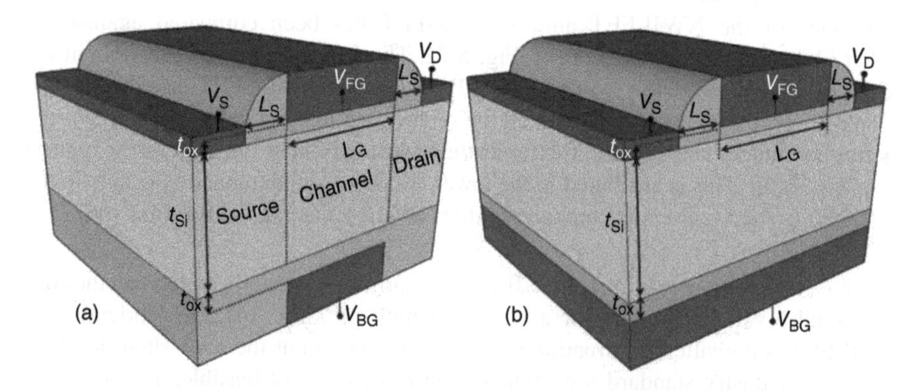

FIGURE 5.73 Schematic view of (a) the conventional DGJLFET and (b) the extended back gate DGJLFET [18].

back gate is only below the channel region. Such an extension in the back gate has been experimentally realized in [67, 68]. An accurate shift in the positioning of the back gate with respect to the front gate in steps of 15 nm using electrical Vernier and E-beam lithography for gate length of 40 nm has been experimentally demonstrated in [68]. Therefore, in addition to the back gate which extends from the source-to-the drain end, a back gate which is underlapped from the source end (back gate extends below the channel and drain regions only) may be experimentally realized [18].

It may also be noted that an EBG is present in the SOI JLFETs with an active gate bias [69–72]. However, the large buried oxide (BOX) thickness (typically 10 nm) restricts the back gate from influencing the device characteristics. Recently, a silicon-on-thin BOX (SOTB) architecture with a BOX thickness of 5 nm was experimentally demonstrated [70]. Replacing the SiO_2 BOX with HfO_2 or any other high-κ dielectric using the process flow described in [4], an effective BOX thickness of ~1 nm may be realized yielding the proposed EBG DGJLFET.

5.8.8.B Transfer Characteristics The transfer characteristics of the DGJLFET, EBG DGJLFET, and DGJAMFET are shown in Fig. 5.74(a). The DGJAMFET exhibits a considerably high OFF-state current and a consequent poor ON-state to OFF-state current ratio (I_{ON}/I_{OFF}) due to the enlarged BTBT-induced parasitic BJT action in JAMFETs as compared -to the JLFETs as discussed in Section 5.7. Although the OFF-state current is reduced in the DGJLFET as compared to the DGJAMFET due to a reduced BTBT-induced parasitic BJT action, it still exhibits a low I_{ON}/I_{OFF}. However, the EBG DGJLFET exhibits a significantly low OFF-state current and a considerably high $I_{ON}/I_{OFF}$of ~10^8. Moreover, the subthreshold swing of the EBG DGJLFET is also improved (~62 mV/decade) as compared to the DGJLFET (~100 mV/decade).

The ON-state current of the EBG DGJLFET is somewhat lower than the DGJLFET. The DGJAMFET exhibits the highest ON-state current owing to the low

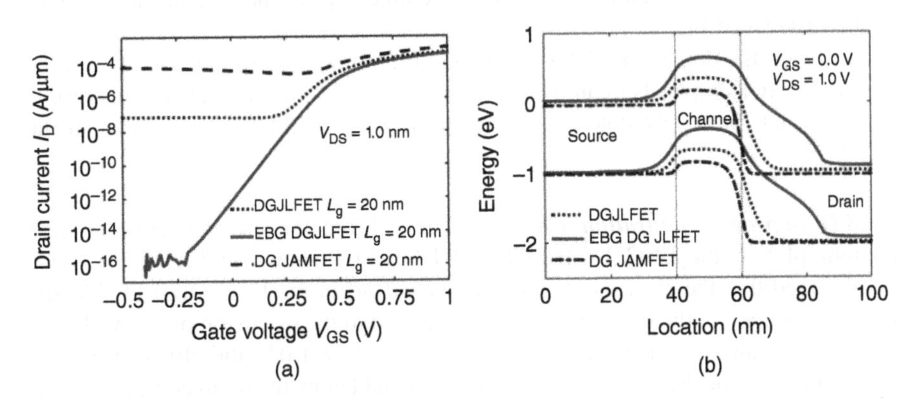

FIGURE 5.74 (a) Transfer characteristics and (b) energy-band profiles taken at a cutline 1 nm below the top Si–SiO_2 interface of the conventional DGJAMFET, DGJLFET, and the EBG DGJLFET (L_g = 20 nm).

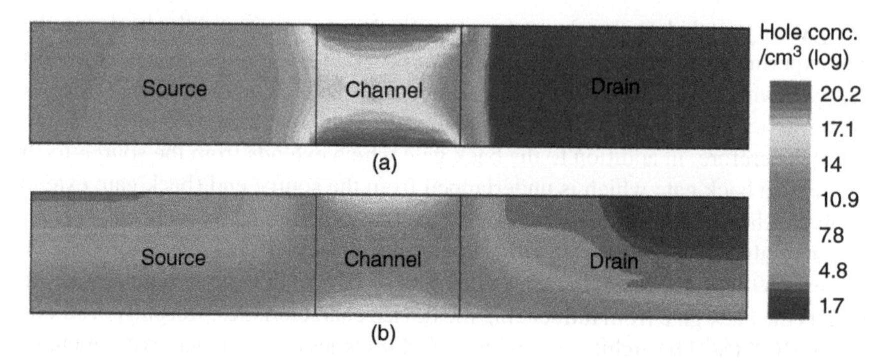

FIGURE 5.75 Hole concentration contour plot of (a) the conventional DGJLFET and (b) the EBG DGJLFET in the OFF-state ($V_{GS} = 0.0$ V and $V_{DS} = 1.0$ V).

source/drain series resistance and higher mobility due to the lightly doped channel region [18].

5.8.8.C Operation The significant reduction in the OFF-state current of the EBG DGJLFET can be understood from Figs. 5.74(b) and 5.75. The extension of the back gate below the drain region facilitates the depletion of the drain region at the channel–drain interface increasing the tunneling width as shown in Fig. 5.74(b). An increase in the tunneling width reduces the BTBT significantly [18]. The low hole concentration in the channel region of the EBG DGJLFET (Fig. 5.75) is attributed to the reduced BTBT. The extension of the back gate below the source region also facilitates the depletion of the source region adjacent to the source–channel interface. This leads to a higher source-to-channel barrier height in the EBG DGJLFET as compared to the DGJLFET. A higher source-to-channel barrier height does not allow the parasitic BJT to turn on. These factors lead to a considerable improvement in the performance of the EBG DGJLFET [18].

However, the depletion of the source and drain regions from bottom due to the back gate extension leads to an increased source/drain series resistance. This results in somewhat lower ON-state current in the EBG DGJLFET as compared to the DGJLFET.

5.8.8.D Improved Saturation Characteristics From the electron concentration contour plot of the EBG DGJLFET and DGJLFET shown in Fig. 5.76, it can be observed that the electron channel is pinched off in the EBG DGJLFET below the drain contact in the ON-state. This is attributed to the extension of the back gate below the drain contact. In the ON-state, $V_{GS} = V_{DS} = 1.0$ V, the effective voltage experienced by the drain region above the EBG and below the drain contact is 0.0 V. As a result, this region is effectively depleted by the EBG due to its high work function. Therefore, the EBG DGJLFET facilitates an efficient pinch-off mechanism and exhibits a better saturation characteristic than the DGJLFET as shown in Fig. 5.77(a).

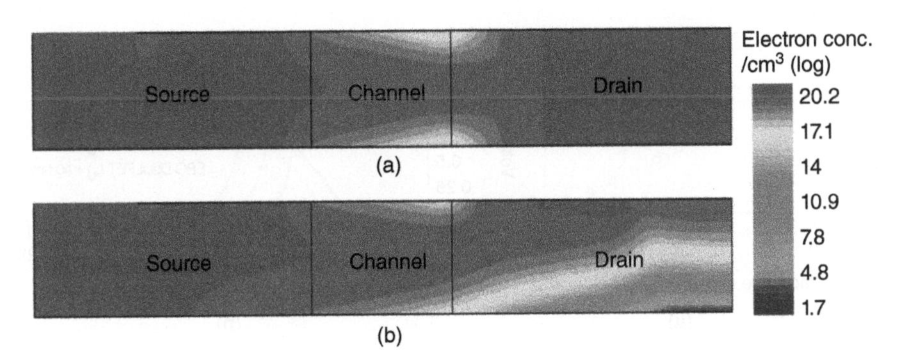

FIGURE 5.76 Electron concentration contour plot of (a) the conventional DGJLFET and (b) the EBG DGJLFET in the ON-state ($V_{GS} = 1.0\,V$ and $V_{DS} = 1.0\,V$).

The drain current in DGJLFET does not saturate and has a weak linear dependence on the drain voltage [22, 62, 73]. However, the drain current saturates efficiently and remains nearly constant in the EBG DGJLFET [18] as shown in Fig. 5.77(a).

5.8.8.E Dynamic Performance Although the EBG DGJLFET exhibits a high I_{ON}/I_{OFF}, a close to ideal 60 mV/decade subthreshold swing and a better saturation behavior, the extension of the back gate from the source end to the drain end is expected to increase the gate capacitance significantly. An increase in the gate capacitance would degrade the dynamic performance of the EBG DGJLFET due to the enlarged delay [18]. Therefore, the EBG DGJLFET with a back gate which is underlapped from the source end was also proposed as shown in Fig. 5.77(b). As shown in Fig. 5.78(a), the gate capacitance reduces with an increase in the back gate underlap length (L_U) owing to the reduction in the effective gate area.

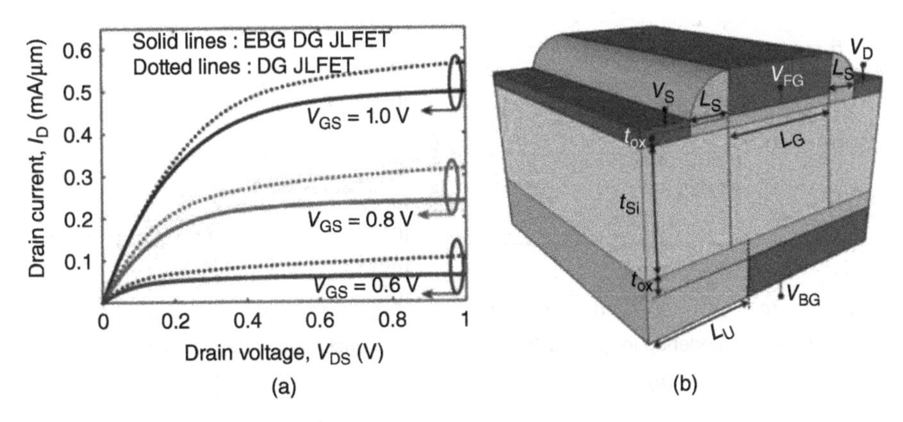

FIGURE 5.77 (a) The output characteristics of the DGJLFET and the EBG DGJLFET and (b) the schematic view of the EBG DGJLFET with a back gate underlap.

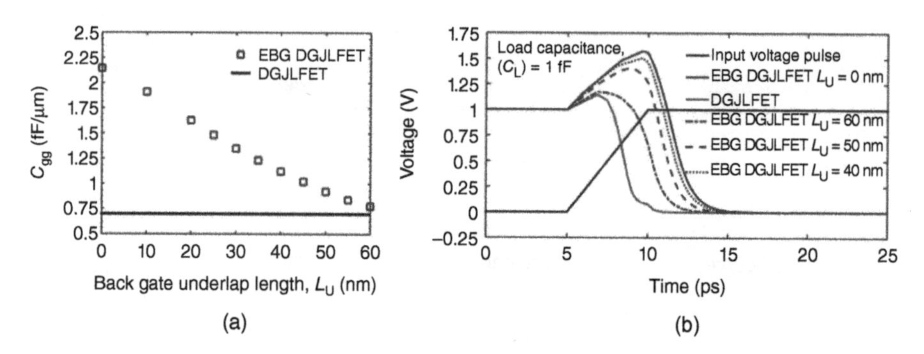

FIGURE 5.78 (a) The total gate capacitance C_{gg} (at $V_{GS} = 1.0$ V) and (b) the transient response of the CMOS inverter of the conventional DGJLFET and the EBG DGJLFET for different back gate underlap lengths.

The transient simulations of a simple inverter obtained using EBG DGJLFET and DGJLFET are also shown in Fig. 5.78(b). The EBG DGJLFET exhibits a larger delay and an increased overshoot. The larger delay is attributed to the increased gate capacitance, and the increased overshoot is due to the large gate-to-drain Miller capacitance owing to the gate-on-drain overlap architecture. However, the overshoot and delay reduces with an increment in the underlap length [18].

Although the delay reduces with increasing underlap length owing to the reduction in the gate capacitance, the OFF-state current and the consequent I_{ON}/I_{OFF} also degrades significantly with increasing L_U as shown in Fig. 5.79. This is due to a reduction in the effective channel region being volume depleted by the back gate. Therefore, the back gate should extend from the source–channel interface to the drain end for optimizing the dynamic as well as logic performance of the EBG DGJLFET.

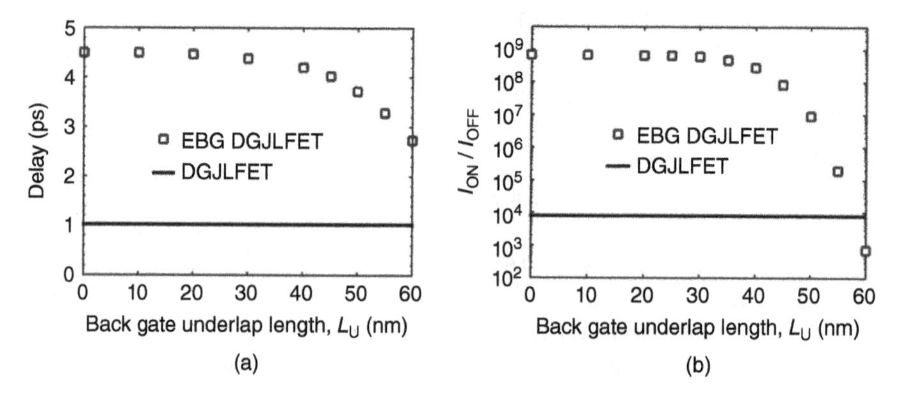

FIGURE 5.79 (a) The intrinsic delay and (b) the I_{ON}/I_{OFF} ratio of the DGJLFET and the EBG DGJLFET for different back gate underlap lengths.

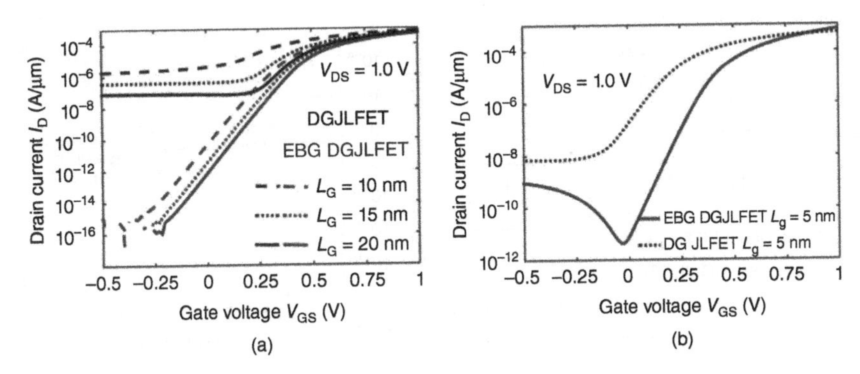

FIGURE 5.80 The impact of gate length scaling on the transfer characteristics of the EBG DGJLFET and the DGJLFET for (a) $L_g \geq 10$ nm and (b) $L_g = 5$ nm.

5.8.8.F Impact of Scaling Figure 5.80 shows the impact of scaling of the gate length on EBG DGJLFETs and DGJLFETs with (a) $L_g \geq 10$ nm with $t_{Si} = 10$ nm and (b) $L_g = 5$ nm with $t_{Si} = 5$ nm. As can be observed from Fig. 5.80(a), the OFF-state current increases for both DGJLFET and the EBG DGJLFET with channel length scaling owing to the increased gain of the parasitic BJT due to the reduced base width (channel length) as discussed in Section 5.3. However, the EBG DGJLFET exhibits a significantly low OFF-state current and a considerably high I_{ON}/I_{OFF} of ~10^6 for a gate length of 10 nm.

However, the OFF-state current reduces significantly when the silicon thickness is reduced to 5 nm for both DGJLFET and the EBG DGJLFET irrespective of scaling the channel length to 5 nm as shown in Fig. 5.80(b). This is attributed to (a) the quantum confinement effects which become dominant as the channel thickness is scaled below 7 nm [29] and (b) the reduction in DIBL and the consequent increase in the source-to-channel barrier height and electrostatic integrity as discussed in Section 5.4.1. The quantization effects become dominant as the silicon thickness is reduced below 7 nm and lead to a significant discretization in the conduction band and the valence band [29–32]. This restricts the tunneling phenomenon to occur between the first subband of the valence band and the first subband of the conduction band leading to an increase in the effective band gap as discussed in Section 5.4.2. Therefore, tunneling probability is reduced due to scaling of the silicon thickness to 5 nm.

Moreover, a reduction in the silicon thickness also improves the electrostatic integrity by reducing the DIBL and increasing the source-to-channel barrier height. If you compare the transfer characteristics of the DGJLFET and the EBG DGJLFET in Fig. 5.80, you may easily comprehend that the performance degradation brought about by the gate length scaling can be compensated by reducing the channel thickness.

Now, if you carefully observe Fig. 5.80(b), you notice that while the drain current becomes independent of the gate voltage (constant) when $V_{GS} < 0.0$ V following the typical GIDL characteristics in DGJLFETs, it continues to increase with negative

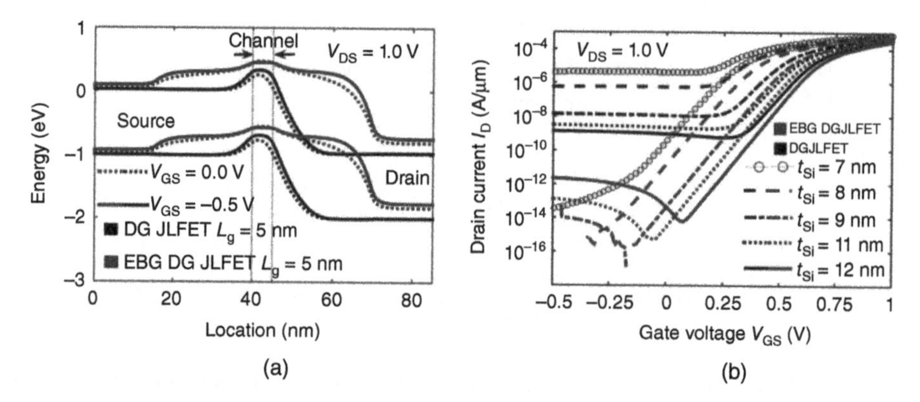

FIGURE 5.81 (a) Energy band profiles at 1 nm below Si–SiO_2 interface for $L_g = 5$ nm (with $t_{Si} = 5$ nm) and (b) the transfer characteristics for different silicon film thickness (t_{Si}) of the conventional DGJLFET and the EBG DGJLFET.

gate voltages in case of EBG DGJLFETs for $L_g = 5$ nm, which is the GIDL characteristic for MOSFET or JAMFETs as discussed in Section 5.7. Figure 5.81(a) can help in understanding this difference in the GIDL behavior of the DGJLFET and the EBG DGJLFET. For large and negative gate voltage ($V_{GS} = -0.5$ V in this case), the tunneling phenomenon occurs close to the drain contact in EBG DGJLFET instead of the channel–drain interface. This leads to the accumulation of holes in the drain region, which spread throughout the active device layer due to the absence of any potential barrier. As a result, the effective base width of the parasitic BJT now spans from the source contact to the drain contact. An increase in the effective base width reduces the gain of the parasitic BJT significantly and leads to a diminished parasitic BJT action. Therefore, as discussed in Section 5.7, the tunneling current starts dominating the drain current similar to the MOSFETs and the drain current increases with the negative gate voltages.

5.8.8.G Impact of Silicon Film Thickness As discussed in Section 5.4, the OFF-state current in JLFETs is highly dependent on the silicon thickness. An increase in the silicon thickness increases the OFF-state current significantly because of a reduction in the electrostatic integrity, which translates into a higher DIBL and the consequent higher DIBT [15]. The process variations due to etching of the active silicon film may result in a different active silicon layer thickness between two devices on a wafer. Since the OFF-state current depends on the silicon film thickness, process variation is a major technological challenge for JLFETs [21]. The DGJLFET exhibits a poor I_{ON}/I_{OFF} of ~10 when the silicon film thickness increases by 20%.

However, the EBG DGJLFET exhibits a significantly high I_{ON}/I_{OFF} of $\sim 10^7$ even when the channel thickness increases by 20% (Fig. 5.81(b)). Therefore, the EBG DGJLFET performs more reliably than the DGJLFET even in the presence of process variations.

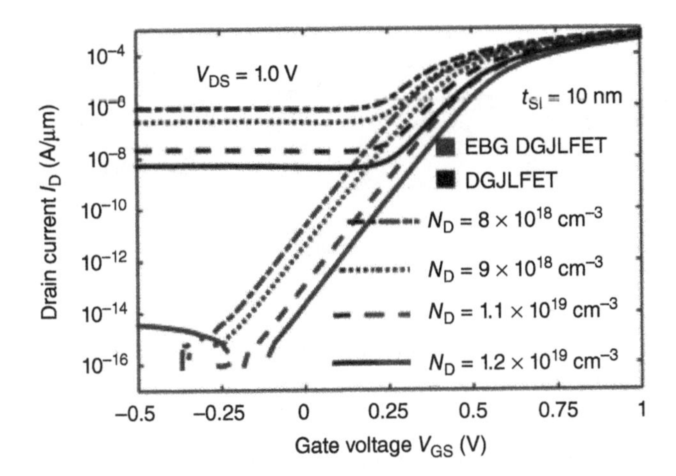

FIGURE 5.82 Transfer characteristics of the conventional DGJLFET and the EBG DGJLFET for different silicon film doping.

5.8.8.H Impact of Silicon Film Doping As per our discussion in Section 5.5, an increase in the silicon film doping increases the short-channel effects, reduces the tunneling width, and also leads to an inefficient volume depletion. However, the conventional doping methodology, i.e. ion implantation, does not guarantee a uniform doping profile and an exact concentration of active dopant atoms as discussed in Section 3.1. Therefore, not all devices in the wafer would have an exact doping of 10^{19} cm^{-3}. Since the performance of the JLFETs degrades significantly with increasing doping concentration, this poses a technological challenge for the conventional JLFETs.

However, as can be observed from Fig. 5.82, the EBG DGJLFET performs significantly better than the DGJLFET even in the presence of doping concentration variation. The EBG DGJLFET exhibits a significantly large I_{ON}/I_{OFF} of ~10^8 even when the doping concentration increases by 20%. Therefore, the EBG DGJLFET is relatively robust against doping concentration variations [18].

5.8.8.I Challenges The EBG DGJLFET may be a promising alternative for ultimate scaling of DGJLFET. However, realizing EBGs via the process flows described in section 5.8.8.A is not easy and requires extremely controlled shifting mechanisms like electrical Vernier [67]. The EBG DGJLFET with an underlap, which shows the optimum performance, is more difficult to realize considering the stringent constraint of exact alignment of the back gate with the source–channel interface.

5.8.9 Nanotube JLFETs

5.8.9.A Structure The 3D view of the nanotube junctionless FET (NTJLFET) is shown in Fig. 5.83(a). It can be simply viewed as a nanowire, which has a core gate in addition to the outer gate. The core gate is electrically connected to the outer gate

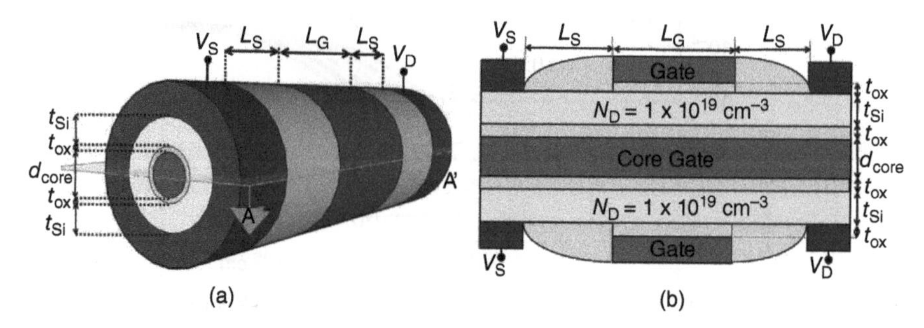

FIGURE 5.83 (a) Three-dimensional view and (b) cross-sectional view at cutline A–A′ of the nanotube JLFET [16].

and extends from the source end to the drain end. Silicon nanotubes (NTs) have been experimentally demonstrated in [74] using high precision plasma-treated nanosphere lithography and reactive-ion etching. In addition, NTJLFETs with a core gate may be fabricated using the similar process steps as outlined in [75–82] for NT MOSFETs. Since the NTJLFET does not require formation of any ultrasteep metallurgical junctions, it can be realized using a low thermal budget. This also enables a larger flexibility in choosing different materials for gate stack. It may be noted that gate sidewall spacers are indispensable for the formation of the NTFETs [75–82].

5.8.9.B Transfer Characteristics The transfer characteristics of the NTJLFET are compared with the NWJLFET and NWJAMFET in Fig. 5.84(a). For a fair comparison, the value of the drain current is normalized by dividing it with the effective silicon film circumference, as done in [75–81], which is $\pi(t_{Si})$ for the NWs and $\pi(d_{core} + t_{Si})$ for the NTs. The NTJLFET exhibits a significantly reduced OFF-state current and

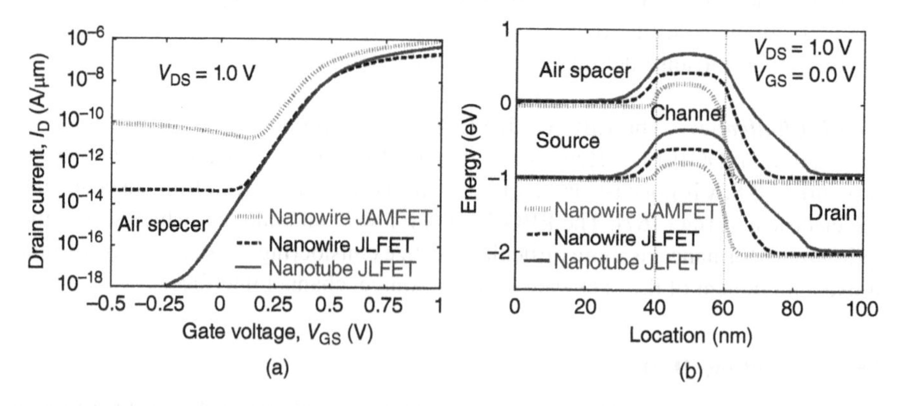

FIGURE 5.84 (a) Transfer characteristics and (b) energy band profiles taken at a cutline 1 nm below the Si–SiO$_2$ interface of the NTJLFET and NWJLFET and NWJAMFET with air and SiO$_2$ spacer.

a considerably high I_{ON}/I_{OFF} ratio of $\sim 10^9$ as compared to the NWJLFET ($\sim 10^6$) and NWJAMFET ($\sim 10^4$). The subthreshold swing is also improved in the NTJLFET (~62 mV/decade) as compared to the NWJLFET (~68 mV/decade) and NWJAMFET (~69 mV/decade). The ON-state current also improves in NWJLFET as compared to the NWJLFET by ~2.5 times.

Another noticeable feature of NTJLFET is that the drain current reduces with negative gate voltages opposed to the NWJLFET where the drain current becomes independent of the gate voltage.

5.8.9.C Operation As discussed in Section 5.6, the OFF-state current of the JLFETs reduces in the presence of gate sidewall spacers. Therefore, the transfer characteristics have been analyzed in the presence of both a high-κ spacer and an air spacer to decouple the improvement brought about by the NT architecture and the spacer. Also, as per our discussion in Section 2.6.4, L-BTBT GIDL originates due to an increased gate control of the channel region. The L-BTBT GIDL and the consequent OFF-state current should increase in the NTJLFETs as compared to the NWJLFET since NT architecture offers a better control as compared to the NW [16].

However, the significantly low OFF-state current observed in NTJLFETs and the reduction in the drain current with negative gate voltage contrary to the NWJLFET and NWJAMFET can be understood from Fig. 5.84(b). The core gate facilitates the depletion of the drain region at the channel–drain interface increasing the tunneling width in NTJLFETs. The depletion of the source region at the source–channel interface owing to the core gate leads to a higher source-to-channel barrier height in the NTJLFETs. In addition, the presence of the core gate, which spans the entire device length, leads to an increased effective channel length (base width) of the parasitic BJT. The significantly reduced OFF-state current in NTJLFETs is attributed to these factors, which lead to a diminished parasitic BJT action [16].

The triggering of the parasitic BJT results in a loss of gate control over the drain current in NWJLFETs and NWJAMFETs for negative gate voltages. However, a reduction in the drain current for negative gate voltages clearly indicates that the parasitic BJT remains inactive in the NTJLFET even when the gate voltage becomes negative. Therefore, the NTJLFET is immune to the parasitic BJT effect [16].

5.8.9.D Impact of Scaling The impact of gate length scaling on the NTJLFETs and NWJLFET and NWJAMFET has been evaluated in Fig. 5.85(a). The performance of NWJLFET and NWJAMFET degrades significantly with scaling due to an increase in the OFF-state current owing to the increased gain of the parasitic BJT due to a lower effective base width as discussed in Section 5.3. However, the NTJLFET performs significantly better and remains immune to the gate length scaling apart from a threshold voltage shift. This can be understood from the increased tunneling width and the higher source-to-channel barrier height in the NTJLFETs even for a gate length of 7 nm as shown in Fig. 5.85(b).

At this juncture, we would also like to discuss about the implication of the direct source-to-drain tunneling leakage current discussed in Section 1.7.4. In FETs

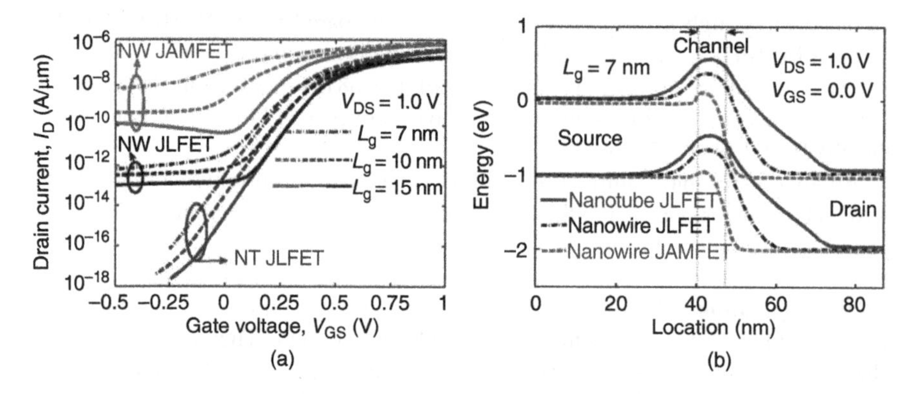

FIGURE 5.85 (a) Transfer characteristics and (b) energy band profiles taken at a cutline 1 nm below the Si–SiO_2 interface.

with ultimately scaled gate length (< 10 nm), the channel poses an ultrathin barrier (< 10 nm) between the source and the drain region. Therefore, the electrons may even tunnel through the channel region through intraband tunneling phenomenon instead of hopping over the source-to-channel barrier height. This phenomenon is called direct source-to-drain tunneling (DSDT) and is dominant short-channel effect in the FETs with the gate length below 10 nm.

As can be observed from Fig. 5.85(b), at a gate length of 7 nm, the barrier is extremely thin for the NWJAMFETs and NWJLFETs, which may facilitate intraband tunneling and DSDT. However, the intraband tunneling barrier is also significantly larger in the NTJLFET as compared to the NWJLFET and NWJAMFETs. Therefore, the NTJLFET is expected to perform better even in the presence of DSDT.

5.8.9.E Impact of Spacer Design Since a gate-sidewall spacer is required to form the NTFETs [75–82], the efficacy of the spacer dielectric constant as a design parameter in improving the performance of the NTJLFETs must be analyzed as done in [15] and shown in Fig. 5.86. An increase in the dielectric constant of the gate sidewall spacer attenuates the transition of the energy bands as shown in Fig. 5.84(b) and results in an increased tunneling width [15]. This leads to suppression of the parasitic BJT action and a reduction in the OFF-state current as the dielectric constant of the spacer is increased.

5.8.9.F Dynamic Performance Although the NTJLFET exhibits a significantly suppressed parasitic BJT action, the presence of the core gate, which spans from the source end to the drain end, should lead to a significant increase in the gate capacitance. As shown in Fig. 5.87(a), the gate capacitance is significantly increased in the NTJLFET as compared with the NWJLFET by ~9 times. The gate capacitance increases as the dielectric constant of the spacer is increased owing to the increased fringing field and associated outer and inner fringe capacitances.

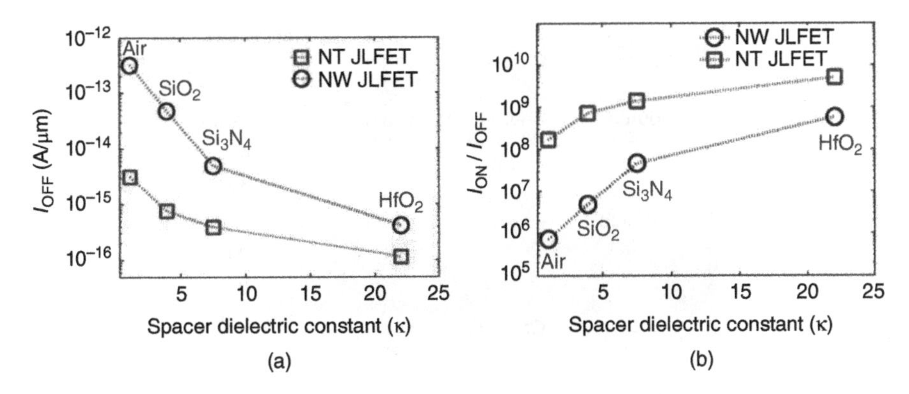

FIGURE 5.86 Variation of (a) OFF-state current (I_{OFF}) and (b) ON-state to OFF-state current ratio (I_{ON}/I_{OFF}) of the NTJLFETs and NWJLFETs with different spacers.

However, the intrinsic delay of the inverter obtained using NTJLFET increases by only ~3.3 times as compared to the NWJLFETs (Fig. 5.87(b)). This is attributed to the higher effective drive current offered by the NT architecture. An increase in the effective drive current somewhat compensates for the significantly increased gate capacitance in NTJLFETs. Although the NTJLFETs exhibit an increased intrinsic delay, the significant reduction in the OFF-state current especially for the ultrascaled gate lengths (< 10 nm) offered by the NTJLFETs clearly indicates its enormous potential for the future technology nodes.

5.8.9.G Impact of Core Diameter The core diameter forms an integral part of the NTJLFET which is responsible for its significant performance improvement. Therefore, it becomes important to analyze the impact of the core diameter on the performance of the NTJLFETs. An increase in the core diameter leads to a larger curvature

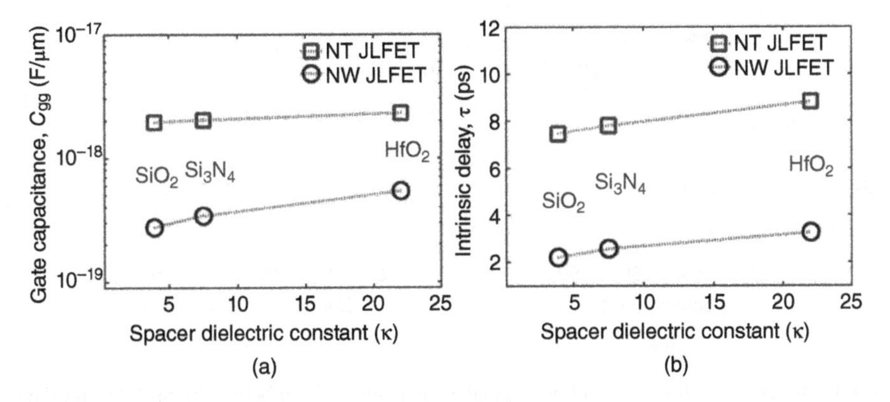

FIGURE 5.87 Variation of (a) gate capacitance (C_{gg}) and (b) intrinsic delay (τ) of the NTJLFETs and NWJLFETs with different spacers.

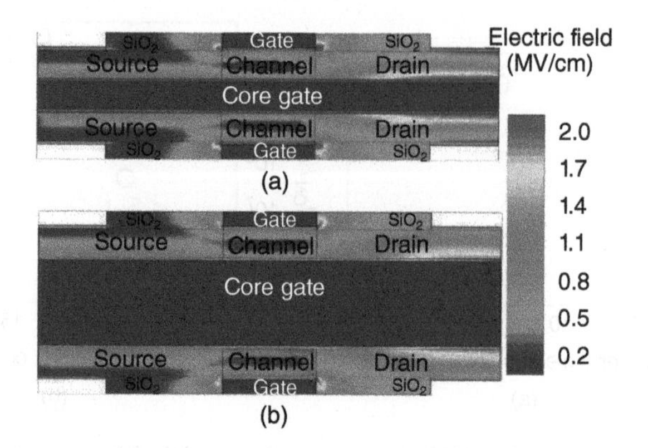

FIGURE 5.88 Electric-field contour plot in the OFF-state ($V_{GS} = 0.0$ V and $V_{DS} = 1.0$ V) of NTJLFET for (a) $d_{core} = 10$ nm and (b) $d_{core} = 30$ nm.

of the core gate and a consequent lower flux of the electric field lines. A smaller curvature due to a lower core diameter would facilitate superimposition of the electric field lines and a consequent increased flux [16]. As a result, an increase in the core diameter leads to a lower electric field as shown in Fig. 5.88. The lower electric field results in an increased tunneling width as shown in Fig. 5.89(a) and a consequent diminished L-BTBT GIDL. The source-to-channel barrier height is also increased in the NTJLFETs with an increased core diameter. Therefore, the OFF-state current reduces with an increase in the core diameter for NTJLFETs as shown in Fig. 5.89(b).

5.8.9.H Impact of Spacer Length Since the NTJLFETs are essentially vertical structures, their foot print is dictated by the core diameter and silicon film thickness unlike lateral nanowires whose scaling is limited by the gate length, spacer length,

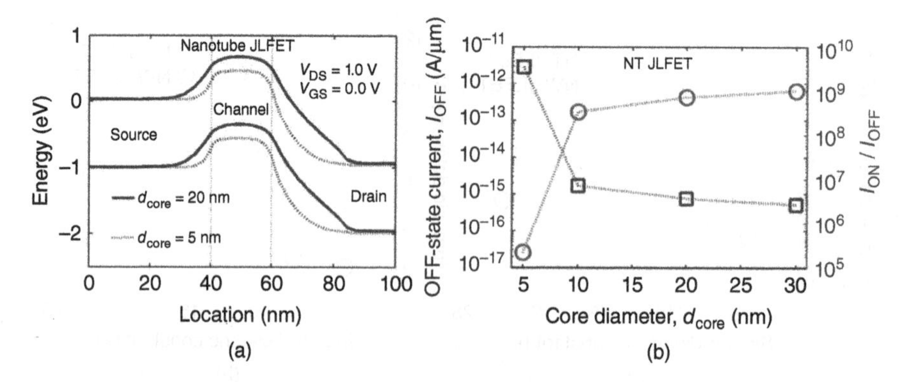

FIGURE 5.89 (a) Energy band profiles taken at a cutline 1 nm below the Si–SiO_2 interface and (b) OFF-state current (I_{OFF}) and I_{ON}/I_{OFF} of the NTJLFETs for different core diameter (d_{core}).

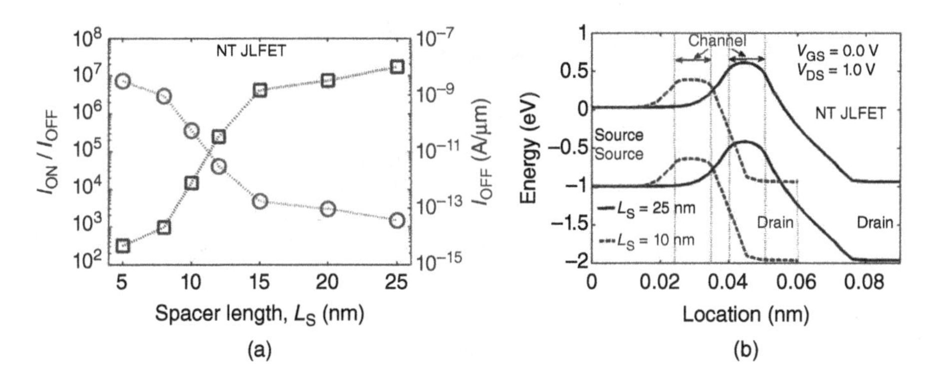

FIGURE 5.90 (a) I_{ON}/I_{OFF} and I_{OFF} and (b) energy band profiles taken at a cutline 1 nm below the Si–SiO_2 interface of the NTJLFETs ($L_g = 10$ nm) for different spacer lengths.

and the length of source/drain contacts. However, the impact of spacer length on the performance of NTJLFETs has also been analyzed in Fig. 5.90. A reduction in the spacer length leads to a larger drain electric field at the channel–drain interface owing to a reduction in the device length. This increases the short-channel effects such as DIBL, and the consequent DIBT leads to a reduction in the tunneling width at the channel–drain interface as shown in Fig. 5.90(b). As a result, the L-BTBT GIDL and the OFF-state current reduces as the spacer length is increased [16]. However, the improvement in the performance brought about by increasing the spacer length tends to saturate close to 15 nm. An increase in the spacer length beyond 15 nm will degrade the dynamic performance severely due to an increase in the fringe capacitances. Therefore, the spacer length should be appropriately chosen in NTJLFETs [16].

5.8.9.1 Challenges Silicon NTs may pave the way for ultimate scaling of nanowire JLFETs. However, realizing silicon NTs with a core gate is very difficult and the large intrinsic delay offered by the NTJLFETs restricts its applications. The formation of a good quality core gate dielectric interface to the silicon film is a major challenge. Apart from this, formation of silicon NTs with precise dimensions is also a fabrication challenge. Therefore, NTJLFETs can be realized only when these fabrication challenges are overcome.

5.8.10 JLFETs with Tunnel Dielectric

5.8.10.A Structure The schematic of the JLFET with a tunnel dielectric (TD) is shown in Fig. 5.91. The tunnel-dielectric junctionless field-effect transistor (TDJLFET) consists of an ultrathin (≤ 1 nm) layer of a dielectric such as Si_3N_4 in the active silicon layer through which the electrons may tunnel easily. This TD may be introduced in the active layer of the JLFET using the process flow described in

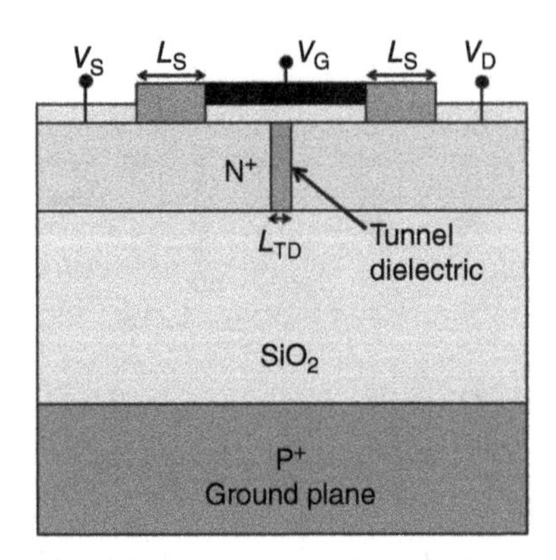

FIGURE 5.91 Schematic view of the JLFET with a tunnel dielectric.

[84–86]. The position of the TD may also be varied in the active silicon layer. For simplicity and realizing a symmetric structure, the TD should be placed at the center of the device.

5.8.10.B Transfer Characteristics The transfer characteristics of the TDJLFET and the conventional SOI JLFET are shown in Fig. 5.92. The TDJLFET exhibits a reduced OFF-state current by ~5 orders of magnitude leading to a considerably high I_{ON}/I_{OFF} of ~10^7 as compared to the SOI JLFET (~10^2). However, the ON-state current is also reduced in the TDJLFET by ~6 times as compared to the SOI JLFET.

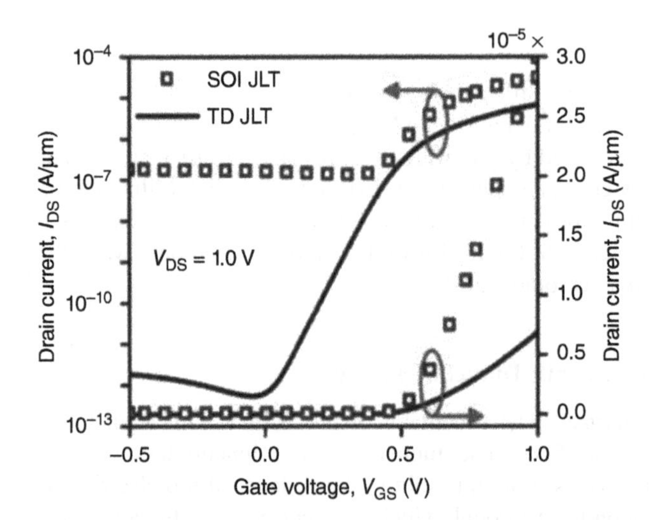

FIGURE 5.92 Transfer characteristics of the TDJLFET and the SOI JLFET [83].

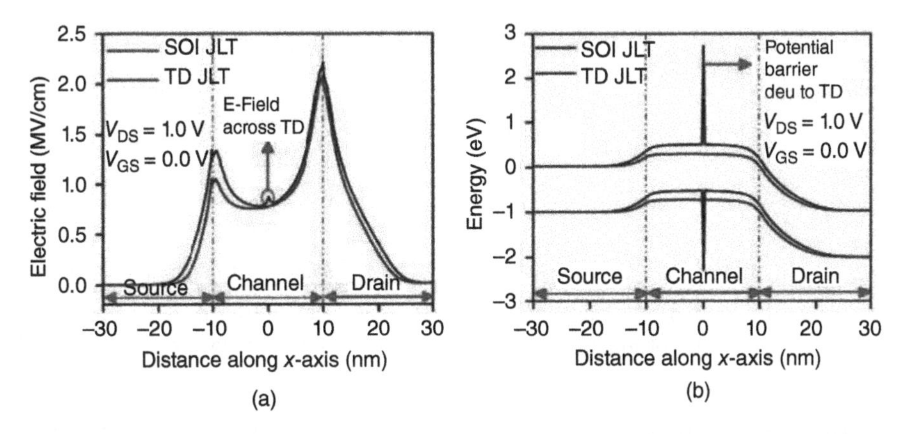

FIGURE 5.93 (a) The electric field profile and (b) the energy band profiles of the TDJLFET and the SOI JLFET at a cutline taken 1 nm below the Si–SiO_2 interface in the OFF-state ($V_{GS} = 0.0$ V, $V_{DS} = 1.0$ V) [83].

5.8.10.C Operation The presence of the TD at the center of the channel leads to a potential drop across it and a consequent electric field across the TD in the OFF-state as shown in Fig. 5.93(a). The increase in the electric field across the TD leads to a reduction in the electric field at the channel–drain interface. This leads to a lower band bending in the TDJLFET as compared to the SOI JLFET and a consequent larger tunneling width which diminishes the BTBT-induced parasitic BJT action in TDJLFET (Fig. 5.93(b)).

The TD also poses a potential barrier for the channel electrons close to the source–channel interface (left of the TD) to tunnel across the barrier and move to the channel–drain interface (right of the TD). The channel electrons close to the channel–drain interface are the ones that tunnel to the drain region and lead to the parasitic BJT action. Therefore, the additional potential barrier facilitated by the presence of the TD also contributes to the reduction in the parasitic BJT action [83].

In addition, the TD also offers a potential barrier for the holes induced into the channel region due to the BTBT. Only the holes that are able to tunnel the TD potential barrier and reach the source–channel interface from the channel–drain interface are injected into the source region leading to the triggering of the parasitic BJT. These factors lead to the significantly reduced parasitic BJT action and the consequent performance improvement in the TD JLFET as compared to the SOI JLFET.

However, in the ON-state, the TD offers a potential barrier for the electrons moving through the channel region. The lower ON-state current exhibited by the TDJLFET is attributed to this potential barrier [83].

5.8.10.D. Impact of TD Material Figure 5.94 shows the impact of the material used for TD in the TDJLFETs. The application of SiO_2 as the TD instead of Si_3N_4results in a higher OFF-state current. The valence band offset reduces for the

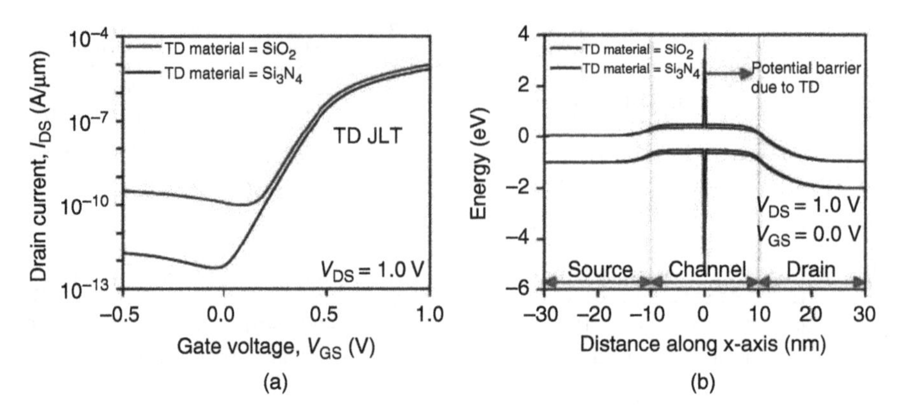

FIGURE 5.94 (a) Transfer characteristics and (b) the energy band profiles of the TDJLFET for different TD [83].

SiO_2 as compared to the Si_3N_4 TD leading to a lower tunneling barrier width [87]. A lower TD barrier width allows more electrons from the channel region close to the source–channel interface (left of TD) to move to the channel region close to the channel–drain interface (right of TD). Therefore, the BTBT is increased in the TDJLFET with SiO_2 TD. A lower BTBT along with a higher source-to-channel barrier width leads to a lower OFF-state current in the TDJLFET with Si_3N_4 TD (Fig. 5.94(b)).

5.8.10.E. Impact of TD Length The impact of the length of TD on the performance of TDJLFET is compared in Fig. 5.95. A larger TD length leads to a higher tunneling barrier width and restricts the flow of channel electrons toward the channel–drain interface. This leads to a lower number of electrons tunneling from the channel region

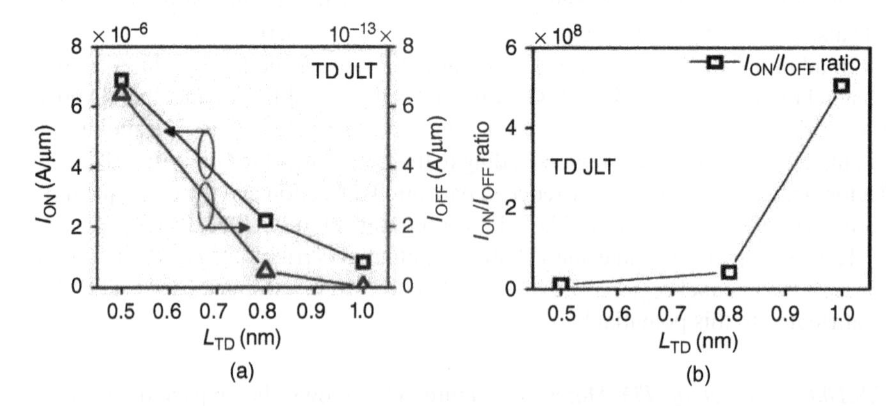

FIGURE 5.95 (a) The ON-state and OFF-state current and (b) the I_{ON}/I_{OFF} ratio for TDJLFET with different TD thickness [83].

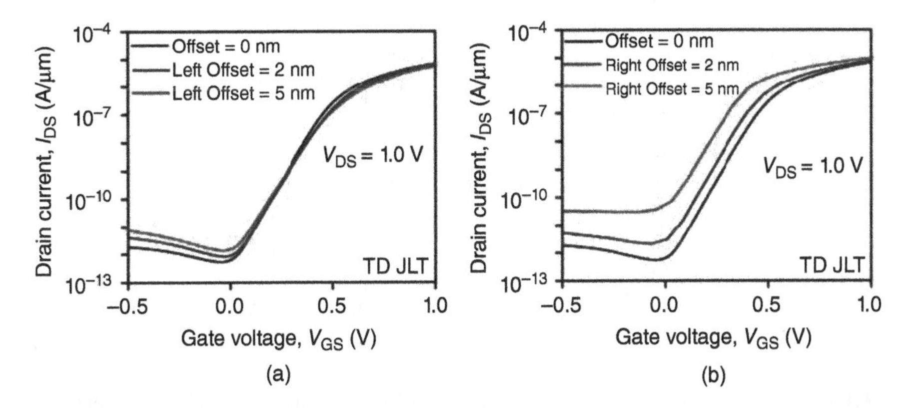

FIGURE 5.96 The impact of misalignment on the TDJLFET for different offset lengths [83].

to the drain region at the channel–drain interface. Therefore, a larger TD length leads to a reduced parasitic BJT action and a lower OFF-state current.

However, the increase in the tunneling barrier width adversely impacts the ON-state current. The ON-state current also reduces with an increase in the TD length as shown in Fig. 5.95 due to a larger barrier width in the path of electrons flowing in the channel region.

5.8.10.F. Impact of Position of TD Accurate placement of the TD at the center of the channel region of the TDJLFET is a technological challenge [83]. The performance of the TDJLFET must be analyzed in the presence of misalignment in the position of the TD from the center of the channel region. As observed from Fig. 5.96, a shift in the position of the TD toward the source–channel interface leads to an increased OFF-state current. A shift in the TD to the source–channel interface leads to a loss of the additional tunneling barrier width offered by the TD which restricted the flow of channel electrons on the left of TD to the right from where they may tunnel into the drain region.

A shift in the TD toward the channel–drain interface facilitates the tunneling at the channel–drain interface. This is because the effective tunneling width on the channel–drain side is now also modulated and dictated by the TD thickness which is extremely thin [84–86]. Therefore, the TD must be placed at the center of the channel region to obtain the optimum performance in TDJLFETs (Fig. 5.96).

5.8.10.G. Impact of Scaling the Gate Length The impact of gate length scaling on the performance of the TDJLFET is shown in Fig. 5.97. The OFF-state current increases in both TDJLFET and the SOI JLFET with gate length scaling due to an increased parasitic BJT action owing to the reduction in the effective base width as discussed in Section 5.3. The TDJLFET exhibits a somewhat high I_{ON}/I_{OFF} ratio ($\sim 10^2$) even when the gate length is reduced to 10 nm. However, it may be noted

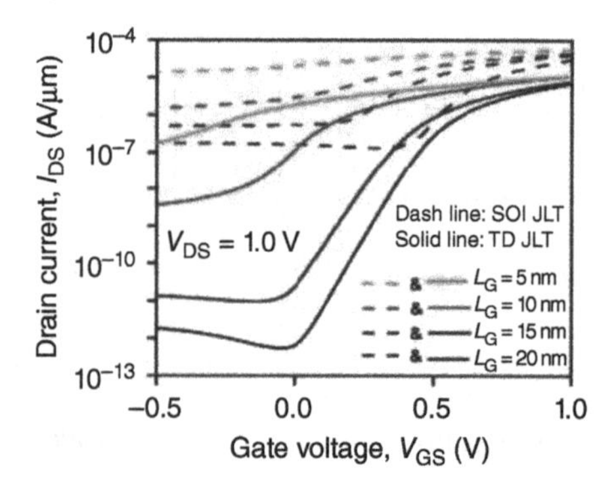

FIGURE 5.97 Impact of gate length scaling on the performance of TDJLFET and SOI JLFET [82].

from Fig. 5.97 that TDJLFET is not scalable beyond the 10-nm regime as it offers an extremely poor $I_{ON}/I_{OFF,}$ which makes it unsuitable for the logic applications.

5.8.10.H. Challenges Although TDJLFETs may appear to be a promising alternative to the conventional SOI JLFETs for gate lengths > 10 nm, the inclusion of the ultrathin < 1 nm TD at the center of the channel of the JLFET with accurate precision is a technological challenge [83]. In addition, the dielectric–semiconductor interface while using TD is bound to have interface defects and traps which requires further analysis. Moreover, the TDJLFET is expected to suffer more severely from reliability concern such as time-dependent dielectric breakdown(TDDB) or bias-temperature instability since the TD is subjugated to high electric fields. Continuous stress of the TD over the time would lead to soft breakdown of the TD. In addition, the TDJLFET is not suitable for scaling beyond the 10-nm regime, which is the dire need of the CMOS industry. The hard breakdown voltage of TDJLFET is also shown to be close to 9 V, which restricts its application in low-power circuits.

5.9 CONCLUSION

In this chapter, we discussed how BTBT leads to the formation of a parasitic BJT in JLFETs. We observed that the parasitic BJT action in JLFETs increases the OFF-state leakage current significantly and hinders their scaling to the sub-10 nm regime. Furthermore, we also analyzed the device architectures proposed to mitigate the parasitic BJT action in JLFETs. In the next chapter, we discuss another important aspect of JLFETs, i.e. the effect of impact ionization phenomenon on the performance of JLFETs. As we shall see, the modified conduction mechanism in JLFETs also leads to a different behavior in terms of the voltage at which impact ionization is triggered and the nature of impact ionization induced steep subthreshold slope in JLFETs as compared to the MOSFETs.

REFERENCES

[1] J.-P. Colinge, C.-W. Lee, A. Afzalian, N. D. Akhavan, R. Yan, I. Ferain, P. Razavi, B. O'Neill, A. Blake, M. White, A.-M. Kelleher, B. McCarthy, and R. Murphy, "Nanowire transistors without junctions," *Nature Nanotechnol.*, vol. 5, no. 3, pp. 225–229, Mar. 2010.

[2] C.-W. Lee, A. Afzalian, N. D. Akhavan, R. Yan, I. Ferain, and J.-P. Colinge, "Junctionless multigate field-effect transistor," *Appl. Phys. Lett.*, vol. 94, no. 5, pp. 053511–053512, 2009.

[3] R. Rios, A. Cappellani, M. Armstrong, A. Budrevich, H. Gomez, R. Pai, N. Rahhal-Orabi, and K. Kuhn, "Comparison of junctionless and conventional trigate transistors with L_g down to 26 nm," *IEEE Electron Device Lett.*, vol. 32, no. 9, pp. 1170–1172, Sept. 2011.

[4] S. Sahay and M. J. Kumar, "Realizing efficient volume depletion in SOI junctionless FETs," *IEEE J. Electron Devices Soc.*, vol. 4, no. 3, pp. 110–115, May 2016.

[5] S. Gundapaneni, S. Ganguly, and A. Kottantharayil, "Bulk planar junctionless transistor (BPJLT): An attractive device alternative for scaling," *IEEE Electron Device Lett.*, vol. 32, no. 3, pp. 261–263, Mar. 2011.

[6] S. Gundapaneni, M. Bajaj, R. K. Pandey, K. V. R. M. Murali, S. Ganguly, and A. Kottantharayil, "Effect of band-to-band tunneling on junctionless transistors," *IEEE Trans. Electron Devices*, vol. 59, no. 4, pp. 1023–1029, Apr. 2012.

[7] J. Fan, M. Li, X. Xu, Y. Yang, H. Xuan, and R. Huang, "Insight into gate-induced drain leakage in silicon nanowire transistors," *IEEE Trans. Electron Devices*, vol. 62, no. 1, pp. 213–219, Jan. 2015.

[8] J. Hur, B.-H. Lee, M.-H. Kang, D.-C. Ahn, T. Bang, S.-B. Jeon, and Y.-K. Choi, "Comprehensive analysis of gate-induced drain leakage in vertically stacked nanowire FETs: Inversion-mode vs. junctionless mode," *IEEE Electron Device Lett.*, vol. 37, no. 5, pp. 541–544, May 2016.

[9] S. Sahay and M. J. Kumar, "A novel gate-stack-engineered nanowire FET for scaling to the sub-10-nm regime," *IEEE Trans. Electron Devices*, vol. 63, no. 12, pp. 5055–5059, Dec. 2016.

[10] S. Sahay and M. J. Kumar, "Comprehensive analysis of gate-induced drain leakage in emerging FET architectures: Nanotube FETs vs. nanowire FETs," *IEEE Access*, vol. 64, no. 7, pp. 18918–18926, Dec. 2017.

[11] S. Sahay and M. J. Kumar, "Spacer design guidelines for nanowire FETs from gate induced drain leakage perspective," *IEEE Trans. Electron Devices*, vol. 64, no. 7, pp. 3007–3015, July 2017.

[12] S. Sahay and M. Jagadesh Kumar, "Physical insights into the nature of gate induced drain leakage in ultra-short channel nanowire field-effect transistors," *IEEE Trans. Electron Devices*, vol. 64, no. 6, pp. 2604–2610, June 2017.

[13] S. Sahay and M. J. Kumar, "Insight into lateral band-to-band-tunneling in nanowire junctionless FETs," *IEEE Trans. Electron Devices*, vol. 63, no. 10, pp. 4138–4142, Oct. 2016.

[14] S. Sahay and M. J. Kumar, "Controlling L-BTBT and volume depletion in nanowire JLFETs using core-shell architecture," *IEEE Trans. Electron Devices*, vol. 63, no. 9, pp. 3790–3794, Sept. 2016.

[15] S. Sahay and M. J. Kumar, "Diameter dependency of leakage current in nanowire junctionless field-effect transistors," *IEEE Trans. Electron Devices*, vol. 64, no. 3, pp. 1330–1335, Mar. 2017.

[16] S. Sahay and M. J. Kumar, "Nanotube junctionless FET: Proposal, design, and investigation," *IEEE Trans. Electron Devices*, vol. 64, no. 4, pp. 1851–1856, Apr. 2017.

[17] M. J. Kumar and S. Sahay, "Controlling BTBT induced parasitic BJT action in junctionless FETs using a hybrid channel," *IEEE Trans. Electron Devices*, vol. 63, no. 8, pp. 3350–3353, Aug. 2016.

[18] S. Sahay and M. J. Kumar, "Symmetric operation in an extended back gate JLFET for scaling to the 5 nm regime considering quantum confinement effects," *IEEE Trans. Electron Devices*, vol. 64, no. 1, pp. 21–27, Jan. 2017.

[19] V. Nathan and N. C. Das, "Gate-induced drain leakage currents in MOS devices," *IEEE Trans. Electron Devices*, vol. 40, no. 10, pp. 1888–1890, Oct. 1993.

[20] T. Hoffmann, G. Doornbos, I. Ferain, N. Collaert, P. Zimmerman, M. Goodwin, R. Rooyackers, A. Kottantharayil, Y. Yim, A. Dixit, K. De Meyer, M. Jurczak, and S. Biesemans "GIDL (gate induced drain leakage) and parasitic Schottky barrier leakage elimination in aggressively scaled HfO_2/TiN FinFET devices," *IEDM Tech. Dig.*, pp. 725–729, 2005.

[21] S.-J. Choi, M. Dong-II, S. Kim, J. P. Duarte, and Y.-K. Choi, "Sensitivity of threshold voltage to nanowire width variation in junctionless transistors," *IEEE Electron Device Lett.*, vol. 32, no. 2, pp. 125–127, Feb. 2011.

[22] C.-H. Park, M.-D. Ko, K.-H. Kim, R.-H. Baek, C.-W. Sohn, C. K. Baek, S. Park, M. J. Deen, Y.-H. Jeong, and J.-S. Lee, "Electrical characteristics of 20-nm Si nanowire transistors," *Solid State Electron.*, vol. 73, pp. 7–10, 2012.

[23] I. Wong, Y. Chen, S. Huang, W. Tu, Y. Chen, and C. W. Liu, "Junctionless gate-all-around PFETs using in-situ boron doped Ge channel on Si," *IEEE Trans. Nanotechnol.*, vol. 14, no. 5, pp. 878–882, Sept. 2015.

[24] S. Migita, Y. Morita, T. Matsukawa, M. Masahara, and Hiroyuki, "Experimental demonstration of ultrashort-channel (3 nm) junctionless FETs utilizing atomically sharp v-grooves on SOI," *IEEE Trans. Nanotechnol.*, vol. 13, no. 2, pp. 208–215, Mar. 2014.

[25] R. Trevisoli, R. T. Doria, M. Souza, and M. A. Pavanello, "Analysis of the leakage current in junctionless nanowire transistors," *J. Appl. Phys.*, vol. 103, no. 20, 202103, Nov. 2013.

[26] K. J. Kuhn, "Considerations for ultimate CMOS scaling," *IEEE Trans. Electron Devices*, vol. 59, no. 7, pp. 1813–1828, July 2012.

[27] L. Liu, D. Mohata, and S. Datta, "Scaling length theory of double-gate interband tunnel field-effect transistors," *IEEE Trans. Electron Devices*, vol. 59, no. 4, pp. 902–908, Apr. 2012.

[28] R. Vishnoi and M. J. Kumar, "Compact analytical model of dual material gate tunneling field-effect transistor using interband tunneling and channel transport," *IEEE Trans. Electron Devices*, vol. 61, no. 6, pp. 1936–1942, June 2014.

[29] J.-P. Colinge, J. C. Alderman, W. Xiong, and C. R. Cleavelin, "Quantum–mechanical effects in trigate SOI MOSFETs," *IEEE Trans. Electron Devices*, Vol. 53, no. 5, pp. 1131–1136, May 2006.

[30] J. L. Padilla, F. Gámiz, and A. Godoy, "A simple approach to quantum confinement in tunneling field-effect transistors," *IEEE Electron Device Lett.*, vol. 33, no. 10, pp. 1342–1344, Oct. 2012.

[31] J. P. Colinge, *FinFETs and Other Multi-Gate Transistors*, vol. 73, Springer, New York, 2008.

[32] D. D. D. Ma, C. S. Lee, F. C. K. Au, S. Y. Tong, and S. T. Lee, "Small-diameter silicon nanowire surfaces," *Science*, vol. 299, no. 5614, pp. 1874–1877, Mar. 2003.

[33] L. Barbut, F. Jazaeri, D. Bouvet, and J. M. Sallese, "Transient off-current in junctionless FETs," *IEEE Trans. Electron Devices*, vol. 60, no. 6, pp. 2080–2083, June 2013.

[34] L. Smith, M. Choi, M. Frey, V. Moroz, A. Ziegler, and M. Luisier, "FinFET to nanowire transition at 5 nm design rules," *Int. Conf. Sim. Semicond. Proc. and Devices (SISPAD)*, 2015, pp. 254–257.

[35] S. Gundapaneni, S. Ganguly, and A. Kottantharayil, "Enhanced electrostatic integrity of short channel junctionless transistor with high-κ; spacers," *IEEE Electron Device Lett.*, vol. 32, no. 10, pp. 1325–1327, Oct. 2011.

[36] J. Y. Choi and J. G. Fossum, "Analysis and control of floating body bipolar effects in FD submicrometer SOI MOSFETs," *IEEE Trans. Electron Devices*, vol. 38, no. 6, pp. 1384–1391, June 1991.

[37] J.-P. Colinge, "Reduction of kink effect in thin-film SOI MOSFETs," *IEEE Electron Device Lett.*, vol. 9, no. 2, pp. 97–99, Feb. 1988.

[38] V. Verma and M. J. Kumar, "Study of the extended P^+ dual source structure for eliminating bipolar induced breakdown in submicron SOI MOSFETs," *IEEE Trans. Electron Devices*, vol. 47, no. 8, pp. 1678–1680, Aug. 2000.

[39] M. J. Kumar and V. Verma, "Elimination of bipolar induced drain breakdown and single transistor latch in submicron PD SOI MOSFETs," *IEEE Trans. Rel.*, vol. 51, no. 3, pp. 367–370, Sept. 2002.

[40] J. Lin, D. A. Antoniadis, and J. A. Alamo, "Off-State leakage induced by band-to-band tunneling and floating-body bipolar effect in InGaAs quantum-well MOSFETs," *IEEE Electron Device Lett.*, vol. 35, no. 12, pp. 1203–1205, Dec. 2014.

[41] J. Knoch, and J. Appenzeller, "Tunneling phenomena in carbon nanotube field-effect transistors," *Phys. Status Solidi (a)*, vol. 205, no. 4, pp. 679–694, Apr. 2008.

[42] J. Knoch, S. Mantl, and J. Appenzeller, "Comparison of transport in carbon nanotube field-effect transistors with Schottky contacts and doped source/drain contacts," *Solid State Electron.*, vol. 49, pp. 73–76, 2005.

[43] J. Appenzeller, Y. M. Lin, J. Knoch, Z. Chen, and P. Avouris, "Comparing carbon nanotube transistors-the ideal choice: A novel tunneling device design," *IEEE Trans. Electron Dev.*, vol. 52, no. 12, pp. 2568–2576, Dec. 2005.

[44] A. Paul, M. Luisier, N. Neophytou, R. Kim, J. Geng, M. McLennan, M. Lundstrom, and G. Klimeck, "Band structure lab," May 2006 [online], Available: http://nanohub.org/resources/1308, Accessed: Dec 23, 2017.

[45] V. Thirunavukkarasu, Y. R. Jhan, Y. B. Liu, E. D. Kurniawan, Y. R. Lin, S. Y. Yang, C. H. Cheng, and Y.-C. Wu, "Gate-all-around junctionless silicon transistors with atomically thin nanosheet channel (0.65 nm) and record sub-threshold slope (43 mV/dec)," *Appl. Phys. Lett.*, vol. 110, no. 3, 032101, Jan. 2017.

[46] J. C. Ho, R. Yerushalmi, Z. A. Jacobson, Z. Fan, R. L. Alley, and A. Javey, "Controlled nanoscale doping of semiconductors via molecular monolayers," *Nature Mater.*, vol. 7, no. 1, pp. 62–67, Jan. 2008.

[47] J. C. Ho, R. Yerushalmi, G. Smith, P. Majhi, J. Bennett, J. Halim, V. N. Faifer, and A. Javey, "Wafer-scale, sub-5 nm junction formation by monolayer doping and conventional spike annealing," *Nano Lett.*, vol. 9, no. 2, pp. 725–730, 2009.

[48] K.-W. Ang, J. Barnett, W.-Y. Loh, J. Huang, B.-G. Min, P. Y. Hung, I. Ok, J. H. Yum, G. Bersuker, M. Rodgers, V. Kaushik, S. Gausepohl, C. Hobbs, P. D. Kirsch, and R. Jammy, "300 mm FinFET results utilizing conformal, damage free, ultra-shallow junctions (X_j $\sim$ 5 nm) formed with molecular monolayer doping technique," in *IEDM Tech. Dig.*, pp. 35.5.1–35.5.4, Dec. 2011.

[49] Y.-J. Lee, T.-C. Cho, S.-S. Chuang, F.-K. Hsueh, Y.-L. Lu, P.-J. Sung, H.-C. Chen, M. I. Current, T.-Y. Tseng, T.-S. Chao, C. Hu, and F.-L. Yang, "Low-temperature microwave annealing processes for future IC fabrication—A review," *IEEE Trans. Electron Devices*, vol. 61, no. 3, pp. 651–665, Mar. 2014.

[50] Y.-J. Lee, T.-C. Cho, K.-H. Kao, P.-J. Sung, F.-K. Hsueh, P.-C. Huang, C.-T. Wu, S.-H. Hsu, W.-H. Huang, H.-C. Chen, Y. Li, M. I. Current, B. Hengstebeck, J. Marino, T. Büyüklimanli, J.-M. Shieh, T.-S. Chao, W.-F. Wu, and W.-K. Yeh, "A novel junctionless FinFET structure with sub-5 nm shell doping profile by molecular monolayer doping and microwave annealing," in *IEDM Tech. Dig.*, pp. 32.7.1–32.7.4, Dec. 2014.

[51] S. Sahay and M. J. Kumar, "Controlling the drain side tunneling width to reduce ambipolar current in Tunnel FETs using heterodielectric BOX," *IEEE Trans. Electron Devices*, vol. 62, no. 11, pp. 3882–3886, Nov. 2015.

[52] Y. Wang, P. H. K. Wei, L. Zeng, X. Liu, G. Du, X. Zhang, and J. Kang, "Impact of random interface traps and random dopants in high-κ/metal gate junctionless FETs," *IEEE Trans. Nanotechnol.*, vol. 13, no. 3, pp. 584–588, May 2014.

[53] Y. Wang, K. Wei, X. Liu, G. Du, and J. Kang, "Random interface trap induced fluctuation in 22 nm high-κ/metal gate junctionless and inversion mode FETs," in *Proc. IEEE Int. Symp. VLSI Tech., Syst., Appl.*, 2013, pp. 177–178.

[54] Y. G. Fedorenko, L. Truong, V. V. Afanas'ev, A. Stesmans, Z. Zhang, and S. A. Campbell, "Impact of nitrogen incorporation on interface states in (100) Si/HfO_2," *J. Appl. Phys.*, vol. 98, no. 12, 123703, Dec. 2005.

[55] B. Ghosh, P. Mondal, M. W. Akram, P. Bal, and A. K. Salimath, "Hetero-gate-dielectric double gate junctionless transistor (HGJLT) with reduced band-to-band tunneling effects in subthreshold regime," *J. Semiconduct.*, vol. 35, no. 6, 064001, June 2014.

[56] W. Y. Choi and W. Lee, "Hetero-gate-dielectric tunneling field-effect transistors," *IEEE Trans. Electron Devices*, vol. 57, no. 9, pp. 2317–2319, Sept. 2010.

[57] M. J. Lee and W. Y. Choi, "Effects of device geometry on hetero-gate dielectric tunneling field-effect transistors," *IEEE Electron Device Lett.*, vol. 33, no. 10, pp. 1459, 2012.

[58] G. Lee, J. S. Jang and W. Y. Choi, "Dual-dielectric-constant spacer hetero-gate-dielectric tunneling field-effect transistors," *Semicond. Sci. Technol.*, vol. 28, pp. 052001, 2013.

[59] N. Ikarashi, K. Watanabe, K. Masuzaki, and T. Nakagawa, "Thermal stability of a HfO_2/SiO_2 interface," *Appl. Phys. Lett.*, vol. 88, no. 10, 101912, 2006.

[60] T. K. Kim, D. H. Kim, Y. G. Yoon, J. M. Moon, B. W. Hwang, D.-I. Moon, G. S. Lee, D. W. Lee, D. E. Yoo, H. C. Hwang, J. S. Kim, Y.-K. Choi, B. J. Cho, and S.-H. Lee,

"First demonstration of junctionless accumulation-mode bulk FinFETs with robust junction isolation," *IEEE Electron Device Lett.*, vol. 34, no. 12, pp. 1479–1481, Dec. 2013.

[61] Y. S. Yu, "A unified analytical current model for n- AND p-type accumulation-mode (junctionless) surrounding-gate nanowire FETs," *IEEE Trans. Electron Devices*, Vol. 61, no. 8, pp. 3007–3010, Aug. 2014.

[62] V. Thirunavukkarasu, Y.-R. Jhan, Y.-B. Liu, and Y.-C. Wu, "Performance of inversion, accumulation, and junctionless mode n-type and p-type bulk silicon FinFETs with 3-nm gate length," *IEEE Electron Device Lett.*, vol. 36, no. 7, pp. 645–647, July 2015.

[63] H. Lou, L. Zhang, Y. Zhu, X. Lin, S. Yang, J. He, and M. Chan, "A junctionless nanowire transistor with a dual-material gate," *IEEE Trans. Electron Devices*, vol. 59, no. 7, pp. 1829–1836, July 2012.

[64] H. Lou, D. Li, Y. Dong, X. Lin, J. He, S. Yang, and M. Chan, "Suppression of tunneling leakage current in junctionless nanowire transistors," *Semicond. Sci. Technol.*, vol. 28, pp. 125016–125022, Nov. 2013.

[65] M. H. Na, E. J. Nowak, W. Haensch, and J. Cai, "The effective drive current in CMOS inverters," in *IEDM Tech. Dig.*, pp. 121–124, 2002.

[66] M. P. Kumar, C. Y. Hu, K. H. Kao, Y. J. Lee, and T. S. Chao, "Impacts of the shell doping profile on the electrical characteristics of junctionless FETs," *IEEE Trans. Electron Devices*, vol. 62, no. 11, pp. 3541–3546, Nov.2015.

[67] M. Vinet, T. Poiroux, J. Widiez, J. Lolivier, B. Previtali, C. Vizioz, B. Guillaumot, P. Besson, J. Simon, F. Martin, S. Maitrejean, P. Holliger, B. Biasse, M. Cassé, F. Allain, A. Toffoli, D. Lafond, J. M. Hartmann, R. Truche, V. Carron, F. Laugier, A. Roman, Y. Morand, D. Renaud, M. Mouis, and S. Deleonibus, "Planar double gate CMOS transistors with 40 nm metal gate for multi-purpose applications," in *Proc. Int. Conf. Solid State Devices Materials*, Tokyo, Japan, Sept. 15–17, 2004, pp. 768–769.

[68] J. Widiez, J. Lolivier, M. Vinet, T. Poiroux, B. Previtali, F. Daugé, M. Mouis, and S. Deleonibus, "Experimental evaluation of gate architecture influence on DG SOI MOSFETs performance," *IEEE Trans. Electron Devices*, vol. 52, no. 8, pp. 1772–1779, Aug. 2005.

[69] N. Sugii, R. Tsuchiya, T. Ishigaki, Y. Morita, H. Yoshimoto, and S. Kimura, "Local v_{th} variability and scalability in silicon-on-thin-BOX (SOTB) CMOS with small random-dopant fluctuation," *IEEE Trans. Electron Devices*, vol. 57, no. 4, pp. 835–845, Apr. 2010.

[70] M. Fujiwara, T. Morooka, N. Yasutake, K. Ohuchi, N. Aoki, H. Tanimoto, M. Kondo, K. Miyano, S. Inaba, K. Ishimaru, and H. Ishiuchi, "Impact of BOX scaling on 30 nm gate length FD SOI MOSFETs," *Proc. IEEE Int. SOI Conf.*, Oct. 2005, pp. 180–182.

[71] S. J. Park, D.-Y. Jeon, L. Montès, S. Barraud, G.-T. Kim, and G. Ghibaudo, "Back biasing effects in tri-gate junctionless transistors," *Solid State Electron.*, vol. 87, pp. 74–79, Sept. 2013.

[72] R. Trevisoli, R. T. Doria, M. D. Souza, and M. A. Pavanello, "Substrate bias influence on the operation of junctionless nanowire transistors," *IEEE Trans. Electron Devices*, vol. 61, no. 5, pp. 1575–1582, May 2014.

[73] C.-W. Lee, A. Borne, I. Ferain, A. Afzalian, R. Yan, N. Dehdashti Akhavan, P. Razavi, and J.-P. Colinge, "High-temperature performance of silicon junctionless MOSFETs," *IEEE Trans. Electron Devices*, vol. 57, no. 3, pp. 620–625, Mar. 2010.

[74] S. S. Amiri, A. Gholizadeh, S. Rajabali, Z. Sanaee, and S. Mohajerzadeh, "Formation of Si nanorods and hollow nanostructures using high precision plasma-treated nanosphere lithography," *RSC Adv.*, vol. 4, pp. 12701–12709, Feb. 2014.

[75] H. M. Fahad, C. E. Smith, J. P. Rojas, and M. M. Hussain, "Silicon nanotube field-effect transistor with core–shell gate stacks for enhanced high performance operation and area scaling benefits," *Nano Lett.*, vol. 11, no. 10, pp. 4393–4399, Oct. 2011.

[76] H. M. Fahad and M. M. Hussain, "Are nanotube architectures advantageous than nanowire architectures for field-effect transistor applications?" *Sci. Rep.*, vol. 2, no. 2, 475, June 2012.

[77] D. Tekleab, H. H. Tran, J. W. Sleight, and D. Chidambarrao, "Silicon nanotube MOSFET," U.S. Patent 0 217 468, Aug. 30, 2012.

[78] D. Tekleab, "Device performance of silicon nanotube field-effect transistor," *IEEE Electron Device Lett.*, vol. 35, no. 5, pp. 506–508, May 2014.

[79] A. N. Hanna, H. M. Fahad, and M. M. Hussain, "InAs/Si hetero-junction nanotube tunnel transistors" *Sci. Rep.*, vol. 9, 9843, Apr. 2015.

[80] H. M. Fahad and M. M. Hussain, "High-performance silicon nanotube tunneling FET for ultralow-power logic applications," *IEEE Trans. Electron Devices*, vol. 60, no. 3, pp. 1034–1039, Mar. 2013.

[81] A. N. Hanna, and M. M. Hussain, "Si/Ge hetero-structure nanotube tunnel field-effect transistor," *J. Appl. Phys.*, vol. 117, no. 1, 014310, Jan. 2015.

[82] A. K. Jain, S. Sahay, and M. J. Kumar, "Controlling L-BTBT in emerging nanotube FETs using dual-material gate," *IEEE J. Electron Dev. Soc.*, vol. 6, pp. 611–621, June 2018.

[83] A. Lahgere and M. J. Kumar, "A tunnel dielectric-based junctionless transistor with reduced parasitic BJT action," *IEEE Trans. Electron Devices*, vol. 64, no. 8, pp. 3470–3475, Aug. 2017.

[84] Z. Luo, H. Wang, N. An, and Z. Zhu, "A tunnel dielectric-based tunnel FET," *IEEE Electron Device Lett.*, vol. 36, no. 9, pp. 966–968, Sept. 2015.

[85] H. Wang, Z. Luo, H. Yin, H. Zhu, J. Liu, and Z. Zhu, "A novel tunnel oxide based tunnel FET," *ECS Trans.*, vol. 34, no. 1, pp. 107–111, 2011.

[86] Z. Luo and H. Wang, "Tunneling field-effect transistor and manufacturing method thereof," CN. Patent 102 544 099, July 2012.

[87] J. Robertson, "Band offsets of high dielectric constant gate oxides on silicon," *J. Non-Cryst. Solids*, vol. 303, pp. 94–100, May 2002.

[88] J. P. Duarte, M. S. Kim, S. J. Choi, and Y. K. Choi, "A compact model of quantum electron density at the subthreshold region for double-gate junctionless transistors," *IEEE Trans. Electron Devices*, vol. 59, no. 4, pp. 1008–1012, Apr. 2012.

6

IMPACT IONIZATION IN JUNCTIONLESS FIELD-EFFECT TRANSISTORS

In Chapter 5, we discussed how band-to-band tunneling (BTBT) leads to the formation of a parasitic bipolar junction transistor (BJT) in junctionless field-effect transistors (JLFETs). We observed that the parasitic BJT action in JLFETs increases the OFF-state leakage current significantly and hinders their scaling to the sub-10 nm regime. Furthermore, we also analyzed the device architectures proposed to mitigate the parasitic BJT action in JLFETs [1–13].

In this chapter, we analyze another important phenomenon: impact ionization and its effect on the performance of the JLFETs. As we see, the different architecture of JLFETs as compared to the conventional metal-oxide-semiconductor field-effect transistors (MOSFETs) leads to an altogether different behavior with respect to the onset of impact ionization induced steep subthreshold swing. This leads to several interesting physical phenomenon in JLFETs such as occurrence of a zero gate oxide thickness coefficient ($Zt_{ox}C$) and single transistor latch-up, which open up a new horizon for application of JLFETs in memories such as capacitorless dynamic random access memory (DRAM) and as biosensors. First, we discuss in detail the physical essence of the impact ionization phenomenon before analyzing its implications on the JLFETs.

Junctionless Field-Effect Transistors: Design, Modeling, and Simulation, First Edition.
Shubham Sahay and Mamidala Jagadesh Kumar.

6.1 IMPACT IONIZATION

As the name suggests, impact ionization is a process in which the electrons impact (collide) with the atoms and if they gain sufficient kinetic energy before collision, they may even ionize the atoms and generate new electron–hole pairs (EHPs). Since this phenomenon requires the electrons to achieve a large kinetic energy before collision, it is observed only when there is a large electric field and the electrons travel sufficient distance under the influence of that field to gain enough energy to ionize the atoms. This phenomenon is visible in long-channel MOSFETs when the drain voltage is large. However, in the short-channel MOSFETs, impact ionization is initiated at a relatively smaller drain voltage.

You may wonder how much is the average lateral electric field in a typical MOSFET? For a MOSFET with a gate length of 1 μm operating at a drain voltage of 5.0 V, an approximate calculation reveals the average lateral electric field to be 50 kV/cm. However, for a MOSFET with a gate length of 10 nm operating at a drain voltage of 1.0 V, the lateral electric field becomes 1 MV/cm. As you can see, the magnitude of the electric field is very high especially for the short-channel MOSFETs. Therefore, the channel electrons gain high momentum and kinetic energy due to this lateral field and collide with the atoms exchanging momentum and energy. The collision may result in generation of an EHP due to the impact ionization mechanism if the electrons transfer sufficient energy to the atoms. The generated EHP may further gain sufficient kinetic energy in the presence of this high field and knock out more EHPs from other atoms. This leads to multiplication of the EHPs resulting in a positive feedback mechanism. As a result, there is an avalanche of carriers which flow from the source region to the drain region increasing the drain current significantly. This process may eventually lead to break down of the MOSFETs [14–17].

6.2 FLOATING BODY EFFECTS IN SILICON-ON-INSULATOR MOSFETs

The silicon-on-insulator (SOI) MOSFETs are segregated into two categories depending upon the extent of depletion of the channel region (body): (a) partially depleted SOI MOSFETs (PDSOI MOSFETs) (Fig. 6.1(a)) in which the SOI channel (body) is so thick that some amount of the channel region remains undepleted and neutral and (b) fully depleted (FD) SOI MOSFETs (FDSOI MOSFET) (Fig. 6.1(b)) in which the SOI film thickness is adjusted such that the entire silicon body is depleted and no neutral region is present.

As shown in Fig. 6.1, in the SOI MOSFETs, there is no contact electrode to the channel. Therefore, the channel region (body) is floating. The impact ionization phenomenon leads to altogether different characteristics in the SOI MOSFETs, known as the floating body effects, since the channel region (body) is floating in the SOI architecture [16]. These floating body effects are dominant in PDSOI MOSFETs and suppressed in FDSOI MOSFETs because of lack of neutral region as we will see in the later sections.

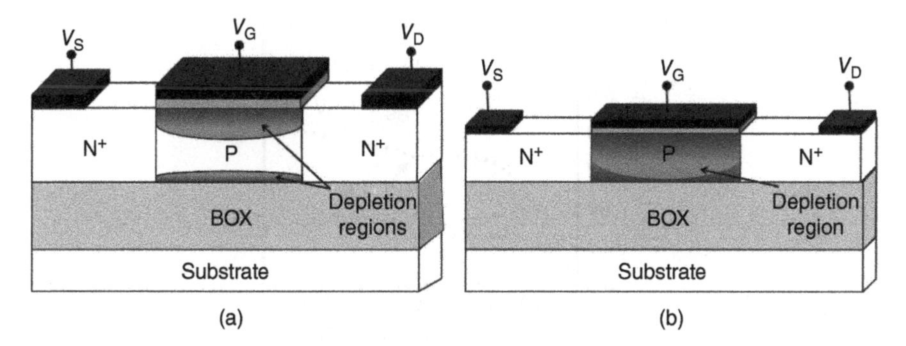

FIGURE 6.1 Three-dimensional view of (a) PDSOI MOSFET and (b) FDSOI MOSFET.

6.2.1 Hole Accumulation

The impact ionization process leads to generation of EHPs. The electrons generated due to the impact ionization process may easily flow through the drain due to the large drain field. The impact ionization generated holes; however, see a barrier equal to the source-to-channel barrier height while moving into the source under the influence of this electric field as shown in Fig. 6.2. Therefore, they are not able to flow to the source and start accumulating in the channel region.

The accumulation of holes leads to an increase in the potential of the channel region as accumulation of positive charge may be perceived as an externally applied positive potential. The application of a positive potential to the channel region reduces the threshold voltage of the MOSFETs. Therefore, the hole accumulation leads to a dynamic reduction in the threshold voltage, which further increases the drain current [16–24]. An increase in the drain current leads to a further increase in the impact ionization process and creates more EHPs. This in turn leads to a significant increase

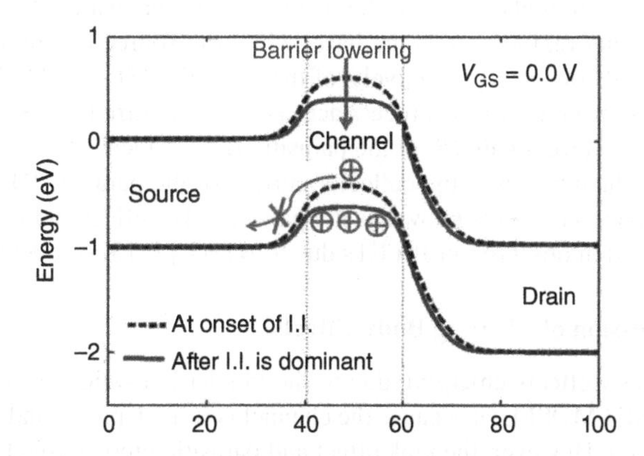

FIGURE 6.2 Energy band profiles of the PDSOI MOSFET at the onset of impact ionization (I.I.) and after the impact ionization becomes the dominant mechanism governing the drain current.

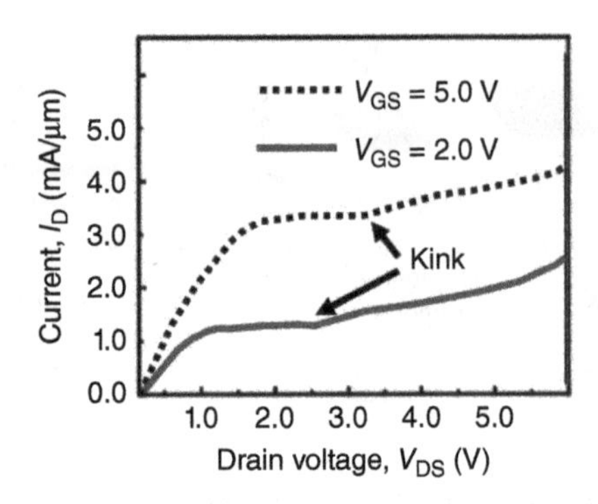

FIGURE 6.3 The output characteristics of a PDSOI MOSFET exhibiting the kink effect [19].

in the accumulation of holes which further lowers the threshold voltage. Therefore, this positive feedback effect leads to a drastic increase in the current through the MOSFET and eventually leads to its breakdown.

6.2.2 Kink Effect

The increase in the channel potential due to the accumulation of holes forward biases the channel–source p–n junction and lowers the potential barrier as shown in Fig. 6.2, leading to a significant injection of holes from the channel to the source region. This injection of holes due to the forward bias of the channel–source p–n junction results in a kink in the output characteristics of the PDSOI MOSFETs as shown in Fig. 6.3 and is known as the "kink" effect [16–24].

6.2.3 Parasitic BJT Action

The hole accumulation also leads to the formation of a parasitic BJT in SOI MOSFETs. The channel region acts as the base, while the source and drain regions act as the emitter and collector, respectively, of the parasitic BJT [18–24]. The hole current due to the impact ionization process acts as the base current. This hole current is multiplied by the current gain (β) of the parasitic BJT and leads to an increase in the drain current, which becomes the collector current of the parasitic BJT as shown in Fig. 6.4. This phenomenon is known as the parasitic BJT effect and is similar to the parasitic BJT action observed in JLFETs due to BTBT [1–13] (see Section 5.2).

6.2.4 Suppression of Floating Body Effects

The floating body effects emerging due to the impact ionization are relatively suppressed in FDSOI MOSFETs because the channel is fully depleted and hole accumulation is hindered. However, the kink effect and parasitic bipolar effect are dominant in PDSOI MOSFETs and result in a degradation of the breakdown voltage [18–24]. Therefore, several architectures like a dual source, SiGe source, lightly doped source

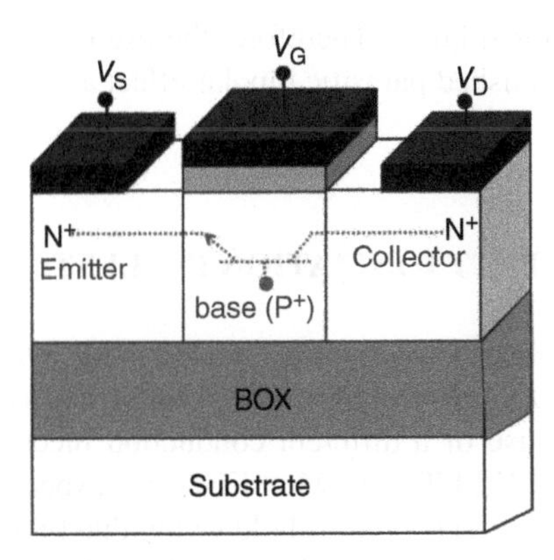

FIGURE 6.4 Parasitic BJT in the PDSOI MOSFET due to impact ionization induced hole accumulation in the body.

along with Ge and proton implantation in the source–drain regions [18–24] have been proposed to mitigate the parasitic bipolar effects in PDSOI MOSFETs. The use of a lightly doped source reduces the emitter injection efficiency of the parasitic BJT, whereas the implantation of Ge in channel and proton implantation in the source/drain regions create recombination centers within the channel and source/drain regions and aids in reducing the life time of the carriers [20]. This reduces their diffusion length and diminishes the parasitic BJT effect. The dual source topology [22–24] utilizes a p^+ Si or SiGe source below the regular n^+ source. The p^+ source (Si) extends into the channel region and is at a lower potential energy compared to the channel region as it is heavily doped. The holes generated due to the impact ionization are carried to the extended p^+ source as it is at a lower potential from where they can be easily collected by the source electrode as shown in Fig. 6.5. Furthermore, due to the band offset, a SiGe-extended source is more efficient than the Si-extended source in attracting the

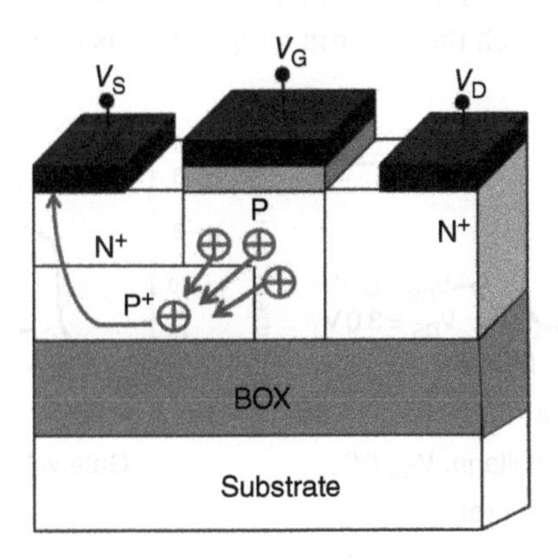

FIGURE 6.5 Three-dimensional view of the dual-source PDSOI MOSFET with a provision for collection of impact ionization generated holes [23].

impact ionization generated holes. Therefore, the use of a dual source architecture [22–24] results in a diminished parasitic bipolar effect and the kink effect in PDSOI MOSFETs.

6.3 NATURE OF IMPACT IONIZATION IN JLFETs

Having examined the impact ionization phenomenon in SOI MOSFETs, let us now discuss the impact ionization mechanism in JLFETs and compare it with respect to the MOSFETs. Because of a different conduction mechanism, the impact ionization properties of the JLFETs and MOSFETs are expected to be different. The impact ionization induced steep subthreshold swing due to the floating body "kink" effect occurs at a low drain voltage (~1.75 V) in JLFETs which is nearly half of the value of drain voltage required to achieve steep subthreshold slope in the MOSFETs (~3.5 V) [25] as shown in Fig. 6.6.

6.3.1 Steep Subthreshold Swing at a Low Voltage in JLFETs

The occurrence of impact ionization induced steep subthreshold swing in JLFETs at a lower drain voltage can be understood from Fig. 6.7. The region where the lateral electric field due to the drain region peaks and accelerates the electrons for impact ionization is located at the channel–drain interface in case of a MOSFET due to the presence of metallurgical junctions. However, the peak of the lateral electric field is located within the drain region in the JLFETs due to the electrostatic pinch-off mechanism rather than metallurgical junction induced depletion of the channel region [25].

The electrons under the influence of such high electric field acquire sufficient energy and their temperature increases. The electron temperature is lower in the JLFETs as compared to MOSFETs close to the drain region (Fig. 6.8) due to the reduced electric field in JLFETs as compared to MOSFETs. However, as shown in Fig. 6.8, the area over which the electron temperature is high in JLFETs exceeds that

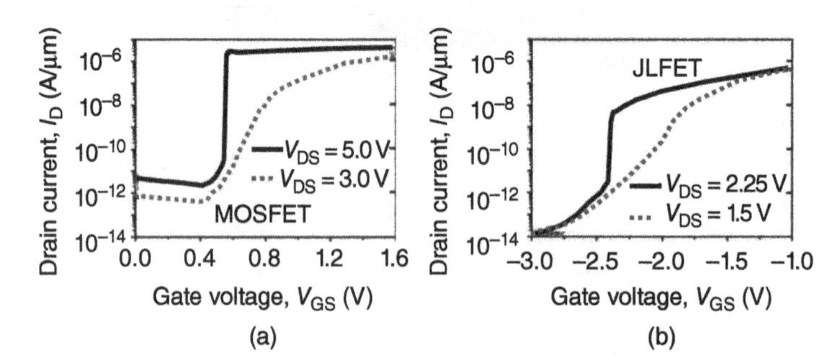

FIGURE 6.6 Transfer characteristics of (a) MOSFETs and (b) JLFETs for different drain voltages [25].

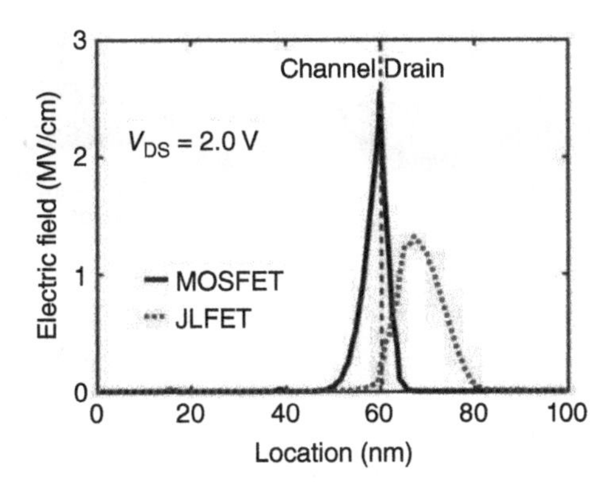

FIGURE 6.7 Lateral electric field profile of the MOSFET and the JLFET in the OFF-state.

of the MOSFETs. Therefore, impact ionization occurs at a lower drain voltage in JLFETs as compared to the MOSFETs [25]. Also, the use of a highly doped silicon film leads to a lower band gap in the JLFETs as compared to the MOSFETs increasing the impact ionization rate [26]. In addition, the impact ionization occurs in the bulk of the silicon film in JLFETs as opposed to the surface inversion layer in case of MOSFETs [26]. The impact ionization coefficients are higher in the bulk than at the surface.

6.3.2 Nature of Floating Body Effect in JLFETs

Another salient feature of the floating body effect in JLFETs is that it is quite dynamic compared to the MOSFETs. Above threshold voltage, the electrons flow in the bulk of the channel in JLFETs. When the drain voltage is large, the high electric field leads to impact ionization and consequent generation of EHP. The electrons generated due

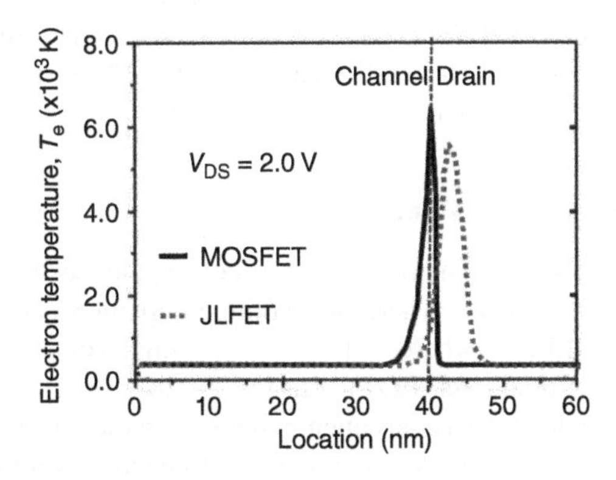

FIGURE 6.8 Electron temperature profile of the MOSFET and the JLFET in the OFF-state [25].

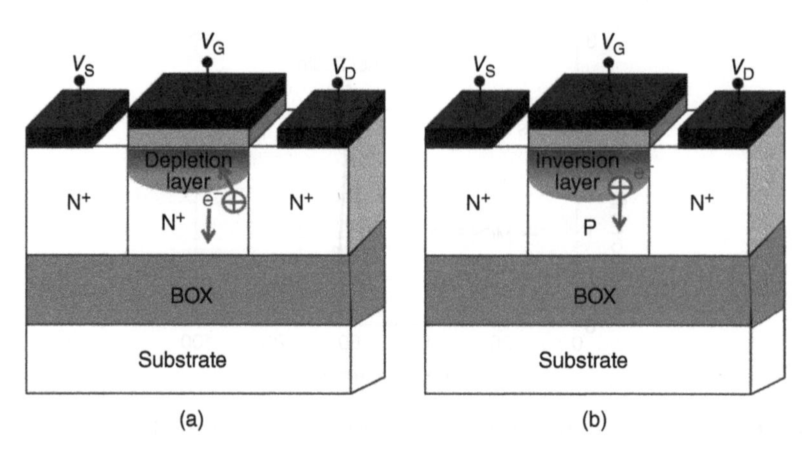

FIGURE 6.9 Transport of impact-generated electrons and holes in (a) JLFETs and (b) MOSFETs.

to the impact ionization are depleted from the surface due to a difference in the gate work function and silicon work function and flow into the bulk, further increasing the electron current [25, 26] as shown in Fig. 6.9(a).

However, the generated holes move close to the Si–SiO_2 surface (which is at lower potential energy) and increase the channel potential leading to a reduction in the threshold voltage and the consequent feedback loop which originates the "kink" effect [26]. However, the "kink" effect is not so pronounced in JLFETs owing to the dynamic recombination of the impact ionization–generated holes with the majority electrons flowing in the bulk. It may be noted that for large overdrive voltages ($V_{GS} - V_{th}$), even at high drain voltages, the JLFETs exhibit drift–diffusion conduction mechanism because of recombination of the excess carriers due to impact ionization [26]. In the MOSFETs, on the other hand, the impact ionization occurs in the inversion layer and the generated holes flow into the center of the channel as opposed to the surface in the case of JLFETs as shown in Fig. 6.9(b). This restricts the interaction between the generated holes and electrons in MOSFETs. Therefore, the impact ionization process is more dynamic in case of JLFETs.

6.3.3 Impact of Gate Oxide Thickness

As the gate oxide thickness is increased, the effective gate electric field that depletes the channel region is reduced. Therefore, volume depletion occurs at a more negative voltage for n-JLFETs whereas it occurs at a high voltage for p-JLFETs if a larger effective oxide thickness (EOT) is used [27, 28]. The shift in the voltage for the volume depletion leads to a consequent reduction in the threshold voltage, which is generally defined as the gate voltage required to achieve volume depletion [27]. Therefore, a larger gate oxide thickness leads to a lower threshold voltage, which results in a higher drain current and further exaggerates the impact ionization process

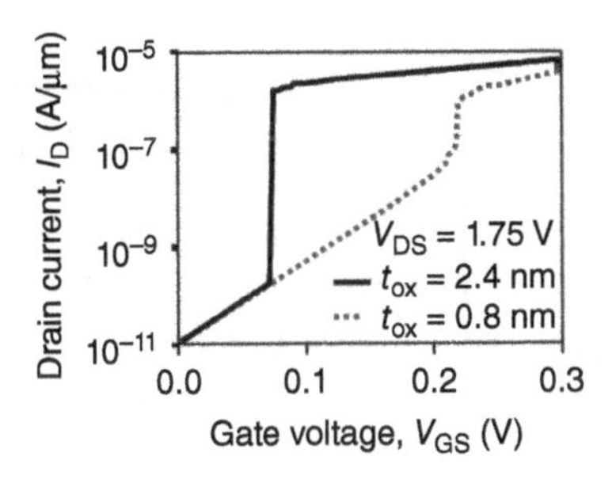

FIGURE 6.10 Impact of gate oxide thickness (t_{ox}) on steep switching of JLFETs [27].

and facilitates bipolar effects in n-JLFETs at lower gate voltages as shown in Fig. 6.10. Therefore, the use of a higher EOT leads to a higher drain current at a lower gate voltage in n-JLFETs unlike MOSFETs in the impact ionization dominant regime. Therefore, scaling the oxide thickness leads to a lower drain current in JLFETs in the impact ionization–dominant regime unlike MOSFETs [27, 28].

6.4 ZERO GATE OXIDE THICKNESS COEFFICIENT

Another implication of the parasitic bipolar effect in JLFETs is the occurrence of a zero gate oxide thickness coefficient $Zt_{ox}C$ [28], which signifies the transition from the bipolar operation regime to the unipolar operation mode in JLFETs. At the gate voltage corresponding to the ZtoxC (V_{ztox}), the JLFETs exhibit the same drain current irrespective of the gate oxide thickness.

As discussed in Section 6.3.3, an increase in the gate oxide thickness exaggerates the impact ionization due to a reduction in the threshold voltage and the consequent increase in the drain current unlike MOSFETs. However, as the impact ionization process is dynamic in case of JLFETs, the excess holes recombine with the conduction electrons at higher gate voltages close to the flat band voltage restoring the unipolar operation in JLFETs prior to a flat band condition. Therefore, the bipolar effects in JLFETs diminish as the gate voltage is increased above the threshold voltage before attaining the flat band condition as discussed in Section 6.4.2.

The JLFETs in the unipolar operation regime (close to the flat band condition) exhibit a reduction in the drain current with increasing gate oxide thickness due to a reduced gate control of the channel region similar to the MOSFETs while the current increases in the bipolar operation regime with increasing gate oxide thickness. Therefore, there exists a gate voltage known as the $Zt_{ox}C$ voltage (V_{ztox}), which signifies the transition from the bipolar mode of operation to the unipolar mode as shown in Fig. 6.11. At gate voltages above the V_{ztox}, drain current decreases upon increasing the oxide thickness owing to the unipolar operation while below V_{ztox} the drain current increases upon increasing the oxide thickness due to the dominance of the bipolar operation mode. At V_{ztox}, the drain current in JLFETs remains same for all the values

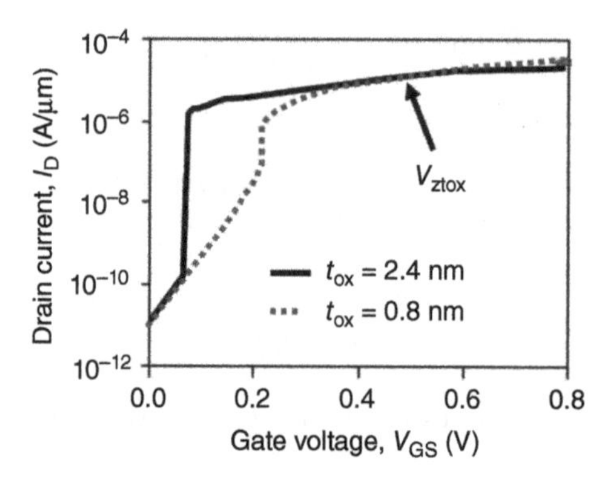

FIGURE 6.11 Transfer characteristics of the JLFETs for different gate oxide thickness (t_{ox}) [28].

of oxide thickness. Therefore, at V_{ztox}, the drain current is insensitive to the variation in the oxide thickness in JLFETs as shown in Fig. 6.11.

However, the Zt$_{ox}$C is not observed in (a) JLFETs with large-channel and gate oxide thickness at high drain voltage where impact ionization–induced bipolar effects are significant or (b) JLFETs with low channel and gate oxide thickness at low drain voltages where the unipolar conduction mode is dominant and the probability of impact ionization is close to zero.

6.4.1 Significance of the Zero Gate Oxide Coefficient

The V_{ztox} signifies the crossover of the majority carriers in the JLFETs. When the gate voltage is reduced below flat band, the electrons concentration decreases due to the gate electrodes with high work function while the holes generated due to impact ionization accumulate at the surface. If we continue to reduce the gate voltage, we arrive at a condition where the peak hole concentration increases above the peak electron concentration [28]. This crossover of the peak hole concentration above the peak electron concentration signifies the change of the operating mode from unipolar (majority electrons) to bipolar mode where the contribution from holes is significant in turning ON the parasitic BJT as shown in Fig. 6.12.

6.4.2 Implication of the Zero Gate Oxide Coefficient on Electric Field

As discussed in Section 3.4.1, the JLFETs exhibit a high transverse electric field in the full depletion mode. The transverse electric field diminishes significantly as the gate voltage is increased and JLFETs operate in partial depletion mode as shown in Fig. 6.13. Subsequently, the transverse electric field attains its minimum value at flat band and again begins to increase as the gate voltage is increased to operate JLFETs in accumulation mode. The transverse electric field in JLFETs with low gate oxide

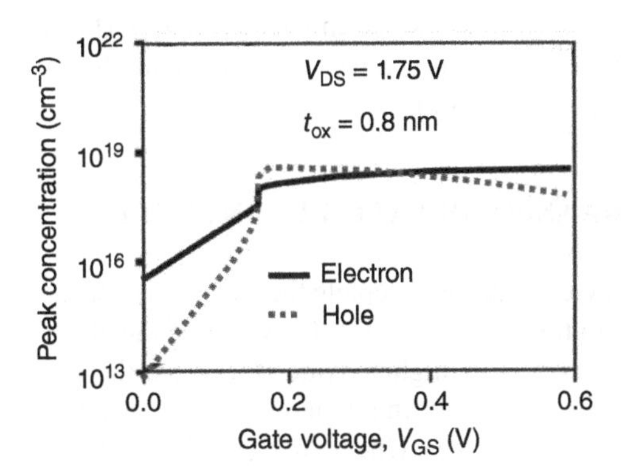

FIGURE 6.12 Peak electron and hole concentration in JLFETs at $V_{DS} = 1.75$ V [28].

thickness is higher as compared to the JLFETs with a higher gate oxide thickness due to a better gate control in JLFETs with low oxide thickness. Therefore, the variation in the electric field is also higher in the JLFETs with a lower gate oxide thickness for the same range of gate voltage.

The electric field at low gate voltages would be higher in the JLFETs with a lower gate oxide thickness (t_{ox}). However, the electric field at high gate voltages would be higher in the JLFETs with a large t_{ox} due to the onset of flat band condition at a higher gate voltage as compared to JLFETs with a low t_{ox} [28]. Therefore, there must exist a gate voltage where the transverse electric field would be same in the JLFETs with high and low t_{ox}. A similar electric field distribution would lead to the same drain current in JLFETs irrespective of parameters such as t_{ox}. This voltage at which same electric field is obtained in JLFETs irrespective of the value of t_{ox} also corresponds to V_{ztox}. It may also be noted that the drain current at V_{ztox} is not the same as the drain

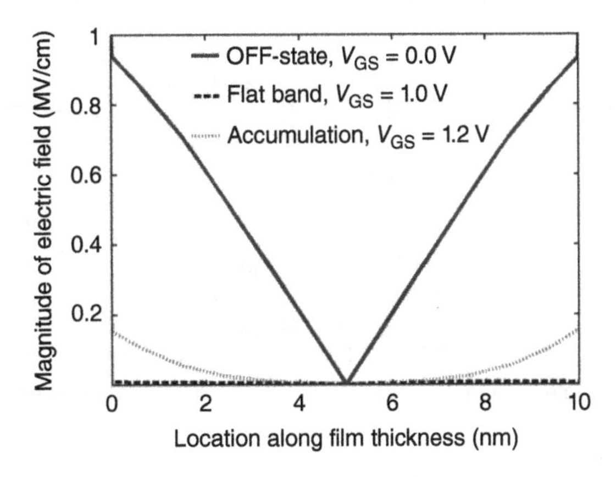

FIGURE 6.13 Transverse electric field along the silicon film thickness of a double gate junctionless field-effect transistor (DGJLFET) in different operating regimes.

current in JLFETs designed to have no impact ionization (low channel thickness and small gate oxide thickness). This is attributed to the complete absence of holes in the JLFETs with no impact ionization.

6.5 SINGLE TRANSISTOR LATCH-UP IN JLFETs

In JLFETs, the gate electrode must deplete the excess carriers generated by the impact ionization process to turn off the device. However, if the JLFET is designed to have a higher channel thickness or higher oxide thickness, the threshold voltage would reduce. Therefore, this will result in an increase in the drain current and impact ionization rate at lower gate voltages. Under such conditions, the gate fails to deplete the excess carriers and the positive feedback loop of impact ionization is not eliminated. Therefore, the impact ionization–induced bipolar effects may lead to the single transistor latch-up phenomena where the JLFET is unable to turn off even if a negative gate voltage is applied [29] as shown in Fig. 6.14.

You may wonder what leads to the single transistor latch-up phenomenon in JLFETs. The excess holes accumulated in the channel region convert the n-type channel into a p-type region. This results in a n^+–p junction at the source–channel interface. As the hole concentration in the channel increases due to the impact ionization, the potential of the channel region increases resulting in an internal forward biasing of this n^+–p junction. The source-to-channel barrier height is significantly reduced because of this internal forward bias mechanism, which indicates that the gate control is lost in JLFETs. Therefore, the excess carriers are not depleted even when the gate voltage is reduced. As a rule of thumb, the product of electric field and current density, which is the measure of impact ionization generated power in the device per unit volume, should not exceed 10^6 AV cm^{-3} to avoid the latch condition [29].

The use of a higher drain voltage or a higher silicon film doping can also result in single transistor latch-up phenomenon as shown in Fig. 6.15. An increase in the

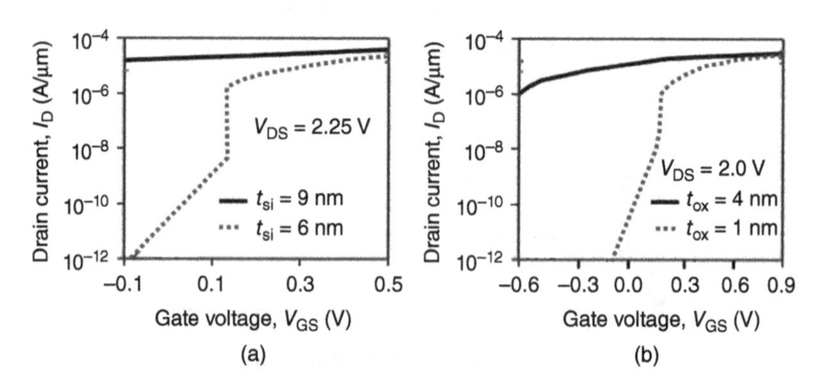

FIGURE 6.14 Single transistor latch-up phenomenon in JLFETs owing to increasing (a) silicon film thickness (t_{Si}) and (b) gate oxide thickness (t_{ox}) [29].

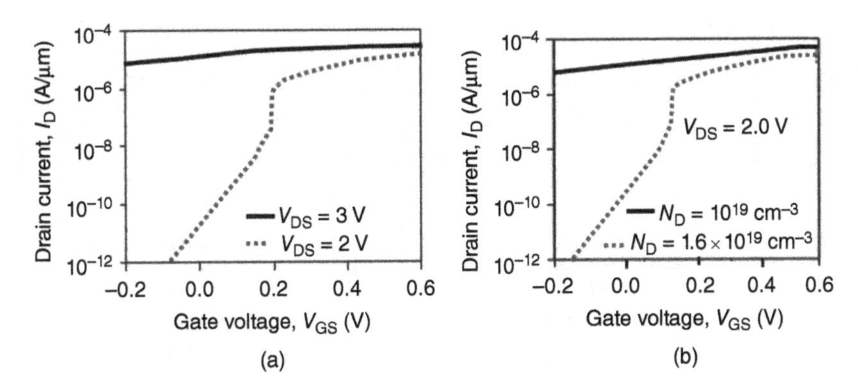

FIGURE 6.15 Single transistor latch-up phenomenon in JLFETs owing to increasing (a) drain voltage (V_{DS}) and (b) silicon film doping (N_D) [29].

silicon film doping leads to an increase in the drain current and increases the impact ionization rate due to band gap narrowing and increased electric field.

6.6 IMPACT OF BODY BIAS ON IMPACT IONIZATION IN JLFETs

The impact ionization–induced floating body effects, which trigger the steep subthreshold swing, are not expected to occur below a drain voltage of 1.5 times the band gap, which is approximately 1.65 V for Si [30]. However, with the aid of a positive substrate bias, impact ionization induced steep subthreshold swing has been observed in JLFETs even at a drain voltage of 1.4 V [30] as shown in Fig. 6.16.

The substrate bias increases the effective electron concentration as well as the electric field as shown in Fig. 6.17, which facilitates more impact ionization and carrier

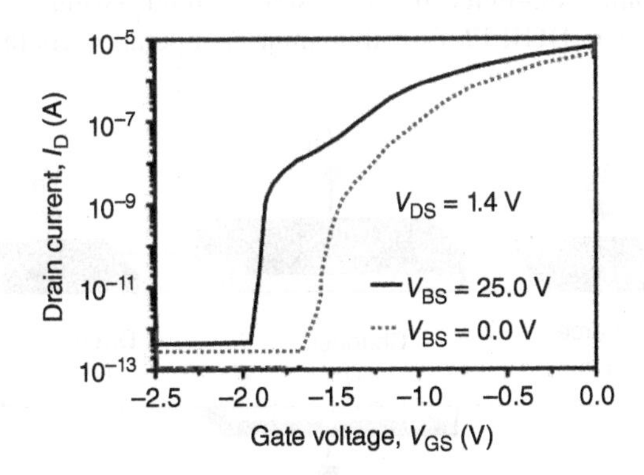

FIGURE 6.16 Transfer characteristics of the JLFETs for different substrate voltage (V_{BS}) [30].

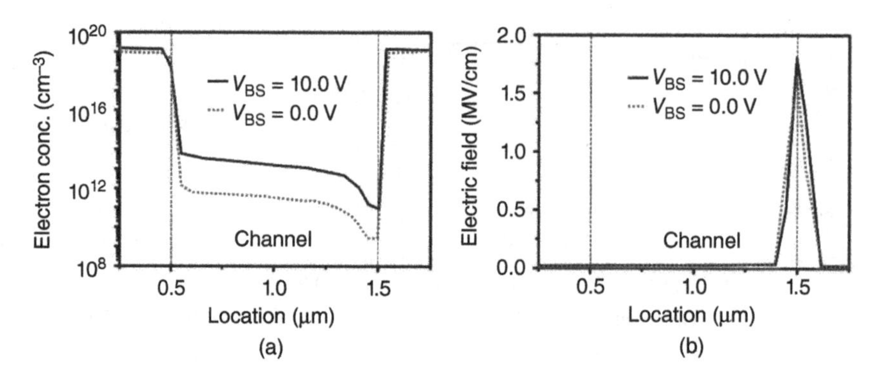

FIGURE 6.17 (a) The electron concentration and (b) electric field profile of the JLFETs with different substrate bias [30].

multiplication [30]. However, impact ionization induced steep subthreshold swing reduces in ultrathin body JLFETs due to increased surface recombination and scattering and the consequent reduction in the drain current and floating body effects [26, 30].

6.7 SUBBAND GAP IMPACT IONIZATION IN DGJLFETS WITH ASYMMETRIC OPERATION

You may wonder whether it is possible to induce impact ionization in JLFETs at subband gap drain voltages. To your surprise, it is indeed possible to achieve impact ionization based steep subthreshold swing in DGJLFETs (Fig. 6.18) at a subband gap drain voltage (< 1.12 V) using asymmetric operation mode (both gates are independent and connected to different voltages, i.e., $V_{Gf} \neq V_{Gb}$) as shown in Fig. 6.19. While impact ionization–induced steep subthreshold swing occurs at a drain voltage of ~2.25 V in DGJLFETs with symmetric operation (both gates tied to the

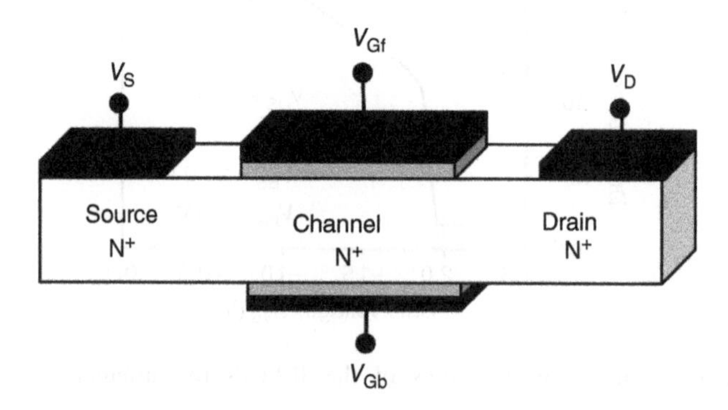

FIGURE 6.18 Three-dimensional view of a DGJLFET. V_{Gf} and V_{Gb} are the applied voltages at the front gate and the back gate, respectively.

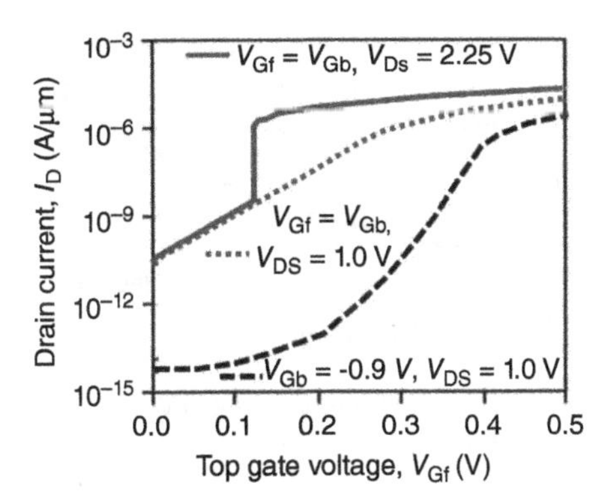

FIGURE 6.19 Transfer characteristics of the DGJLFET in symmetric mode and asymmetric mode [31].

same voltage, i.e., $V_{Gf} = V_{Gb}$), it can be achieved at a drain voltage of 1 V when the back gate is fixed to a negative gate bias ≤ −0.5 V [31].

6.7.1 Operation

The maximum energy (W) gained by an electron can be obtained as an integral over the electric field E for a differential length dl as [31]

$$W = \int \vec{E}.\vec{dl} = -q(V_{DS} + \Delta\varphi) \tag{6.1}$$

where V_{DS} is the drain voltage and $\Delta\varphi$ is the source-to-channel barrier height. This energy W along with the product of current density and electric field is responsible for impact ionization in JLFETs [31]. While the magnitude of $\Delta\varphi$ is low for the symmetric operation mode of DGJLFETs, the source-to-channel barrier height can be modulated and increased using a negative back gate voltage in asymmetric operation mode as shown in Fig. 6.20. Therefore, the energy gained by the electron can be

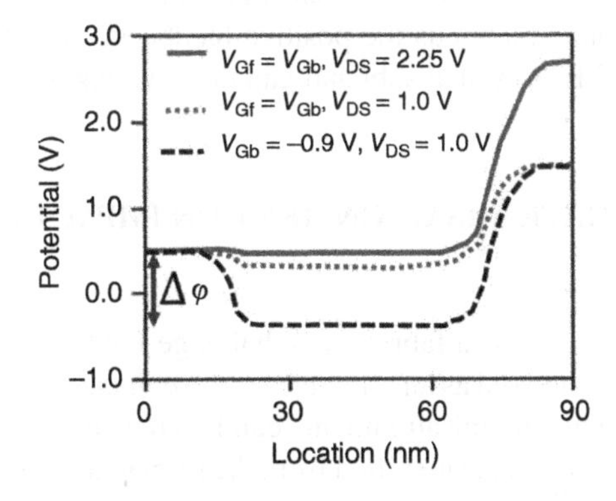

FIGURE 6.20 The potential profile of the JLFETs in symmetric and asymmetric mode [31].

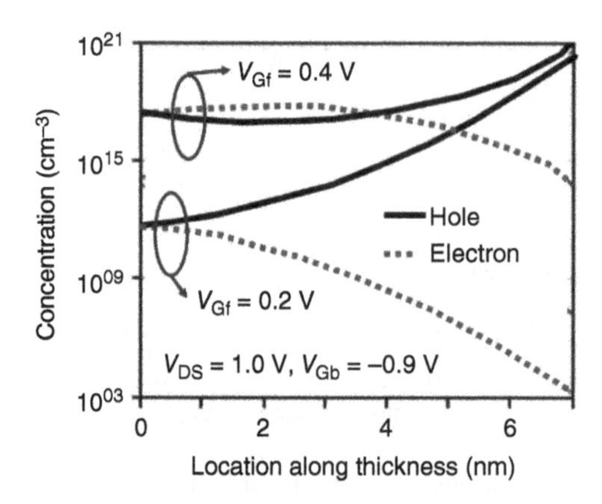

FIGURE 6.21 The carrier concentration profile of the JLFETs in the asymmetric mode at different front gate voltages [31].

increased in DGJLFETs exploiting the independent operation of the two gates even at a low drain voltage (1.0 V).

Another salient feature of the asymmetric operation mode in JLFETs with a negative back gate voltage is that the conduction channel is formed above the center of the silicon film due to the increased depletion from the back gate owing to negative gate voltages as shown in Fig. 6.21. If we continue to reduce the back gate voltage, the silicon film should get depleted further and the conduction path is expected to move toward the front surface. This is indeed the case when impact ionization is not considered. However, when the impact ionization is considered, the conduction path remains fixed at a distance $t_{Si}/4$ (where t_{Si} is the channel thickness) from the front surface irrespective of the negative gate voltage applied on the back gate owing to the carriers generated due to impact ionization [31]. The formation of conduction channel away from the front surface not only leads to a lower surface scattering, which increases the energy relaxation length leading to enhancement in the electron temperature and the impact ionization rate, but it also prevents the loss of holes due to surface recombination, which aids the positive feedback. Therefore, impact ionization is possible in JLFETs even at subband gap drain voltages.

6.8 IMPACT OF GATE MISALIGNMENT ON IMPACT IONIZATION IN DGJLFETs

Alignment of the two gates is a fabrication challenge for DGJLFETs. However, the gate misalignment can be used as an effective tool to enhance the impact ionization in DGJLFETs. Intentional gate misalignments can be created by shifting the electrical Vernier [32, 33]. If the back gate is shifted from the front gate by a distance equal to the

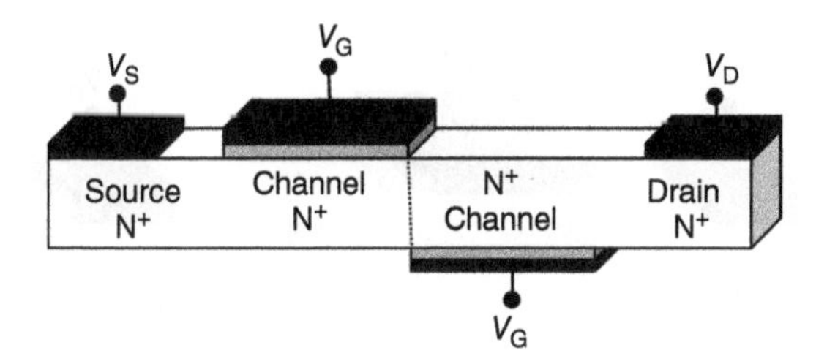

FIGURE 6.22 Three-dimensional view of a DGJLFET with the back gate misaligned with respect to the front gate by a distance equal to the gate length.

channel length (Fig. 6.22), steep subthreshold swing can be obtained in DGJLFETs even for a drain voltage of 0.9 V [34] as shown in Fig. 6.23.

6.8.1 Operation

The operation of the misaligned DGJLFET can be understood from Fig. 6.24. The front gate depletes the electrons from the surface and pushes them toward the back gate. The electrons pushed by the top gate then move toward the drain under the influence of the lateral drain field. However, when they encounter the back gate, it depletes the electrons from the bottom surface and pushes them toward the front surface. Therefore, the electrons not only move toward the drain but are also pushed toward the surface which increases the number of collisions and leads to an enhancement in the impact ionization. This inclined motion of electrons facilitates steep subthreshold swing in misaligned DGJLFETs at subunity drain voltage [34].

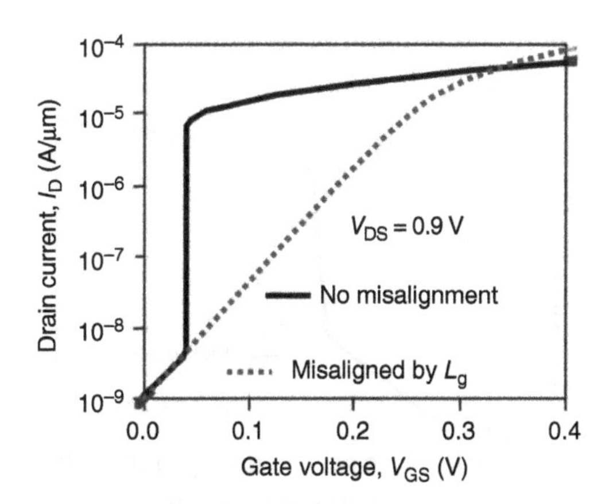

FIGURE 6.23 Transfer characteristics of the DGJLFET with and without misalignment between the front and back gates [34].

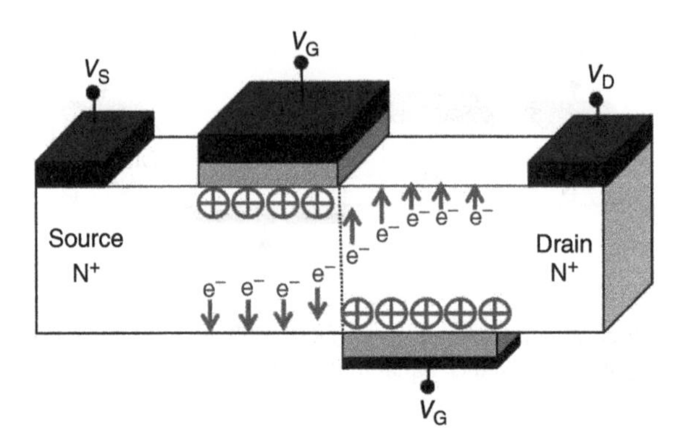

FIGURE 6.24 Movement of electrons in an inclined way in the misaligned DGJLFET.

6.9 SPACER DESIGN GUIDELINE FROM IMPACT IONIZATION PERSPECTIVE

As discussed in Section 4.9, the inclusion of the high-κ spacer enhances the effective channel length of the JLFETs which in turn reduces the short-channel effects. The inclusion of high-κ spacer leads to a redistribution in the electric field such that the lateral drain electric field is reduced at the channel–drain interface while the fringing fields increase through the spacer. This leads to a larger effective channel length compared to the physical gate length owing to the enhanced fringing fields and consequent reduction in the current density due to lower lateral drain field which induces impact ionization. Therefore, a high-κ spacer is not favorable for impact ionization–induced steep subthreshold swing in JLFETs [35] as shown in Fig. 6.25.

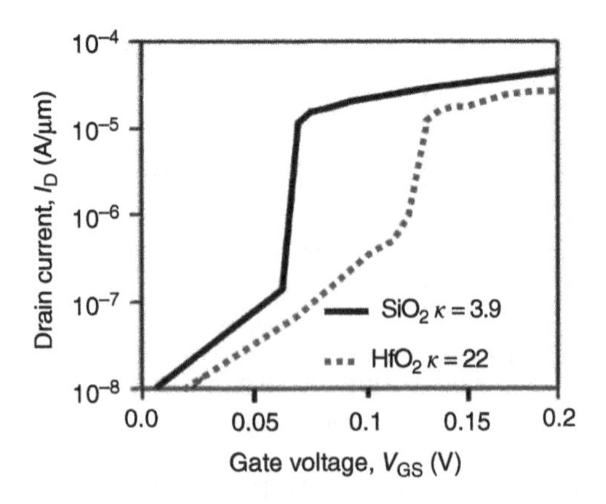

FIGURE 6.25 Transfer characteristics of the DGJLFET with different spacers [35].

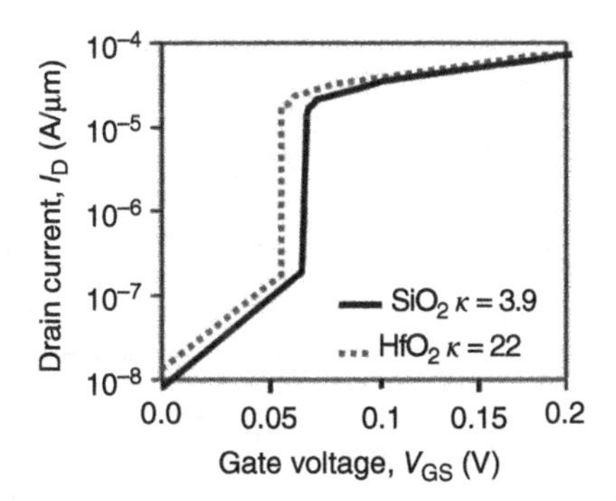

FIGURE 6.26 Transfer characteristics of the DGJLFET with different gate dielectrics with thickness adjusted to obtain the same EOT [35].

For same EOT, a high-κ gate dielectric needs to be thicker. The larger thickness of the high-κ gate dielectric for the same EOT leads to somewhat lower gate control since most of the fringing field lines terminate at the source/drain regions. A reduced gate control increases the influence of the lateral drain field which leads to an enhanced impact ionization in JLFETs with high-κ gate dielectric as shown in Fig. 6.26. Therefore, the use of a high-κ gate dielectric and a low-κ gate sidewall spacer can improve the performance of JLFETs from impact ionization triggered bipolar action perspective [35].

6.9.1 Impact of Temperature

The impact ionization increases in JLFETs at low temperature owing to the increase in mobility, which leads to an increase in the current density. As a result, the product of current density and electric field increases and governs the impact ionization phenomenon in JLFETs. However, such a low-temperature operation may even lead to single transistor latch-up of JLFETs [35] with improperly designed channel thickness or gate oxide thickness. The use of a wider spacer with low-κ dielectric or a narrow spacer with high-κ dielectric can reduce the impact ionization. Therefore, spacer engineering can be used as an effective design parameter to mitigate the single transistor latch-up in JLFETs [35] with a thicker channel or gate oxide at low temperatures.

6.10 HYSTERESIS AND SNAPBACK IN JLFETs

The impact ionization–induced steep switching in JLFETs is attributed to the enhanced bipolar effects. Another salient feature of this steep switching is that the gate voltage at which the impact ionization–induced bipolar action diminishes during the reverse sweep may not be same as the gate voltage at which impact ionization and

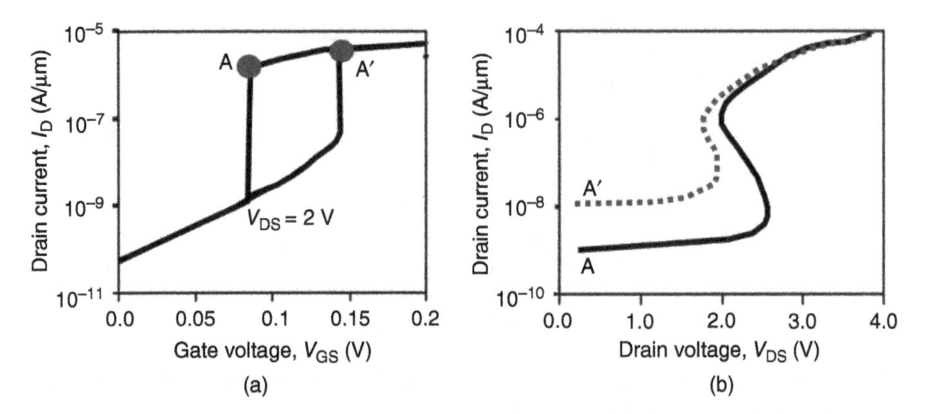

FIGURE 6.27 (a) The hysteresis in the transfer characteristics and (b) the snapback in the output characteristics for the DGJLFETs [36].

bipolar effects are initiated in the forward sweep. Generally, the gate voltage required to completely deplete the excess carriers for turning OFF the JLFETs is more negative as compared to the gate voltage at which impact ionization is triggered. This results in a hysteresis in the transfer characteristics of the JLFETs in the impact ionization–dominant regime, a typical bipolar signature.

In JLFETs, the hysteresis in the transfer characteristics is accompanied with a snapback effect in the output characteristics similar to the PDSOI MOSFETs as shown in Fig. 6.27 [36]. Also, the largest snapback in the output characteristics occurs at the lowest gate voltage in the hysteresis observed in the transfer characteristics. Just before the snapback effect, the impact ionization starts in the JLFETs and the electron and hole concentrations increase in the middle of the silicon film. However, the electrons get depleted by the gate electrode and are confined to flow in the middle of the film while the holes flow toward the Si–SiO_2 surface as it is at a lower potential energy as discussed in Section 6.3.2. At the onset of snapback, the hole concentration becomes more than the electron concentration at the center of the channel in JLFETs. The hole concentration in the middle of the silicon film needs to be at least 10^{17} cm^{-3} to trigger the snapback phenomenon.

The accumulated holes reduce the potential of the channel region and lower the source-to-channel potential barrier height. This internally forward biases the source–channel electrostatic junction (formed as a result of hole accumulation in the channel region) leading to injection of minority carriers, i.e. electrons in the channel region. This in turn results in a larger impact ionization generating more holes and reducing the barrier height further. Therefore, the bipolar action leads to a positive feedback mechanism, which increases the electron concentration significantly. The channel resistance reduces considerably due to the accumulation of a large number of carriers. This leads to the negative differential resistance characteristics in JLFETs leading to the snapback. The snapback simply means that once appreciable amount of carriers is accumulated in the channel region, the same drain current may be obtained even

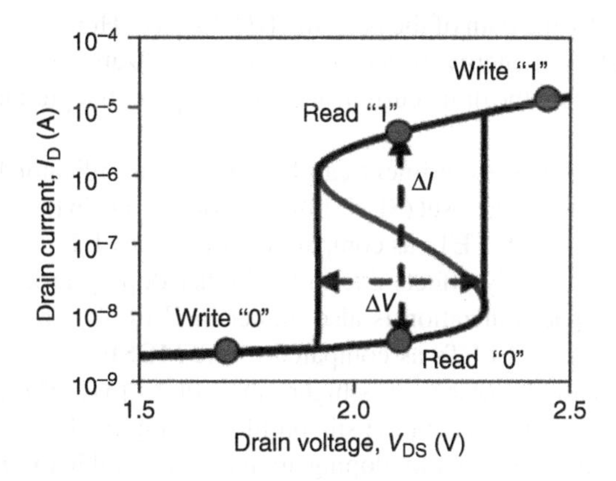

FIGURE 6.28 The hysteresis in the output characteristics of the DGJLFET obtained during voltage-controlled measurements and the different programming voltages for the application of DGJLFETs as capacitorless DRAM [36].

at a lower drain voltage. The snapback is obtained in current-controlled measurements where current is ramped (and voltage is measured) as shown in Fig. 6.27(b). In voltage-controlled measurements (voltage is swept and the current is measured), instead of snapback, a hysteresis is observed in the output characteristics as shown in Fig. 6.28.

However, as discussed in Section 6.3.2, the bipolar action in JLFETs is more dynamic than the MOSFETs and recombination process sets in once the electron concentration rises to a very large value. The recombination process restores the unipolar mode of conduction in the JLFETs leading to an increase in the drain current with drain voltage and terminates the snapback characteristics.

It may, however, be noted that the snapback in the output characteristics depends on the silicon film and gate oxide thickness. Increasing the silicon film thickness and the gate oxide thickness not only leads to a larger floating body effect but also increases the snapback window. Furthermore, the snapback window is more susceptible to the silicon film thickness as compared to the gate oxide thickness. Therefore, these parameters must be adequately chosen while designing JLFETs for applications such as capacitorless DRAM to achieve a large snapback window.

6.11 IMPACT OF HEAVY-ION IRRADIATION ON JLFETs

Ion bombardment in MOSFETs leads to the generation of EHPs in the channel region, which may create additional EHPs due to impact ionization [37]. The generated holes accumulate in the channel region and result in parasitic bipolar action, which is similar to the impact ionization–induced floating body effect. Therefore, the channel potential is increased owing to the ion bombardment and the drain current shoots up due to the generated EHPs. The current generated by these excess carriers is further multiplied by the gain of the parasitic BJT formed. However, the excess carriers do not get sufficient energy to sustain the impact ionization process and generate more EHPs and, hence, the drain current and channel potential gradually drop to their initial values.

The ion bombardment experiment can be used as an indicator for immunity of devices against single event upset (SEU) [38]. The decay rate in the drain current after ion strike is lower in the JLFETs as compared to the MOSFETs. This is attributed to the increased floating body effects due to the higher doping in the channel region. In addition, the impact ionization is also higher in JLFETs. This leads to a larger collected charge in the JLFETs as compared to the MOSFETs, which clearly indicates that the parasitic bipolar gain is higher in JLFETs [37]. The bipolar gain further increases with an increase in the silicon film doping in JLFETs. Therefore, the JLFETs with increased silicon film doping are more vulnerable to SEU phenomenon and the immunity of JLFET to SEU is lower than the MOSFETs [37].

6.12 CONCLUSIONS

In this chapter, we discussed the implications of the impact ionization phenomenon on the JLFETs. We studied the different physical phenomena induced by the impact ionization of carriers in PDSOI MOSFETs. Using this understanding, we saw how the floating body effects in JLFETs are quite different and dynamic as compared to the MOSFETs. We also observed that the JLFETs tend to exhibit a high degree of impact ionization and the consequent steep subthreshold slope at a lower drain voltage as compared to the MOSFETs. This leads to several interesting phenomena in JLFETs such as the occurrence of ZtoxC, single transistor latch-up, and so on. The ZtoxC voltage signifies the change from the bipolar to unipolar conduction mode and a fixed drain current irrespective of the value of gate oxide thickness. In addition, we also discussed the impact of substrate bias on JLFETs. The floating body effects in JLFETs were also found to be more dynamic as compared to the MOSFETs. The efficacy of asymmetric mode of operation in DGJLFETs and gate misalignment in DGJLFETs in reducing the drain voltage required for impact ionization–induced steep subthreshold swing were also discussed. The spacer design guidelines were also given for the JLFETs. Although the JLFETs exhibit a steep subthreshold slope at a lower drain voltage, the heavily doped silicon film still leads to a large spatial (device to device) variability due to RDF as discussed in Chapter 3. Therefore, an altogether new class of dopantless junctionless devices was proposed to alleviate the need for chemical doping and the consequent RDF. We discuss these junctionless devices in the next chapter.

REFERENCES

[1] S. Sahay and M. J. Kumar, "Physical insights into the nature of gate-induced drain leakage in ultrashort channel nanowire FETs," *IEEE Trans. Electron Devices*, vol. 64, no. 6, pp. 2604–2610, June 2017.

[2] S. Sahay and M. J. Kumar, "A novel gate-stack-engineered nanowire FET for scaling to the sub-10-nm regime," *IEEE Trans. Electron Devices*, vol. 63, no. 12, pp. 5055–5059, Dec. 2016.

[3] S. Sahay and M. J. Kumar, "Spacer design guidelines for nanowire FETs from gate-induced drain leakage perspective," *IEEE Trans. Electron Devices*, vol. 64, no. 7, pp. 3007–3015, July 2017.

[4] S. Sahay and M. J. Kumar, "Insight into lateral band-to-band-tunneling in nanowire junctionless FETs," *IEEE Trans. Electron Devices*, vol. 63, no. 10, pp. 4138–4142, Oct. 2016.

[5] S. Sahay and M. J. Kumar, "Controlling L-BTBT and volume depletion in nanowire JLFETs using core-shell architecture," *IEEE Trans. Electron Devices*, vol. 63, no. 9, pp. 3790–3794, Sept. 2016.

[6] S. Sahay and M. J. Kumar, "Diameter dependency of leakage current in nanowire junctionless field-effect transistors," *IEEE Trans. Electron Devices*, vol. 64, no. 3, pp. 1330–1335, Mar. 2017.

[7] S. Sahay and M. J. Kumar, "Nanotube junctionless FET: Proposal, design, and investigation," *IEEE Trans. Electron Devices*, vol. 64, no. 4, pp. 1851–1856, Apr. 2017.

[8] M. J. Kumar and S. Sahay, "Controlling BTBT induced parasitic BJT action in junctionless FETs using a hybrid channel," *IEEE Trans. Electron Devices*, vol. 63, no. 8, pp. 3350–3353, Aug. 2016.

[9] S. Sahay, and M. J. Kumar, "Realizing efficient volume depletion in SOI junctionless FETs," *IEEE J. Electron Devices Soc.*, vol. 4, no. 3, pp. 110–115, May 2016.

[10] S. Sahay and M. J. Kumar, "Controlling the drain side tunneling width to reduce ambipolar current in tunnel FETs using heterodielectric BOX," *IEEE Trans. Electron Devices*, Vol. 62, no. 11, pp. 3882–3886, Nov. 2015.

[11] S. Sahay and M. J. Kumar, "Comprehensive analysis of gate-induced drain leakage in emerging FET architectures: Nanotube FETs vs. nanowire FETs," *IEEE Access*, vol. 5, pp. 18918–18926, Dec. 2017.

[12] S. Sahay and M. J. Kumar, "Symmetric operation in an extended back gate JLFET for scaling to the 5 nm regime considering quantum confinement effects," *IEEE Trans. Electron Devices*, vol. 64, no. 1, pp. 21–27, Jan. 2017.

[13] A. K. Jain, S. Sahay, and M. J. Kumar, "Controlling L-BTBT in emerging nanotube FETs using dual-material gate," *IEEE J. Electron Dev. Soc.*, vol. 6, pp. 611–621, June 2018.

[14] T. Toyabe, K. Yamaguchi, S. Asai, and M. S. Mock, "A numerical model of avalanche breakdown in MOSFET's," *IEEE Trans. Electron Devices*, vol. 25, no. 7, pp. 825–832, 1978.

[15] S. Tam, F. C. Hsu, P. K. Ko, C. Hu, and R. S. Muller, "Hot-electron induced excess carriers in MOSFET's," *IEEE Electron Device Lett.*, vo. 3, no. 12, pp. 376–378, 1982.

[16] N. O. Kotani, S. A. Kawazu, and S. H. Komori, "Behavior of holes generated by impact ionization in n-channel MOSFET's," *IEEE Trans. Electron Devices*, vol. 30, no. 12, pp. 1678–1680, 1983.

[17] A. O. Adan and K. Higashi, "OFF-state leakage current mechanisms in bulk Si and SOI MOSFETs and their impact on CMOS ULSIs standby current," *IEEE Trans. Electron Devices*, vol. 48, no. 9, pp. 2050–2057, 2001.

[18] J. G. Fossum, M. M. Pelella, and S. Krishnan, "Scalable PD/SOI CMOS with floating bodies," *IEEE Electron Device Lett.*, vol. 19, no. 11, pp. 414–416, 1998.

[19] J P. Colinge, "Reduction of kink effect in thin-film SOI MOSFETs," *IEEE Electron Device Lett.*, vol. 9, no. 2, pp. 97–99, 1988.

[20] K. R. Mistry, J. W. Sleight, G. Grula, R. Flatley, B. Miner, L. A. Bair, and D. A. Antoniadis, "Parasitic bipolar gain reduction and the optimization of 0.25 μm partially depleted SOI MOSFETs," *IEEE Trans. Electron Devices*, vol. 46, no. 11, pp. 2201–2209, 1999.

[21] J. Y. Choi and J. G. Fossum, "Analysis and control of floating-body bipolar effects in fully depleted submicrometer SOI MOSFET's," *IEEE Trans. Electron Devices*, vol. 38, no. 6, pp. 1384–1391, 1991.

[22] V. Ploeg, "Elimination of bipolar-induced breakdown in fully-depleted SOI MOSFETs," in *IEDM*, pp. 337–340, 2012.

[23] V. Verma and M. J. Kumar, "Study of the extended P^+ dual source structure for eliminating bipolar induced breakdown in submicron SOI MOSFETs," *IEEE Trans. Electron Devices*, vol. 47, no. 8, pp. 1678–1680, Aug. 2000.

[24] M. J. Kumar and V. Verma, "Elimination of bipolar induced drain breakdown and single transistor latch in submicron PD SOI MOSFETs," *IEEE Trans. Rel.*, vol. 51, no. 3, pp. 367–370, Sept. 2002.

[25] C W. Lee, A. N. Nazarov, I. Ferain, N. D. Akhavan, R. Yan, P. Razavi, R. Yu, R. T. Doria, and J. P. Colinge, "Low subthreshold slope in junctionless multigate transistors," *Appl. Phys. Lett.*, vol. 96, no. 10, 102106, 2010.

[26] R. Yu, A. N. Nazarov, V. S. Lysenko, S. Das, I. Ferain, P. Razavi, M. Shayesteh, A. Kranti, R. Duffy, and J. P. Colinge, "Impact ionization induced dynamic floating body effect in junctionless transistors," *Solid-State Electron.*, vol. 90, pp. 28–33, 2013.

[27] M. S. Parihar, D. Ghosh, G. A. Armstrong, R. Yu, P. Razavi, and A. Kranti, "Bipolar effects in unipolar junctionless transistors," *Appl. Phys. Lett.*, vol. 101, no. 9, 093507, 2012.

[28] M. S. Parihar, D. Ghosh, and A. Kranti, "Occurrence of zero gate oxide thickness coefficient in junctionless transistors," *Appl. Phys. Lett.*, vol. 102, no. 20, 203509, 2013.

[29] M. S. Parihar, D. Ghosh, and A. Kranti, "Single transistor latch phenomenon in junctionless transistors," *J. Appl. Phys.*, vol. 113, no. 18, 184503, 2013.

[30] S. M. Lee and J. T. Park, "The impact of substrate bias on the steep subthreshold slope in junctionless MuGFETs," *IEEE Trans. Electron Devices*, vol. 60, no. 11, pp. 3856–3861, 2013.

[31] M. S. Parihar and A. Kranti, "Back bias induced dynamic and steep subthreshold swing in junctionless transistors," *Appl. Phys. Lett.*, vol. 105, no. 3, 033503, 2014.

[32] M. Vinet, T. Poiroux, J. Widiez, J. Lolivier, B. Previtali, C. Vizioz, B. Guillaumot, P. Besson, J. Simon, F. Martin, S. Maitrejean, P. Holliger, B. Biasse, M. Cassé, F. Allain, A. Toffoli, D. Lafond, J. M. Hartmann, R. Truche, V. Carron, F. Laugier, A. Roman, Y. Morand, D. Renaud, M. Mouis, and S. Deleonibus, "Planar double gate CMOS transistors with 40 nm metal gate for multi-purpose applications," in *Proc. Int. Conf. Solid State Devices Materials*, Tokyo, Japan, Sept. 15–17, 2004, pp. 768–769.

[33] J. Widiez, J. Lolivier, M. Vinet, T. Poiroux, B. Previtali, F. Daugé, M. Mouis, and S. Deleonibus, "Experimental evaluation of gate architecture influence on DG SOI MOSFETs performance," *IEEE Trans. Electron Devices*, vol. 52, no. 8, pp. 1772–1779, Aug. 2005.

[34] M. Gupta and A. Kranti, "Transforming gate misalignment into a unique opportunity to facilitate steep switching in junctionless nanotransistors," *Nanotechnology*, vol. 27, no. 5, 455204, 2012.

[35] M. Gupta and A. Kranti, "Sidewall spacer optimization for steep switching junctionless transistors," *Semicond. Sci. Technol.*, vol. 31, no. 6, 065017, 2016.

[36] M. S. Parihar, D. Ghosh, G. A. Armstrong, and A. Kranti, "Bipolar snapback in junctionless transistors for capacitorless dynamic random access memory," *Appl. Phys. Lett.*, vol. 101, no. 26, 263503, 2012.

[37] D. Munteanu and J. L. Autran, "3-D Numerical simulation of bipolar amplification in junctionless double-gate MOSFETs under heavy-ion irradiation," *IEEE Trans. Nucl. Sci.*, vol. 59, no. 4, pp. 773–780, 2012.

[38] I. Y. Chung, H. Jang, J. Lee, H. Moon, S. M. Seo, and D. H. Kim, "Simulation study on discrete charge effects of SiNW biosensors according to bound target position using a 3D TCAD simulator," *Nanotechnology*, vol. 23, no. 6, 065202, 2012.

7

JUNCTIONLESS DEVICES WITHOUT ANY CHEMICAL DOPING

From our discussions in Chapter 3–5, you may comprehend that the root cause of all the challenges faced by the device designers in scaling the device technology to the sub-10 nm node is the chemical doping. Realizing highly doped source/drain regions using conventional doping processes such as ion implantation with high dopant activation, formation of ultrasteep doping profiles at the source–channel and channel–drain metallurgical junction [1–3], and threshold voltage variation due to random dopant fluctuations (RDF) owing to arbitrary placement of the dopant atoms in the channel region are the current technological challenges [4–16]. Moreover, the stringent requirement of high dopant activation in the source/drain region requires a high-temperature annealing process. However, the formation of ultrasteep doping profile requires the absence of annealing process to inhibit dopant diffusion, which is accelerated at high temperatures. Therefore, the chemical doping process leads to a complex thermal budget.

Even in the modern day Fin field-effect transistors (FinFETs) with the undoped channel region to eliminate RDF, the dopant diffusion from the source/drain regions leads to stochastic distribution of dopants in the channel region. Therefore, RDF is still prominent in the sub-10 nm technology node [4–16].

The chemical doping techniques restrict the scaling of the conventional metal–oxide–semiconductor field-effect transistors (MOSFETs). Therefore, our life as a device designer would have been much simpler if there were an alternative

Junctionless Field-Effect Transistors: Design, Modeling, and Simulation, First Edition.
Shubham Sahay and Mamidala Jagadesh Kumar.

doping procedure, which could eliminate the detrimental impact of the chemical doping. Recently, two alternative electrostatic doping procedures have been proposed: (a) electrostatic doping, which employs gates over the source/drain regions to induce the required concentration of electron and holes in the source/drain regions in the intrinsic silicon film by utilizing the work function difference between the intrinsic silicon film and the source/drain metal electrode (known as charge plasma doping), and (b) electrostatic doping, which employs additional gates over the source/drain regions (known as polarity gate) to induce the required concentration of electron and holes in the source/drain regions in the intrinsic silicon film by the application of appropriate gate voltage. Furthermore, since the application of an external bias induces carriers, it is possible to reconfigure the type of carrier (electron or hole) depending on the applied bias. This unique property translates into an opportunity to realize dynamically configurable devices with electrostatic doping technique.

At this juncture, we would also like to point out that although the electrostatic doping induces carriers in the silicon film, it does not lead to the creation of any metallurgical junction although an electrical junction may be formed where the induced electron and hole concentrations are equal. The absence of any metallurgical junction in devices realized using these alternative doping schemes allows us to classify them as "junctionless" devices.

In the subsequent sections, we discuss in detail about the electrostatic doping techniques. We present a brief overview of the potential benefits, trends, and the challenges for the nanoscale field-effect transistors realized using these alternative doping procedures.

7.1 CHARGE PLASMA DOPING

The charge plasma doping utilizes the induced carriers in the semiconductor when it is brought in contact with a metal electrode of a different work function [17–32]. This can be simply understood from the energy band profiles of a metal–semiconductor (M–S) junction with the work function of metal electrode (ϕ_M) being lower than that of the work function of the semiconductor (ϕ_S), i.e., $\phi_M < \phi_S$ as shown in Figs. 7.1(a) and 7.1(b). In this case, at equilibrium, for the Fermi levels to align at the metal semiconductor interface, the electrons from the metal (which are at a higher potential energy) will flow into the semiconductor (Fig. 7.1(a)). Therefore, the concentration of electrons in the semiconductor at the M–S junction increases and the conduction band comes closer to the Fermi level at the M–S interface (a characteristic of n-type doped semiconductor). Now, if the semiconductor film thickness is less than the Debye length, i.e. the length of the depletion region induced due to the M–S contact, the induced electrons are distributed across the semiconductor film with the peak concentration at the interface.

Therefore, the use of a metal with a small work function as compared to the semiconductor may induce a near uniform concentration of electrons within the Debye length of the semiconductor. Similarly, as shown in Figs. 7.1(c) and 7.1(d), if a metal

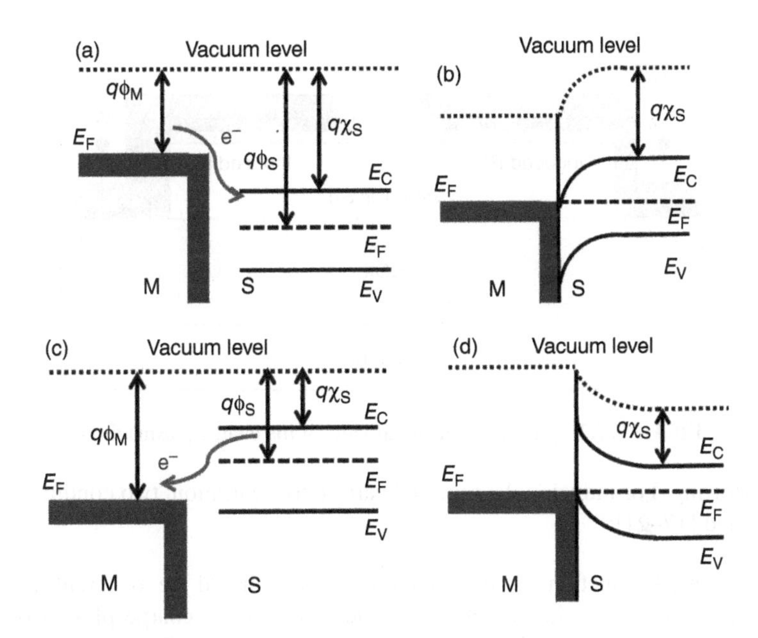

FIGURE 7.1 Energy band profiles of a M–S junction for $\phi_M < \phi_S$: (a) before contact and (b) after contact in thermal equilibrium and for $\phi_M > \phi_S$: (c) before contact and (d) after contact in thermal equilibrium.

electrode with $\phi_M > \phi_S$ is chosen for contacting the semiconductor, it will induce a near uniform concentration of holes within the Debye length.

An extremely thin insulator may also be used between the metal electrode and the semiconductor to further enhance the uniformity of the induced carrier concentration within the semiconductor film. The insulator just reduces the peak electric field at the interface and also minimizes the possibility of silicide formation between the metal and semiconductor [17–21]. In this case, the structure no longer remains a M–S junction but starts behaving like a metal–oxide–semiconductor (M–O–S) structure with a gate electrode with a different work function as compared to the semiconductor. This property of the M–S or M–O–S junction is utilized in the charge plasma–based doping.

7.2 CHARGE PLASMA BASED p–n DIODE

7.2.1 Structure

The three-dimensional (3D) view of the charge plasma based p–n diode is shown in Fig. 7.2. It consists of an ultrathin silicon-on-insulator (SOI) layer over which two different metal electrodes are deposited with an appropriate spacing between them to

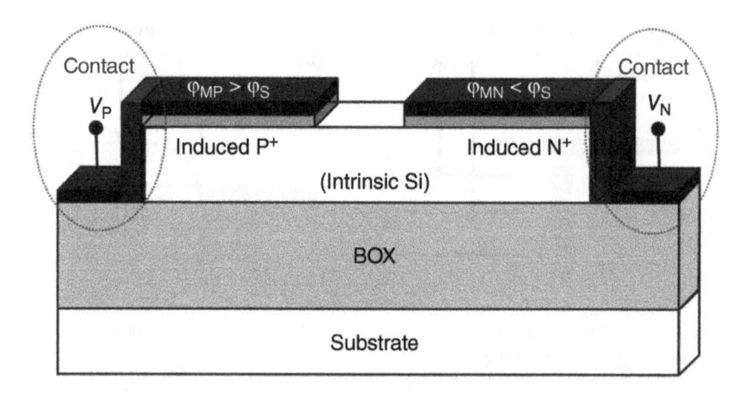

FIGURE 7.2 Three-dimensional view of the charge plasma diode.

avoid shorting. For inducing the required carrier concentration, two conditions must be satisfied [17–21]:

a) The work function of the metal electrode should be different than the semiconductor to induce charge carriers. To induce charge plasma of electrons in the semiconductor film, the work function of the metal electrode should be lower than that of the semiconductor so that electrons may flow from metal to semiconductor to align the Fermi levels in equilibrium, i.e. $\phi_M < \phi_S (= \chi_S + (E_g/2))$, where χ_S is the electron affinity of the semiconductor and E_g is the band gap of the semiconductor. Similarly, to induce a charge plasma of holes, a metal electrode with a work function greater than the semiconductor work function should be chosen such that the electrons from the semiconductor flow into the metal to align the Fermi levels at equilibrium, i.e. $\phi_M > \phi_S (= \chi_S + (E_g/2))$.

   Table 7.1 summarizes the work function of some of the metals. It should also be noted that not only the availability of the metals with a particular work

TABLE 7.1 Metals with Their Work Function

Metal	Work function (ϕ_M in eV) [19]
Gadolinium (Gd)	3.1
Hafnium (Hf)	3.9
Zirconium (Zr)	4.05
Tantalum (Ta)	4.25
Aluminum (Al)	4.28
Tungsten titanium alloy (TiW)	4.3
Cobalt (Co)	5.0
Gold (Au)	5.1
Palladium (Pd)	5.12
Nickel (Ni)	5.15
Platinum (Pt)	5.65–5.93

function is essential but the metal should also be compatible with the complementary metal-oxide-semiconductor (CMOS) process. For instance, for silicon with an electron affinity of 4.05 eV and a band gap of 1.12 eV at room temperature, the work function becomes ~4.6 eV. Therefore, to induce electron concentration in Si, metals such as hafnium (Hf) or aluminum (Al) with a lower work function than Si should be used. Similarly, for inducing a hole plasma in silicon, metals such as platinum (Pt) and palladium (Pd) with a work function more than Si should be used.

b) The silicon film thickness must be less than the Debye length given as: $L_D = \sqrt{(\varepsilon_{Si} \cdot V_T)/qN}$, where L_D is the Debye length, ε_{Si} is the permittivity of silicon, V_T is the thermal voltage (25.9 mV), q is the elementary charge of an electron, and N is the concentration of the SOI layer. Considering an undoped or unintentionally doped SOI layer with a doping concentration of $\sim 10^{15}$ cm^{-3}, a Debye length of 130 nm is obtained. This condition simply implies that the charge carrier distribution inside the silicon film is dominated by the carriers induced owing to the difference in the work function between the metal electrode and the semiconductor. If this criterion is met, the depletion charges or the charges due to the background doping of the semiconductor may be neglected compared to those induced by the metal electrode [17–19] as shown in Fig. 7.3(b).

Generally, an ultrathin insulator layer of SiO_2 is also used on top of the SOI film below the metal electrodes. This serves two purposes: (a) It provides a uniform concentration of the induced carrier concentration as shown in Figs. 7.3(b), and (b) it eliminates the possibility of silicide formation at the M–S interface [17–26].

Now, you may wonder that in conventional SOI devices, the contacts are taken on the top surface of the SOI layer. Since an insulator is used between the SOI layer and the top metal electrodes, no contact is formed on the top. The extension of the metal electrodes on top of the SOI layer in charge plasma–based devices is just to induce the

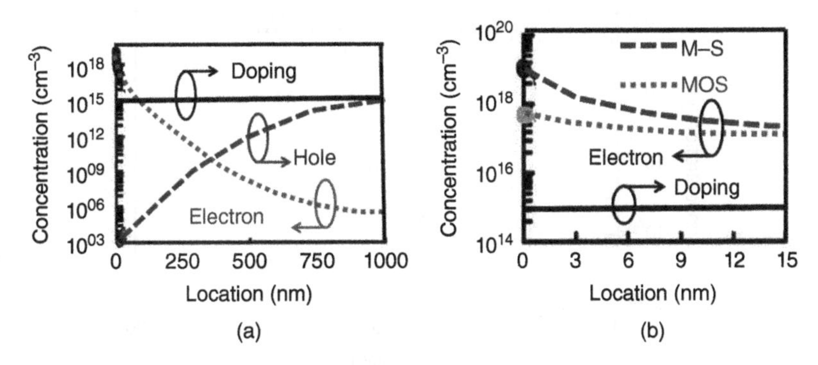

FIGURE 7.3 The electron concentration as a function of distance from the hafnium–silicon interface when (a) silicon film thickness is large and (b) when film thickness is within the Debye length. The electron concentration when an ultrathin insulator is placed between hafnium and silicon is also reported in part (b) [19].

required carrier concentration. So, an immediate question that arises at this juncture is how to contact the underlying SOI layer.

As shown in Fig. 7.2, the contacts to the SOI layer are taken on the sides of the SOI film. Therefore, the purpose of the metal electrode on the top of the SOI layer is to induce sufficient carriers inside the SOI film while the extended metal electrode on the sides of the SOI layer facilitates contact formation as also indicated in Fig. 7.2.

7.2.2 Operation and Characteristics

As shown in Fig. 7.4, an electron plasma is induced below the metal electrode with $\phi_M < \phi_S$ converting this region into n-type and a hole plasma is induced below the metal electrode with $\phi_M > \phi_S$ transforming this region into p-type in the equilibrium state (anode voltage, $V_p = 0.0$ V) although the silicon film is intrinsic or undoped. Also, there is net zero charge at the center of the device as the electron and hole concentrations are same.

The output characteristics of the conventional p–n diode and the charge plasma (CP) diode are shown in Fig. 7.5. Similar to the conventional p–n diodes, the current in a CP diode is also dictated by the diffusion of the minority carriers at low forward bias voltages and increases exponentially with the anode voltage.

In symmetrically doped conventional p–n diodes ($N_A = N_D = 10^{17}$ cm^{-3}), at high forward bias voltages ($V_p > 0.5$ V), the minority carrier concentration increases as compared to the majority carrier concentration and the high injection effects become dominant. Therefore, the rate of increase of the anode current with the anode voltage is reduced as shown in Fig. 5. However, for the CP diode realized on a substrate with the same doping concentration ($N_A = 10^{17}$ cm^{-3}), owing to the large electron and hole concentrations induced by the metal electrodes with different work functions,

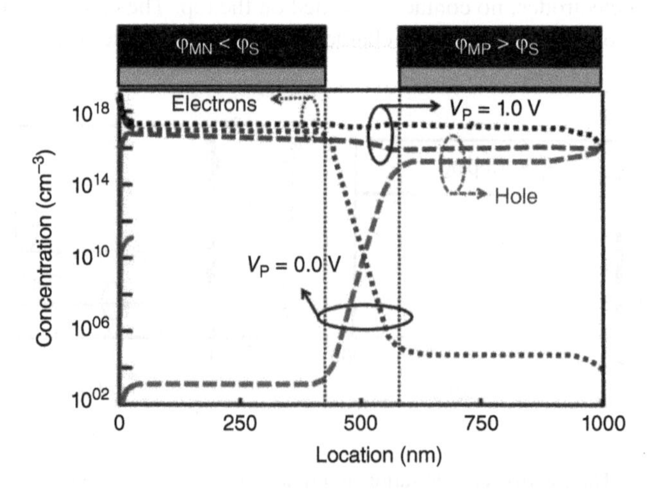

FIGURE 7.4 The electron and hole concentration along the length of the CP diode at different anode voltages [19].

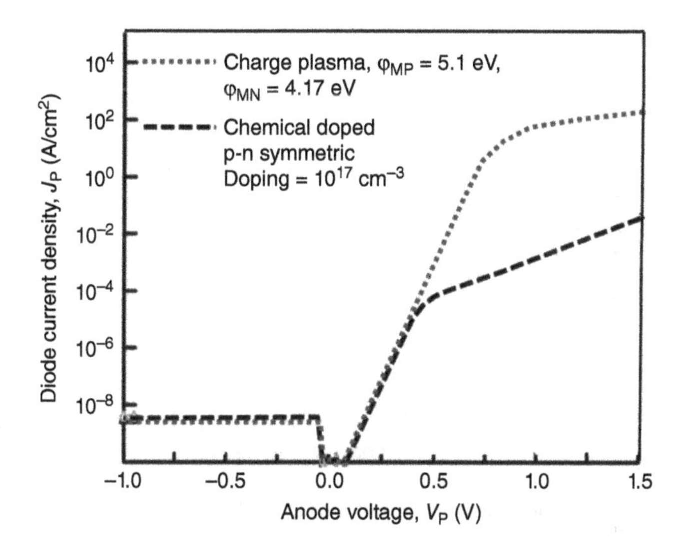

FIGURE 7.5 Output characteristics of the CP diode and the conventional chemically doped diode with symmetrical p–n doping [19].

the voltage at which high injection effects dominate increases significantly (noticeable close to $V_p = 1.0$ V). As you may observe from Fig. 7.4, at $V_p = 1.0$ V, the concentration of electrons (minority carrier) injected into the anode via diffusion owing to the forward bias becomes significantly larger than the induced hole (majority carrier) concentration at equilibrium. This leads to the high injection effect in the CP diode as well.

However, an increase in the electron and hole concentrations increases the majority carrier concentration and reduces the possibility of high injection effects. This can be achieved by increasing the doping concentration in the conventional p–n diodes and by selecting a metal electrode with a lower work function for inducing an enhanced electron concentration at the cathode end and a higher work function for inducing an enlarged hole concentration at the anode end [19].

The anode current in the CP diode increases with an increase in the temperature similar to the conventional chemically doped p–n diode as shown in Fig. 7.6. This clearly indicates that the current mechanism remains same in the conventional diode and the CP diode irrespective of whether the carriers are induced by chemical doping or electrostatic doping utilizing metal electrode with different work functions [18].

In the subsequent sections, we analyze the performance of different emerging devices such as junctionless field-effect transistors (JLFETs), tunnel field-effect transistors (TFETs), and impact ionization MOSFET (IMOS) FETs proposed using the charge plasma doping. We see how the charge plasma concept improves the performance of these devices apart from eliminating the need for high-temperature annealing process for dopant activation and mitigating defects after ion implantation, reducing the thermal budget, and opening up a possible route for realization of these

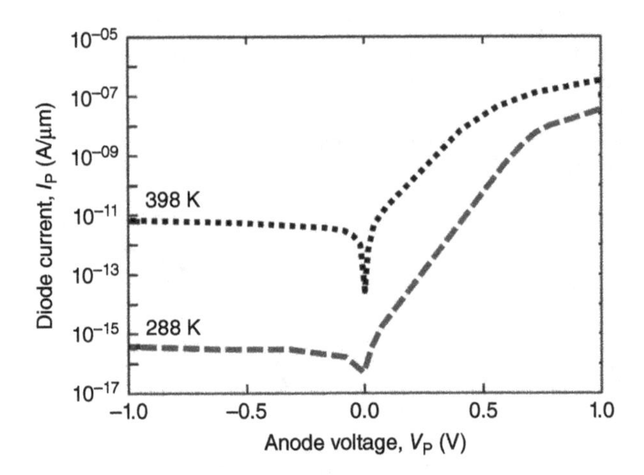

FIGURE 7.6 Output characteristics of the CP diode at different temperatures [18].

devices on single crystal silicon-on-glass substrates. Therefore, the charge plasma doping concept may lead to a new domain of exciting possibilities whereby these devices could be employed in display devices and for biocompatible and optoelectronic applications.

7.3 JUNCTIONLESS IMOS FET

As discussed in Section 2.2, the IMOS device is essentially a reverse biased p–i–n diode and consists of a gate which does not overlap the entire intrinsic silicon region and is underlapped from the source side [33–39]. The gate-modulated region in IMOS is the channel region. A high voltage is applied to the source terminal which drops across the underlapped intrinsic region, the channel region, and the depletion regions at the source–intrinsic region and the channel–drain interface. However, the application of a gate voltage modulates the conductivity of the channel region. For instance, at a high gate voltage, the channel region becomes inverted and offers a low series resistance. Therefore, the entire source voltage drops across the small intrinsic region leading to impact ionization. The impact ionization leads to abrupt turn on of the IMOS devices.

However, it suffers from two major problems [33–39]: (a) It requires the formation of abrupt metallurgical junctions at the source–intrinsic region and the channel–drain interface, which increases the fabrication complexity and the thermal budget, and (b) the operating voltage required to achieve impact ionization is very high (>5 V).

7.3.1 Structure

The schematic view of the junctionless IMOS [40] is shown in Fig. 7.7. For obtaining a carrier concentration profile similar to the conventional p–i–n IMOS, while hafnium

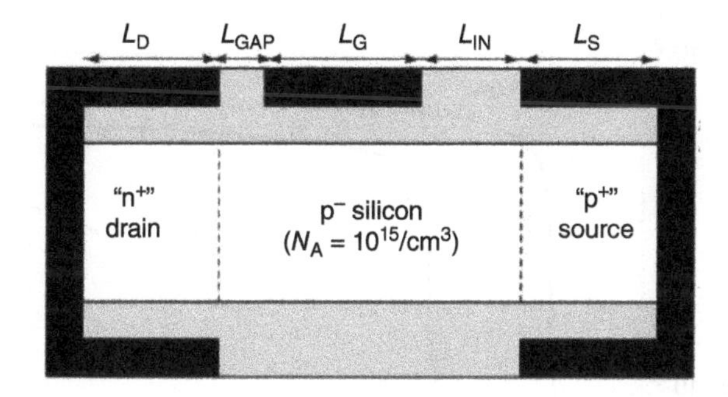

FIGURE 7.7 Schematic view of the junctionless IMOS [40].

has been used to induce electron plasma, platinum with a work function of 5.93 eV has been used to induce the hole plasma. The junctionless IMOS (JIMOS) obviates the need for the formation of metallurgical junction and eliminates the need for high thermal budget [40]. The performance of the JIMOS has been compared against the conventional chemically doped IMOS.

7.3.2 Characteristics

The output characteristics of the JIMOS and the conventional IMOS are compared in Fig. 7.8. You may observe that the breakdown voltage is reduced significantly in the JIMOS as compared to the conventional IMOS by ~0.2 V. Therefore, the JIMOS not only facilitates the realization of IMOS using a low thermal budget but also exhibits a lower breakdown voltage indicating that JIMOS may be operated at lower voltages as compared to the conventional IMOS. Furthermore, the tunneling current dominates the drain current in the conventional IMOS for source voltage, V_S

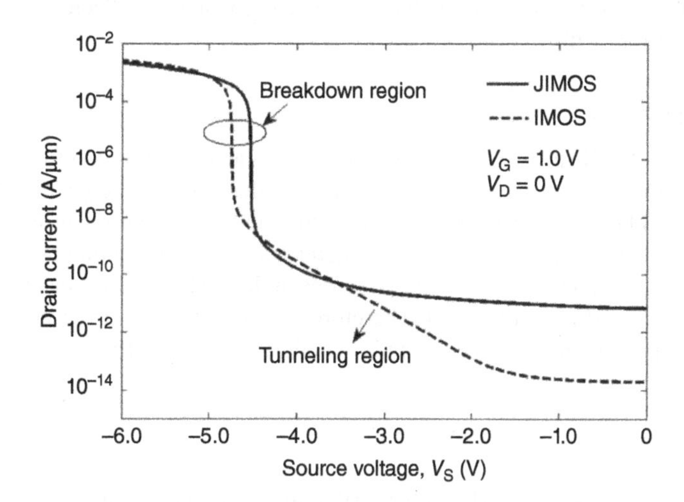

FIGURE 7.8 Output characteristics of the JIMOS and the conventional IMOS [40].

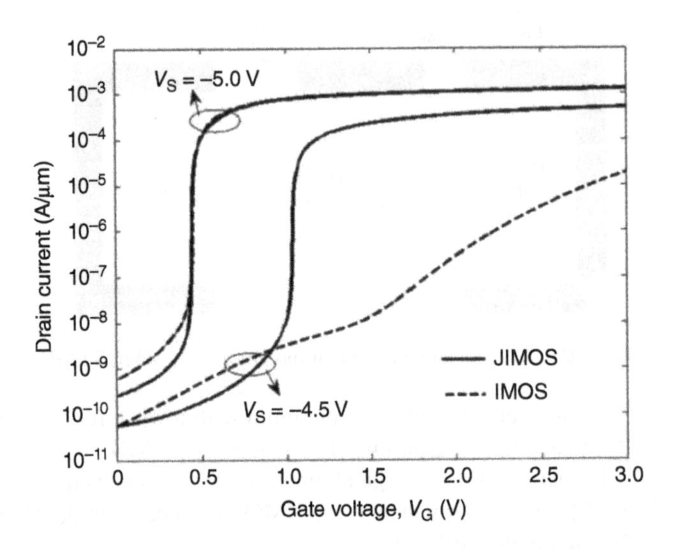

FIGURE 7.9 Transfer characteristics of the JIMOS and conventional IMOS [40].

between −2 and −4.7 V. However, the generation current dominates the tunneling current in a JIMOS and, therefore, the tunneling regime is not observed in the output characteristics of the JIMOS. In addition, the transfer characteristics of the JIMOS and conventional IMOS are compared in Fig. 7.9. It can be observed that both IMOS and JIMOS exhibit nearly identical transfer characteristics at $V_S = -5.0$ V. However, while the conventional IMOS fails to operate at $V_S = -4.5$ V due to insufficient energy for impact ionization, the JIMOS shows steep switching behavior even at a reduced operating voltage. However, the impact ionization is reduced even in the JIMOS when the operating voltage is reduced, which is reflected as the lower ON-state current with $V_S = -4.5$ V.

7.3.3 Operation

The ability of the JIMOS to operate at even −4.5 V can be understood from Fig. 7.10. The nonuniform carrier concentration induced due to the charge plasma effect leads to an inherent electric field in the source and drain regions in the direction along the thickness of the device. This inherent electric field facilitates impact ionization even in the source and the drain regions in the JIMOS unlike the conventional IMOS as observed from Fig. 7.10. Therefore, the nonuniform carrier concentration induced by the charge plasma concept is beneficial for IMOS devices [40].

7.4 JUNCTIONLESS TUNNEL FETs

As discussed in Section 2.1.3, chemical doping is one of the major concerns affecting the performance of TFETs. The tunneling efficiency of the TFETs depends

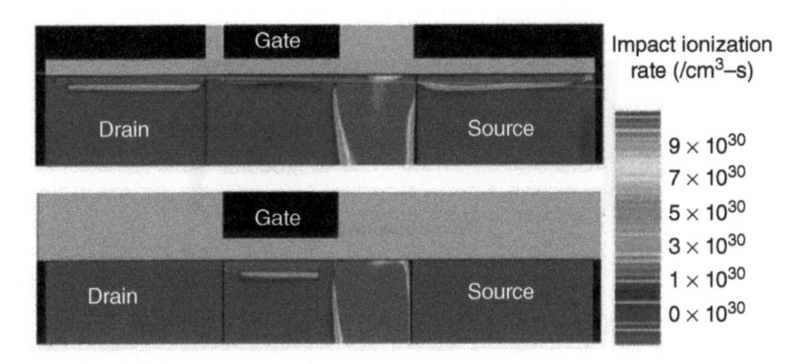

FIGURE 7.10 Impact generation rate contour plot of (a) the JIMOS and (b) conventional IMOS [40].

significantly on the abruptness of the doping profile at the source–channel interface. For efficient tunneling, the source–channel junction doping should be as abrupt as possible. Any gradient in the doping profile tends to increase the tunneling distance which reduces the tunneling efficiency exponentially.

In addition, high doping concentration is also required in the source region in TFETs. The consequent dopant activation and damage mitigation owing to the ion implantation process necessitates a high-temperature annealing. However, the stringent requirement of an abrupt junction simply means that the TFETs cannot be exposed to any high-temperature process which could lead to dopant diffusion from the source region to the channel region. Therefore, this presents a complex thermal budget requirement for TFETs [41].

Since the tunneling efficiency is exponentially dependent upon the doping profile, even a small change in the position of dopants in the silicon film may lead to a large variation in the tunneling current. Therefore, the sensitivity of the threshold voltage to RDF is also larger in TFETs as compared to the MOSFETs as discussed in Section 3.6.3. Chemical doping is, therefore, not favorable for realizing efficient TFETs. Hence, we discuss the proposals for realizing TFETs without chemical doping using the charge plasma concept. First, we look at the junctionless tunnel FETs (JLTFET) on the intrinsic silicon film and then we focus on the JLTFET on a highly doped silicon film which is a characteristic of the JLFETs.

7.4.1 Structure

The schematic view of the JLTFET is shown in Fig. 7.11. A TFET can also be viewed as a gated reverse biased p–i–n diode. Therefore, to induce the hole concentration in the source region, a platinum electrode has been employed while hafnium has been used to induce electrons in the drain region. Since the JLTFET does not require formation of abrupt junctions or dopant activation, it may be realized using low-temperature processes. Also, the JLTFET is immune to the RDF since no chemical doping is used.

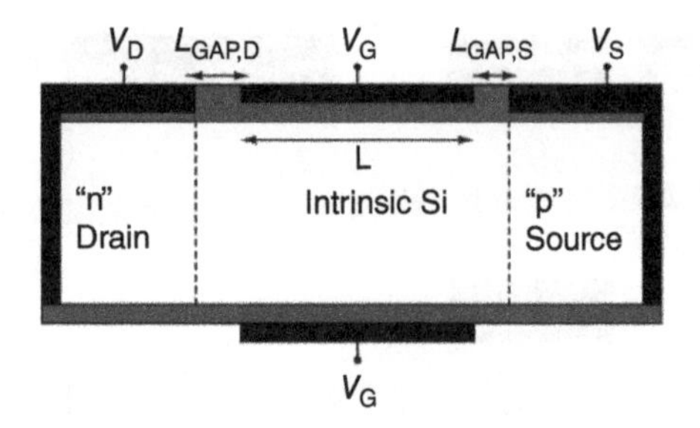

FIGURE 7.11 Schematic view of a JLTFET.

7.4.2 Operation and Characteristics

Figure 7.12 shows the energy band profiles of the JLTFET in the OFF-state ($V_{GS} = 0.0$ V) and ON-state ($V_{GS} = 1.0$ V). The energy band profiles clearly indicate that the operation of the JLTFET is similar to the conventional chemically doped TFET. In the OFF-state, due to the large potential barrier at the source–channel interface, the drain current corresponds to the leakage current in a reverse biased p–i–n diode. However, in the ON-state, there exists a considerable spatial proximity between the valence band of the source region and the conduction band of the channel region that leads to a gate-modulated tunneling current.

The transfer characteristics of the JLTFET are similar to the conventional TFET (Fig. 7.13), which corroborates with our observation that the operating principle of the JLTFET is same as that of the conventional TFET. Therefore, the JLTFET exhibits a performance similar to the conventional TFET without the need of well-controlled chemical doping using molecular beam epitaxy/plasma implantation/laser annealing, or complex thermal budget [41].

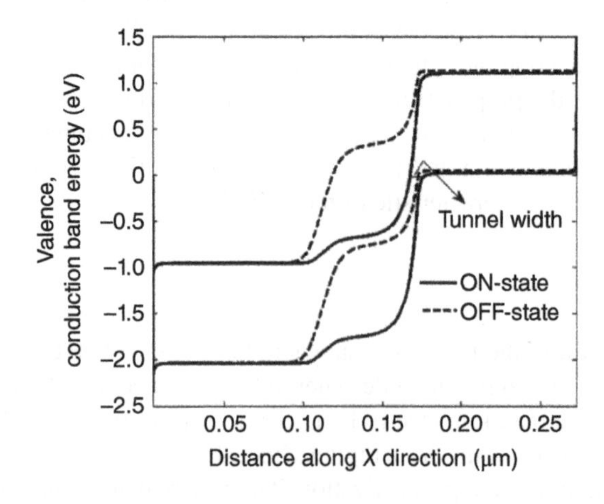

FIGURE 7.12 Energy band profiles of the JLTFET in the ON-state and the OFF-state [41].

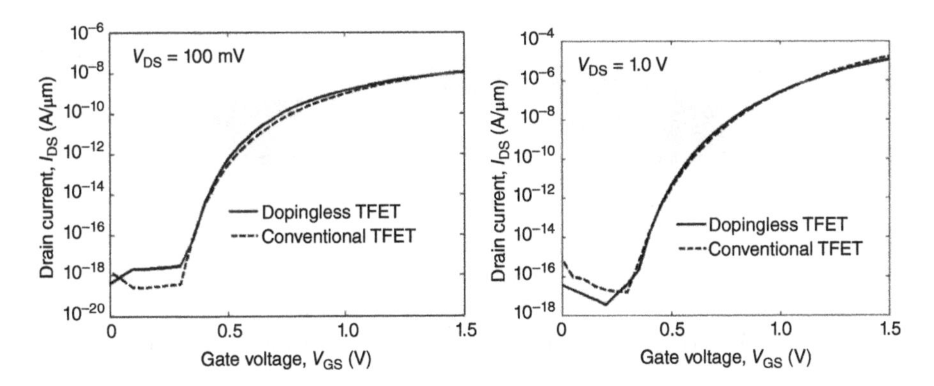

FIGURE 7.13 Transfer characteristics of the JLTFET and the conventional TFET for (a) $V_{DS} = 100\,\text{mV}$ and (b) $V_{DS} = 1.0\,\text{V}$ [41].

Furthermore, the gap between the source electrode and the gate electrode determines (a) the proximity between the gate electrode and the induced source–channel interface and (b) the abruptness of the induced P^+ source–channel interface [41]. Therefore, the gap length should be kept as minimum as possible to achieve an enhanced tunneling efficiency and consequently an increased ON-state current as shown in Fig. 7.14. However, close spacing between two electrodes may lead to coupling and large parasitic capacitances, which may degrade the dynamic performance of the JLTFET.

The ON-state current of the JLTFET can be increased by utilizing the in-built n^+ pocket using the charge plasma doping [42] or by using a metal inside the gate dielectric on the source side [43].

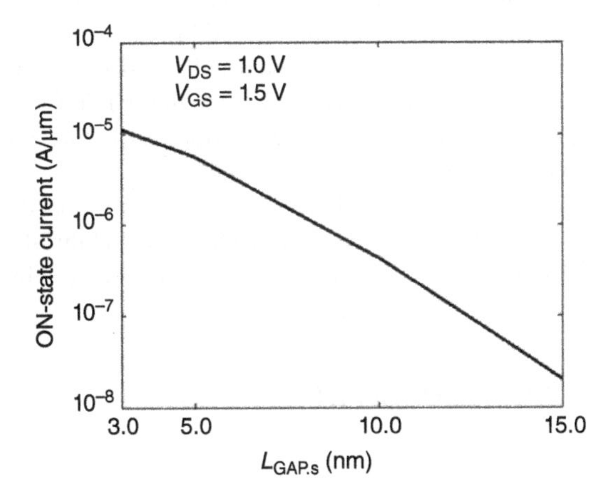

FIGURE 7.14 ON-state current of the JLTFET as a function of the gap length between source electrode and gate electrode [41].

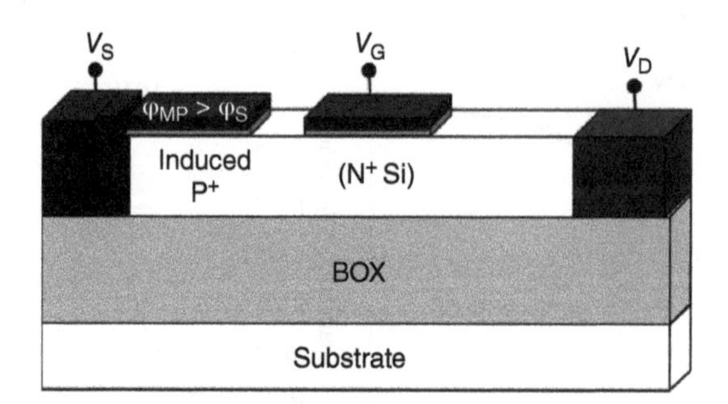

FIGURE 7.15 Three-dimensional view of a JLTFET on a heavily doped SOI film.

7.5 JLTFET ON A HIGHLY DOPED SILICON FILM

As discussed in Chapter 3, the JLFETs consist of a heavily doped ultrathin silicon film. However, the subthreshold swing of the JLFETs is also limited to the Boltzmann tyranny of 60 mV/decade owing to the thermionic injection mechanism. The performance of the conventional JLFETs may be enhanced by utilizing the band-to-band tunneling (BTBT) mechanism in the ON-state of JLFETs. This can be achieved by employing the charge plasma concept and converting the n^+ source into a p^+ source using a platinum electrode over the source region and using a gate electrode with a work function such that the carrier concentration resembles that of a conventional TFET (p–i–n) [44].

7.5.1 Structure and Characteristics

The structure of such a JLTFET realized on a highly doped silicon film is shown in Fig. 7.15. Compared to the JLTFET on an intrinsic silicon film (discussed in Section 7.4.1), this configuration does not require an additional metal electrode to realize the drain region. Therefore, the JLTFET on the heavily doped silicon film is expected to exhibit lower parasitic capacitance as compared to the JLTFET on an intrinsic substrate. However, since the silicon film is heavily doped, the threshold voltage variability owing to the RDF would remain prominent.

The transfer characteristics of the JLTFET on a heavily doped silicon film are shown in Fig. 7.16. The ON-state current as well as the subthreshold slope of the JLTFET on a heavily doped SOI film may be enhanced by utilizing a high-κ gate dielectric.

7.6 BIPOLAR ENHANCED JLTFET

7.6.1 Structure

The 3D view of the bipolar enhanced JLTFET (BEJLTFET) is shown in Fig. 7.17. The only difference between the BEJLTFET and the JLTFET on a heavily doped SOI film

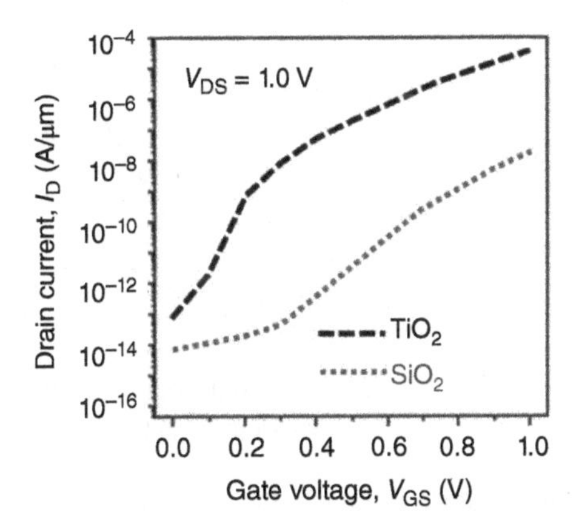

FIGURE 7.16 Transfer characteristics of the JLTFET on a heavily doped SOI film with a different gate dielectric [44].

is that in BEJLTFET the platinum electrode does not induce holes in the entire source region [45]. The N$^+$-doped source region near the source contact is intentionally left in the BEJLTFET for the reasons explained in the subsequent sections.

7.6.2 Transfer Characteristics

The transfer characteristics of the BEJLTFET have been compared against the JLTFET and the conventional TFET are shown in Fig. 7.18. The BEJLTFET exhibits a higher ON-state current by one order of magnitude as compared to the JLTFET or conventional TFET. Furthermore, a reduction in the drain doping leads to a significant reduction in the OFF-state current as observed in Fig. 7.19.

7.6.3 Operation

The introduction of a doped N$^+$ source adjacent to the induced P$^+$ source region in the BEJLTFET leads to an altogether different current transport phenomenon in the

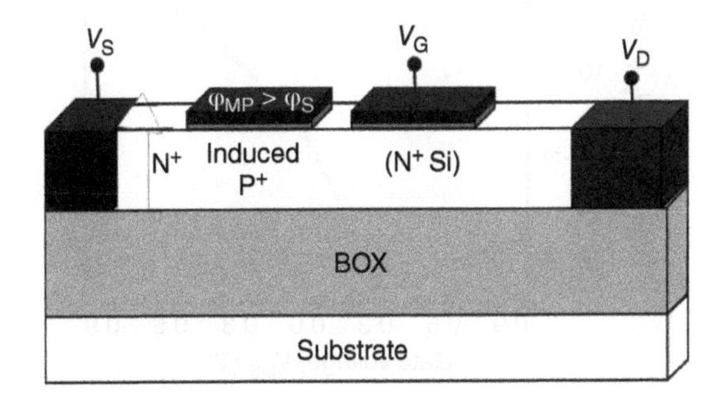

FIGURE 7.17 Three-dimensional view of a BEJLTFET.

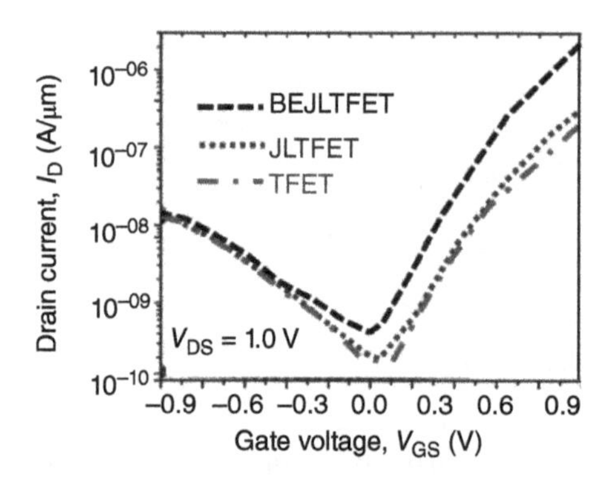

FIGURE 7.18 Transfer characteristics of the conventional TFET, JLTFET, and the BEJLTFET [45].

BEJLTFET [45]. As can be observed from Fig. 7.20, the BTBT of electrons occurs at the induced P^+ source–channel interface. The electrons tunneling from the P^+ source leave behind holes which leads to hole accumulation in the induced P^+ source region. This accumulation of hole increases the potential of the induced P^+ source region and lowers the doped N^+ source-induced P^+ source potential barrier height which internally forward biases this junction [46–57].

Similar to the parasitic BJT action discussed in Section 5.2, a parasitic BJT is formed in the BEJLTFET with the doped N^+ source acting as the emitter, induced P^+ source acting as the base, and the channel region through drain acting as the collector of the parasitic BJT. The tunneling current acts like the base current and gets amplified by the gain of the parasitic BJT. Therefore, the bipolar gain offered by the BEJLTFET to the tunneling current increases the ON-state current significantly.

However, the OFF-state current is also larger in the BEJLTFET. Therefore, techniques to reduce the peak electric field at the drain region such as lightly doped drain, heterodielectric BOX, gate drain overlap, and so on [58–64] may be effectively

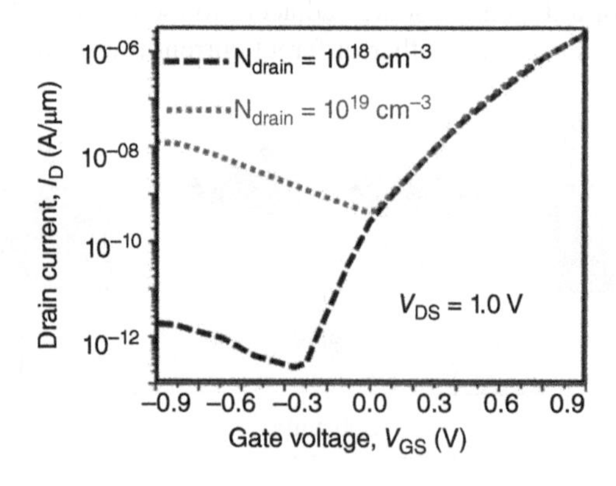

FIGURE 7.19 Transfer characteristics of the BEJLTFET with different drain dopings [45].

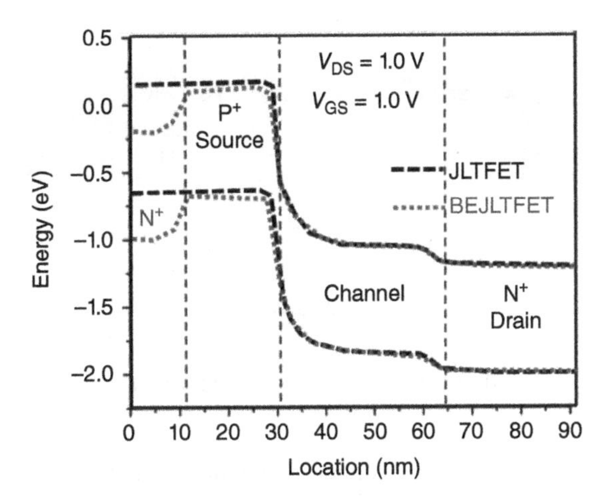

FIGURE 7.20 Energy band profiles of the JLTFET and the BEJLTFET in the ON-state [45].

utilized to reduce the OFF-state current. A reduction in the OFF-state current with a lower drain doping in the BEJLTFET is attributed to a wider depletion width and a lower electric field at the channel–drain interface owing to the lightly doped drain region [45].

7.7 JUNCTIONLESS FETS WITHOUT ANY CHEMICAL DOPING

As discussed in Section 3.6.3, the chemically doped JLFETs suffer from severe threshold voltage fluctuations owing to RDF and a variation in the silicon film thickness or nanowire width. Furthermore, as discussed in Section 3.2.2, the threshold voltage as a function of silicon film doping and the silicon film thickness is given as

$$V_{\text{th}} = V_{\text{FB}} - \frac{qN_{\text{D}}{T_{\text{Si}}}^2}{8\varepsilon_{\text{S}}} - \frac{qN_{\text{D}}T_{\text{Si}}T_{\text{ox}}}{2\varepsilon_{\text{ox}}} \tag{7.1}$$

Therefore, the sensitivity to threshold voltage variation due to doping and thickness may be given as

$$\frac{\partial V_{\text{th}}}{\partial N_{\text{D}}} = -\frac{q{T_{\text{Si}}}^2}{8\varepsilon_{\text{S}}} - \frac{qT_{\text{Si}}T_{\text{ox}}}{2\varepsilon_{\text{ox}}} \tag{7.2}$$

$$\frac{dV_{\text{th}}}{dT_{\text{Si}}} = -\frac{qN_{\text{D}}T_{\text{Si}}}{4\varepsilon_{\text{S}}} - \frac{qN_{\text{D}}T_{\text{ox}}}{2\varepsilon_{\text{ox}}} \tag{7.3}$$

Equation (7.3) clearly indicates that the variation is larger in JLFETs with higher silicon film doping, which is required for a reduced source/drain series resistance.

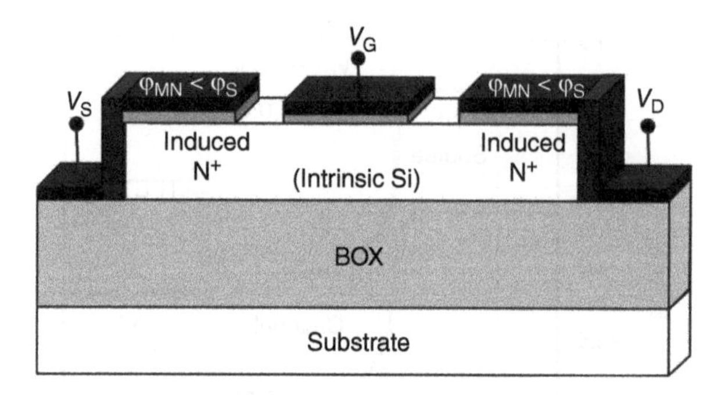

FIGURE 7.21 Three-dimensional view of a CPJLFET without any chemical doping.

The significantly large threshold voltage fluctuation in JLFETs is the major challenge, hindering the replacement of conventional MOSFETs by JLFETs. Therefore, if somehow the JLFETs could be realized without chemical doping while preserving the advantages such as immunity against short-channel effects and lower reliability issues, it would be a promising alternative to the MOSFETs for the future technology nodes. The charge plasma concept may be extended to realize JLFETs without any chemical doping as discussed in the subsequent sections.

7.7.1 Structure

The 3D view of the JLFET realized using the charge plasma concept (CPJLFET) without any chemical doping is shown in Fig. 7.21. Hafnium metal is employed to induce electron plasma in the source and drain regions on an intrinsic silicon film and to obtain an electron concentration close to that of conventional JLFETs. The application of an intrinsic silicon film in the CPJLFET results in an intrinsic channel, which may be volume depleted utilizing even gate electrodes with a midgap ~4.5 eV work function. Therefore, realizing volume depletion in CPJLFET does not pose a stringent requirement of extremely high work functions or ultrathin films like JLFETs and significantly relaxes the fabrication complexity for the conventional JLFETs [65].

7.7.2 Transfer Characteristics and Operation

The transfer characteristics of the conventional JLFET and the CPJLFET with source/drain electrodes with different work functions are shown in Fig. 7.22. The CPJLFET with a hafnium source/drain electrode ($\phi_{MS/D} = 3.9\,eV$) exhibits both a high ON-state current as well as a low OFF-state current as compared to the conventional JLFET [65]. However, when a source/drain electrode with a midgap work function (4.5 eV) is chosen, both the OFF-state current and the ON-state current reduce significantly. This is attributed to the lower electron plasma induced by the metal electrodes with midgap work function.

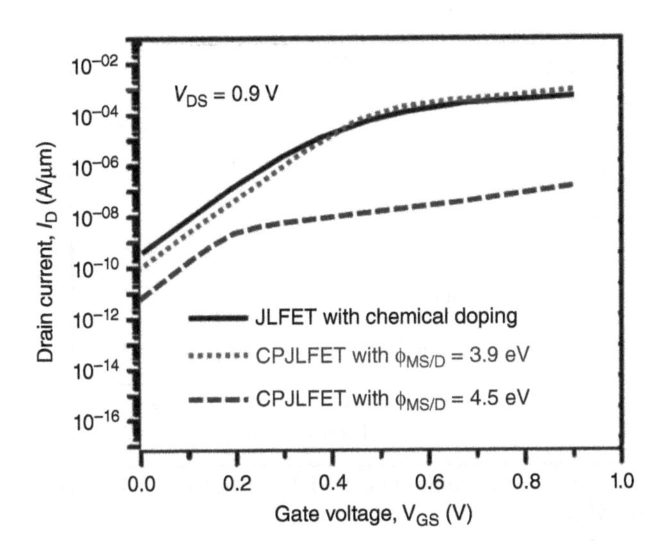

FIGURE 7.22 Transfer characteristics of the JLFET and CPJLFET for different source/drain work functions [65].

Another concern with the chemically doped conventional JLFETs is the dopant activation. Even at room temperature, FETs with a high doping close to 10^{19} cm^{-3} are susceptible to incomplete ionization phenomena where the dopants are inactivated and the effective carrier concentration is lower than the film doping [66]. The incomplete ionization results in an increased series resistance and a lower current driving capability in the conventional doped JLFETs as shown in Fig. 7.23(a).

Furthermore, the increased ON-state current in the CPJLFET is attributed to the higher mobility due to the lower ionized impurity scattering mechanism owing to the intrinsic channel region. However, a distinguishing feature of the CPJLFET as

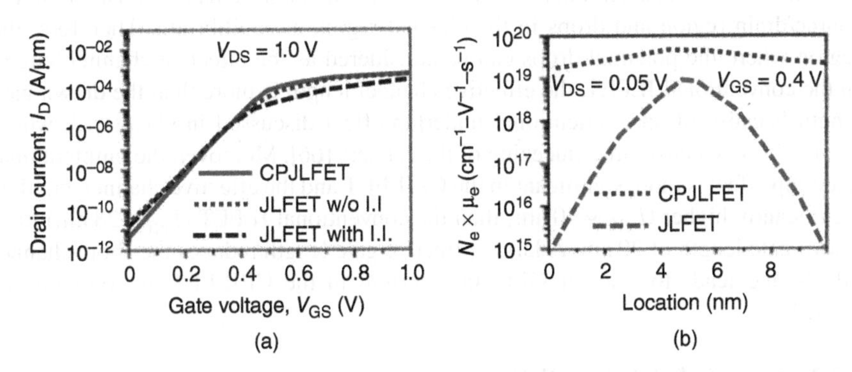

FIGURE 7.23 (a) Transfer characteristics and (b) product of electron mobility and the electron concentration of the JLFET with and without incomplete ionization (I.I.) model and the CPJLFET [66].

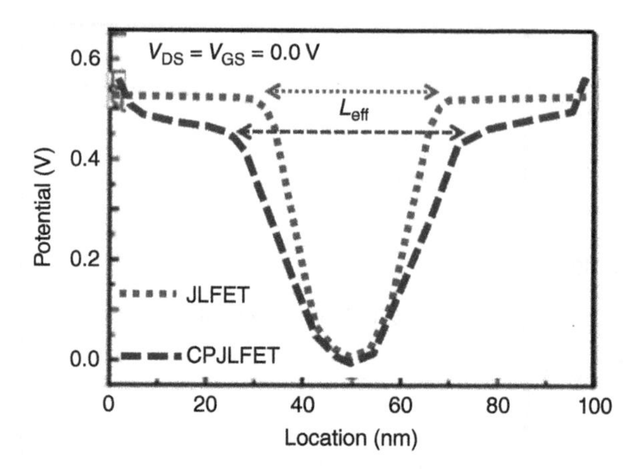

FIGURE 7.24 The potential profile of the JLFET and the CPJLFET in the equilibrium state ($V_{GS} = V_{DS} = 0.0$ V) [66].

compared to the JLFETs is that it operates in the accumulation mode in the ON-state as compared to the flat band conditions in a JLFET. Therefore, the surface scattering mechanism forms the dominating mechanism affecting the mobility of CPJLFETs. The product of electron concentration and electron mobility is significantly higher in the CPJLFET as compared to the JLFETs as shown in Fig. 7.23(b). At this juncture, we would also like to point out that the value of $n_e \times \mu_e$, where n_e is the electron concentration in the channel region and μ_e is the electron mobility peaks at the center in conventional JLFETs, clearly indicating current conduction at the center while in CPJLFET it remains high throughout the film thickness indicating volume accumulation. The increased ON-state current in the CPJLFETs is attributed to these factors [66].

The potential profile of the CPJLFET and the conventional JLFET are shown in Fig. 7.24. In conventional MOSFETs, the potential remains constant in the source/drain region and drops in the channel region at equilibrium. Therefore, the region where the potential drops can be considered as the effective channel length. In the conventional JLFET, the effective channel length is more than the drawn gate length because of an unintentional underlap effect discussed in Chapter 3, which occurs due to electrostatic squeezing of the carriers [66]. Moreover, the unintentional underlap effect is more significant in the CPJLFET and the effective channel length is significantly higher ($L_{eff} \sim 40$ nm) than the conventional JLFET ($L_{eff} = 28$ nm) for a drawn gate length of 20 nm. A larger effective gate length reduces the short-channel effects and leads to a lower OFF-state current in the CPJLFET as compared to the JLFET.

7.7.3 Impact of Gate Length Scaling

The impact of gate length scaling on the JLFETs and the CPJLFETs is shown in Fig. 7.25. While the OFF-state current increases with a reduction in the gate length

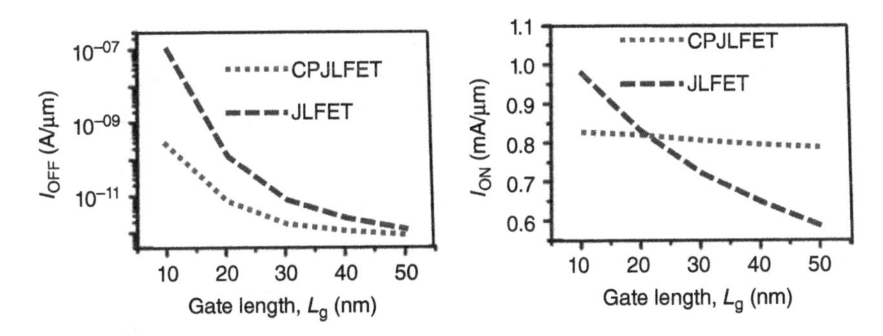

FIGURE 7.25 Impact of gate length scaling on the OFF-state current (I_{OFF}) and the ON-state current (I_{ON}) of the JLFET and the CPJLFET [66].

for both devices due to increased short-channel effects, the increment is lower for the CPJLFETs and it exhibits a lower OFF-state current than JLFETs by more than two orders of magnitude even when the gate length is 10 nm.

The ON-state current also increases with gate length scaling for both configurations. However, the rate of increase is higher for the JLFETs. This clearly indicates that CPJLFET is less sensitive to the gate length scaling than JLFETs [66].

7.7.4 Sensitivity to Process Variation

The sensitivity of the JLFET and the CPJLFET toward silicon film doping is compared in Fig. 7.26. The transfer characteristics of the CPJLFET remain the same even when the silicon film doping is changed by 20%. This is because within the Debye length, as discussed in Section 7.2, the induced electron concentration is independent of the silicon film doping and the depletion charges. Therefore, the CPJLFET is also immune to the doping fluctuation caused during the silicon film doping process

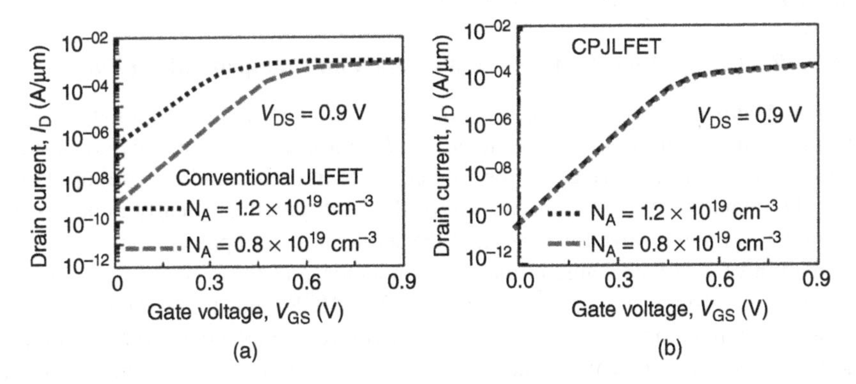

FIGURE 7.26 Sensitivity of the JLFET and the CPJLFET to the variation in the silicon film doping [65].

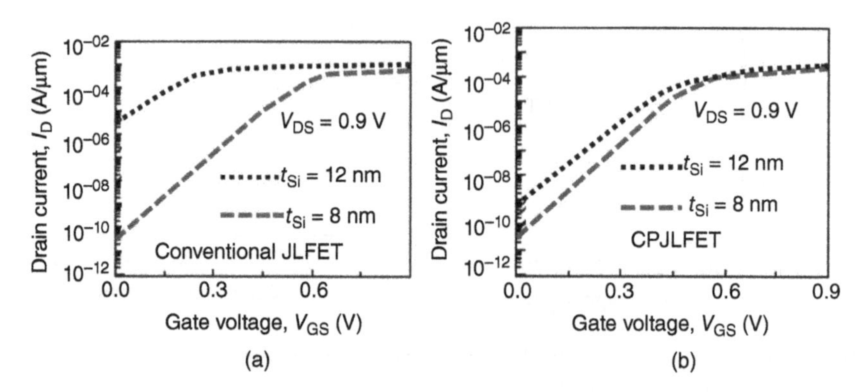

FIGURE 7.27 Sensitivity of the JLFET and the CPJLFET to the variation in the silicon film thickness [65].

[65]. Furthermore, since no chemical doping is utilized, the CPJLFET is inherently immune to RDF.

Since the JLFETs also exhibit a large sensitivity to the variation in the silicon film thickness, the transfer characteristics of the CPJLFET and the JLFET have also been compared when the silicon film thickness changes by 20%. As can be observed from Fig. 7.27, the variation in the transfer characteristics of the CPJLFET is lower as compared to the changes in the JLFETs. Therefore, the CPJLFET is relatively immune to the process variation even in the silicon film thickness [65]. This also corroborates with equation (7.3), which clearly indicates that the variation in threshold voltage would be lower in JLFETs if a low silicon film doping is used which is the case with the CPJLFET.

7.8 CHALLENGES FOR CPJLFETs

7.8.1 Realizing Ohmic Source/Drain Contact on Undoped Silicon Film

7.8.1.A Metal-Induced Gap States All the junctionless implementations without chemical doping discussed in Sections 7.2–7.8 have assumed ohmic source/drain contacts at the M–S interface on the side contacts in their simulation set up. However, the quantum mechanical treatment of the M–S junctions clearly indicates the penetration of metal wave functions into the semiconductor upon contact. This leads to creation of interface states in the semiconductor, and these states are popularly known as metal-induced gap states (MIGS). The MIGS tends to pin the Fermi level of the semiconductor at the M–S junction. Therefore, the effective work function of the metal seen by the semiconductor at the M–S junction is significantly different from its value when it is isolated. The MIGS leads to parasitic Fermi level pinning at the midgap for silicon and close to the valence band in germanium [67–68].

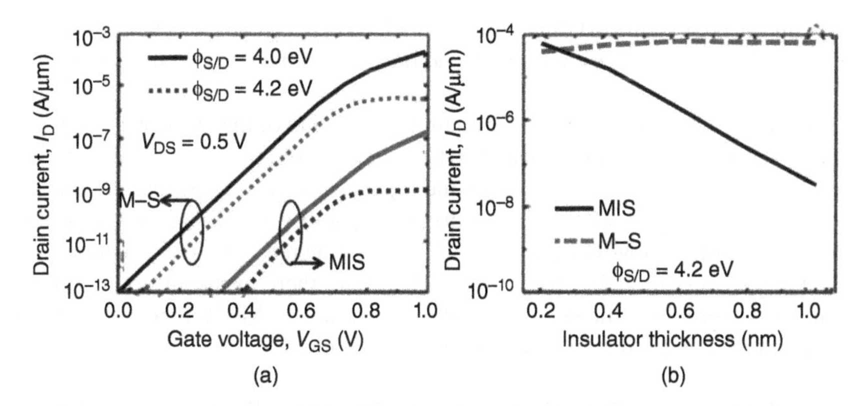

FIGURE 7.28 (a) The transfer characteristics of the CPJLFET with metal–semiconductor (M–S) and metal–insulator–semiconductor (MIS) contact with varying work function of the source/drain electrode ($\phi_{S/D}$) and (b) the current components through the top MIS contact and the side M–S contact in a CPJLFET with M–S contact at side and MIS contact at the top [67].

7.8.1.B Fermi-Level Pinning Experimental investigation of the M–S contacts over lightly doped or undoped silicon (which is the case for the junctionless devices discussed in this chapter) has revealed that the work function of most metals (even though their actual work function ranges between 3 and 5 eV in isolation) is pinned at ~4.6 eV [69–75]. A work function of 4.6 eV at the source/drain contact implies a Schottky barrier height of ~0.5 eV, which would significantly reduce the injection of electrons from source/drain contacts into a undoped or lightly doped semiconductor. Therefore, the current-driving capability of the JLFETs over the undoped silicon film would be extremely poor.

The impact of MIGS-induced Fermi level pinning can be incorporated by specifying the effective work function at which the metal electrode at the source/drain contact is pinned. Figure 7.28 shows the impact of varying the work function of the source/drain electrode ($\phi_{S/D}$) taken at the sides of the silicon film. The ON-state current decreases significantly when the $\phi_{S/D}$ increases to 4.2 eV. The ON-state current reaches in the range of fA/μm when the $\phi_{S/D}$ is increased to 4.6 eV [67]. Therefore, Fermi level pinning is detrimental for operation of JLFETs on the undoped silicon film and should be properly analyzed.

7.8.1.C Mitigating MIGS However, the MIGS can be considerably reduced at the M–S interface by incorporating a thin insulator yielding metal–insulator–semiconductor (MIS) contacts. An insulator with a high band gap increases the potential barrier for the electrons in the metal, attenuating the penetration of the electron wave function into the semiconductor. Moreover, an insulator with low-κ material inhibits the coupling between the metal and the semiconductor. Furthermore, the thickness of the insulator dictates the barrier width that the wave has to penetrate. Therefore, a thick insulator can minimize the MIGS. Using a MIS contact,

$\phi_{S/D}$ same as the metal work function in isolation may be realized. Therefore, ohmic contacts may be realized even on undoped or lightly doped silicon films using MIS contacts [67].

However, in the MIS contacts, the quantum tunneling of the electrons through the insulator is the dominant mechanism for the current flow at the source/drain interface. The thickness of the insulator dictates the tunneling width. Therefore, an ultrathin insulator is required for a large current driving capability. However, a thick insulator is needed to diminish MIGS and the MIGS would increase with a reduction in the insulator thickness reducing the current. Therefore, there exists an optimal insulator thickness for which the decrease in the current due to MIGS is minimal and the increase in the current due to quantum tunneling is large.

7.8.1.D Revisiting the Top MIS Contacts in the Charge Plasma Based FETs In all the charge plasma concept based devices discussed in Sections 7.2–7.7, an ultrathin insulator was used between the top metal electrode and the semiconductor for enhancing the uniformity of the induced charge carrier concentration and to eliminate the possibility of silicide formation. However, the impact of tunneling through the ultrathin insulator was not analyzed in the top MIS structures. Figure 7.28(a) analyzes the decoupled current components through the top MIS contact and the side M–S contacts when different $\phi_{S/D}$ are used. Irrespective of whether M–S or MIS contacts are utilized, the current reduces significantly as the $\phi_{S/D}$ is increased. However, the diminished MIGS in MIS contacts may enable the designers to realize the exact work function of the metal electrode for $\phi_{S/D}$ rather than the effective $\phi_{S/D}$ due to MIGS in M–S contacts [67].

In addition, the ON-state current decoupled for metal–semiconductor (M–S) and metal–insulator–semiconductor (MIS) contacts is also shown in Fig. 7.28(b) for different insulator thicknesses. It can be observed that the ON-state current component through MIS increases significantly with a reduction in the insulator thickness owing to the reduced tunneling barrier width. Moreover, the current through MIS contact may even be larger than the current through M–S contact as the insulator thickness is significantly reduced. This clearly indicates that, for undoped silicon films, MIS contacts may yield larger ON-state current if an ultrathin insulator is used [67]. Therefore, MIS source/drain contacts were proposed even for the side contacts in junctionless devices without chemical doping as shown in Fig. 7.29. Furthermore, Fig. 7.28(b) also reveals that in junctionless devices utilizing the charge plasma concept, the tunneling current through the top MIS contact is significant and dominates when the insulator thickness is low. Therefore, quantum tunneling through the top MIS contacts should not be ignored while analyzing the charge plasma based devices.

7.8.1.E JLFETs with MIS Contacts The structure of the CPJLFET with MIS contacts even on the side source/drain contacts is shown in Fig. 7.29. The transfer characteristics of the CPJLFET with MIS contacts for different insulator thicknesses are shown in Fig. 7.30(a). The current driving ability of the MIS CPJLFET increases with a reduction in the insulator thickness owing to the lower tunneling barrier

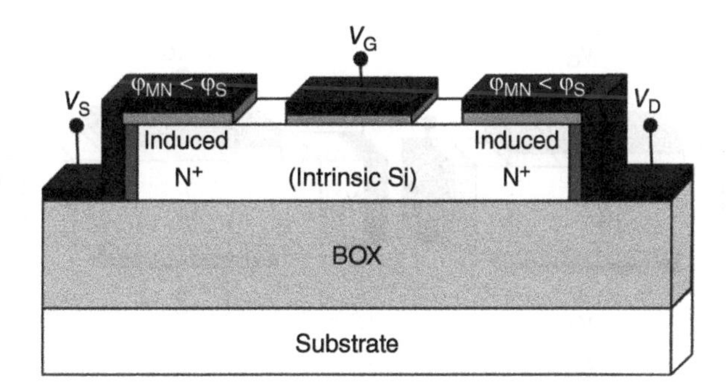

FIGURE 7.29 Schematic view of the CPJLFET with a MIS source/drain contact.

width facilitated by the ultrathin insulator. Furthermore, the use of a high-κ insulator increases the electric field at the MIS interface and results in a larger tunneling probability increasing the ON-state current as shown in Fig. 7.30(b). Furthermore, the use of MIS contact can make $\phi_{S/D}$ similar to the metal in isolation, i.e. the true work function of hafnium (3.9 eV) can be achieved in MIS JLFETs. This would increase the ON-state current significantly in the MIS CPJLFET as shown in Fig. 7.30(c).

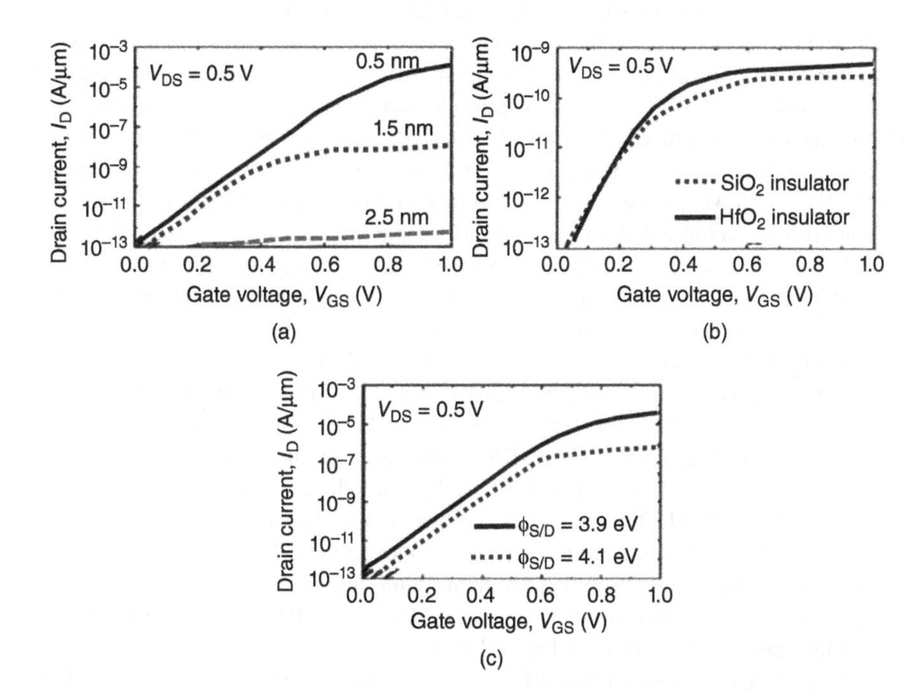

FIGURE 7.30 Transfer characteristics of the CPJLFET with MIS contact for (a) different insulator thicknesses, (b) different dielectric constants of the insulator and (c) different work functions of the source/drain electrode ($\phi_{S/D}$) [67].

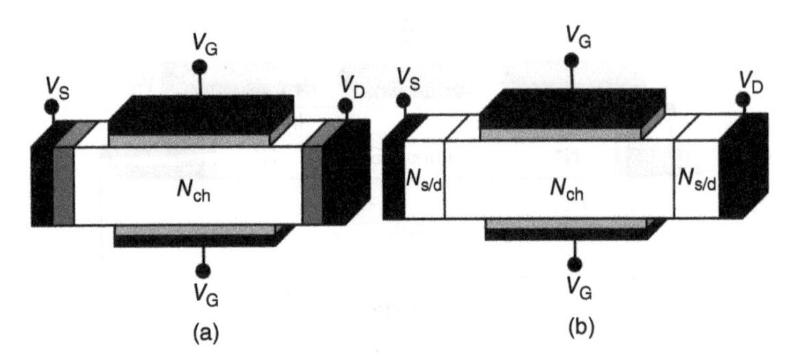

FIGURE 7.31 Schematic view of a JLFET with (a) MIS contact and (b) M–S contact.

7.8.1.F MIS Contacts versus M–S Contacts in JLFETs The contact resistance of M–S and MIS configurations was compared experimentally in [75], and it was observed that Ti/TiO_2/n-Si (MIS) contact exhibits a lower contact resistivity as compared to Ti/n-Si (M–S) contact when the doping concentration of silicon film is lower than 4×10^{19} cm^{-3}. Therefore, MIS configuration appears to be an appropriate contacting strategy for JLFETs, which are inherently doped in the range of 10^{19} cm^{-3}.

To have a fair comparison between the performance of JLFETs with M–S and MIS contacting strategy, (a) the work function of the source/drain contacts in M–S configuration which are pinned at the midgap due to MIGS was taken as 4.5 eV and (b) the work function of the source/drain electrodes in the MIS configuration was chosen as 4.0 eV. Furthermore, both M–S and MIS contacts were taken only on the side of the silicon film as shown in Fig. 7.31. Since the Schottky barrier height is larger in JLFETs with M–S contacts, an additional highly doped source/drain extension region (similar to the JAMFETs with doping $N_{S/D} = 10^{20}$ cm^{-3}) has also been used in JLFETs with M–S contacts to increase the Schottky tunneling current and reduce the potential barrier height as done in dopant-segregated Schottky FETs [68].

Figure 7.32 shows the transfer characteristics of the JLFETs with MIS contact and different silicon film dopings. It can be observed that irrespective of the nature of doping (p or n) and the doping concentration of the initial silicon film, the transfer characteristics are identical as long as the doping is below 10^{17} cm^{-3}. This can be understood from Fig. 7.33(a). The MIS contact induces the same electron concentration in the JLFETs irrespective of the background doping. Therefore, the transfer characteristics of the JLFETs with MIS contacts remain the same as long as the channel doping is lower than the induced electron concentration. Furthermore, the energy band profile of the conduction band remains same in MIS JLFETs with a different doping since it is the electric field emanating from the MIS contact which dictates the surface potential as shown in Fig. 7.33(b).

The contact resistivity of the MIS contacts is independent of the channel doping and dependent only on the thickness of the insulator as shown in Fig. 7.33(c). Also, the contact resistivity of the MIS contacts is lower than that of the M–S contacts even when a highly doped source/drain extension region of 3 nm is used. As

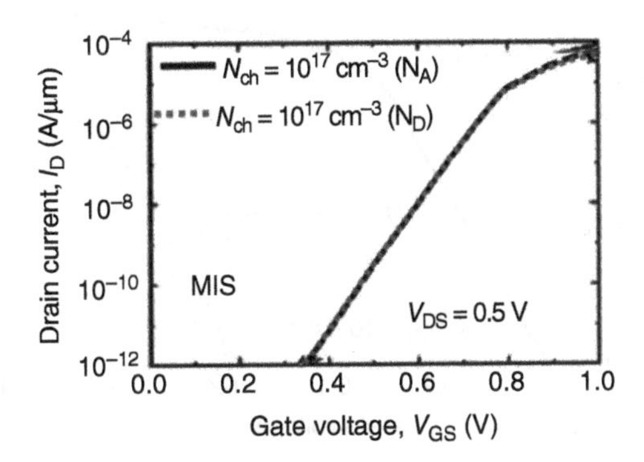

FIGURE 7.32 Transfer characteristics of the JLFETs with MIS contacts for different silicon film dopings [68].

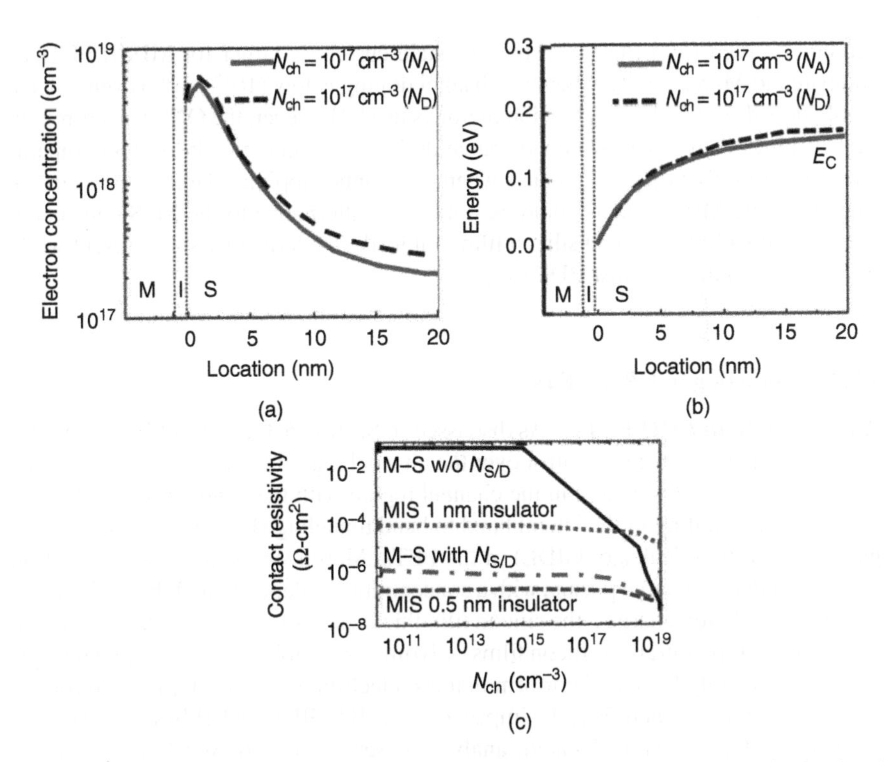

FIGURE 7.33 (a) The electron density and (b) energy band profile of the conduction band taken at a cutline 1 nm below the top Si–SiO_2 interface and (c) the contact resistivity (ρ_c) of the JLFETs with M–S and MIS contacts for different insulator thicknesses [68].

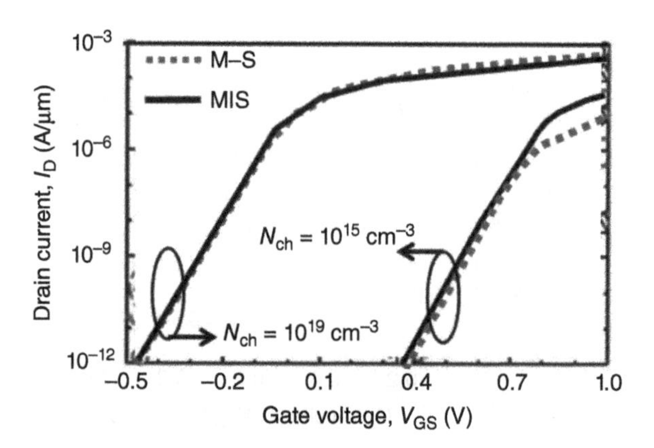

FIGURE 7.34 Transfer characteristics of JLFETs with MIS contacts M–S contacts for different channel dopings [68].

can be observed from Fig. 7.34, the ON-state current is larger for MIS JLFETs as compared to M–S JLFETs when the channel doping is low (10^{15} cm^{-3}) even when a highly doped source/drain extension region is used. However, the ON-state current in M–S JLFETs with a heavily doped source/drain extension region becomes comparable to the JLFETs with MIS contacts when the channel doping is large ($>10^{19}$ cm^{-3}). Therefore, the MIS contacts could be a lucrative alternative to the M–S contacts in JLFETs with a lightly doped silicon film as it leads to a large ON-state current while providing immunity against RDF [68].

7.8.2 Tunneling in CPJLFETs

7.8.2.A GIDL in CPJLFETs As discussed in Section 5.1, an ultrathin silicon film not only increases the gate control over the channel region but also leads to a significant overlap of valence band in the channel region with the conduction band in the drain region, resulting in the lateral band to band tunneling (L-BTBT) component of gate-induced drain leakage (GIDL) in the FETs. Moreover, as discussed in Section 7.2.1, for inducing charge plasma with a concentration independent of the background doping or depletion charges, the silicon film thickness should be less than the Debye length. Therefore, ultrathin silicon films (<10 nm) are employed in CP-based devices to have sufficiently large and uniform induced electron/hole concentrations throughout the silicon film. Therefore, the impact of L-BTBT GIDL on CP-based FETs must be analyzed. It may be noted that the analysis presented in Sections 7.1–7.7 is without including the BTBT models to account for the L-BTBT GIDL.

You may also notice that since the gate electrode is always separated from the drain electrode used for inducing charges with a spacer, the CP-based FETs inherently have a gate-drain underlapped architecture. Therefore, the possibility of transverse BTBT

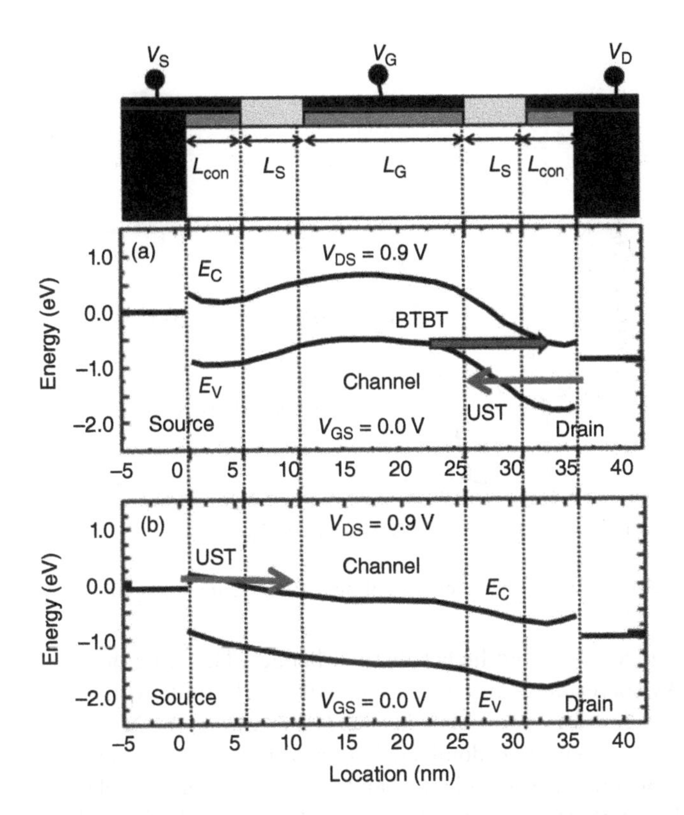

FIGURE 7.35 Energy band profiles of the CPJLFET in (a) the OFF-state and (b) the ON-state [76].

component of GIDL, which occurs as a result of tunneling in the gate-drain overlap region, is eliminated in the CP-based FETs.

However, as can be observed from Fig. 7.35(a), there exists a significant spatial proximity between the valence band of the channel region and conduction band of the drain region, which facilitates L-BTBT even in CPJLFETs in the OFF-state ($V_{GS} = 0.0$ V). Therefore, the drain current increases for negative gate voltages ($V_{GS} < 0.0$ V) when the BTBT models are included unlike the case when BTBT models are not included as shown in Fig. 7.36. This ambipolar behavior is attributed to the L-BTBT.

7.8.2.B Tunneling in the ON-State As discussed in Section 7.6.1, the MIGS lead to Fermi-level pinning at M–S interface making it difficult to realize ohmic contacts on lightly doped silicon films. The MIGS also increases the Schottky barrier height at the M–S junction as compared to the actual difference between the metal work function in vacuum and electron affinity value of the semiconductor. Furthermore, the CPFETs also exhibit a higher contact resistance due to the absence of any silicide

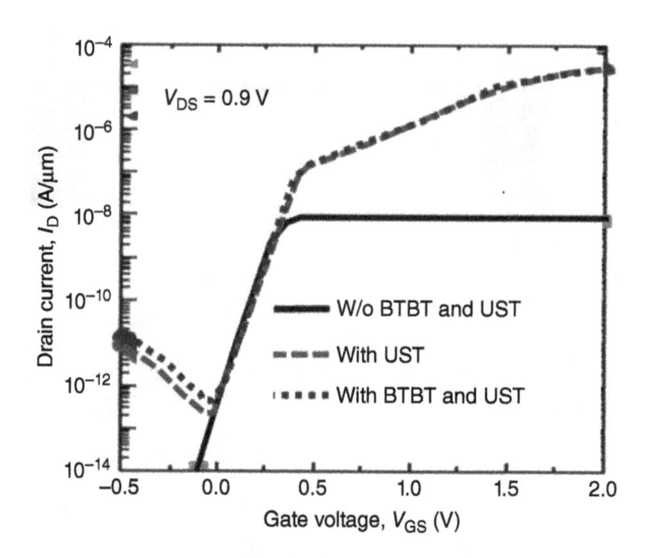

FIGURE 7.36 Transfer characteristics of the CPJLFET including tunneling models [76].

annealing process to avoid higher thermal budget. Therefore, to take into account the Schottky source/drain contacts, a source/drain electrode with a work function of 4.31 eV and a Schottky barrier height of 0.3 eV was used in [76].

The thermionic emission dominated drain current saturates in the CPJLFETs due to the large Schottky barrier height when BTBT models are not taken into account as shown in Fig. 7.36. However, the drain current continues to rise with increasing gate voltage once the universal Schottky tunneling model is considered. This is attributed to the barrier thinning with the increasing gate voltage, which facilitates the BTBT from the metal to the semiconductor as shown in Fig. 7.35(b). The BTBT leads to a large ON-state current in the CPJLFETs. However, the ON-state current in the CPJLFETs is still considerably less than the ON-state current in the conventional MOSFETs and JLFETs (Fig. 7.37) owing to the low concentration of the carriers in the source/drain region.

7.8.2.C Impact of Spacer Length Since the spacer length dictates the separation between the source/drain and the channel region, it is an important design parameter for CPJLFETs. A smaller spacer length leads to a larger band bending at the M–S interface at source increasing the tunneling probability. Furthermore, it also reduces the series resistance offered by the undoped silicon film below the spacer region owing to the smaller length of the undoped regions. Therefore, a small spacer length boosts the ON-state current as shown in Fig. 7.38.

However, the spacer length dictates the tunneling width at the channel drain interface. Therefore, a smaller spacer length increases the L-BTBT GIDL significantly leading to a large OFF-state current. This clearly indicates that there exists a trade-off between the increment in the ON-state current and the reduction in the ON-state to

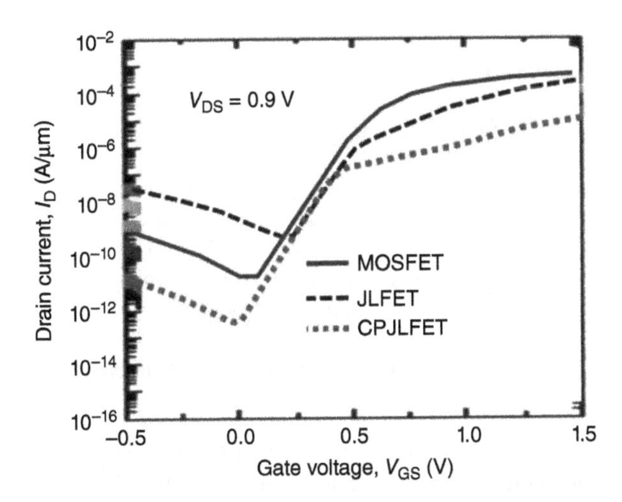

FIGURE 7.37 Transfer characteristics of the conventional MOSFETs, JLFETs, and the CPJLFET with Schottky barrier tunneling model included [76].

OFF-state current ratio with decreasing spacer lengths [76]. Therefore, the spacer length must be appropriately chosen to optimize the ON-state to OFF-state current ratio in CPJLFETs.

7.8.2.D Impact of Source/Drain Electrode Length The length of the source/drain electrodes over the silicon film inducing the charge concentrations while forming MIS

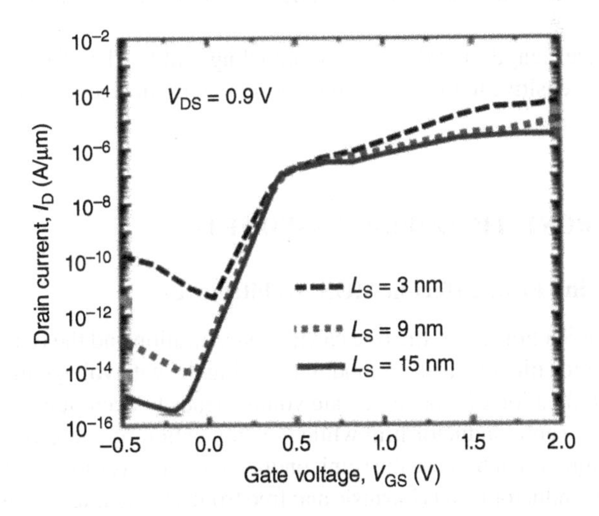

FIGURE 7.38 Transfer characteristics of the CPJLFET with different spacer lengths (L_S) [76].

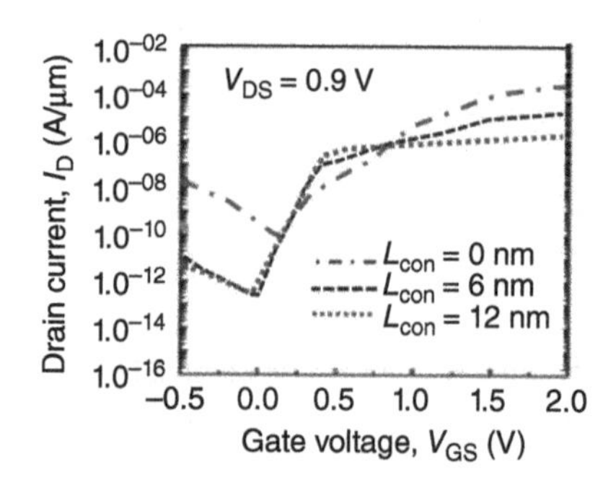

FIGURE 7.39 Transfer characteristics of the CPJLFET with different contact lengths [76].

junction is also an important design parameter for CPJLFETs. Ideally, to reduce the gate pitch, the contact length should be as small as possible. The impact of changing the length of the source/drain electrodes over the silicon film is shown in Fig. 7.39. CPJLFETs with a lower length of source/drain electrodes exhibit a lower drain current at small gate voltages and a higher drain current at higher gate voltages. This is attributed to the interplay between the electric field from the Schottky source contact and the gate electrode at low gate voltages. The gate electric field is sufficient to suppress the impact of the electric fields emanating from the source Schottky contact when a smaller electrode length is chosen. However, as the gate is placed far away from the source–Schottky contact when a large electrode length is used, the efficacy of the gate electrode in suppressing the electric fields at the source–Schottky contact diminishes significantly increasing the drain current. However, a larger electrode thickness increases the Schottky tunneling width and leads to a lower ON-state current [76].

It is the spacer length that dictates the tunneling width. Therefore, the GIDL current is nearly insensitive to the variation in the source/drain electrode length as shown in Fig. 7.39.

7.9 ELECTROSTATIC DOPING BASED FETs

7.9.1 Revisiting Field-Effect in MOS Architectures

As discussed in Section 1.2.1, the free carrier concentration and the type of carriers in the semiconductor film can also be modulated using the gate voltage in a MOS architecture. The application of a positive gate voltage leads to accumulation of electrons at the surface of semiconductor film while the application of a negative gate voltage leads to accumulation of holes at the semiconductor surface as shown in Fig. 7.40 even when the semiconductor film is intrinsic and free from any chemical doping. The electrostatic doping technique utilizes the gate voltage as an effective tool for modulating the carrier concentration in the semiconductor film.

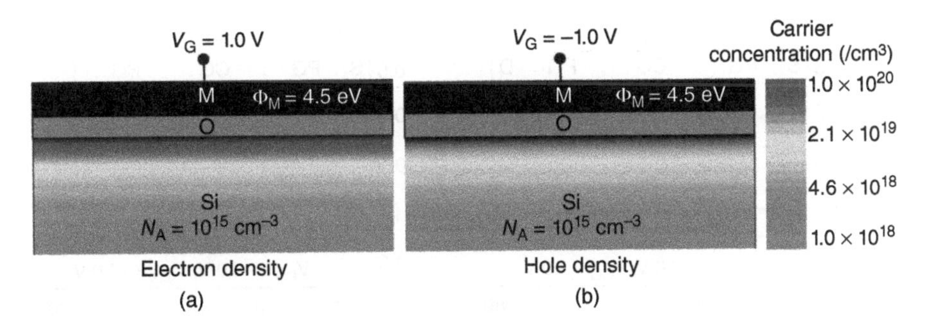

FIGURE 7.40 Induced (a) electron and (b) hole concentration in the silicon film due to the field-effect in a MOS capacitor.

Another interesting observation that can be made from Fig. 7.40 is that the same MOS architecture can induce both electron concentration and hole concentration in the semiconductor film depending upon the applied gate voltage. This property of the MOS architecture renders the electrostatic-doped FETs their unique property of reconfigurability, i.e. the same FET structure can be programmed as n-FET or p-FET depending upon the gate voltages.

Also, since no chemical doping is involved, the FETs with electrostatic doping are also more robust against RDF.

7.9.2 Structure

The structure of an electrostatically doped nanowire NWFET (EDFET) is shown in Fig. 7.41. It consists of a metal–semiconductor–metal (MSM) heterostructure [78–81]. It may be noted that although there are two M–S junctions in the EDFET, an intrinsic silicon film is used and no chemical doping process is used for inducing carriers. Nickel silicide with a Schottky barrier height of 0.45 eV for hole injection and 0.66 eV for electron injection is taken for source/drain contacts [77]. A control gate is employed at the center of the semiconductor, and two polarity gates responsible for inducing free electron/hole concentrations are placed at the side of the control gate.

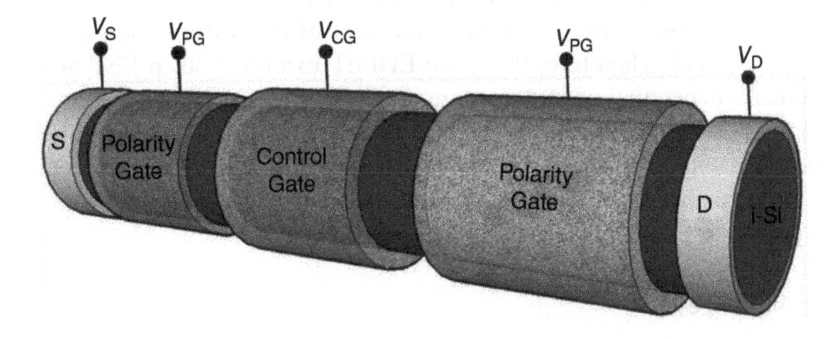

FIGURE 7.41 Three-dimensional view of the electrostatically doped NWFET (EDFET).

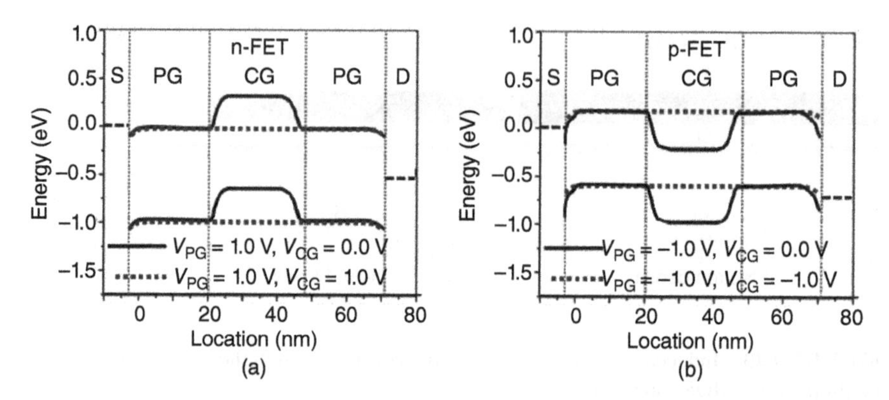

FIGURE 7.42 Energy band profiles of the EDFET in (a) n-FET configuration in the ON-state ($V_{CG} = 1.0$ V) and OFF-state ($V_{CG} = 0.0$ V) and (b) p-FET configuration in the ON-state ($V_{CG} = -1.0$ V) and OFF-state ($V_{CG} = 0.0$ V) taken at a cutline 1 nm below the Si–SiO_2 interface for different polarity gate and control gate voltages [78].

7.9.3 Operation

The application of a voltage over the polarity gates not only induces carriers in the intrinsic silicon nanowire but also modulates the injection of carriers through the Schottky barriers. For instance, the application of a positive polarity gate voltage induces a large concentration of electrons adjacent to the M–S Schottky junction. The induced n^+ silicon and the high electric field facilitate the injection of electrons through the Schottky barrier via the tunneling mechanism due to a reduction in the Schottky tunneling width as shown in Fig. 7.42. The control gate can then be utilized to modulate the potential barrier and the consequent conduction of electrons through the channel region. In the OFF-state, the potential barrier at the source–channel interface is large for electrons to surmount the barrier and conduct current; whereas in the ON-state the potential barrier is lowered and the electrons can flow easily from the source to the drain [78].

Similarly, application of a negative polarity gate voltage not only induces holes but also leads to injection of holes at the drain electrode via the tunneling mechanism due to the large electric field. The control gate then effectively modulates the potential barrier of the channel region controlling the flow of holes from drain to source as shown in Fig. 7.42. Therefore, the same EDFET exhibits both p-FET and n-FET characteristics depending upon the nature of applied voltages at the polarity gates as shown in Fig. 7.43. At this juncture, it may be noted that the EDFET exhibits the p-type operation without utilizing a negative voltage on the drain electrode, i.e. for both n-type or p-type operations, a positive drain voltage is utilized unlike the conventional CMOS operation [78].

7.9.4 Challenges

Since the MSM junctions require formation of metal silicides, the thermal budget is not low for EDFETs. For instance, nickel silicidation takes place at around 800°C.

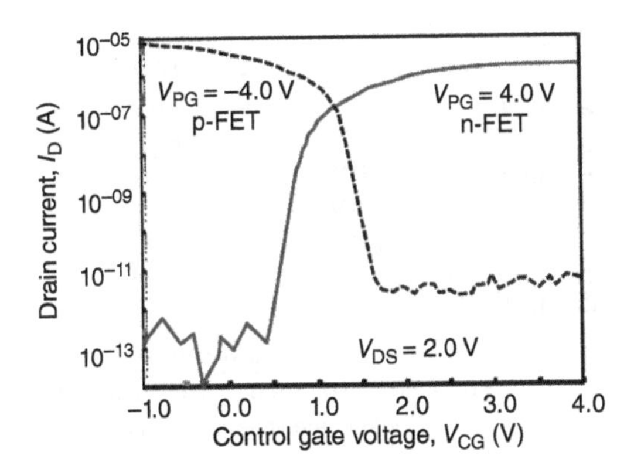

FIGURE 7.43 Transfer characteristics of the EDFET for different polarity gate voltages [78].

Although the electrostatic doping eliminates the impact of RDF, it does not alleviate the need for high thermal budget due to the formation of metal silicides.

The proximity of polarity gates and control gates may lead to a large parasitic capacitance. Also, intergate coupling and cross-talk would be unavoidable at ultra-scaled technological nodes. However, the uniquc property of configurability may allow realization of even complex functional blocks like XOR using only four FETs instead of the conventional CMOS XOR, which requires eight FETs as shown in [75]. The reconfigurable FET architecture may facilitate realization of large functional blocks utilizing few EDFETs or just by altering the interconnection or the routing of the different inputs as done in the field-programmable gate arrays [78–82].

7.9.5 Reconfigurability Utilizing Single Polarity Gate

The parasitic capacitance would be decreased, and the scaling of the EDFETs would be facilitated if the number of polarity gates is reduced or the distance between control gate and polarity gate is increased. Therefore, another interesting realization of the reconfigurable electrostatic doped FETs utilizing only two gates: one polarity gate and one control gate instead of three gates was proposed in [83] as shown in Fig. 7.44.

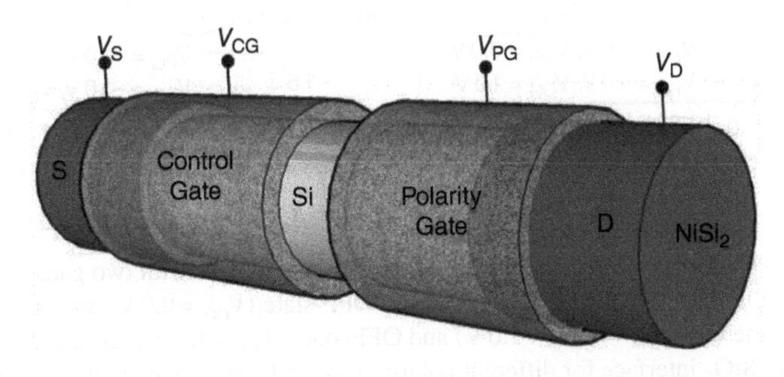

FIGURE 7.44 Three-dimensional view of the reconfigurable FET utilizing only two gates.

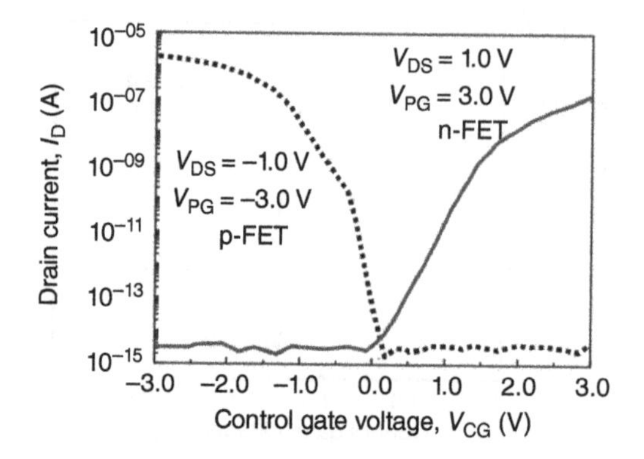

FIGURE 7.45 Transfer characteristics of the reconfigurable FET with two gates [83].

The polarity gate is placed over the channel–drain Schottky contact, whereas the control gate is placed over the source–channel Schottky contact.

The transfer characteristics of the EDFET with two gates are shown in Fig. 7.45. The operation of the EDFET with two gates can be understood from Fig. 7.46. Since nickel silicide exhibits a high Schottky barrier height for both electrons and holes, in the OFF-state, there is no injection of either holes or electrons in the silicon nanowire. However, for instance, in the n-program mode, when a positive voltage is applied to the control gate, the Schottky barrier reduces and results in thermionic emission of electrons over the barrier as shown in Fig. 7.46(a). If the control gate voltage is further increased, the Schottky barrier width becomes significantly narrow and facilitates injection of electrons via tunneling as shown in Fig. 7.46. It may be noted that, in this case, the p-FET operation occurs when a negative voltage is applied to the drain,

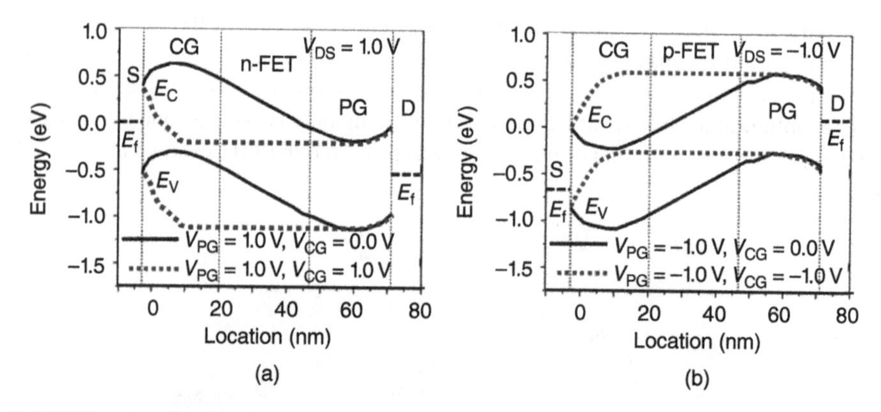

FIGURE 7.46 Energy band profiles of the reconfigurable FET with two gates in (a) n-FET configuration at the ON-state (V_{CG} = 1.0 V) and OFF-state (V_{CG} = 0.0 V) and (b) p-FET configuration at the ON-state (V_{CG} = −1.0 V) and OFF-state (V_{CG} = 0.0 V) taken at a cutline 1 nm below the Si–SiO_2 interface for different polarity gate and control gate voltages [83].

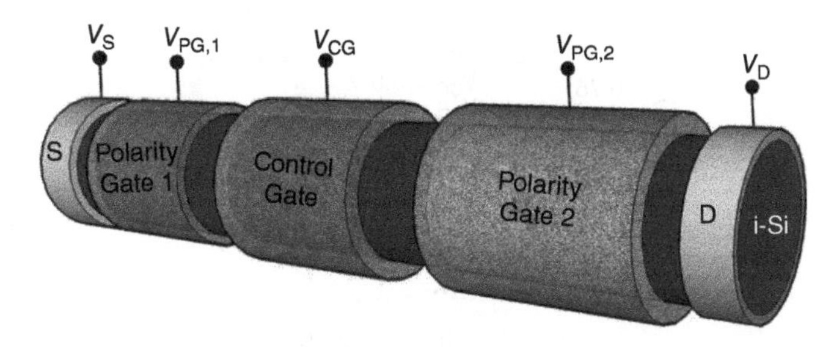

FIGURE 7.47 Three-dimensional view of the electrostatic doped TFET (EDTFET).

whereas n-FET operation requires application of a positive drain voltage unlike the three gated EDFET [83, 84].

7.9.6 Electrostatically Doped TFET

Although the EDFETs offer a reconfigurable alternative to the CMOS devices, their subthreshold swing is still limited to the Boltzmann tyranny of 60 mV/decade. This hinders their application in ultralow power circuits targeting sub-0.5 V supply voltages. Therefore, if somehow the thermionic injection mechanism of current conduction is replaced with the tunneling, sub-60 mV/decade subthreshold swing may be obtained even in EDFETs. This can be done by inducing p^+ instead of n^+ in the source region resulting in a TFET p–i–n configuration [85–88].

7.9.6.A Structure The 3D view of the electrostatic doped TFET (EDTFET) is similar to the EDFET as shown in Fig. 7.47. However, for realizing TFET, instead of connecting the two polarity gates together, the polarity gate on the source side (PG 1) is connected to a negative gate voltage to induce holes and form p^+ source whereas the polarity gate on the drain side (PG 2) is biased independently and connected to a positive potential to induce electrons forming the n^+ drain. Therefore, EDTFET essentially exploits electrostatic doping for formation of p–i–n architecture without the need for chemical doping. Nickel silicide is used as the source/drain contact.

7.9.6.B Operation Figure 7.48 shows the energy band profiles of the EDTFET in the ON-state and the OFF-state. In the OFF-state the EDTFET resembles a reverse biased p–i–n diode, whereas in the ON-state the inversion of the channel region results in significant spatial proximity between the valence band of the source and the conduction band of the channel which facilitates BTBT. The transfer characteristics of EDTFET are also shown in Fig. 7.49. The EDTFET exhibits a lower ON-state current owing to the increased series resistance of the regions separating the gate electrodes and a low OFF-state current.

7.9.6.C Sensitivity to process–voltage–temperature Variations As discussed in Section 2.1, TFETs are more sensitive to process-induced variations in oxide

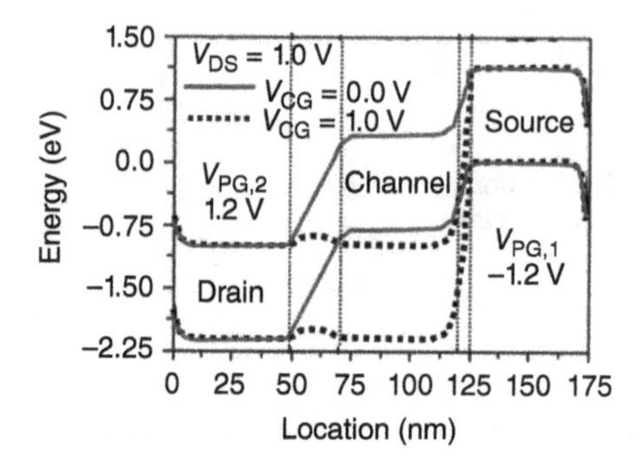

FIGURE 7.48 Energy band profiles of the EDTFET in the ON-state (V_{CG} = 1.0 V) and OFF-state (V_{CG} = 0.0 V) taken at a cutline 1 nm below the Si–SiO_2 interface for different control gate voltages [89].

thickness, silicon film thickness, doping gradient, and so on as compared to the conventional MOSFETs. This is because tunneling probability is exponentially dependent upon the tunneling distance which in turn depends on the electric field and the doping gradient dictated by these parameters. Furthermore, TFETs are also more sensitive to RDF as compared to MOSFETs or JLFETs as discussed in Section 3.6.2.

However, EDTFETs are immune to RDF since they are not chemically doped. In addition, the relative change in the performance of the EDTFETs due to process variations is also significantly small as compared to the TFETs [89]. Furthermore, an increase in temperature leads to an increase in the intrinsic concentration due to a reduction in the band gap. The band gap reduction due to temperature leads to an increase in the tunneling probability and a consequent increment in the ON-state current. However, the increment in the ON-state current per kelvin change in the temperature is also lower in the EDTFETs [89]. Therefore, the EDTFETs are robust against process–voltage–temperature induced variations as compared to the conventional TFETs.

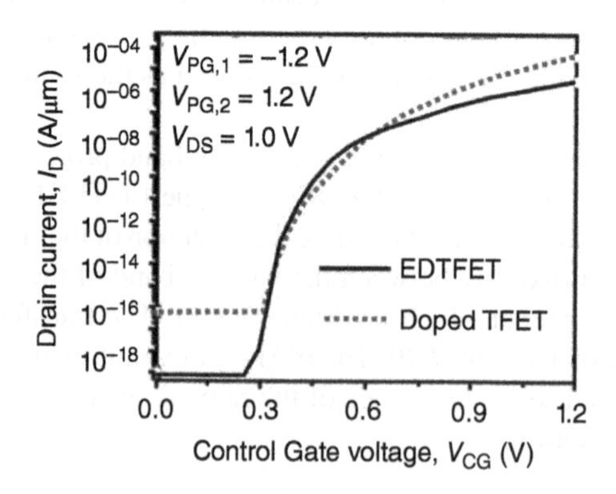

FIGURE 7.49 Transfer characteristics of the EDTFET and the conventionally doped TFET [89].

7.9.6.D Challenges The EDTFET also suffers from the same challenges faced by the EDFETs. For instance, the proximity of the three gates would increase the parasitic capacitances and degrade the dynamic performance of the EDTFETs significantly. Furthermore, the ON-state current of EDTFETs is low and the tunneling distance in EDTFETs is dictated by the gap between the control gate and the source polarity gate. However, reducing the gap between the two electrodes is limited by lithographic challenges as well as intergate coupling and cross-talks. Therefore, these challenges must be mitigated to make EDTFETs a lucrative alternative to the conventional TFETs and MOSFETs.

7.10 CONCLUSIONS

In this chapter, we discussed about the alternate approaches to the chemical doping, i.e., charge plasma concept and the electrostatic doping technique. We saw how the charge plasma doping can be used to fabricate all the emerging FET devices like TFET, IMOS, JLFETs, ans so on utilizing a low thermal budget opening up a new horizon for realizing these devices on glass substrate while at the same time improving their performance. We also discussed how electrostatic doping makes it possible to realize FETs with reconfigurability whereby the same FET exhibits both n-FET and p-FET characteristics by simply altering the polarity gate bias. The challenges in realizing the junctionless devices without any chemical doping such as fabrication complexity while designing additional gates and the intergate coupling and cross-talk, which eventually lead to poor dynamic performance, were also highlighted.

Having discussed the design of the JLFETs for optimum performance with and without chemical doping, we would now switch to the other important aspect of JLFETs, i.e. modeling the surface potential and the drain current in JLFETs. In the next chapter, we discuss about modeling of the surface potential and drain current of JLFETs using simple approaches. We also discuss in detail the charge-based modeling approach for JLFETs.

REFERENCES

[1] J. F. Gibbons, "Ion implantation in semiconductors—Part I: Range distribution theory and experiments," *Proc. IEEE*, vol. 56, no. 3, pp. 295–319, Mar. 1968.

[2] J. F. Gibbons, "Ion implantation in semiconductors—Part II: Damage production and annealing," *Proc. IEEE*, vol. 60, no. 9, pp. 1062–1096, Sept. 1972.

[3] S. Furukawa, H. Matsumura, and H. Ishiwara, "Theoretical considerations on lateral spread of implanted ions," *Jap. J. Appl. Phys.*, vol. 11, no. 2, pp. 134, Feb. 1972.

[4] J. T. Watt and J. D. Plummer, "Surface potential fluctuations in MOS devices induced by the random distribution of channel dopant ions," *IEEE Trans. Electron Devices*, vol. 35, no. 12, pp. 2431, Dec. 1988.

[5] J. T. Watt and J. D. Plummer, "Dispersion of MOS capacitance-voltage characteristics resulting from the random channel dopant ion distribution," *IEEE Trans. Electron Devices*, vol. 41, no. 11, pp. 2222–2232, Nov. 1994.

[6] M. J. Pelgrom, A. C. Duinmaijer, and A. P. Welbers, "Matching properties of MOS transistors," *IEEE J. Solid-State Circuits*, vol. 24, no. 5, pp. 1433–1439, Oct. 1989.

[7] A. R. Brown, A. Asenov, and J. R. Watling, "Intrinsic fluctuations in sub 10-nm double-gate MOSFETs introduced by discreteness of charge and matter," *IEEE Trans. Nanotechnol.*, vol. 99, no. 4, pp. 195–200, Dec. 2002.

[8] A. Gnudi, S. Reggiani, E. Gnani, and G. Baccarani, "Analysis of threshold voltage variability due to random dopant fluctuations in junctionless FETs," *IEEE Electron Device Lett.*, vol. 33, no. 3, pp. 336–338, Mar. 2012.

[9] S. M. Nawaz, S. Dutta, A. Chattopadhyay, and A. Mallik, "Comparison of random dopant and gate-metal workfunction variability between junctionless and conventional FinFETs," *IEEE Electron Device Lett.*, vol. 35, no. 6, pp. 663–665, June 2014.

[10] G. Ghibaudo, "Evaluation of variability performance of junctionless and conventional trigate transistors," *Solid-State Electron.*, vol. 75, pp. 13–15, Sept. 2012.

[11] Y. Taur, H. P. Chen, W. Wang, S. H. Lo, and C. Wann, "On–off charge–voltage characteristics and dopant number fluctuation effects in junctionless double-gate MOSFETs," *IEEE Trans. Electron Devices*, vol. 59, no. 3, pp. 863–866, Mar. 2012.

[12] M. Aldegunde, A. Martinez, and J. R. Barker, "Study of discrete doping-induced variability in junctionless nanowire MOSFETs using dissipative quantum transport simulations," *IEEE Electron Device Lett.*, vol. 33, no. 2, pp. 194–196, Feb. 2012.

[13] G. Leung and C. O. Chui, "Variability impact of random dopant fluctuation on nanoscale junctionless FinFETs," *IEEE Electron Device Lett.*, vol. 33, no. 6, pp. 767–769, June 2012.

[14] G. Giusi and A. Lucibello, "Variability of the drain current in junctionless nanotransistors induced by random dopant fluctuation," *IEEE Trans. Electron Devices*, vol. 61, no. 3, pp. 702–706, Mar. 2014.

[15] S. M. Nawaz and A. Mallik, "Effects of device scaling on the performance of junctionless FinFETs due to gate-metal work function variability and random dopant fluctuations," *IEEE Electron Device Lett.*, vol. 37, no. 8, pp. 958–961, Aug. 2016.

[16] S. M. Nawaz, S. Dutta, and A. Mallik, "A comparison of random discrete dopant induced variability between Ge and Si junctionless p-FinFETs," *Appl. Phys. Lett.*, vol. 107, no. 3, 033506, July 2015.

[17] B. Rajasekharan, C. Salm, R. J. E. Hueting, T. Hoang, and J. Schmitz, "Dimensional scaling effects on transport properties of ultrathin body pin diodes," in *ULIS*, pp. 195–198, 2008.

[18] B. Rajasekharan, R. J. E. Hueting, C. Salm, T. V. Hemert, R. A. Wolters, and J. Schmitz, "Fabrication and characterization of the charge-plasma diode," *IEEE Electron Device Lett.*, vol. 31, no. 6, pp. 528–530, 2010.

[19] R. J. Hueting, B. Rajasekharan, C. Salm, and J. Schmitz, "The charge plasma PN diode," *IEEE Electron Device Lett.*, vol. 29, no. 12, pp. 1367–1369, 2008.

[20] J. Li, Z. B. Zhang, Z. Qiu, and S. L. Zhang, "Contact-electrode insensitive rectifiers based on carbon nanotube network transistors," *IEEE Electron Device Lett.*, vol. 29, no. 5, pp. 500–502, 2010.

[21] G. Gupta, B. Rajasekharan, and R. J. Hueting, "Electrostatic Doping in Semiconductor Devices," *IEEE Trans. Electron Devices*, vol. 64, no. 8, pp. 3044–3055, 2017.

[22] M. J. Kumar and K. Nadda, "Bipolar charge-plasma transistor: A novel three terminal device," *IEEE Trans. Electron Devices*, vol. 59, no. 4, pp. 962–967, 2012.

[23] K. Nadda and M. J. Kumar, "Thin-film bipolar transistors on recrystallized polycrystalline silicon without impurity doped junctions: Proposal and investigation," *J. Display Technol.*, vol. 10, no. 7, pp. 590–594, 2014.

[24] K. Nadda and M. J. Kumar, "Schottky collector bipolar transistor without impurity doped emitter and base: Design and performance," *IEEE Trans. Electron Devices*, vol. 60, no. 9, pp. 2956–2959, 2013.

[25] M. J. Kumar, M. Maheedhar, and P. P. Varma, "Junctionless biristor: A bistable resistor without chemically doped PN junctions," *IEEE J. Electron Dev. Soc.*, vol. 3, no. 4, pp. 311–315, 2015.

[26] K. Nadda and M. J. Kumar, "Vertical bipolar charge plasma transistor with buried metal layer," *Sci. Rep.*, vol. 5, pp. 7860, 2015.

[27] A. Sahu, L. K. Bramhane, and J. Singh, "Symmetric lateral doping-free BJT: A novel design for mixed signal applications," *IEEE Trans. Electron Devices*, vol. 63, no. 7, pp. 2684–2690, 2016.

[28] M. S. Ram and D. B. Abdi, "Single grain boundary dopingless PNPN tunnel FET on recrystallized polysilicon: Proposal and theoretical analysis," *IEEE J. Electron Dev. Soc.*, vol. 3, no. 3, pp. 291–296, 2015.

[29] M. Panchore, J. Singh, and S. P. Mohanty, "Impact of channel hot carrier effect in junction-and doping-free devices and circuits," *IEEE Trans. Electron Devices*, vol. 63, no. 12, pp. 5068–5071, 2016.

[30] C. Shan, Y. Wang, and M. T. Bao, "A charge-plasma-based transistor with induced graded channel for enhanced analog performance," *IEEE Trans. Electron Devices*, vol. 63, no. 6, pp. 2275–2281, 2016.

[31] S. Ramaswamy and M. J. Kumar, "Raised source/drain dopingless junctionless accumulation mode FET: Design and analysis," *IEEE Trans. Electron Devices*, vol. 63, no. 11, pp. 4185–4190, 2016.

[32] F. Bashir, S. A. Loan, M. Rafat, A. R. M. Alamoud, and S. A. Abbasi, "A high-performance source engineered charge plasma-based Schottky MOSFET on SOI," *IEEE Trans. Electron Devices*, vol. 62, no. 10, pp. 3357–3364, 2015.

[33] K. Gopalakrishnan, P. B. Griffin, and J. D. Plummer, "Impact ionization MOS (IMOS)—Part I: Device and circuit simulations," *IEEE Trans. Electron Devices*, vol. 52, no. 1, pp. 69–76, Jan. 2005.

[34] K. Gopalakrishnan, R. Woo, C. Jungemann, P. B. Griffin, and J. D. Plummer, "Impact ionization MOS (IMOS)—Part II: Experimental results," *IEEE Trans. Electron Devices*, vol. 52, no. 1, pp. 77–84, Jan. 2005.

[35] C. Onal, R. Woo, H.-Y. S. Koh, P. B. Griffin, and J. D. Plummer, "A novel depletion-IMOS (DIMOS) device with improved reliability and reduced operating voltage," *IEEE Electron Device Lett.*, vol. 30, no. 1, pp. 64–67, Jan. 2009.

[36] E.-H. Toh, G. H. Wang, L. Chan, G. Samudra, and Y.-C. Yeo, "A double spacer IMOS transistor with shallow source junction and lightly doped drain for reduced operating

voltage and enhanced device performance," *IEEE Electron Device Lett.*, vol. 29, no. 2, pp. 189–191, Feb. 2008.

[37] Q. Huang, R. Huang, Z. Wang, Z. Zhan, and Y. Wang, "Schottky barrier impact-ionization metal–oxide–semiconductor device with reduced operating voltage," *Appl. Phys. Lett.*, vol. 99, no. 8, pp. 083507, Aug. 2011.

[38] E.-H. Toh, G. H. Wang, L. Chan, G.-Q. Lo, G. Samudra, and Y.-C. Yeo, "Strain and materials engineering for the IMOS transistor with an elevated impact-ionization region," *IEEE Trans. Electron Devices*, vol. 54, no. 10, pp. 2778–2785, Oct. 2007.

[39] D. Sarkar, N. Singh and K. Banerjee, "A novel enhanced electric-field impact-ionization MOS transistor," *IEEE Electron Devices Lett.*, vol. 31, no. 11, pp. 1175, 2010.

[40] S. Ramaswamy and M. J. Kumar, "Junctionless impact ionization MOS: Proposal and investigation," *IEEE Trans. Electron Devices*, vol. 61, no. 12, pp. 4295–4298, Dec. 2014.

[41] M. J. Kumar and S. Janardhanan, "Doping-less tunnel field-effect transistor: Design and investigation," *IEEE Trans. Electron Devices*, vol. 60, no. 10, pp. 3285–3290, 2013.

[42] D. B. Abdi and M. J. Kumar, "In-built N+ pocket p-n-p-n tunnel field-effect transistor," *IEEE Electron Device Lett.*, vol. 35, no. 12, pp. 1170–1172, 2014.

[43] B. R. Raad, S. Tirkey, D. Sharma, and P. Kondekar, "A new design approach of dopingless tunnel FET for enhancement of device characteristics," *IEEE Trans. Electron Devices*, vol. 64, no. 4, pp. 1830–1836, 2017.

[44] B. Ghosh and M. W. Akram, "Junctionless tunnel field-effect transistor," *IEEE Electron Device Lett.*, vol. 34, no. 5, pp. 584–586, 2013.

[45] M. Rahimian and M. Fathipour, "Junctionless nanowire TFET with built-in NPN bipolar action: Physics and operational principle," *J. Appl. Phys.*, vol. 120, no. 22, 225702, 2016.

[46] S. Sahay and M. J. Kumar, "Physical insights into the nature of gate-induced drain leakage in ultrashort channel nanowire FETs," *IEEE Trans. Electron Devices*, vol. 64, no. 6, pp. 2604–2610, June 2017.

[47] S. Sahay and M. J. Kumar, "A novel gate-stack-engineered nanowire FET for scaling to the sub-10-nm regime," *IEEE Trans. Electron Devices*, vol. 63, no. 12, pp. 5055–5059, Dec. 2016.

[48] S. Sahay and M. J. Kumar, "Spacer design guidelines for nanowire FETs from gate-induced drain leakage perspective," *IEEE Trans. Electron Devices*, vol. 64, no. 7, pp. 3007–3015, July 2017.

[49] S. Sahay and M. J. Kumar, "Insight into lateral band-to-band-tunneling in nanowire junctionless FETs," *IEEE Trans. Electron Devices*, vol. 63, no. 10, pp. 4138–4142, Oct. 2016.

[50] S. Sahay and M. J. Kumar, "Controlling L-BTBT and volume depletion in nanowire JLFETs using core-shell architecture," *IEEE Trans. Electron Devices*, vol. 63, no. 9, pp. 3790–3794, Sept. 2016.

[51] S. Sahay and M. J. Kumar, "Diameter dependency of leakage current in nanowire junctionless field-effect transistors," *IEEE Trans. Electron Devices*, vol. 64, no. 3, pp. 1330–1335, Mar. 2017.

[52] S. Sahay and M. J. Kumar, "Nanotube junctionless FET: proposal, design, and investigation," *IEEE Trans. Electron Devices*, vol. 64, no. 4, pp. 1851–1856, Apr. 2017.

[53] M. J. Kumar and S. Sahay, "Controlling BTBT induced parasitic BJT action in junctionless FETs using a hybrid channel," *IEEE Trans. Electron Devices*, vol. 63, no. 8, pp. 3350–3353, Aug. 2016.

[54] S. Sahay, and M. J. Kumar, "Realizing efficient volume depletion in SOI junctionless FETs," *IEEE J. Electron Devices Soc.*, vol. 4, no. 3, pp. 110–115, May 2016.

[55] S. Sahay and M. J. Kumar, "Comprehensive analysis of gate-induced drain leakage in emerging FET architectures: Nanotube FETs vs. nanowire FETs", *IEEE Access*, vol. 5, pp. 18918–18926, Dec. 2017.

[56] S. Sahay and M. J. Kumar, "Symmetric operation in an extended back gate JLFET for scaling to the 5 nm regime considering quantum confinement effects," *IEEE Trans. Electron Devices*, vol. 64, no. 1, pp. 21–27, Jan. 2017.

[57] A. K. Jain, S. Sahay, and M. J. Kumar, "Controlling L-BTBT in emerging nanotube FETs using dual-material gate," *IEEE J. Electron Devices Soc.*, vol. 6, pp. 611–621, June 2018.

[58] A. Hraziia, A. Gupta, A. Vladimirescu, A. Amara, and C. Anghel, "30-nm Tunnel FET with improved performance and reduced ambipolar current,", *IEEE Trans. Electron Devices*, vol. 58, pp. 1649–1654, June 2011.

[59] W. Y. Choi and W. Lee, "Hetero-gate-dielectric tunneling field-effect transistors," *IEEE Trans. Electron Devices*, vol. 57, no. 9, pp. 2317–2319, Sept. 2010.

[60] A. S. Verhulst, W. G. Vandenberghe, K. Maex, and G. Groeseneken, "Tunnel field-effect transistor without gate–drain overlap," *Appl. Phys. Lett.*, vol. 91, no. 5, pp. 053102–053103, July 2007.

[61] J. Wan, C. Le Royer, A. Zaslavsky, and S. Cristoloveanu, "SOI TFETs: Suppression of ambipolar leakage and low-frequency noise behavior", in *Proc. European Solid-State Device Research Conference (ESSDERC)*, 2010, pp. 341–344.

[62] J. Wana, C. L. Royer, A. Zaslavsky, and S. Cristoloveanu, "Tunneling FETs on SOI: Suppression of ambipolar leakage, low-frequency noise behavior, and modeling," *Solid-State Electron.*, vol. 65, pp. 226–233, Nov. 2011.

[63] D. B. Abdi and M. J. Kumar, "Controlling ambipolar current in tunneling FETs using overlapping gate-on-drain," *IEEE J. Electron Devices Soc.*, vol. 2, no. 6, pp. 187–190, Nov. 2014.

[64] S. Sahay and M. J. Kumar, "Controlling the drain side tunneling width to reduce ambipolar current in tunnel FETs using heterodielectric BOX," *IEEE Trans. Electron Devices*, vol. 62, no. 11, pp. 3882–3886, Nov. 2015.

[65] C. Sahu and J. Singh, "Charge-plasma based process variation immune junctionless transistor," *IEEE Electron Device Lett.*, vol. 35, no. 3, pp. 411–413, 2014.

[66] C. Sahu, and J. Singh, "Potential benefits and sensitivity analysis of dopingless transistor for low power applications," *IEEE Trans. Electron Devices*, vol. 62, no. 3, pp. 729–735, 2015.

[67] K. H. Kao and L. Y. Chen, "A dopingless FET with metal–insulator–semiconductor contacts," *IEEE Electron Device Lett.*, vol. 38, no. 1, pp. 5–8, 2017.

[68] L. Y. Chen, Y. F. Hsieh, and K. H. Kao, "Undoped and doped junctionless FETs: Source/drain contacts and immunity to random dopant fluctuation," *IEEE Electron Device Lett.*, vol. 38, no. 6, pp. 708–711, 2017.

[69] A. Dimoulas, P. Tsipas, A. Sotiropoulos, and E. K. Evangelou, "Fermi-level pinning and charge neutrality level in germanium," *Appl. Phys. Lett.*, vol. 89, no. 25, p. 252110, 2006.

[70] D. Connelly, C. Faulkner, P. A. Clifton, and D. E. Grupp, "Fermi-level depinning for low-barrier Schottky source/drain transistors," *Appl. Phys. Lett.*, vol. 88, no. 1, 012105, 2006.

[71] M. K. Husain, X. V. Li, and C. H. De Groot, "High-quality Schottky contacts for limiting leakage currents in Ge-based Schottky barrier MOSFETs," *IEEE Trans. Electron Devices*, vol. 56, no. 3, pp. 499–504, Mar. 2009.

[72] V. Heine, "Theory of surface states," *Phys. Rev.*, vol. 138, A1689, June 1965.

[73] W. Mönch, *Electronic Properties of Semiconductor Interfaces*. New York: Springer, 2004.

[74] T. Nishimura, K. Kita, and A. Toriumi, "Evidence for strong Fermi-level pinning due to metal-induced gap states at metal/germanium interface," *Appl. Phys. Lett.*, vol. 91, no. 12, 123123, 2007.

[75] A. Agrawal, J. Lin, M. Barth, R. White, B. Zheng, S. Chopra, S. Gupta, K. Wang, J. Gelatos, S. E. Mohney, and S. Data, "Fermi-level depinning and contact resistivity reduction using a reduced titania interlayer in n-silicon metal-insulator-semiconductor ohmic contacts," *Appl. Phys. Lett.*, vol. 104, no. 11, 112101, 2014,

[76] J. Hur, D. I. Moon, J. W. Han, G. H. Kim, C. H. Jeon, and Y. K. Choi, "Tunneling effects in a charge-plasma dopingless transistor," *IEEE Trans. Nanotechnol.*, vol. 16, no. 2, pp. 315–320, 2017.

[77] H. Yu, M. Schaekers, K. Barla, N. Horiguchi, N. Collaert, A. V.-Y. Thean, and K. De Meyer, "Contact resistivities of metal-insulator-semiconductor contacts and metal-semiconductor contacts," *Appl. Phys. Lett.*, vol. 108, no. 17, 171602, 2016.

[78] M. D. Marchi, D. Sacchetto, S. Frache, J. Zhang, P. E. Gaillardon, Y. Leblebici, and G. D. Micheli, "Polarity control in double-gate, gate-all-around vertically stacked silicon nanowire FETs," in *IEDM*, pp. 8–12, 2012.

[79] M. D. Marchi, J. Zhang, S. Frache, D. Sacchetto, P. E. Gaillardon, Y. Leblebici, and G. D. Micheli, "Configurable logic gates using polarity-controlled silicon nanowire gate-all-around FETs," *IEEE Electron Device Lett.*, vol. 35, no. 8, pp. 880–882, 2014.

[80] J. Zhang, M. D. Marchi, D. Sacchetto, P. E. Gaillardon, Y. Leblebici, and G. D. Micheli, "Polarity-controllable silicon nanowire transistors with dual threshold voltages," *IEEE Trans. Electron Devices*, vol. 61, no. 11, pp. 3654–3660, 2014.

[81] A. Bhattacharjee and S. Dasgupta, "Impact of gate/spacer-channel underlap, gate oxide EOT, and scaling on the device characteristics of a DG-RFET," *IEEE Trans. Electron Devices*, vol. 64, no. 8, pp. 3063–3070, 2017.

[82] A. Bhattacharjee, M. Saikiran, A. Dutta, B. Anand, and S. Dasgupta, "Spacer engineering-based high-performance reconfigurable FET with low off current characteristics," *IEEE Electron Device Lett.*, vol. 36, no. 5, pp. 520–522, 2015.

[83] A. Heinzig, S. Slesazeck, F. Kreupl, T. Mikolajick, and W. M. Weber, "Reconfigurable silicon nanowire transistors," *Nano Lett.*, vol. 12, no. 1, pp. 119–124, 2011.

[84] T. Baldauf, A. Heinzig, J. Trommer, T. Mikolajick, and W. M. Weber, "Stress-dependent performance optimization of reconfigurable silicon nanowire transistors," *IEEE Electron Device Lett.*, vol. 36, no. 10, pp. 991–993, 2015.

[85] J. Knoch and J. Appenzeller, "A novel concept for field-effect transistors—The tunneling carbon nanotube FET," in *Proc. Device Res. Conf. Dig.*, 2005, pp. 153–156.

[86] W. Y. Choi, B-G. Park, J.D. Lee, and T-J. K. Liu "Tunneling field-effect transistors (TFETs) with subthreshold swing (SS) less than 60 mV/dec," *IEEE Electron Device Lett.*, vol. 28, no. 8, pp. 743–745, Aug. 2007.

[87] A. C. Seabaugh and Q. Zhang, "Low-voltage tunnel transistors for beyond CMOS logic," *Proc. IEEE*, vol. 98, no. 12, pp. 2095–2110, Dec. 2010.

[88] A. M. Ionescu and H. Riel, "Tunnel field-effect transistors as energy efficient electronic switches," *Nature*, vol. 479, no. 7373, pp. 329–337, Nov. 2011.

[89] A. Lahgere, C. Sahu, and J. Singh, "PVT-aware design of dopingless dynamically configurable tunnel FET," *IEEE Trans. Electron Devices*, vol. 62, no. 8, pp. 2404–2409, 2015.

8

MODELING JUNCTIONLESS FIELD-EFFECT TRANSISTORS

From our discussions in Chapters 3–7, it becomes quite clear that junctionless FETs (JLFETs) have been extensively studied through experimental characterization and numerical simulations using technology computer aided design (TCAD) owing to their excellent electrostatics [1–13]. Several novel architectures have been proposed to mitigate the challenges faced by the JLFETs and to further improve their performance. In addition, electrostatic doping techniques have also been employed to get rid of the chemical doping processes and the metallurgical junctions yielding a new class of dopantless junctionless devices.

However, to facilitate the application of JLFETs, it becomes necessary to develop appropriate models for JLFETs. The theoretical models are essential for investigating the foundation of JLFETs and to better understand its operating principles. JLFETs can then be explored for designing analog or digital circuits by using these models in circuit simulators. Developing accurate, robust, and computationally-efficient models for JLFETs not only helps in exploring the design space of individual devices but also helps to analyze the performance of the systems utilizing JLFETs. Therefore, in this chapter, we cover the modeling aspect of JLFETs. We begin with a discussion of the different modeling approaches used for field-effect transistors (FETs) in general followed by a comprehensive analysis of the different analytical modeling techniques used for JLFETs.

Junctionless Field-Effect Transistors: Design, Modeling, and Simulation, First Edition.
Shubham Sahay and Mamidala Jagadesh Kumar.

8.1 INTRODUCTION TO FET MODELING

In general, FET modeling involves development of analytical relations that describe the electrical behavior of FETs based on the fundamental physics. Models are not only necessary to provide physical insight into the working of the FETs but also help in exploring their design space, optimization, and circuit-level analysis. There are two well-known approaches for theoretical modeling of FETs: (a) analytical modeling and (b) compact modeling.

In analytical modeling, different known physical equations such as the Poisson equation, continuity equation, Schrödinger equation, and so on are solved with appropriate approximations to yield a relation between the input voltages and the output current. Analytical models are accurate since they use first principle analysis and augment the information that may be derived from experimental results. They can be used to investigate the theoretical foundations of the FETs and help in better understanding of the microscopic properties, which are not intuitive from the experimental data. However, they are computationally inefficient and consume a lot of time. Therefore, they are not useful for system level or circuit simulations, which may involve numerous such devices. This calls for the need of a fast, computationally efficient, and robust model for circuit simulations which led to the development of the compact models. Analytical modeling is important for the device physicists and technologists, whereas compact models are more useful for the circuit designers.

Compact models are usually empirical and generated using crude curve fitting methods to match the experimental characteristics. They treat the device under test as a black box and parameterize the experimental behavior based on the terminal voltages and some empirical constants that can be tweaked to match experimental data. Therefore, the empirical compact models do not give any useful information about the microscopic properties. However, they are of utmost importance as they form a bridge between the foundry and the circuit designers.

There is yet another class of compact analytical models that are based on fundamental physics but are computationally efficient. The compact analytical models are obtained from the analytical models by using intelligent approximations and some empirical relations that have a physical basis. The number of parameters that need to be adjusted to match the experimental results is significantly lower in these models as compared to the compact models.

A quantum mechanics based atomistic modeling approach also exists, which is more accurate and relevant for the ultrathin silicon films where quantum confinement effects are prominent. The analytical models based on classical physics fail to account for subband quantization and the change in carrier profile due to quantization effects. Therefore, for the future technology nodes where the silicon film thickness reaches the quantum limit of 7 nm and below, the use of atomistic modeling and simulation is inevitable. The atomistic modeling approach relies on self-consistent solution of the Poisson equation and the Schrödinger equation as discussed in Section 5.4.2. The nonequilibrium Green's function (NEGF) formalism is generally utilized for this purpose. NEGF is a quantum domain counterpart of the Boltzmann transport equation (BTE) used for the analysis of semiclassical transport. It provides a method to solve the Schrödinger equation while keeping in mind the thermodynamic processes involved in the channel and at the source/drain contacts. The first process in this method is calculation of the Hamiltonian operator, utilizing a suitable basis such as effective mass for analysis of energy band profiles. The results are then used along with the ballistic quantum transport model derived using the Landaur's approach [14–19] to give final transport expressions. The NEGF formalism enables us to analyze the wave functions, energy Eigen values, which provide information about the allowed energy states and the energy subbands, transport effective mass in each subband, and so on. The atomistic modeling approach is most accurate for ultrathin silicon film FETs such as those used in the sub-10 nm technology nodes or the JLFETs where ultrathin film is required to achieve volume depletion. However, this method is computationally inefficient and may take even a month or more for converging to the final results even on a supercomputing cluster. This restricts widespread use of such an approach.

Once we have understood the importance and the different types of modeling approaches, let us discuss in detail the different analytical modeling approaches to get some more physical insight into the operation of JLFETs. As we see, most of these approaches follow the same principles used for modeling the metal–oxide–semiconductor field-effect transistors (MOSFETs) [20–29].

The general approach in the analytical modeling of MOSFETs is to first find a solution of the Poisson equation using appropriate assumptions, boundary conditions, approximations, and mathematical methods to get a relationship between the surface potential (φ_S) and the gate voltage (V_G). This step is known as surface potential modeling [25–29]. The surface potential can then be used to obtain the threshold voltage using the classical definition, i.e. the gate voltage at which the inversion layer carrier density is same as the bulk carrier concentration. At the threshold voltage (V_{Th}), the surface potential is twice the Fermi potential (φ_f), i.e. $\varphi_S = 2\varphi_f$, where φ_f is defined as $V_t \ln N_A/n_i$ and V_t is the thermal voltage, N_A is the channel doping, and n_i is the intrinsic carrier concentration. The surface potential model is extremely useful as it can also be used to predict the threshold voltage variation in the presence of oxide charges or traps [30–32].

Once a closed form surface potential model is obtained, the next step is to model the drain current. The gradual channel approximation is then used, which simply states that the potential drops gradually from the drain end to the source end along the channel length [25–29]. Using this approximation, the drain current (I_D) can be found out by simply using Ohm's law and integrating the current over the entire channel length between the source and the drain ends as

$$I_D = \mu \frac{W}{L} \int_{V_S}^{V_D} Q_{in}(x, V)\, dV \tag{8.1}$$

where $Q_{in}(x, V)$ is the inversion layer charge density, which depends on the location along the channel direction (x) as well as on the quasi-electron Fermi-level in the channel (V), W is the width of the channel region, L is the channel length, and μ is the effective mobility.

Following a similar approach for modeling JLFETs, we discuss the different techniques used for surface potential modeling, i.e. obtaining an analytical expression relating surface potential and the gate voltage in Section 8.2. We deal with the long-channel JLFETs first since the two-dimensional (2D) Poisson equation can be reduced to a single-dimension problem using the gradual channel approximation as discussed in Section 8.2.

8.2 SURFACE POTENTIAL MODELING OF JLFETs

The three-dimensional (3D) view of a long-channel single-gate JLFET is shown in Fig. 8.1. The electric field distribution of the JLFET operating in the partial depletion regime in the *x*- and *y*-directions (along the cutline A–A' in Fig. 8.1) is shown in Fig. 8.2. As can be observed from Fig. 8.2, the variation of the electric field component along the *y*-direction is negligible in the channel region of a long-channel

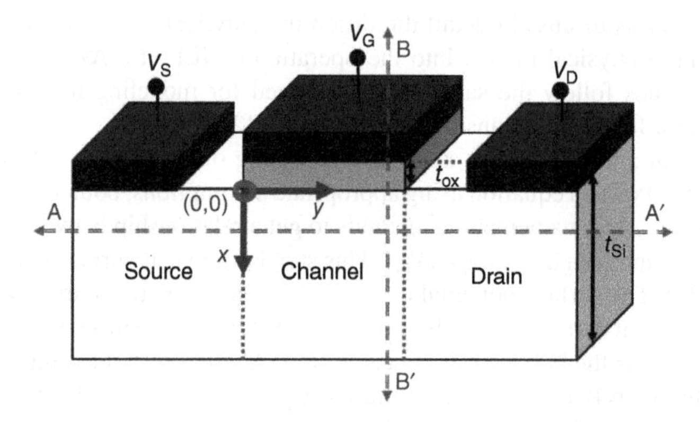

FIGURE 8.1 Three-dimensional view of the single-gate JLFET.

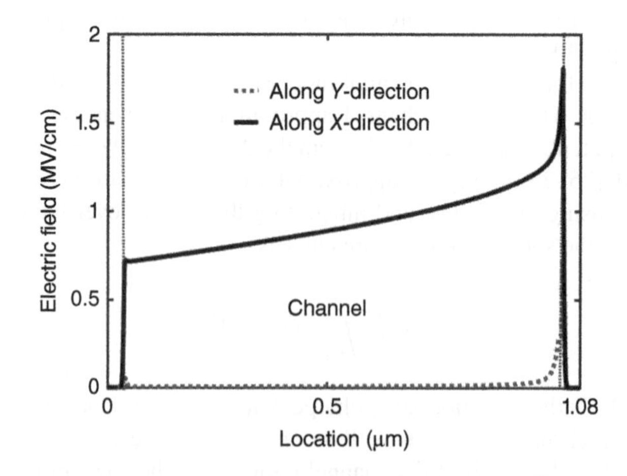

FIGURE 8.2 Electric field profile in the *x*- and *y*-directions along the cutline A-A' in the OFF-state ($V_{GS} = 0.0$ V) of JLFETs.

JLFET as compared to the electric field along the x-direction. Therefore, the gradual channel approximation is valid for JLFETs similar to the MOSFETs [25–29]. The gradual channel approximation allows us to neglect the electric field in the y-direction along the channel length and simplifies the solution of the original 2D Poisson equation given as

$$\frac{d^2\varphi(x,y)}{dx^2} + \frac{d^2\varphi(x,y)}{dy^2} = -\frac{\rho}{\varepsilon_{\mathrm{Si}}} \tag{8.2}$$

by reducing it to a single-dimensional problem in the x-direction along the channel thickness given as

$$\frac{d^2\varphi(x)}{dx^2} = -\frac{\rho}{\varepsilon_{\mathrm{Si}}} \tag{8.3}$$

where $\varphi(x)$ is the potential distribution, ρ is the charge density in the silicon film, and $\varepsilon_{\mathrm{Si}}$ is the permittivity of silicon.

Now, as discussed in Section 3.2.2, to find the surface potential, we use the depletion approximation (Fig. 8.3) and solve equation (8.3) for the potential distribution inside the silicon film. The charge density in the silicon channel can be simply given as $\rho = qN_{\mathrm{D}}$, where q is the electronic charge and N_{D} is the donor-doping concentration. Now, integrating equation (8.3) with respect to x, we have

$$E(x) = \frac{qN_{\mathrm{D}}x}{\varepsilon_{\mathrm{Si}}} + c \tag{8.4}$$

where $E(x)$ is the electric field distribution within the silicon film and c is the constant of integration. Now, with the assumption that the electric field emanating from the gate diminishes to 0 at $x = x_{\mathrm{dep}}$, where x_{dep} is the depletion region width in the silicon film, we can obtain the constant as

$$c = -\frac{qN_{\mathrm{D}}x_{\mathrm{dep}}}{\varepsilon_{\mathrm{Si}}} \tag{8.5}$$

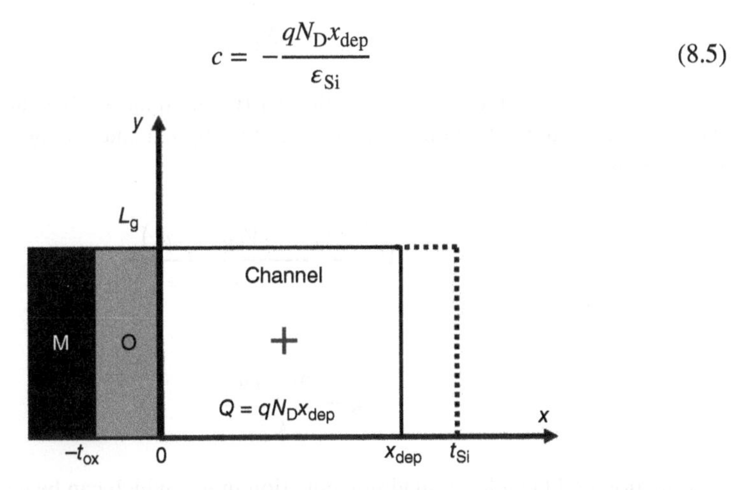

FIGURE 8.3 Depletion charge profile assumed in the channel region in JLFETs.

Therefore, the surface potential can be found by integrating equation (8.4) with respect to x from $x = 0$ (Si–SiO_2) interface to $x = x_{dep}$, which yields

$$\varphi\left(x_{dep}\right) - \varphi(0) = \frac{qN_D x_{dep}{}^2}{2\varepsilon_{Si}} \tag{8.6}$$

Now, assuming that the potential at $x = x_{dep}$ is the electron quasi-Fermi potential V, we can obtain an expression for the surface potential at the Si–SiO_2 interface as

$$\varphi_s = V - \frac{qN_D x_{dep}{}^2}{2\varepsilon_{Si}} \tag{8.7}$$

Now, applying the Gauss's law at the Si–SiO_2 interface relates the charge density in the semiconductor film (Q_{SC}) to the surface electric field (E_S) as

$$Q_{SC} = \varepsilon_{Si} E_S \tag{8.8}$$

Also, the electrical displacement vector should be continuous at the Si–SiO_2 interface, which means that

$$\varepsilon_{Si} E_S = \varepsilon_{ox} E_{ox} = \varepsilon_{ox} \frac{\left(V_G - V_{FB} - \varphi_s\right)}{t_{ox}} = C_{ox}\left(V_G - V_{FB} - \varphi_s\right) = Q_{SC} \tag{8.9}$$

where E_{ox} is the electric field in the oxide region, t_{ox} is the gate oxide thickness, and ε_{ox} is the permittivity of SiO_2. Equation (8.9) relates the charge density in the semiconductor film to the surface potential. The charge density in the semiconductor film can be approximated as the depletion charge and given as

$$Q_{SC} = qN_D x_{dep} \tag{8.10}$$

Using the value of x_{dep} from equation (8.10) in equation (8.7) and simplifying, we obtain an analytical relationship between the applied gate voltage and the surface potential as

$$\varphi_s = V - \frac{\left(V_G - V_{FB} - \varphi_s\right)^2}{\alpha} \tag{8.11}$$

where

$$\alpha = \frac{2\varepsilon_{Si} q N_D}{C_{ox}^2} \tag{8.12}$$

Equation (8.11) yields a quadratic equation in φ_S, which can be easily solved.

Although we have formulated a surface potential model for the single-gate JLFETs, from our discussions in Chapter 3, we know that the single-gate JLFETs suffer from several challenges such as inefficient volume depletion, and so on. The multigate architectures such as double-gate (DG), trigate (TG), and the gate-all-around nanowire (GAANW) JLFETs were proposed to improve the performance of the JLFETs. Apart from an improvement in the performance, symmetric structures like DG and GAANW JLFETs are better even from an analytical modeling perspective. The symmetric structure leads to additional boundary conditions, which ease the solution of the Poisson equation [18–21]. For instance, in the GAA NWJLFET with a circular cross section in cylindrical coordinates, owing to the symmetrical structure, the potential along the azimuth direction is same, i.e. the potential profile along the azimuth consists of equipotential lines. Therefore, the variation of the electric field in the azimuth direction is zero. This allows us to solve a 2D Poisson equation in radial (r) and the height (h) rather than solving a 3D Poisson equation in radial, azimuth, and height directions. Moreover, in the double-gate junctionless field-effect transistor (DGJLFETs), the presence of two gates and the symmetric structure yields three boundary conditions instead of only two used for the silicon-on-insulator (SOI) single-gated JLFETs [18–21]. Therefore, owing to the ease in modeling, we discuss the modeling techniques used for the long-channel DGJLFETs in rest of this chapter.

8.2.1 Operating Regionwise Approximation Technique

Let us consider the case of a DGJLFET as shown in Fig. 8.4. Following the Boltzmann statistics, the mobile electron concentration (n) in the channel region can be given as [33]

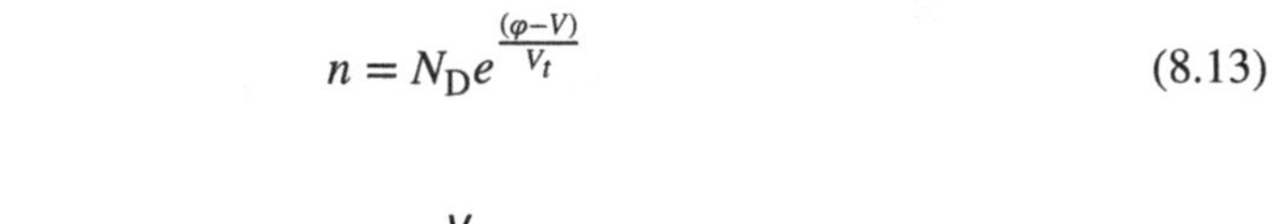

$$n = N_{\mathrm{D}} e^{\frac{(\varphi - V)}{V_t}} \tag{8.13}$$

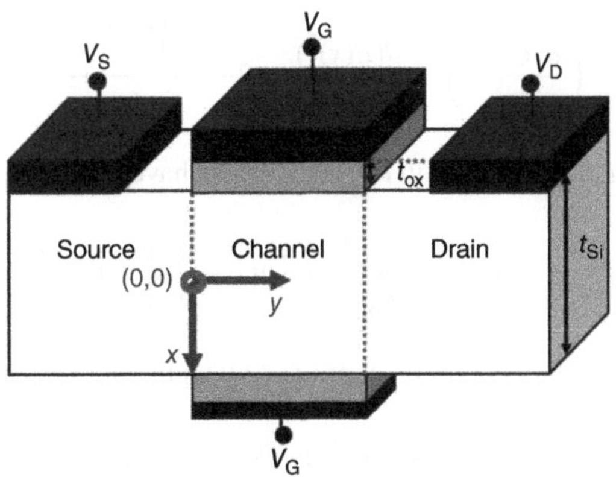

FIGURE 8.4 Three-dimensional view of the DGJLFET.

Ignoring the contribution due to the holes, the Poisson equation may be written as

$$\frac{d^2\varphi(x)}{dx^2} = -\frac{\rho}{\varepsilon_{\mathrm{Si}}} = -\frac{\left[qN_{\mathrm{D}} - qN_{\mathrm{D}}e^{\frac{(\varphi-V)}{V_{\mathrm{t}}}}\right]}{\varepsilon_{\mathrm{Si}}} = \frac{qN_{\mathrm{D}}\left[e^{\frac{(\varphi-V)}{V_{\mathrm{t}}}} - 1\right]}{\varepsilon_{\mathrm{Si}}} \tag{8.14}$$

The boundary conditions in a DGJLFET are

1. The electric field at the center should be zero due to the symmetric nature, i.e.

$$\frac{d\varphi}{dx} = 0 \text{ at } x = 0 \tag{8.15}$$

2. The potential at the Si–SiO_2 interface should be the surface potential, i.e. $\varphi\left(\pm\frac{t_{\mathrm{Si}}}{2}\right) = \varphi_{\mathrm{S}}$.
3. The electric displacement vector at the Si–SiO_2 interface adjacent to the top gate must be continuous, i.e.

$$\varepsilon_{\mathrm{Si}}E_{S,t} = C_{\mathrm{ox}}\left(V_{\mathrm{G}} - V_{\mathrm{FB}} - \varphi_{\mathrm{s}}\right) \tag{8.16}$$

4. The electric displacement vector at the Si–SiO_2 interface adjacent to the bottom gate must be continuous, i.e.

$$\varepsilon_{\mathrm{Si}}E_{S,b} = -C_{\mathrm{ox}}\left(V_{\mathrm{G}} - V_{\mathrm{FB}} - \varphi_{s}\right) \tag{8.17}$$

Multiplying both sides of equation (8.14) with $2(d\varphi/dx)$, we have

$$2\frac{d\varphi}{dx}\left(\frac{d^2\varphi(x)}{dx^2}\right) = \frac{d(E(x)^2)}{dx} = 2\frac{qN_{\mathrm{D}}\left[e^{\frac{(\varphi-V)}{V_{\mathrm{t}}}} - 1\right]}{\varepsilon_{\mathrm{Si}}}\frac{d\varphi}{dx} \tag{8.18}$$

Integrating equation (8.18) with respect to x, we have

$$E(x)^2 = \frac{2qN_{\mathrm{D}}V_{\mathrm{t}}}{\varepsilon_{\mathrm{Si}}}\left[e^{\frac{(\varphi-V)}{V_{\mathrm{t}}}} - \frac{\varphi}{V_{\mathrm{t}}}\right] + C \tag{8.19}$$

Utilizing the boundary condition expressed by equation (8.15), we have

$$C = -\frac{2qN_{\mathrm{D}}V_{\mathrm{t}}}{\varepsilon_{\mathrm{Si}}}\left[e^{\frac{(\varphi_0-V)}{V_{\mathrm{t}}}} - \frac{\varphi_0}{V_{\mathrm{t}}}\right] \tag{8.20}$$

where φ_0 is the potential at the center of the silicon film known as the central potential. Now, the electric field at the bottom surface ($E_{S,b}$) can be obtained by using equation (8.20), boundary condition (2) and $x = t_{\mathrm{Si}}/2$ in (8.19) as

$$E_{S,b}^2 = \frac{2qN_{\mathrm{D}}V_{\mathrm{t}}}{\varepsilon_{\mathrm{Si}}}\left[e^{\frac{(\varphi_{\mathrm{S}}-V)}{V_t}} - e^{\frac{(\varphi_0-V)}{V_{\mathrm{t}}}} - \frac{\varphi_{\mathrm{S}}-\varphi_0}{V_{\mathrm{t}}}\right] \tag{8.21}$$

Although equation (8.21) appears very simple, it has no closed form analytical solution. Therefore, we use approximations that are valid in different operating regimes of DGJLFET and obtain analytical solutions for equation (8.21). As discussed in Section 3.2, DGJLFETs operate mainly in three regions: (a) accumulation, (b) partial depletion, and (c) full depletion. First, let us consider the case of DGJLFET in accumulation.

8.2.1.A Case A: DGJLFET Operating in the Accumulation Region As discussed in Section 3.2, the electric field along the x-direction ceases to be zero at the flat band condition and begins to increase as the gate voltage is increased forcing the DGJLFETs into the accumulation regime. However, the electric field is still very low in the accumulation regime. The accumulation layer appears only at the surface, whereas the entire silicon film remains neutral. Therefore, the assumption of a constant potential (at least in the neutral regions) in the silicon film is valid. This constant potential can be approximated to be the electron quasi-Fermi potential V [33]. Approximating $\varphi(x) \approx V$ in the silicon film apart from the surface in equation (8.21), we have

$$E_{S,b} \approx \sqrt{\frac{2qN_{\mathrm{D}}V_{\mathrm{t}}}{\varepsilon_{\mathrm{Si}}}\left[\left\{e^{\frac{(\varphi_{\mathrm{S}}-V)}{V_{\mathrm{t}}}} - 1\right\} - \frac{\varphi_{\mathrm{S}}-V}{V_{\mathrm{t}}}\right]} \tag{8.22}$$

Now, close to the flat band conditions, i.e. near accumulation, $\varphi_S > V$. Therefore, the ratio

$$\frac{e^{\frac{(\varphi_{\mathrm{S}}-V)}{V_{\mathrm{t}}}} - 1}{\frac{\varphi_{\mathrm{S}}-V}{V_{\mathrm{t}}}} \tag{8.23}$$

is much higher than unity for the accumulation mode and becomes unity only at the flat band condition. Therefore, by neglecting the denominator term of equation (8.23) in equation (8.22), we obtain

$$E_{S,b} \approx \sqrt{\frac{2qN_{\mathrm{D}}V_{\mathrm{t}}}{\varepsilon_{\mathrm{Si}}}\left[e^{\frac{(\varphi_{\mathrm{S}}-V)}{V_{\mathrm{t}}}} - 1\right]} \tag{8.24}$$

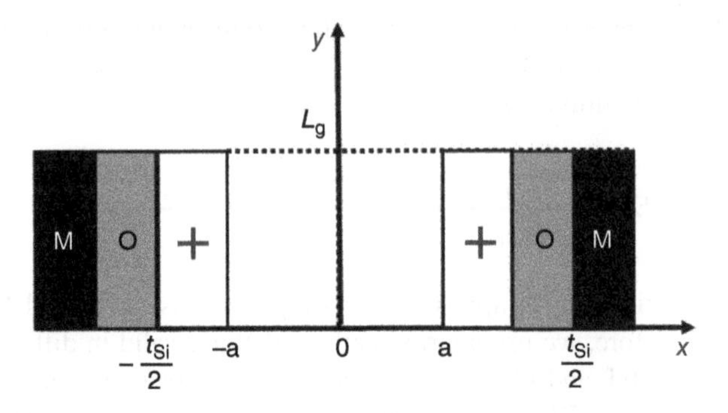

FIGURE 8.5 Depletion charges in the DGJLFET in the partially depletion operating mode.

Now, using the boundary condition (4), i.e. equation (8.17) in equation (8.24), we have

$$\left(V_{\mathrm{G}} - V_{\mathrm{FB}} - \varphi_{\mathrm{S}}\right)^2 = \alpha V_{\mathrm{t}} \left[e^{\frac{(\varphi_{\mathrm{S}} - V)}{V_{\mathrm{t}}}} - 1\right] \tag{8.25}$$

where $\alpha = \left(2q\varepsilon_{\mathrm{Si}} N_{\mathrm{D}}\right) / C_{\mathrm{ox}}^2$ is already defined in equation (8.12)

Equation (8.25) relates the surface potential to the gate voltage in the accumulation regime.

8.2.1.B. Case B: DGJLFET Operating in the Partial Depletion Region In the partial depletion regime, a part of the silicon film thickness at the center is uncovered and remains neutral while the silicon film close to the surface remains depleted as shown in Fig. 8.5. As done in Section 8.2, we take the depletion approximation to solve the Poisson equation in the partial depletion regime. The charge density ρ can be given as

$$\rho = \begin{cases} 0, \ 0 \le x \le a \\ qN_{\mathrm{D}}, \ a \le x \le \frac{t_{\mathrm{Si}}}{2} \end{cases} \tag{8.26}$$

where $a = \left(t_{\mathrm{Si}}/2\right) - x_{\mathrm{dep}}$ and x_{dep} is the depletion region width close to the bottom gate. Therefore, the Poisson equation gets modified as

$$\frac{d^2\varphi(x)}{dx^2} = \begin{cases} 0, \ \ 0 \le x \le a \\ -\frac{qN_{\mathrm{D}}}{\varepsilon_{\mathrm{Si}}}, \ a \le x \le \frac{t_{\mathrm{Si}}}{2} \end{cases} \tag{8.27}$$

Integrating equation (8.27) with respect to x and using the boundary condition (1), i.e. equation (8.15), we have

$$E(x) = 0; \quad 0 \le x \le a \tag{8.28}$$

Integrating equation (8.28) with respect to x and using the boundary condition that $\varphi(x) = \varphi_0$ at $x = 0$, we have

$$\varphi(x) = \varphi_0; \quad 0 \leq x \leq a \tag{8.29}$$

Now, integrating equation (8.27) with respect to x once for $a \leq x \leq 0.5\, t_{\mathrm{Si}}$, we get

$$E(x) = \frac{qN_{\mathrm{D}}x}{\varepsilon_{\mathrm{Si}}} + C_1, \quad a \leq x \leq \frac{t_{\mathrm{Si}}}{2} \tag{8.30}$$

Now, the electric field diminishes to zero at the boundary of the depletion region, i.e. at $x = a$, we obtain

$$C_1 = -\frac{qN_{\mathrm{D}}a}{\varepsilon_{\mathrm{Si}}} \tag{8.31}$$

Integrating equation (8.30) with respect to x, we get

$$\varphi(x) = -\frac{qN_{\mathrm{D}}}{2\varepsilon_{\mathrm{Si}}}\left(x^2 - 2ax\right) + C_2, \quad a \leq x \leq \frac{t_{\mathrm{Si}}}{2} \tag{8.32}$$

Now, the potential should be continuous ($\varphi(x) = \varphi_0$) at $x = a$. Using this boundary condition, we get

$$C_2 = \varphi_0 - \frac{qN_{\mathrm{D}}a^2}{2\varepsilon_{\mathrm{Si}}} \tag{8.33}$$

The potential distribution in the silicon film for $a \leq x \leq t_{\mathrm{Si}}/2$ can be obtained by using equation (8.33) in equation (8.32) as

$$\varphi(x) = \varphi_0 - \frac{qN_{\mathrm{D}}(x-a)^2}{2\varepsilon_{\mathrm{Si}}} \tag{8.34}$$

Since the center of the silicon film is neutral and undepleted, the central potential is equal to the electron quasi-Fermi potential $\varphi_0 = V$ [33]. Using this, the potential at the bottom surface ($x = t_{\mathrm{Si}}/2$) can be calculated as

$$\varphi_{S,b} = V - \frac{qN_{\mathrm{D}}x_{\mathrm{dep}}^2}{2\varepsilon_{\mathrm{Si}}} \tag{8.35}$$

Also, the total charge density in the silicon film can be obtained by applying the Gauss's law at the Si–SiO_2 interface adjacent to the bottom gate as

$$Q_{SC} = 2\varepsilon_{Si} E_{S,b} = -2C_{ox}(V_G - V_{FB} - \varphi_s) \tag{8.36}$$

For DGJLFETs operating in partial depletion regime, $Q_{SC} = 2qN_D x_{dep}$, which relates x_{dep} to the gate voltage as

$$x_{dep} = -\frac{C_{ox}\left(V_G - V_{FB} - \varphi_s\right)}{qN_D} \tag{8.37}$$

Using the value of x_{dep} in equation (8.35), we obtain a relationship between gate voltage and surface potential as

$$\varphi_{S,b} = V - \frac{\left(V_G - V_{FB} - \varphi_s\right)^2}{\alpha} \tag{8.38}$$

which is same as that obtained for a single-gate SOI JLFET working in the partial depletion regime.

Another interesting observation is that using the threshold conditions, i.e. $x_{dep} = t_{Si}/2$ and $V_G = V_{Th}$, where V_{Th} is the threshold voltage, we obtain the same result as obtained in Section 3.2.2 as

$$V_{Th} = V_{FB} - \frac{qN_D t_{Si}^2}{8\varepsilon_{Si}} - \frac{qN_D t_{Si} t_{ox}}{2\varepsilon_{ox}} \tag{8.39}$$

8.2.1.C. Case C: DGJLFET Operating in the Full Depletion (Subthreshold) Region
The depletion approximation used earlier is not suitable for subthreshold condition where the depletion regions merge completely and overlap each other. Therefore, the deep depletion approximation, which provides accurate results for the subthreshold full depletion regime, should be used [33]. Under the full depletion operating regime (Fig. 8.6), integrating the Poisson equation (8.3) once, we have

$$E(x) = \frac{qN_D x}{\varepsilon_{Si}} + C_3 \tag{8.40}$$

Now, due to the symmetric structure, $E(0) = 0$, which implies $C_3 = 0$. Now, integrating equation (8.40) with respect to x, we have

$$\varphi(x) = -\frac{qN_D}{2\varepsilon_{Si}} x^2 + C_4 \tag{8.41}$$

At $x = 0$, $\varphi(x) = \varphi_0$. Using this in equation (8.41) yields $C_4 = \varphi_0$.

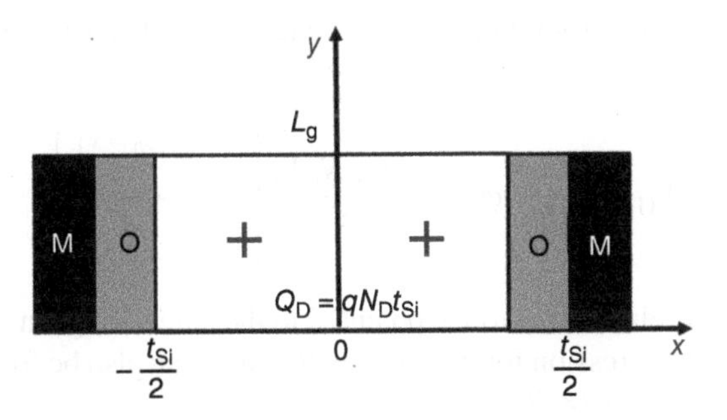

FIGURE 8.6 Depletion charges in the DGJLFETs in the full depletion regime.

At $x = \frac{t_{Si}}{2}$, $\varphi(x) = \varphi_{S,b}$. Using this condition, we have a relation between the central potential and the surface potential in the full depletion mode:

$$\varphi_{S,b} = \varphi_0 - \frac{qN_D t_{Si}^2}{8\varepsilon_{Si}} \tag{8.42}$$

From equation (8.17) and equation (8.21), we have

$$C_{ox}^2 \left(V_G - V_{FB} - \varphi_{S,b}\right)^2 = -2qN_D\varepsilon_{Si}V_t \left[e^{\frac{(\varphi_S - V)}{V_t}} - e^{\frac{(\varphi_0 - V)}{V_t}} - \frac{\varphi_S - \varphi_0}{V_t}\right] \tag{8.43}$$

Now, to simplify, we define $\beta = \frac{\varphi_0 - \varphi_S}{V_t}$, which yields:

$$V_G - V_{FB} - \varphi_{S,b} = -\sqrt{\frac{2q\varepsilon_{Si}N_D V_t \beta}{C_{ox}^2}} \sqrt{1 - \left[\frac{1 - e^{-\beta}}{\beta}\right] e^{\frac{(\varphi_0 - V)}{V_t}}} \tag{8.44}$$

The value of β can be obtained from (8.42) and used in equation (8.44) to give an analytical relation between the surface potential and the gate voltage. However, equation (8.44) is transcendental and needs to be simplified using mathematical techniques. In the full depletion mode, $\varphi_0 > \varphi_{S,b}$. Therefore, β is a positive quantity more than unity, which implies

$$\frac{1 - e^{-\beta}}{\beta} < 1 \tag{8.45}$$

Expanding the square root on the RHS of equation (8.44) using the binomial theorem, we have

$$V_{\mathrm{G}} - V_{\mathrm{FB}} - \varphi_{S,b} = \frac{q\varepsilon_{\mathrm{Si}} N_{\mathrm{D}} V_{\mathrm{t}}}{2C_{\mathrm{ox}}^2} \left[1 - \frac{e^{\frac{(\varphi_0 - V)}{V_{\mathrm{t}}}}}{2} \right] \tag{8.46}$$

Equation (8.46) relates the surface potential to the gate voltage in the subthreshold region. An explicit expression for the surface potential can also be found by rearranging the terms in equation (8.46) as

$$\varphi_{S,b} = V_{\mathrm{G}} - V_{\mathrm{Th}} - \frac{qN_{\mathrm{D}} t_{\mathrm{Si}}^2}{8\varepsilon_{\mathrm{Si}}} - V_{\mathrm{t}} W \left[\frac{qN_{\mathrm{D}} t_{\mathrm{Si}}}{4C_{\mathrm{ox}} V_{\mathrm{t}}} e^{\frac{(V_{\mathrm{G}} - V_{\mathrm{Th}} - V)}{V_{\mathrm{t}}}} \right] \tag{8.47}$$

where W is the Lambert W function [34], which is the inverse of the function $z = W(z)\, e^{W(z)}$. The Lambert W function is a well-known function used for circuit analysis of the bipolar junction transistor (BJT) [35]. The proposed model is in well agreement with the TCAD results for various ranges of silicon film doping, silicon film thickness, and gate oxide thickness as shown in Fig. 8.7.

In this section, we presented analytical expressions for the surface potential of DGJLFETs utilizing suitable approximations in different operating regions. However, a regionwise approach may cause a convergence problem in circuit simulators. Therefore, a continuous model covering all the operating regions is desired, which will be discussed in the next section.

8.2.2 Parabolic Approximation Technique

The parabolic approximation technique or the pseudo-2D method is widely used for surface potential modeling in MOSFETs and tunnel FETs [30, 31, 36–40]. The basis of this method can be understood from Fig. 8.8. The energy band profile ($-e\varphi$) and, hence, the potential variation along any point along the channel length direction (y-axis) is monotonous in the direction along the channel thickness (x-direction). However, the nature of the potential distribution differs as we move along the channel length. Therefore, the potential distribution can be approximated as a quadratic polynomial in x as [36]

$$\varphi(x, y) = a_0(y) + a_1(y)\, x + a_2(y)\, x^2 \tag{8.48}$$

The reason for limiting the order of polynomial to 2 is that utilizing the available boundary conditions for double-gate metal-oxide-semiconductor field-effect transistor (DGMOSFETs), only a quadratic polynomial may be solved. Therefore, higher order terms are ignored [36]. By using such an approximation, the original 2D Poisson equation reduces to a second-order linear differential equation in a single dimension, which may be solved easily.

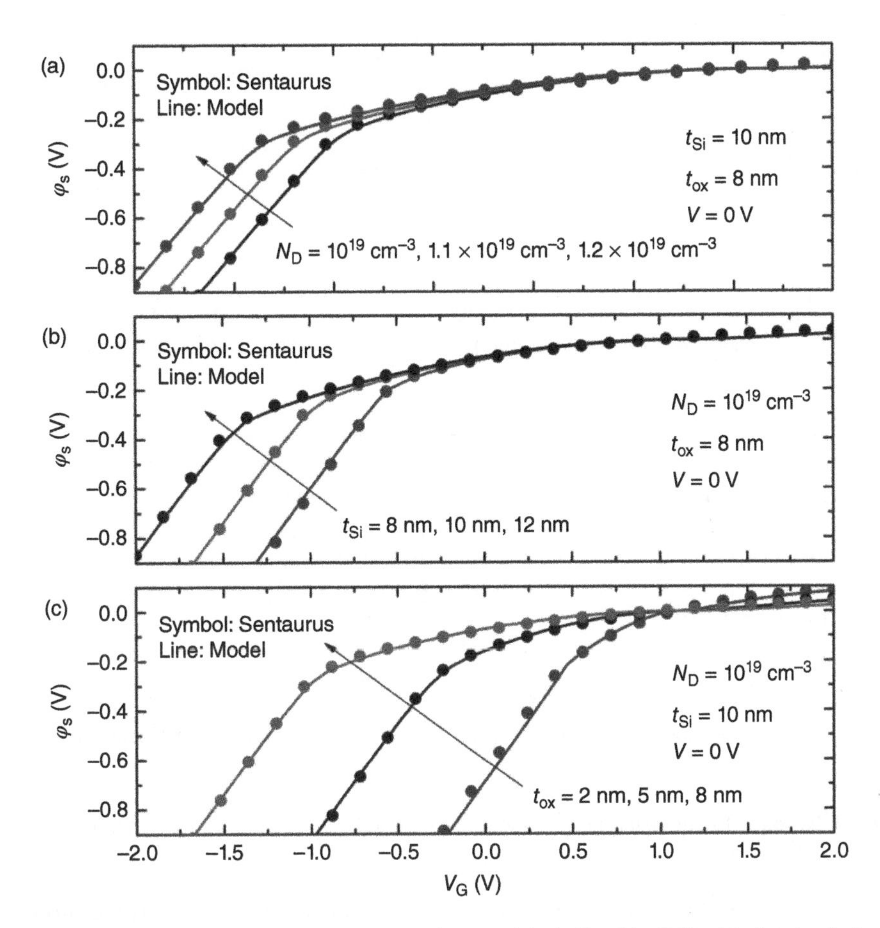

FIGURE 8.7 Comparison between the surface model obtained in [33] with the simulation data for different (a) silicon film doping, (b) silicon thickness, and (c) gate oxide thickness.

Utilizing the boundary conditions (equations 8.15–8.17) and the coordinate system used for the operating regionwise approximation in Section 8.2.1, we will try to solve equation (8.48) and obtain a relationship between the surface potential and the gate voltage, which is continuous and valid for all operating regions of interest [42].

Utilizing the first boundary condition, i.e. the electric field at the center is 0 $[E(0, y) = 0]$ in equation (8.48), we obtain, $a_1(y) = 0$. Now, equation (8.48) reduces to

$$\varphi(x, y) = a_0(y) + a_2(y)x^2 \tag{8.49}$$

Using the boundary condition (2) that the potential at the center is φ_0, i.e. $[\varphi(0, y) = \varphi_0]$, we obtain $a_0(y) = \varphi_0$, where φ_0 is the central potential. Now,

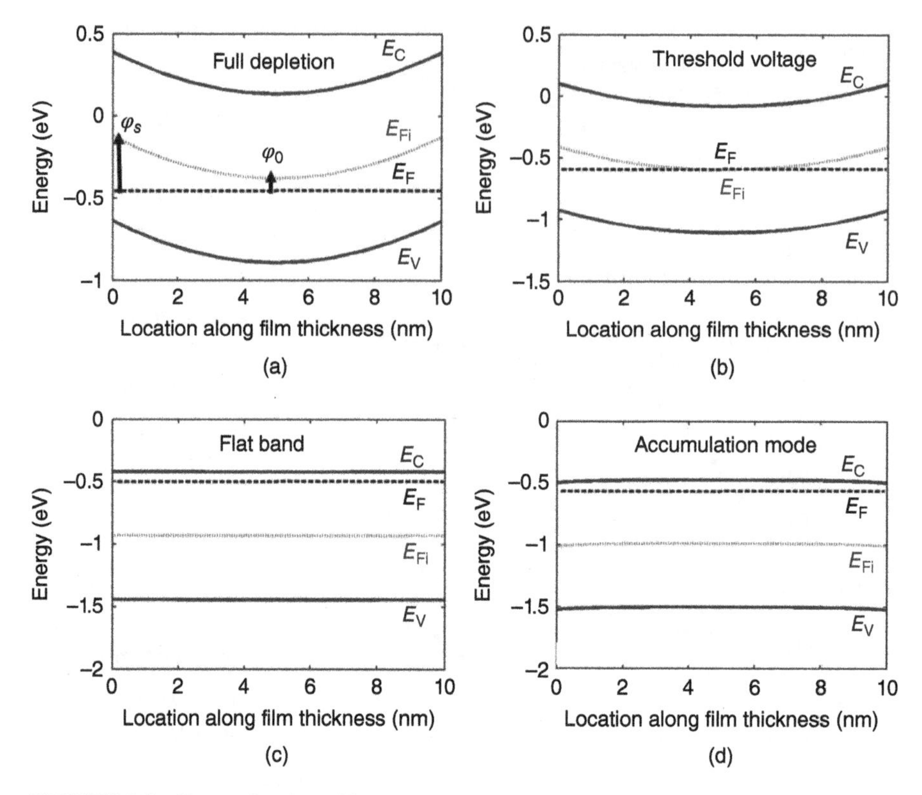

FIGURE 8.8 Energy band profiles along the cutline B–B' (Fig. 8.1) in different operating modes of JLFETs.

utilizing the third boundary condition, i.e. the potential at the surface is the surface potential $\left[\varphi\left(t_{Si}/2, y\right) = \varphi_{S,b}\right]$, we obtain

$$a_2(y) = \left(\varphi_{S,b} - \varphi_0\right)\frac{4}{t_{\mathrm{Si}}^2} \tag{8.50}$$

Therefore, the potential distribution can be expressed as

$$\varphi(x, y) = \varphi_0 + \left(\varphi_{S,b} - \varphi_0\right)\frac{4x^2}{t_{\mathrm{Si}}^2} \tag{8.51}$$

Now, using the boundary condition that the electric displacement vector at the bottom Si–SiO_2 interface should be continuous, we have $\varepsilon_{\mathrm{ox}}E_{\mathrm{ox}} = \varepsilon_{\mathrm{Si}}E_{\mathrm{Si}}$, which can be elaborated using equation (8.17) as

$$-C_{\mathrm{ox}}(V_{\mathrm{G}} - V_{\mathrm{FB}} - \varphi_{S,b}) = -\varepsilon_{\mathrm{Si}}\frac{d\varphi\left(\frac{t_{\mathrm{Si}}}{2}, y\right)}{dx} = \frac{4\varepsilon_{\mathrm{Si}}\Delta\varphi}{t_{\mathrm{Si}}} \tag{8.52}$$

where $\Delta\varphi = \left(\varphi_0 - \varphi_{S,b}\right)$. Although equation (8.52) relates the gate voltage to the surface potential, the central potential is an unknown term. Therefore, another relationship between $\varphi_{S,b}$ and φ_0 is required. For this, we can relate the charge density in the bottom half silicon film using Gauss's law at the bottom Si–SiO_2 interface and utilize the symmetric conditions in DGJLFET to find the total charge density as

$$Q_{SC} = -2\varepsilon_{Si}\frac{d\varphi\left(\frac{t_{Si}}{2}, y\right)}{dx} = \frac{8\varepsilon_{Si}\Delta\varphi}{t_{Si}} \tag{8.53}$$

Now, the total charge density in the silicon film consists of both mobile electrons and the depletion charge, i.e.

$$\rho = qN_D\left[e^{\frac{(\varphi-V)}{V_t}} - 1\right] \tag{8.54}$$

Integrating equation (8.54) throughout the silicon film with respect to x, we get the total charge density inside the silicon film as

$$Q_{Sc} = qN_D t_{Si} - qN_D \int_{-t_{Si}/2}^{t_{Si}/2} e^{\frac{(\varphi(x,y)-V)}{V_t}} dx \tag{8.55}$$

Replacing $\varphi(x, y)$ by equation (8.51), we have

$$Q_{Sc} = qN_D t_{Si} - qN_D e^{\frac{(\varphi_0-V)}{V_t}} \int_{-t_{Si}/2}^{t_{Si}/2} e^{-\left(\frac{4\Delta\varphi}{t_{Si}^2 V_t}\right)x^2} dx \tag{8.56}$$

The integral on the right side of equation (8.56) takes the form of the error function which can be solved using

$$\sqrt{\frac{c}{\pi}}\int_p^q e^{-Cx^2} dx = \frac{\left[\text{erf}\left(q\sqrt{c}\right) - \text{erf}\left(p\sqrt{c}\right)\right]}{2} \tag{8.57}$$

Using the above integral, we obtain total charge density in the silicon film as

$$Q_{Sc} = qN_D t_{Si}\left[1 - \frac{e^{\frac{(\varphi_0-V)}{V_t}}}{2}\sqrt{\frac{\pi V_t}{\Delta\varphi}}\left\{\text{erf}\left(\sqrt{\frac{\Delta\varphi}{V_t}}\right)\right\}\right] \tag{8.58}$$

Although equation (8.58) relates the total charge density inside the silicon film to the surface and central potential, the obtained expression is not analytical. However, if we carefully observe, the difference between the surface potential and the central

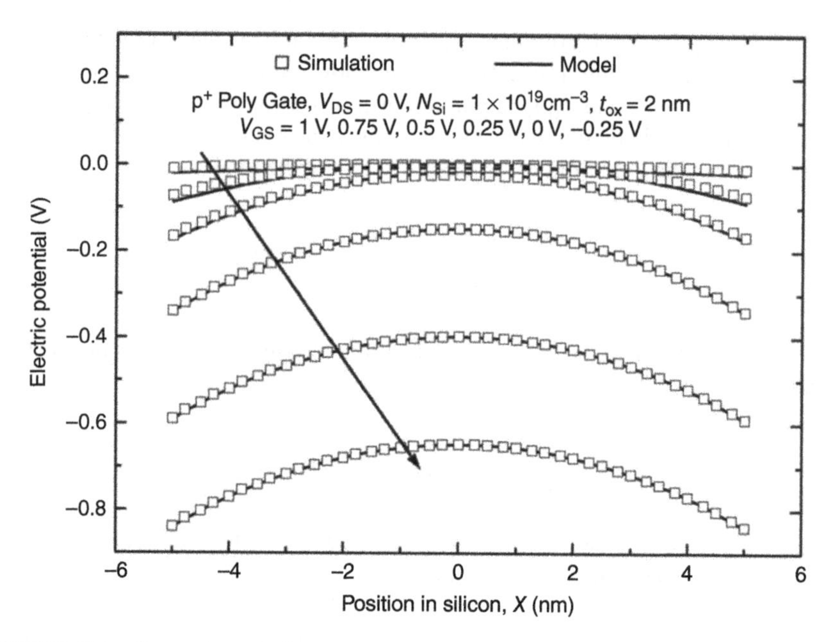

FIGURE 8.9 Comparison of the potential in the channel region along the channel thickness for different gate voltages calculated utilizing the analytical model [42] and the simulations.

potential $\Delta\varphi$ is more than the thermal voltage apart from the flat band and weak accumulation conditions [42]. Therefore, the error function term in equation (8.58) tends to unity and equation (8.58) can be simplified into an analytical relation. Now, using equation (8.58), a second relationship between the surface potential and the central potential may be found as

$$\frac{8\varepsilon_{\mathrm{Si}}\Delta\varphi}{t_{\mathrm{Si}}} = qN_{\mathrm{D}}t_{\mathrm{Si}}\left[1 - \frac{e^{\frac{(\varphi_0 - V)}{V_t}}}{2}\sqrt{\frac{\pi V_t}{\Delta\varphi}}\right] \tag{8.59}$$

Equations (8.52) and (8.59) can be solved in a self-consistent manner to obtain the surface potential in terms of the gate voltage.

A numerical example calculated using the above equations is shown in Fig. 8.9 and compared with simulation results obtained using a commercial TCAD silvaco Atlas. The model shows a good agreement with the simulations and provides a continuous model for relating surface potential to the applied gate voltage [42]. However, it may be noted that the assumption of a parabolic potential profile is true when the JLFETs operate in the full depletion or partial depletion mode as shown in Fig. 8.8. Close to the flat band condition, in the weak accumulation regime, the potential profile does not resemble a parabola. However, the JLFETs are generally designed to operate in the flat band condition only once the ON-state is reached. Therefore, even the parabolic approximation works well for modeling JLFETs.

8.2.3 Initial Guess technique

As discussed in Section 8.2.1, although the Poisson equation for JLFETs appears simple, it has got no closed form analytical solution unlike the undoped MOSFETs [43–46]. We may find an analytical relationship between the gate voltage and the surface potential if we solve the Poisson equation for undoped DGMOSFETs and use it as an initial guess [47]. In the next iteration, the Poisson equation for the DGJLFETs may be updated based on the initial guess. The surface potential model can then be developed utilizing the boundary conditions. We begin our analysis from the simple Poisson equation for DGJLFETs, which we formulated as equation (8.14).

Now, in an undoped DGMOSFET, the depletion charge density can be neglected. Therefore, by including only the mobile electron charge density in equation (8.14), we obtain the Poisson equation for DGMOSFETs as

$$\frac{d^2\varphi(x)}{dx^2} = -\frac{\rho}{\varepsilon_{\mathrm{Si}}} = \frac{qn_{\mathrm{i}}e^{\frac{(\varphi-V)}{V_{\mathrm{t}}}}}{\varepsilon_{\mathrm{Si}}} \tag{8.60}$$

where n_{i} is the intrinsic carrier concentration.

Note that we ignored the hole contribution as done in Section 8.2.1. Using the similar steps as done in Section 8.2.1, we obtain the expression for the electric field by the following equation:

$$2\frac{d\varphi}{dx}\left(\frac{d^2\varphi(x)}{dx^2}\right) = \frac{d(E(x)^2)}{dx} = 2\frac{qn_{\mathrm{i}}\,e^{\frac{(\varphi-V)}{V_{\mathrm{t}}}}}{\varepsilon_{\mathrm{Si}}}\frac{d\varphi}{dx} \tag{8.61}$$

Integrating equation (8.61), we get

$$\frac{d\varphi}{dx} = \sqrt{\frac{2n_{\mathrm{i}}kT}{\varepsilon_{\mathrm{Si}}}e^{\frac{(\varphi-V)}{V_{\mathrm{t}}}} + C} \tag{8.62}$$

where C is the constant of integration. Rearranging equation (8.62), we get

$$\frac{d\varphi}{\sqrt{\frac{2n_{\mathrm{i}}kT}{\varepsilon_{\mathrm{Si}}}e^{\frac{(\varphi-V)}{V_{\mathrm{t}}}} + C}} = dx \tag{8.63}$$

Integrating both sides, we may get an expression for the potential as a function of x. The integral on the LHS can be performed using the substitution method. We may use the integration constant C in two ways, yielding two different expressions. We can rearrange equation (8.63) as shown below:

$$\frac{d\varphi}{\sqrt{\frac{2n_{\mathrm{i}}kT}{\varepsilon_{\mathrm{Si}}}}\sqrt{\left[e^{\frac{(\varphi-V)}{V_{\mathrm{t}}}} + C_1\right]}} = dx \tag{8.64}$$

where C_1 is another constant. If we substitute $e^{(\varphi-V)/2V_t} = u$ and perform the integration of equation (8.64) using the appropriate boundary conditions, we arrive at the standard expressions given in [43]. However, for finding the initial guess, we should not use any boundary condition as the initial guess is only an approximate solution to the Poisson equation and not the final solution. Therefore, we use a different approach without changing the constant C and rearrange equation (8.63) as

$$\frac{d\varphi}{2V_t\sqrt{\frac{q^2 n_i}{2\varepsilon_{Si}kT}\, e^{\frac{(\varphi-V)}{V_t}} + C'}} = dx \tag{8.65}$$

where $V_t = kT/q$ and C' is yet another constant. Now, substituting $\sqrt{(q^2 n_i)/(2\varepsilon_{Si}kT)}e^{(\varphi-V)/2V_t} = u$, we have, $(2V_t du)/u = d\varphi$. Substituting this in equation (8.65), we get

$$\frac{du}{u\sqrt{u^2 + C'}} = dx \tag{8.66}$$

Now, using $C' = -(2\beta/t_{Si})^2$, where β is a constant, and substituting $y = u(t_{Si}/2\beta)$, we obtain

$$\frac{dy}{y\sqrt{y^2-1}} = \frac{2\beta}{t_{Si}}dx \tag{8.67}$$

Integrating both sides, we get, $\sec^{-1}y = 2\beta x/t_{Si} + k$, where k is a constant. Substituting the value of y and u in integral of equation (8.67), we get

$$\varphi(x) = V - 2V_t \ln\left[\frac{t_{Si}}{2\beta}\sqrt{\frac{q^2 n_i}{2\varepsilon_{Si}kT}}\cos\left(\frac{2\beta x}{t_{Si}} + k\right)\right] \tag{8.68}$$

Now, utilizing the boundary condition $d\varphi/dx = 0$ at $x = 0$, we get $\tan(k) = 0$ yielding $k = 0$. Therefore, for an undoped DGMOSFET, we can solve analytically to obtain a closed form solution of the potential as [44, 45]

$$\varphi(x) = V - 2V_t \ln\left[\frac{t_{Si}}{2\beta}\sqrt{\frac{q^2 n_i}{2\varepsilon_{Si}kT}}\cos\left(\frac{2\beta x}{t_{Si}}\right)\right] \tag{8.69}$$

Now, the Poisson equation for the DGJLFETs can also be written as

$$\frac{d^2\varphi(x)}{dx^2} = -\frac{\rho}{\varepsilon_{Si}} = \frac{q\left[n_i\, e^{\frac{(\varphi-V)}{V_t}}\right]}{\varepsilon_{Si}} - \frac{qN_D}{\varepsilon_{Si}} \tag{8.70}$$

Therefore, we may use the solution for the first term in the RHS of equation (8.70) as an initial guess to find an approximate solution. Differentiating equation (8.69) twice with respect to x and substituting into equation (8.70), we get

$$\frac{d^2\varphi(x)}{dx^2} = \frac{q}{\varepsilon_{\mathrm{Si}}}\left[\frac{8\beta^2\varepsilon_{\mathrm{Si}}kT}{q^2t_{\mathrm{Si}}^2}\sec^2\left(\frac{2\beta x}{t_{\mathrm{Si}}}\right) - N_{\mathrm{D}}\right] \tag{8.71}$$

Now, integrating equation (8.71) once with respect to x, we get

$$E(x) = -\frac{d\varphi}{dx} = -\left[\frac{q}{\varepsilon_{\mathrm{Si}}}\left\{\frac{4\beta\varepsilon_{\mathrm{Si}}kT}{q^2t_{\mathrm{Si}}}\tan\left(\frac{2\beta x}{t_{\mathrm{Si}}}\right) - N_{\mathrm{D}}x\right\} + C_1\right] \tag{8.72}$$

where C_1 is the constant of integration. Applying the boundary condition due to symmetry in DGJLFETs that $E(x) = 0$ at $x = 0$, we get $C_1 = 0$. Integrating equation (8.72) again with respect to x, we obtain

$$\varphi(x) = \frac{q}{\varepsilon_{\mathrm{Si}}}\left[-\frac{2\varepsilon_{\mathrm{Si}}kT}{q^2}\ln\left\{\cos\left(\frac{2\beta x}{t_{\mathrm{Si}}}\right)\right\} - \frac{N_{\mathrm{D}}x^2}{2}\right] + C_2 \tag{8.73}$$

where C_2 is another constant, which can be evaluated by using the boundary condition that the potential at the center $\varphi(x) = \varphi_0$ at $x = 0$, yielding $C_2 = \varphi_0$. Now, utilizing the boundary condition that the electric displacement vector must be continuous at the bottom Si–SiO_2 interface, i.e. equation (8.17), gives the required relationship between the gate voltage and the surface potential in terms of the design parameters silicon film doping (N_{D}), silicon film thickness (t_{Si}), and the integration constant obtained from the solution of the Poisson equation for DGMOSFETs as

$$C_{\mathrm{ox}}\left(V_{\mathrm{G}} - V_{\mathrm{FB}} - \varphi_{\mathrm{s}}\right) = \left[\frac{4\beta\varepsilon_{\mathrm{Si}}kT}{qt_{\mathrm{Si}}}\tan(\beta) - \frac{qN_{\mathrm{D}}t_{\mathrm{Si}}}{2}\right] \tag{8.74}$$

However, we need to find an explicit expression for β to relate the gate voltage and the surface potential using equation (8.74). For this, we use the Gauss's law, which gives the charge density (Q_{SC}) inside the silicon film as

$$Q_{\mathrm{SC}} = 2\varepsilon_{\mathrm{Si}}E_{S,b} \tag{8.75}$$

where $E_{S,b}$ is the electric field at the bottom Si–SiO_2 interface. Now, the charge density can be expressed as the sum of mobile charge density (Q_{m}) and the depletion charge (Q_{D}) as $Q_{\mathrm{SC}} = Q_{\mathrm{m}} + Q_{\mathrm{D}}$, where

$$Q_{\mathrm{m}} = -2\int_0^{t_{\mathrm{Si}}/2} n_{\mathrm{i}}\, e^{\frac{(\varphi(x)-V)}{V_{\mathrm{t}}}}\, dx \tag{8.76}$$

$$Q_{\mathrm{D}} = 2N_{\mathrm{D}}t_{\mathrm{Si}} \tag{8.77}$$

Using equations (8.75) –(8.77), we obtain a relationship between the integral constant β and the electron quasi-Fermi potential V as

$$V = -V_t \ln\left[\frac{4\beta\varepsilon_{\mathrm{Si}}kT\tan(\beta)}{t_{\mathrm{Si}}q^2\int_0^{t_{\mathrm{Si}}/2} n_{\mathrm{i}}\, e^{\frac{\varphi(x)}{V_t}}\, dx}\right] \tag{8.78}$$

The value of the integral constant β can be obtained from equation (8.78) and used in equation (8.74) to yield a surface potential model for DGJLFETs [47].

8.2.4 Finite Difference Approach

The parabolic approximation method, discussed in Section 8.2.2, assumes that the potential profile along the direction of channel thickness (x-axis) at any point along the channel length (y-axis) is parabolic with some arbitrary coefficients, which can be obtained utilizing the boundary conditions. Although this assumption is true when the JLFETs operate in the full depletion or partial depletion mode, close to the flat band condition, in the weak accumulation regime, the potential profile does not resemble a parabola. Hence, the parabolic approximation is not valid in this regime. Also, in the JLFETs, the channel thickness needs to be ultrathin to achieve volume depletion. Therefore, we may follow a discretization step and use the well-known finite difference technique for modeling JLFETs. This discretization step not only relaxes any preassumption of the potential profile but also provides a computationally efficient model [48, 49].

The modeling approach utilizing the finite difference method uses three discrete points as shown in Fig. 8.10: first at the top Si–SiO_2 interface, second at the center of the channel region, and third at the bottom Si–SiO_2 interface to solve the Poisson

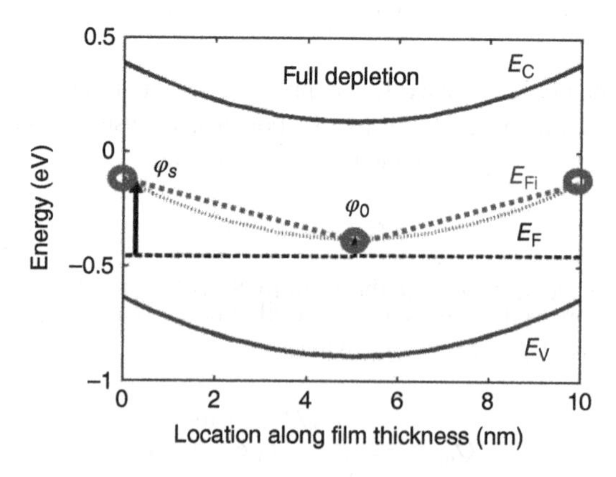

FIGURE 8.10 The points considered for the solution of the Poisson equation using the finite difference method.

equation along the channel thickness. For the case of JLFETs, it is a valid approximation since the points are located only 5 nm apart for a channel thickness of 10 nm, which is typically used while designing JLFETs. As the channel thickness reduces, the discretization step tends to be more accurate.

The Poisson equation for the DGJLFETs, i.e. equation (8.70), may be expressed as

$$\frac{d^2\varphi(x)}{dx^2} = -\frac{\rho}{\varepsilon_{\mathrm{Si}}} = \frac{qn_{\mathrm{i}}\left[e^{\frac{(\varphi-V)}{V_{\mathrm{t}}}} - \frac{N_{\mathrm{D}}}{n_i}\right]}{\varepsilon_{Si}} \tag{8.79}$$

Following the same approach as in Sections 8.2.1 and 8.2.3, i.e. multiplying both sides by $2\,(d\varphi/dx)$, we have

$$\frac{d}{dx}\left(\frac{d\varphi}{dx}\right)^2 = \frac{d}{dx}E(x)^2 = \frac{2qn_{\mathrm{i}}\left[e^{\frac{(\varphi-V)}{V_{\mathrm{t}}}} - \frac{N_{\mathrm{D}}}{n_i}\right]}{\varepsilon_{\mathrm{Si}}}\frac{d\varphi}{dx} \tag{8.80}$$

Now, integrating equation (8.80) from the center of the channel with $E(0) = 0$ and $\varphi(0) = \varphi_0$ to the bottom Si–SiO$_2$ surface with $E\left(t_{\mathrm{Si}}/2\right) = E_{S,b}$ and $\varphi\left(t_{\mathrm{Si}}/2\right) = \varphi_S$, we have

$$\int_0^{E_{S,b}} \frac{d}{dx}E(x)^2 = \int_{\varphi_0}^{\varphi_S} \frac{2qn_{\mathrm{i}}\left[e^{\frac{(\varphi-V)}{V_{\mathrm{t}}}} - \frac{N_{\mathrm{D}}}{n_{\mathrm{i}}}\right]}{\varepsilon_{\mathrm{Si}}}\frac{d\varphi}{dx} \tag{8.81}$$

Solving, we get

$$E_{S,b}^2 = \frac{2qn_{\mathrm{i}}V_{\mathrm{t}}}{\varepsilon_{\mathrm{Si}}}\left[e^{\frac{(\varphi_{\mathrm{S}}-V)}{V_{\mathrm{t}}}} - e^{\frac{(\varphi_0-V)}{V_{\mathrm{t}}}} - \frac{N_{\mathrm{D}}}{n_{\mathrm{i}}}\frac{\left(\varphi_{\mathrm{S}} - \varphi_0\right)}{V_{\mathrm{t}}}\right] \tag{8.82}$$

Now, using the finite-difference method along the three points discussed earlier, the first derivative of the potential can be expressed as

$$\frac{d\varphi\,(x = 0)}{dx} = \frac{\varphi\left(-\frac{t_{\mathrm{Si}}}{2}\right) - \varphi\left(\frac{t_{\mathrm{Si}}}{2}\right)}{t_{\mathrm{Si}}} = \frac{\varphi_{\mathrm{S}} - \varphi_{\mathrm{S}}}{t_{\mathrm{Si}}} = 0 \tag{8.83}$$

which is also evident from the boundary condition that the electric field at the center is zero due to the symmetric operation in DGJLFET. Now, the second derivative of the potential using the finite difference method along the three chosen points can be obtained as

$$\frac{d^2\varphi(x=0)}{dx^2} = \frac{1}{\frac{t_{Si}}{2}}\left[\frac{\varphi\left(-\frac{t_{Si}}{2}\right) - \varphi(0)}{\frac{t_{Si}}{2}} - \frac{\varphi(0) - \varphi\left(\frac{t_{Si}}{2}\right)}{\frac{t_{Si}}{2}}\right] \tag{8.84}$$

Solving equation (8.84) and using equation (8.79), we get

$$\varphi_S - \varphi_0 = \frac{qt_{Si}^2}{8\varepsilon_{Si}}\left[n_i\, e^{\frac{(\varphi_0 - V)}{V_t}} - N_D\right] \tag{8.85}$$

Equation (8.85) relates the surface potential to the central potential without any assumption about the shape of the potential profile. Moreover, ignoring the contribution due to the mobile electrons, it converges to the results obtained using depletion approximation for the JLFETs operating in the full depletion mode in Section 8.2.1.

Equation (8.85) alone cannot be used to find a relationship between the gate voltage and the surface potential. We need another expression relating the gate voltage to the central or the surface potential for a self-consistent solution. Following the approach in Section 8.2.1–8.2.3, we use the boundary condition that the electric displacement vector at the bottom Si–SiO_2 interface is continuous and the Gauss's law as

$$\varepsilon_{Si}E_{S,b} = \frac{Q_{SC}}{2} = -C_{ox}\left(V_G - V_{FB} - \varphi_s\right) \tag{8.86}$$

Using equation (8.82) in equation (8.86), we get the following expression relating the surface potential, central potential, and the gate voltage:

$$C_{ox}\left(V_G - V_{FB} - \varphi_s\right) = -\sqrt{2q\varepsilon_{Si}n_iV_t\left[e^{\frac{(\varphi_S - V)}{V_t}} - e^{\frac{(\varphi_0 - V)}{V_t}} - \frac{N_D}{n_i}\frac{(\varphi_S - \varphi_0)}{V_t}\right]} \tag{8.87}$$

Equations (8.85) and (8.87) can be solved in a self-consistent way to get the surface potential model for DGJLFETs.

At this juncture, we point out that equation (8.85) could have been directly derived by using the Maclaurin series expansion of the potential about the center as

$$\varphi(x) = \varphi(0) + \frac{d\varphi(x=0)}{dx}x + \frac{d^2\varphi(x=0)}{dx^2}x^2 + \frac{d^3\varphi(x=0)}{dx^3}x^3 + \ldots \tag{8.88}$$

Because of the symmetric structure of DGJLFETs, the odd power derivatives would cease to be zero. Putting the values from equation (8.79), we get

$$\varphi(x) = \varphi_0 + \frac{qn_i\left[e^{\frac{(\varphi_0 - V)}{V_t}} - \frac{N_D}{n_i}\right]}{\varepsilon_{Si}}x^2 \tag{8.89}$$

Substituting $x = t_{Si}/2$ in equation (8.89), we obtain equation (8.85) without using any sophisticated technique.

8.3 CHARGE-BASED MODELING APPROACH

Until now, we have discussed about the surface potential modeling approach for the JLFETs. There is yet another modeling approach, which relates the charges in the semiconductor film directly to the gate voltage, known as the charge-based model. The charge-based modeling approach is more practical compared to the surface potential–based models as it is indeed the charge density, which determines the drain current and the behavior of the FETs [25–29]. However, obtaining an analytical and computationally efficient charge-based model is a challenge. Therefore, charge-based models are not commonly used in circuit simulators. Fortunately, computationally efficient charge-based models have been derived for JLFETs like the École polytechnique fédérale de Lausanne (EPFL) charge-based model (EPFL JL-1.0) [25]. We discuss the charge-based modeling approach in detail in this section.

Throughout Sections 8.2.1–8.2.4, we have related the charge in the semiconductor to the gate voltage and the surface potential by utilizing the Gauss's law as equation (8.86). However, we need to eliminate the surface potential term from equation (8.86) to get an explicit relationship between the charge density in the silicon film and the gate voltage to get the charge-based model for DGJLFETs [48]. For this, we proceed by using equation (8.82) in equation (8.86) which relates the charge density to the gate voltage, surface potential, and the surface potential as

$$\begin{aligned} Q_{SC} &= 2\,\text{sign}\left(\varphi_0 - \varphi_S\right)\sqrt{2q\varepsilon_{Si}n_iV_t\left[e^{\frac{(\varphi_S - V)}{V_t}} - e^{\frac{(\varphi_0 - V)}{V_t}} - \frac{N_D}{n_i}\frac{\left(\varphi_S - \varphi_0\right)}{V_t}\right]} \\ &= -2C_{ox}\left(V_G - V_{FB} - \varphi_s\right) \end{aligned} \tag{8.90}$$

The signum function has been used in equation (8.90) to determine the exact nature of the channel charge as the operation mode changes from full depletion to accumulation. In the full depletion and partial depletion mode, $\varphi_0 > \varphi_S$, the total charge density is dominated by the depletion charges and the semiconductor charge density is positive. However above flat band conditions, $\varphi_S > \varphi_0$, the mobile electrons dominate the total charge density and the net charge density becomes negative. However, at the flat band condition, the net charge density inside the semiconductor is zero and

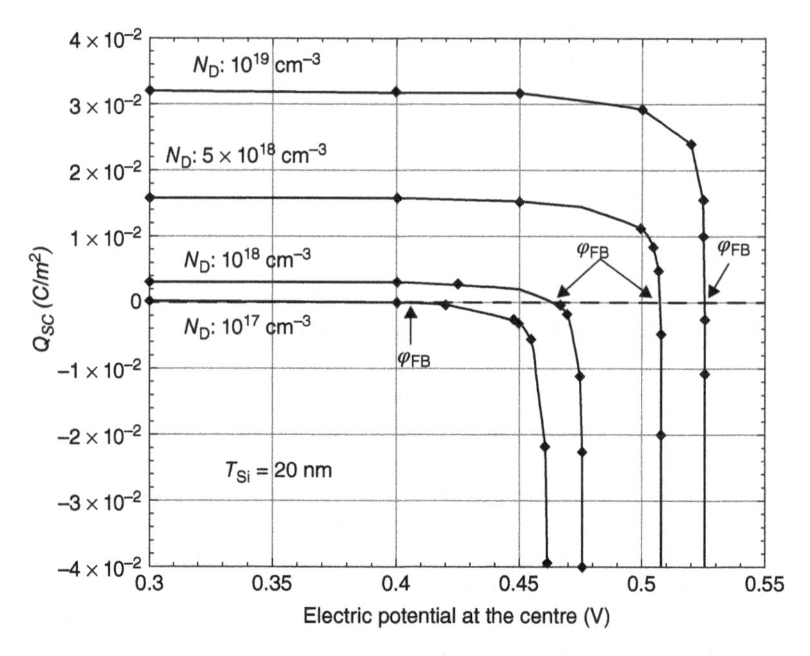

FIGURE 8.11 The charge density as a function of the central potential for different doping concentrations obtained by numerical solution of equation (8.79) [48].

the central potential is equal to the surface potential, which may be obtained by using $\left(d^2\varphi(x)\right)/dx^2 = 0$ in equation (8.79) yielding

$$\varphi_0 = \varphi_S = \varphi_{FB} = V + V_T \ln\left(\frac{N_D}{n_i}\right) \tag{8.91}$$

Now, the numerical simulations of equation (8.79) shown in Fig. 8.11 indicate that the central potential remains stuck at the flat band potential obtained by equation (8.91) even when the DGJLFETs are biased in the accumulation regime [48]. This asymptote in the central potential is attributed to the shielding of the electric field by the surface-accumulated electrons. Therefore, to relate the charge density to the gate voltage in the accumulation mode, we may use the value of the central potential obtained from equation (8.91) in equation (8.90) as

$$Q_{SC}\,(\text{accu}) \approx -2\sqrt{2q\varepsilon_{Si}n_iV_t}\sqrt{\left[\underbrace{\left\{e^{\frac{(\varphi_S - V)}{V_t}} - \frac{N_D}{n_i}\right\}}_{1} - \underbrace{\frac{N_D}{n_i}\left\{\frac{(\varphi_S - V)}{V_t} - \ln\left(\frac{N_D}{n_i}\right)\right\}}_{2}\right]} \tag{8.92}$$

We can clearly observe in equation (8.92) that the ratio of the terms denoted by 1 to that of the terms denoted by 2 is larger than unity for the accumulation mode of operation. Therefore, we can omit the term denoted by 2 and approximate equation (8.92) as

$$Q_{SC}\,(\text{accu}) \approx -2\sqrt{2q\varepsilon_{Si}n_iV_t}\sqrt{e^{\frac{(\varphi_S - V)}{V_t}} - \frac{N_D}{n_i}} \tag{8.93}$$

Now, using equation (8.93), we obtain a relationship between the surface potential and the charge density as

$$\varphi_S \approx V + V_t \ln\frac{N_D}{n_i}\left(1 + \frac{Q_{SC}^2}{8\varepsilon_{Si}qN_DV_t}\right) \tag{8.94}$$

Utilizing the value of surface potential obtained from equation (8.94) in equation (8.86), we obtain an analytical expression relating the charge density in the accumulation mode of operation to the gate voltage:

$$V_G - V_{FB} - V \approx -\frac{Q_{SC}}{2C_{ox}} + V_t \ln\frac{N_D}{n_i}\left(1 + \frac{Q_{SC}^2}{8\varepsilon_{Si}qN_DV_t}\right) \tag{8.95}$$

In general, the charge density can be expressed as a sum of the mobile charge density and the depletion charge density, i.e. $Q_{SC} = Q_m + Q_D = Q_m + qN_Dt_{Si}$. The calculation of mobile charge density is also essential as it can be directly used to get an analytical expression for the drain current using the Pao–Sah integral, which we discuss in Section 8.4.2. A general approach may also be utilized for finding a charge-based model for JLFET. Equation (8.90) may be reexpressed after defining $C = e^{(\varphi_0 - V)/V_t}$ and $K = \left(qn_it_{Si}^2\right)/8V_t\varepsilon_{Si}$ as

$$V_G - V_{FB} - V = V_t \ln C + KV_t\left(C - \frac{N_D}{n_i}\right) - \frac{2\,\text{sign}\,(\varphi_0 - \varphi_S)}{C_{ox}} \times\sqrt{2q\varepsilon_{Si}n_iV_t}\sqrt{C\left[\underbrace{e^{K\left(C - \frac{N_D}{n_i}\right)}}_{1} - \frac{KN_D}{n_i} - 1\right] + K\left(\frac{N_D}{n_i}\right)^2} \tag{8.96}$$

Equation (8.96) may be used to calculate C using standard numerical techniques once the gate voltage and the central potential are known [48]. The mobile charge density may be known once C is obtained from equation (8.96). Combining equations (8.85) and (8.86), the generalized charge-based model for DGJLFETs may be obtained as

$$Q_{\mathrm{SC}} = -2C_{\mathrm{ox}}\left(V_{\mathrm{G}} - V_{\mathrm{FB}} - V + \frac{KV_{\mathrm{t}}N_{\mathrm{D}}}{n_{\mathrm{i}}} - KV_{\mathrm{t}}C - V_{\mathrm{t}}\ln C\right) \tag{8.97}$$

Although equation (8.97) represents a generalized charge-based model for DGJLFETs, the numerical approach of calculating C should be avoided for compact models used in the circuit simulators. However, using valid approximations, we obtained an analytical relation for charge density in the accumulation regime [48]. We extend this approach to find the charge-based model for DGJLFETs in the depletion mode of operation.

In the depletion regime, the central potential is larger than the surface potential and the mobile charge density is lower than the depletion charge density. This implies that the exponential term in equation (8.96) indicated by 1 is less than unity [48]. Assuming $C = N_{\mathrm{D}}/n_{\mathrm{i}}$, which appears like the flat band relationship results in the best possible case for the exponential term being unity. Under this assumption, the charge density can be expressed as

$$Q_{\mathrm{SC}}\,(\mathrm{dep}) = 2\sqrt{2q\varepsilon_{\mathrm{Si}}n_{\mathrm{i}}V_{\mathrm{t}}}\sqrt{\frac{KN_{\mathrm{D}}}{n_{\mathrm{i}}}\left(\frac{N_{\mathrm{D}}}{n_{\mathrm{i}}} - C\right)} \tag{8.98}$$

From equation (8.98), C may be extracted as

$$C\,(\mathrm{dep}) = \frac{N_{\mathrm{D}}}{n_{\mathrm{i}}}\left[1 - \left(\frac{Q_{\mathrm{SC}}}{qN_{\mathrm{D}}t_{\mathrm{Si}}}\right)^2\right] \tag{8.99}$$

Putting the value of C obtained from equation (8.99) in equation (8.96), we get the charge-based model for the DGJLFETs even in the depletion regime as

$$V_{\mathrm{G}} - V_{\mathrm{FB}} - V = V_{\mathrm{t}}\ln\frac{N_{\mathrm{D}}}{n_{\mathrm{i}}}\left[1 - \left(\frac{Q_{\mathrm{SC}}}{qN_{\mathrm{D}}t_{\mathrm{Si}}}\right)^2\right] - \frac{Q_{\mathrm{SC}}^2}{8qN_{\mathrm{D}}\varepsilon_{\mathrm{Si}}} - \frac{Q_{\mathrm{SC}}}{2C_{\mathrm{ox}}} \tag{8.100}$$

The results obtained by the approximations based on the operating regions shows a good match with the TCAD simulations (Fig. 8.12.), indicating the accuracy of the proposed charge-based model [48].

It is also possible to find an analytical expression for the mobile charge density in the DGJLFETs utilizing the parabolic approximation method [42] used in Section 8.2.2. From equation (8.52), we have

$$\Delta\varphi = \frac{\left(Q_{\mathrm{m}} + qN_{\mathrm{D}}t_{\mathrm{Si}}\right)t_{\mathrm{Si}}}{8\varepsilon_{\mathrm{Si}}} \tag{8.101}$$

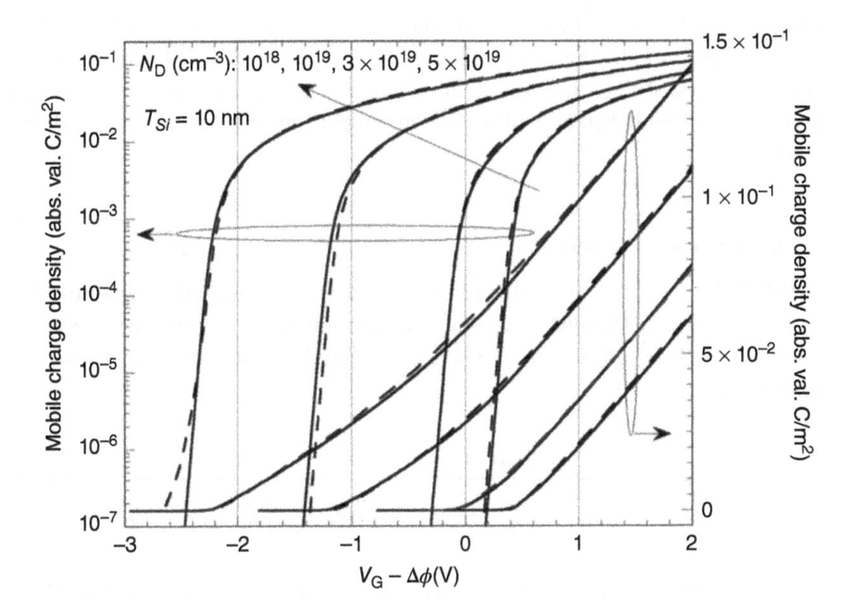

FIGURE 8.12 Mobile charge density as a function of effective gate voltage, i.e. gate voltage – work function of the gate electrode, for different silicon film doping. The TCAD simulations are shown with dotted lines, whereas results obtained by the model are shown as a solid line [48].

Replacing the value of $\Delta\varphi$ in equation (8.59) and utilizing equation (8.52), after rearrangement, we obtain a closed form relationship between the mobile charge density and the gate voltage as

$$V_G - V_{Th} - V = -\frac{\beta'}{2C_{ox}}Q_m + V_t \ln\left[-Q_m\sqrt{\frac{qN_D t_{Si} + Q_m}{2\varepsilon_{Si}\pi V_t q^2 N_D^2 t_{Si}}}\right] \tag{8.102}$$

where $\beta' = 1 + \frac{C_{ox}t_{Si}}{4\varepsilon_{Si}}$ and V_{Th} is the threshold voltage given by equation (8.39).

8.4 DRAIN CURRENT MODELING APPROACH

As discussed in Section 8.1, the main motivation for developing surface potential-based or charge-based models is to get more physical insight into the working of the FETs. These models can be used to obtain the drain current of the FETs for analyzing its static behavior such as the transfer characteristics, the output characteristics, subthreshold swing, DIBL, and so on. In this section, we discuss the different approaches for modeling the drain current in DGJLFETs utilizing the insights gained from Sections 8.2.1–8.2.4. First, we take the simplest bulk current model, which utilizes Ohm's law to calculate the drain current in Section 8.4.1. Then, we discuss the classical work of Pao–Sah, which allows us to calculate the drain current accurately considering both drift and diffusion components once the mobile charge density is known.

8.4.1 Bulk Current Model

In DGJLFETs (Fig. 8.13), above the threshold voltage, the current flows through the center of the channel. Therefore, we may calculate the drain current by simply using Ohm's law, i.e. $dV = IdR$, where dR is the differential channel resistance, which can be expressed as

$$dR = \frac{\rho dl}{A} = \frac{dy}{\sigma W2\left(\frac{t_{\mathrm{Si}}}{2} - x_{\mathrm{dep}}\right)} = \frac{dy}{2qN_{\mathrm{D}}\mu_n W\left(\frac{t_{\mathrm{Si}}}{2} - x_{\mathrm{dep}}\right)} \tag{8.103}$$

where ρ is the sheet resistivity and σ is the sheet conductivity given by $\sigma = qN_{\mathrm{D}}\mu_n$. However, the depletion region width x_{dep} is an unknown in equation (8.103). However, x_{dep} can be easily calculated from our analysis in Section 8.2 utilizing the depletion approximation. By combining equations (8.35) and (8.36), we obtain a quadratic equation in x_{dep} as

$$qN_{\mathrm{D}}x_{\mathrm{dep}} = C_{ox}\left(V_{\mathrm{G}} - V_{\mathrm{FB}} - V + \frac{qN_{\mathrm{D}}x_{\mathrm{dep}}^2}{2\varepsilon_{\mathrm{Si}}}\right) \tag{8.104}$$

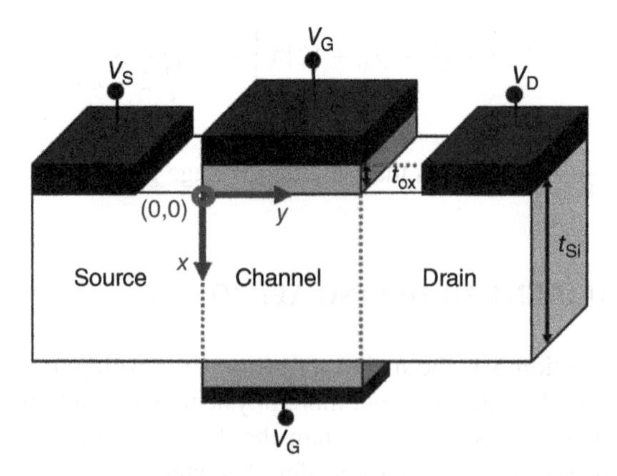

FIGURE 8.13 The DGJLFET with a modified coordinate system used for the bulk current modeling in [50].

Solving equation (8.104), we get a simple expression for the depletion region width as

$$x_{\text{dep}}\left(V_{\text{G}}, V\right) = \frac{\varepsilon_{\text{Si}}}{C_{\text{ox}}}\left[-1 + \sqrt{1 - \frac{2C_{\text{ox}}^2}{qN_{\text{D}}\varepsilon_{\text{Si}}}\left(V_{\text{G}} - V_{\text{FB}} - V\right)}\right] \tag{8.105}$$

Now, to get a simplified expression for the depletion region width, we expand $x_{\text{dep}}\left(V_{\text{G}}, V\right)$ using the Taylor series around the threshold voltage V_{Th} (where $x_{\text{dep}} = t_{\text{Si}}/2$) given by equation (8.39), which yields

$$x_{\text{dep}}\left(V_{\text{G}}, V\right) = x_{\text{dep}}\left(V_{\text{Th}}, V\right) + \frac{dx_{\text{dep}}\left(V_{\text{Th}}, V\right)}{dV_{\text{G}}}\left(V_{\text{G}} - V_{\text{Th}}\right) + \ldots \tag{8.106}$$

Ignoring the higher order terms, we obtain

$$x_{\text{dep}}\left(V_{\text{G}}, V\right) \approx -\frac{C_{\text{eq}}}{qN_{\text{D}}}\left(V_{\text{G}} - V_{\text{FB}} - V\right) + \frac{t_{\text{Si}}}{2}\left[1 - \underbrace{\frac{\frac{C_{\text{ox}}}{2} + C_{\text{dep}}}{C_{\text{ox}} + C_{\text{dep}}}}_{1}\right] \tag{8.107}$$

where C_{dep} is the half depletion capacitance given by $C_{\text{dep}} = \left(2\varepsilon_{\text{Si}}\right)/t_{\text{Si}}$ and C_{eq} is the net series capacitance of C_{ox} and C_{dep} given as $C_{\text{eq}} = \frac{C_{\text{dep}}C_{\text{ox}}}{C_{\text{dep}} + C_{\text{ox}}}$. Equation (8.107) can be further simplified by assuming that the bracketed term designated by 1 is unity for DGJLFETs, which are essentially designed with a ultrathin channel to achieve volume depletion such that $t_{\text{Si}}/2$ is comparable to the t_{ox} [50]. Using this approximation, equation (8.107) simplifies to

$$x_{\text{dep}}\left(V_{\text{G}}, V\right) \approx -\frac{C_{\text{eq}}}{qN_{\text{D}}}\left(V_{\text{G}} - V_{\text{FB}} - V\right) \tag{8.108}$$

Using equation (8.108), Ohm's law expressed as $dV = IdR$ can be integrated utilizing the gradual channel approximation to yield:

$$I\int_0^{L_{\text{g}}} dy = \int_0^{V_{\text{DS}}} 2qN_{\text{D}}\mu_n W\left(\frac{t_{\text{Si}}}{2} + \frac{C_{\text{eq}}}{qN_{\text{D}}}\left(V_{\text{G}} - V_{\text{FB}} - V\right)\right)dV \tag{8.109}$$

which gives a closed form analytical expression for the drain current in the partial depletion regime as

$$I = \frac{2qN_{\text{D}}\mu_n W}{L_{\text{g}}}\left[\left(\frac{t_{\text{Si}}}{2} + \frac{C_{\text{eq}}}{qN_{\text{D}}}\left(V_{\text{G}} - V_{\text{FB}}\right)\right)V_{\text{DS}} - \frac{C_{\text{eq}}}{2qN_{\text{D}}}V_{\text{DS}}^2\right] \tag{8.110}$$

Equation (8.110) indicates that the JLFETs and MOSFETs behave in a similar way with a linear current dependence on V_{DS} in the triode region, where V_{DS} is small. The saturation current can also be found from equation (8.110) by using the saturation voltage given as $V_{\mathrm{DS,sat}} = V_{\mathrm{GS}} - V_{\mathrm{Th}}$.

Once analytical expression for the current in linear and saturation regimes is found, the next step is to find an analytical expression for the drain current in the subthreshold regime to obtain a complete drain current model. Since the current in the subthreshold regime is dictated by the diffusion of mobile carriers from the source to the drain, the drain current in subthreshold may be expressed as the diffusion current [51]:

$$I = qWD_n \frac{dn\,(y)}{dy} = \frac{qWD_n\left[n(y=0) - n(y=L_{\mathrm{g}})\right]}{L_{\mathrm{g}}} \tag{8.111}$$

where $n(y=0)$ and $n(y=L_{\mathrm{g}})$ are the areal electron density at the source end and the drain end, respectively. Since the electrons at the drain end encounter an additional potential barrier equal to the drain voltage as compared to the electrons at the source, we have, $n(y=L_{\mathrm{g}}) = n(y=0)e^{-\frac{V_{\mathrm{DS}}}{V_t}}$. Using this in equation (8.111), we have

$$I = \frac{qWD_n n\,(y=0)\left[1 - e^{-\frac{V_{\mathrm{DS}}}{V_t}}\right]}{L_{\mathrm{g}}} \tag{8.112}$$

The areal electron density at the source end may be written as

$$n\,(y=0) = N_{\mathrm{D}} \int_0^{t_{\mathrm{Si}}} e^{\frac{\varphi(x)}{V_t}}\,dx \tag{8.113}$$

Now, in the coordinate system defined in Fig. 8.13 and used in [50], in the subthreshold mode of operation, using the depletion approximation, the Poisson equation may be integrated with the condition that electric field is zero at $x = t_{\mathrm{si}}/2$ to yield

$$\frac{d\varphi\,(x)}{dx} = -\frac{qN_{\mathrm{D}}}{\varepsilon_{\mathrm{Si}}}\left(x - \frac{t_{\mathrm{Si}}}{2}\right) \tag{8.114}$$

Integrating equation (8.114) and using the boundary condition that $\varphi\,(x=0) = \varphi_{\mathrm{S}}$, we obtain

$$\varphi\,(x) = -\frac{qN_{\mathrm{D}}}{2\varepsilon_{\mathrm{Si}}}\left(x - \frac{t_{\mathrm{Si}}}{2}\right)^2 + \frac{qN_{\mathrm{D}}t_{\mathrm{Si}}^2}{8\varepsilon_{\mathrm{Si}}} + \varphi_S \tag{8.115}$$

Also, using the Gauss's law, we have, $2C_{\mathrm{ox}}\left(V_{\mathrm{G}} - V_{\mathrm{FB}} - \varphi_{\mathrm{S}}\right) = Q_{\mathrm{SC}} = qN_{\mathrm{D}}t_{\mathrm{Si}}$.

Using equations (8.86) and (8.115) in equation (8.113) and defining

$$\alpha' = 2N_D\, e^{\left[\frac{qN_D t_{Si}^2}{8\varepsilon_{Si}} + V_G - V_{FB} + \frac{qN_D t_{Si}}{2C_{ox}}\right]/V_t} \tag{8.116}$$

and

$$\beta'' = \frac{qN_D}{2\varepsilon_{Si} V_t}, \tag{8.117}$$

we get

$$n(y=0) = \alpha' \int_0^{t_{Si}/2} e^{-\beta'' x^2} dx = \frac{\alpha'}{2}\sqrt{\frac{\pi}{\beta''}}\operatorname{erf}\left(\frac{\sqrt{\beta''}t_{Si}}{2}\right) \tag{8.118}$$

Although the expression obtained for the areal density at the source end is not analytic, the channel region is highly doped in DGJLFETs, which implies that β'' is much larger than unity [50]. Since the error function reduces to unity if the argument is more than one, equation (8.118) can be simplified.

Substituting simplified equation (8.118) in equation (8.112) after replacing the error function by unity, we get a simplified analytic expression for the subthreshold current as

$$I = \frac{qWD_n \frac{\alpha'}{2}\sqrt{\frac{\pi}{\beta''}}\left[1 - e^{-\frac{V_{DS}}{V_t}}\right]}{L_g} \tag{8.119}$$

The proposed model shows a good agreement with the TCAD simulations for different silicon film doping, thicknesses, and gate oxide thicknesses [50].

8.4.2 The Pao–Sah Integral

In the conventional MOSFETs, modeling the drain current requires the inversion layer charge density as discussed in the introduction of this chapter. However, calculation of the inversion layer charge density is a tedious task as it does not yield a closed form analytical expression and involves numerical integration which is time consuming and cannot be used in circuit simulators [27, 52]. In their classical work, Pao and Sah derived an exact solution for the inversion layer charge density as a function of the electron quasi-Fermi potential only [53]. Their analysis involved an assumption that the inversion layer charge density is negligible beyond a position $x = x_f$ where the band bending approaches the Fermi-potential, i.e. $\varphi = \varphi_f$. This is a valid assumption as the classical definition of the weak inversion regime also starts once the surface potential is equal to the Fermi-potential, i.e. $\varphi_S = \varphi_f$. The inversion layer charge density can be found analytically as

$$Q_{\text{in}}(x, V) = q \int_{x_c}^{0} n(x, V)\, dx = \int_{\varphi_{\text{f}}}^{\varphi_{\text{S}}} \frac{n_{\text{i}}^2}{N_{\text{A}}} e^{\frac{\varphi(x)-V}{V_{\text{t}}}} \frac{dx}{d\varphi} d\varphi \tag{8.120}$$

where $n(x, V)$ is the electron concentration at location x and electron quasi-Fermi level (V) in the channel, φ is the potential, N_{A} is the channel doping, and n_{i} is the intrinsic carrier concentration. Using equation (8.120) to get the inversion layer charge in equation (8.1) gives us an analytical expression for the drain current. The Pao–Sah method is the most comprehensive model and takes both drift and diffusion transport into account. Therefore, it is also considered as the benchmark model for modeling drain current in MOSFETs [27]. The Pao–Sah integral approach for calculating the drain current may be generalized for any FET considering only drift and diffusion transport as

$$I = -\mu \frac{W}{L_{\text{g}}} \int_{V_{\text{S}}}^{V_{\text{D}}} Q_{\text{m}} dV = -\mu \frac{W}{L_{\text{g}}} \int_{Q_{\text{S}}}^{Q_{\text{D}}} Q_{\text{m}} \frac{dV}{dQ_{\text{m}}} dQ_{\text{m}} = \mu \frac{W}{L_{\text{g}}} \int_{V_{\text{S}}}^{V_{\text{D}}} (Q_{\text{D}} - Q_{\text{SC}}) dV \tag{8.121}$$

where Q_m is the mobile charge density.

For DGJLFETs, an analytical relationship between the mobile charge density and the gate voltage is presented in Section 8.3 as equation (8.102) [42]. Below the flat band condition, the mobile charge density is less than the depletion charge density. Therefore, equation (8.102) may be simplified as

$$Q_{\text{m}} - \frac{2C_{\text{ox}}}{\beta'} V_{\text{t}} \ln \left[-\frac{Q_{\text{m}}}{\sqrt{2\varepsilon_{\text{Si}} \pi V_{\text{t}} q N_{\text{D}}}} \right] = -\frac{2C_{\text{ox}}}{\beta'} \left(V_{\text{G}} - V_{\text{Th}} - V \right) \tag{8.122}$$

From equation (8.122), we may write

$$\frac{dV}{dQ_{\text{m}}} = \frac{\beta'}{2C_{\text{ox}}} - \frac{V_{\text{t}}}{Q_{\text{m}}} \tag{8.123}$$

Using equation (8.123) in equation (8.121) and solving the integral, we get

$$I = -\mu \frac{W}{L_{\text{g}}} \left[\frac{\beta'}{4C_{\text{ox}}} Q_{\text{m}}^2 - V_{\text{t}} Q_{\text{m}} \right] \Bigg|_{Q_{\text{D}}}^{Q_{\text{S}}} \tag{8.124}$$

where Q_{S} and Q_{D} are the mobile charge densities at the source and the drain end, respectively, and may be calculated by using $V = 0$ for Q_{S} and $V = V_{\text{DS}}$ in equation (8.122). Equation (8.124) yields a continuous drain current model for DGJLFETs, which is valid in all the operating regimes below the flat band [42]. In the linear region, the mobile charges are dominant at both the source and drain ends. Therefore, the first term of equations (8.122) and (8.124) is dominant and governs the current [42]. Ignoring the second term, we get the following expression for the drain current in the linear region:

$$I = \frac{2C_{ox}\mu W}{\beta L_g}\left(V_G - V_{Th} - \frac{V_{DS}}{2}\right)V_{DS} \tag{8.125}$$

Similarly, in the saturation regime, although the mobile charges dominate on the source side, the channel region is pinched off on the drain side. Therefore, the first term of equations (8.122) and (8.124) dominates for the source side whereas the second term of these equations dominate and should be used for calculating Q_D and the drain current [42]. Using this simplification, the drain current is obtained as

$$I = \frac{C_{ox}\mu W}{\beta L_g}\left[(V_G - V_{Th})^2 - \frac{V_t\beta}{C_{ox}}\sqrt{2\varepsilon_{Si}\pi V_t q N_D}\, e^{\left(\frac{V_G - V_{Th} - V_{DS}}{V_t}\right)}\right. \tag{8.126}$$

In the subthreshold regime, the mobile charges are negligible at both source and drain end. Therefore, the second term of equations (8.122) and (8.124) is dominant and should be used for calculation of both Q_S and Q_D as well as the drain current [42]. The simplified expression for the drain current based on this assumption can be given as

$$I = \frac{\mu W V_t}{L_g}\sqrt{2\varepsilon_{Si}\pi V_t q N_D}\, e^{\left(\frac{V_G - V_{Th}}{V_t}\right)}\left[1 - e^{-\frac{V_{DS}}{V_t}}\right] \tag{8.127}$$

Therefore, using the mobile charge model developed in Section 8.3, we can simply obtain the drain current model. The model shows a good agreement with the TCAD simulation results [42]. However, this model does not account for the current in the accumulation regime. Therefore, we use the charge-based model developed in Section 8.3 for accumulation and depletion regimes for calculating the drain current [48]. Equation (8.121) can also be expressed as

$$I = \mu\frac{W}{L_g}\int_{V_S}^{V_D}(qN_D t_{Si} - Q_{SC})\,dV = \mu\frac{W}{L_g}qN_D t_{Si}V_{DS} - \mu\frac{W}{L_g}\int_{V_S}^{V_D}Q_{SC}\,dV \tag{8.128}$$

Now, $Q_{SC}\,dV$ in the accumulation mode can be easily found using equation (8.95) as

$$Q_{SC}\,dV\,(\text{accu}) \approx \frac{Q_{SC}dQ_{SC}}{2C_{ox}} - \frac{Q_{SC}^2\,dQ_{SC}}{4qN_D\varepsilon_{Si} + \frac{Q_{SC}^2}{2V_t}} \tag{8.129}$$

The integration of the second term in equation (8.128) in the accumulation regime can then be expressed as

$$\int_{V_S}^{V_D} Q_{SC}\, dV\,(\text{accu})$$
$$\approx \frac{1}{4C_{ox}} Q_{SC}^2 \Big|_{Q_S}^{Q_D} - 2V_t \left[Q_{SC} - \sqrt{8\varepsilon_{Si} q N_D V_t} \tan^{-1} \frac{Q_{SC}}{\sqrt{8\varepsilon_{Si} q N_D V_t}} \right] \Bigg|_{Q_S}^{Q_D} \tag{8.130}$$

Q_S and Q_D may be obtained by putting $V=0$ and $V=V_{DS}$ in equation (8.95), respectively. Now, when the entire channel is in accumulation, we have, $V_G - V_{FB} - V_{DS} > V_t \ln(N_D/ni)$ which ensures that the channel at the drain end is in accumulation. Once this is satisfied, the channel is bound to be in accumulation mode at the source end [48]. In this regime, the drain current can be expressed as

$$I = \mu \frac{W}{L_g} q N_D t_{Si} - \mu \frac{W}{L_g} \int_{V_S}^{V_D} Q_{SC}\, dV\ (\text{accu}) \tag{8.131}$$

Now, $Q_{SC}\, dV$ in the depletion mode can be obtained from equation (8.100) as

$$Q_{SC}\, dV\,(\text{dep}) \approx \frac{Q_{SC}^2 dQ_{SC}}{4\varepsilon_{Si} q N_D} + \frac{Q_{SC} dQ_{SC}}{2C_{ox}} + \frac{V_t 2 Q_{SC}^2\, dQ_{SC}}{(qN_D t_{Si})^2 - Q_{SC}^2} \tag{8.132}$$

The integration of the second term in equation (8.128) for the depletion region can be obtained as

$$\int_{V_S}^{V_D} Q_{SC}\, dV\,(\text{dep}) \approx \frac{1}{12\varepsilon_{Si} q N_D} Q_{SC}^3 \Big|_{Q_S}^{Q_D} + \frac{1}{4C_{ox}} Q_{SC}^2 \Big|_{Q_S}^{Q_D} - 2V_t\, Q_{SC} \Big|_{Q_S}^{Q_D}$$
$$+ V_t q N_D t_{Si} \left[\ln \frac{(qN_D t_{Si} + Q_{SC})}{(qN_D t_{Si} - Q_{SC})} \right] \tag{8.133}$$

When the entire channel is depleted, we have, $V_G - V_{FB} < V_t \ln(N_D/ni)$ at the source end of the channel, which ensures that the channel is depleted at the source end. If this condition is met, the drain end of the channel region is also depleted [48]. Therefore, in the depletion regime, the drain current can be expressed as

$$I = \mu \frac{W}{L_g} q N_D t_{Si} - \mu \frac{W}{L_g} \int_{V_S}^{V_D} Q_{SC}\, dV\ (\text{dep}) \tag{8.134}$$

However, in the partial depletion mode of operation, the source end of the channel is in the accumulation mode whereas the drain end is in the depletion mode and the conditions $V_G - V_{FB} > V_t \ln(N_D/ni)$ and $V_G - V_{FB} - V_{DS} < V_t \ln(N_D/ni)$ are satisfied. In this case of a hybrid channel, the integral of the second term is divided into two parts: the accumulated channel at from the source end to the flat band position in the channel region (where $Q_{SC} = 0$) and the depleted region from the flat band position to the drain end as given below:

$$I_{hybrid} = \mu \frac{W}{L_g} qN_D t_{Si} - \mu \frac{W}{L_g} \int_{V_S}^{V_{FB}} Q_{SC}\, dV\, (\text{accu}) - \mu \frac{W}{L_g} \int_{V_{FB}}^{V_D} Q_{SC}\, dV\, (\text{dep}) \tag{8.135}$$

Therefore, the drain current in all the operating regimes can be calculated using the charge-based model obtained in Section 8.3 [48].

Now, we discuss the methodology to obtain the drain current model from the surface potential modeling approach using the Pao–Sah integral. For this, we use a modified version of the Pao–Sah integral as

$$I = -\mu \frac{W}{L_g} \int_{\varphi_S(0)}^{\varphi_S(L_g)} Q_m \frac{dV}{d\varphi_S} d\varphi_S = -\mu \frac{W}{L_g} \int_{\varphi_S(0)}^{\varphi_S(L_g)} \left(Q_{SC} - Q_D\right) \frac{dV}{d\varphi_S} d\varphi_S \tag{8.136}$$

Now, utilizing the relationship between the surfacc potential and the electron quasi-Fermi potential derived in Section 8.2.1 using the regional approximation approach, we may develop a drain current model for the DGJLFETs [41]. Using equation (8.86) for Q_{SC}, equation (8.136) may be expressed as

$$I = \mu \frac{W}{L_g} \int_{\varphi_S(0)}^{\varphi_S(L_g)} \left[2C_{ox}\left(V_G - V_{FB} - \varphi_S\right) + qN_D t_{Si}\right] \frac{dV}{d\varphi_S} d\varphi_S \tag{8.137}$$

For $V_{GS} > V_{FB} + V_{DS}$, the mobile electrons accumulate in the entire channel region and the relationship between the surface potential and the electron quasi-Fermi potential given by equation (8.25) should be differentiated to obtain $dV/d\varphi_S$ as

$$\frac{dV}{d\varphi_S} = 1 + \frac{2V_t\left(V_G - V_{FB} - \varphi_S\right)}{\left(V_G - V_{FB} - \varphi_S\right)^2 + \alpha V_t} \tag{8.138}$$

Using this value of $dV/d\varphi_S$ in equation (8.137) and integrating, we obtain the drain current for the DGJLFETs in the accumulation region as

$$I_{accu} = \mu \frac{W}{L_g} \Bigg[qN_D t_{Si} V - 4C_{ox} V_t (V_G - V_{FB} - \varphi_S) - C_{ox}(V_G - V_{FB} - \varphi_S)^2 + 4C_{ox} V_t \sqrt{\alpha V_t} \tan^{-1} \frac{(V_G - V_{FB} - \varphi_S)}{\sqrt{\alpha V_t}} \Bigg] \Bigg|_{\varphi_S(0)}^{\varphi_S(L_g)} \tag{8.139}$$

where $\varphi_S(0)$ and $\varphi_S(L_g)$ are the surface potential values at the source end and drain end, which may be calculated by using $V=0$ and $V=V_{DS}$ in equation (8.25), respectively.

Similarly, for the partial depletion regime ($V_{Th} < V_{GS} < V_{FB}$), differentiating equation (8.38), we get,

$$\frac{dV}{d\varphi_S} = 1 - \frac{2}{\alpha}\left(V_G - V_{FB} - \varphi_S\right) \tag{8.140}$$

Utilizing this relationship in equation (8.137) and evaluating the integral, we obtain the drain current in the partially depletion mode as

$$I_{dep} = \mu\frac{W}{L_g}\left[qN_D t_{Si} V - C_{ox}\left(V_G - V_{FB} - \varphi_S\right)^2 + \frac{4}{3\alpha}\left(V_G - V_{FB} - \varphi_S\right)^3\right]\Bigg|_{\varphi_S(0)}^{\varphi_S(L_g)} \tag{8.141}$$

Again, $\varphi_S(0)$ and $\varphi_S(L_g)$ are the surface potential values at the source end and drain end which may be calculated by using $V=0$ and $V=V_{DS}$ in equation (8.38), respectively.

Now, for $V_{FB} < V_{GS} < V_{FB} + V_{DS}$, the channel is accumulated near the source end while it is depleted near the drain end. Therefore, for this case of hybrid channel, the integral must be separated for the accumulated channel from source end to the flat band point in the channel, which follows equation (8.25), and for the depleted channel until the drain end, which follows equation (8.38), as

$$I_{hybrid} = \mu\frac{W}{L_g}[qN_D t_{Si} V + 2C_{ox}\int_0^{V_G - V_{FB}}\left(V_G - V_{FB} - \varphi_S\right)dV\Big|(\text{accu})$$

$$+2C_{ox}\int_{V_G - V_{FB}}^{V_{DS}}\left(V_G - V_{FB} - \varphi_S\right)dV\Big|(\text{dep}) \tag{8.142}$$

For $V_{GS} < V_{Th}$, the entire channel is depleted. Using equation (8.47) for the full-depletion mode, we obtain the drain current as

$$I_{sub} = \mu\frac{W}{L_g}V_t 2C_{ox}\int_0^{V_{DS}} W\left[\frac{qN_D t_{Si}}{4C_{ox}\beta}e^{\frac{V_G - V_{Th} - V}{V_t}}\right]dV \tag{8.143}$$

The subthreshold current obtained by equation (8.143) saturates when the operating region changes to the partial depletion mode [54]. At the point of saturation, to improve the accuracy in the partial depletion regime, a simple interpolation may be used to find the total drain current as

$$I = \left[I_{sub}^2 + I_{dep}^2\right]^{1/2} \tag{8.144}$$

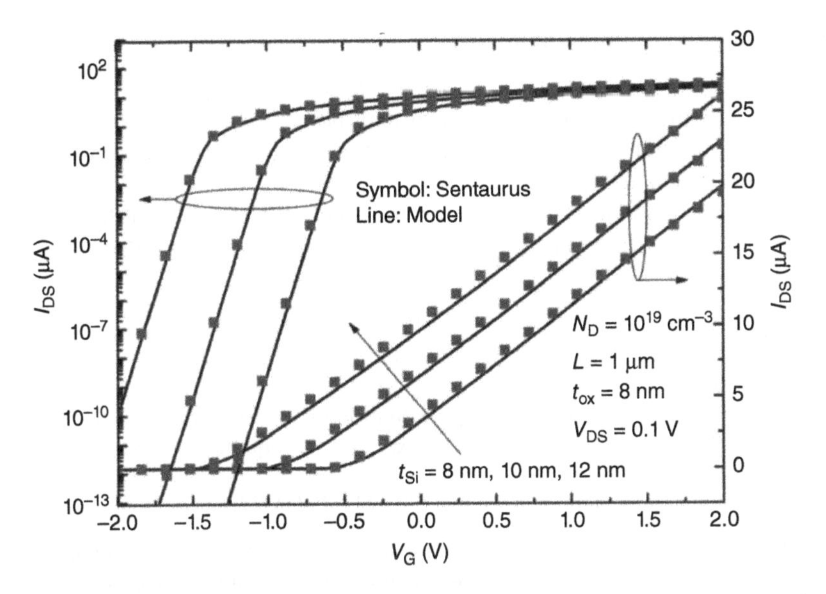

FIGURE 8.14 Transfer characteristics of the DGJLFETs for different silicon film thickness using the interpolation technique [41].

As can be observed from Fig. 8.14, the drain current model obtained using the interpolation technique shows a close agreement with the TCAD simulations [41].

8.5 MODELING SHORT-CHANNEL JLFETs

In Sections 8.1–8.4, we have discussed the different approaches used for modeling the surface potential, the charge, and the drain current in DGJLFETs. However, the approaches discussed in these sections are accurate only for the long-channel JLFETs where there is a negligible impact of the drain voltage on the surface potential at the source–channel interface as the lateral electric field is low. In this case only, we may separately solve the Poisson equation in the direction along the channel thickness and use it to find the drain current using the Pao–Sah integral or numerical integration. However, in the short-channel JLFETs, due to the proximity of the source–channel interface to the drain, the interaction between the surface potential and the drain electric field is inevitable. This interaction leads to a perturbation in the surface potential at the source–channel interface giving rise to short-channel effects such as drain-induced barrier lowering (DIBL), and so on. Therefore, the 2D Poisson equation must be solved for modeling the short-channel JLFETs. One of the simplest techniques to solve the 2D Poisson equation and to model the threshold voltage and the short-channel effects such as threshold voltage roll-off and DIBL in JLFETs is discussed in the next section.

8.5.1 Quasi-2D Scaling Equation

The quasi-2D scaling equation simplifies the 2D Poisson equation into an ordinary second-order differential equation, which may be solved analytically. Before discussing the quasi-2D approach, we would like to discuss about the concept of scaling factor and natural length [55].

In the subthreshold regime of operation, the contribution to the total charge density due to the mobile carriers may be ignored and the 2D Poisson equation for MOSFETs may be expressed as

$$\frac{d^2\varphi(x,y)}{dx^2} + \frac{d^2\varphi(x,y)}{dy^2} = \frac{qN_A}{\varepsilon_{Si}} \tag{8.145}$$

Assuming a parabolic potential approximation along the x-direction as presented in Section 8.2.2, the 2D potential may be expressed as equation (8.48):

$$\varphi(x,y) = a_0(y) + a_1(y)x + a_2(y)x^2$$

Utilizing the boundary condition that the electric field is zero at the center and the potential at $x = 0$ is the central potential, equation (8.48) simplifies to equation (8.49):

$$\varphi(x,y) = \varphi_0(y) + a_2(y)x^2$$

Now, from the continuity of the electric displacement vector at the bottom Si–SiO_2 interface, we get a_2

$$-C_{ox}\left(V_G - V_{FB} - \varphi_{S,b}\right) = -\varepsilon_{Si}\frac{d\varphi\left(\frac{t_{Si}}{2},y\right)}{dx} = -\varepsilon_{Si}t_{Si}a_2(y) \tag{8.146}$$

Putting the value of a_2 in equation (8.49), we obtain a relation for the 2D potential as

$$\varphi(x,y) = \varphi_0(y) + \frac{C_{ox}}{\varepsilon_{Si}t_{Si}}(V_G - V_{FB} - \varphi_S(y))y^2 \tag{8.147}$$

From equation (8.147), a relation between the surface potential and the central potential can be obtained as

$$\varphi\left(\frac{t_{Si}}{2},y\right) = \varphi_S(y) = \varphi_0(y) + \frac{C_{ox}t_{Si}}{4\varepsilon_{Si}}(V_G - V_{FB} - \varphi_S(y)) \tag{8.148}$$

Rearranging equation (8.148), we get a relationship between the surface potential and the central potential as

$$\varphi_S(y) = \frac{\varphi_0(y) + \frac{C_{ox}t_{Si}}{4\varepsilon_{Si}}\left(V_G - V_{FB}\right)}{1 + \frac{C_{ox}t_{Si}}{4\varepsilon_{Si}}} \tag{8.149}$$

Now, putting equation (8.147) into equation (8.145) after utilizing the value of central potential obtained from (8.149), we get

$$\frac{d^2\varphi_S(y)}{dy^2} - \frac{4C_{ox}}{\varepsilon_{Si}t_{Si}}\left(\varphi_S(y) - V_G - V_{FB} + \frac{qN_A t_{Si}}{4C_{ox}}\right) = 0 \tag{8.150}$$

Now, defining $1/\lambda_1^2 = (4C_{ox})/\varepsilon_{Si}t_{Si}$, $\emptyset_S = V_G - \omega_1$, and $\omega_1 = V_{FB} - (qN_A t_{Si})/4C_{ox}$, where λ_1 has the dimensions of length and is referred to as the natural length and $\emptyset_S$ is defined as the long-channel surface potential, we get a simplified version of equation (8.150) as a simple second-order differential equation:

$$\frac{d^2\varphi_S(y)}{dy^2} - \frac{1}{\lambda_1^2}\left(\varphi_S(y) - \emptyset_S\right) = 0 \tag{8.151}$$

This simplified differential equation has the solution of the form:

$$\varphi_S(y) = \emptyset_S + a\,e^{\frac{y}{\lambda_1}} + b\,e^{-\frac{y}{\lambda_1}} \tag{8.152}$$

It may be noted from equation (8.152) that λ_1 represents the spread in the potential profile along the lateral direction and depends upon the channel thickness and the gate oxide thickness [55]. A smaller channel or gate oxide thickness improves the effective gate control, leading to a reduced natural length and a lower influence of the drain region on the electrostatics of the channel region. Therefore, the natural length is also an indicator of the influence of the drain electric field on the channel region. The gate length should be much larger than the natural length for mitigating the short-channel effects. According to the quasi-2D scaling theory [55], the FETs should be designed to have a large value of the scaling factor α given as $\alpha = L_g/2\lambda$ to suppress the short-channel effects.

Now, for the short-channel DGJLFETs (Fig. 8.15), the Poisson equation is modified as equation (8.2):

$$\frac{d^2\varphi(x,y)}{dx^2} + \frac{d^2\varphi(x,y)}{dy^2} = -\frac{qN_D}{\varepsilon_{Si}}$$

Now, following the approach similar to the DGMOSFETs and utilizing the parabolic potential approximation and the boundary conditions, we obtain

$$\varphi(x,y) = \varphi_0(y) + \frac{C_{ox}}{\varepsilon_{Si}t_{Si}}\left(V_G - V_{FB} - \varphi_S(y)\right)y^2 \tag{8.153}$$

Now, we develop the scaling equations for the JLFETs considering the central potential since the current flows at the center at the verge of threshold conditions in a DGJLFET as discussed in Section 3.2.2 [56]. Therefore, replacing $\varphi_S(y)$ in equation

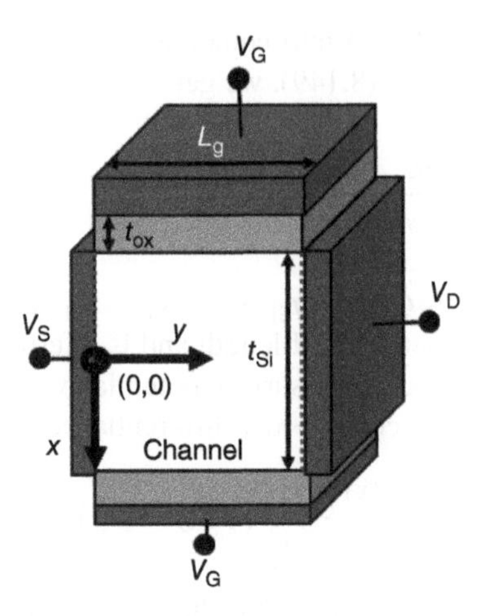

FIGURE 8.15 Three-dimensional view of the short-channel DGJLFET used for obtaining the quasi-2D scaling equation in [56]. For simplicity, the source/drain thickness is assumed to be zero and contacts are abutted on the sides.

(8.153) utilizing the expression relating the central potential and surface potential from equation (8.149), we have

$$\varphi(x,y) = \varphi_0(y) + \frac{C_{ox}}{\varepsilon_{Si} t_{Si}} \left(V_G - V_{FB} - \frac{\varphi_0(y) + \frac{C_{ox} t_{Si}}{4\varepsilon_{Si}} (V_G - V_{FB})}{1 + \frac{C_{ox} t_{Si}}{4\varepsilon_{Si}}} \right) y^2 \quad (8.154)$$

Utilizing equation (8.154) in equation (8.2), we get

$$\frac{d^2 \varphi_0(y)}{dy^2} - \frac{8C_{ox}}{4\varepsilon_{Si} t_{Si} + C_{ox} t_{Si}^2} \left(\varphi_0(y) - V_G - V_{FB} + \frac{qN_D t_{Si}}{2C_{ox}} + \frac{qN_D t_{Si}^2}{8\varepsilon_{Si}} \right) = 0 \quad (8.155)$$

Now, defining $1/\lambda_2^2 = 8C_{ox}/(4\varepsilon_{Si} t_{Si} + C_{ox} t_{Si}^2)$, $\emptyset_0 = V_G - \omega_2$, and $\omega_2 = V_{FB} - \left(qN_D t_{Si}\right)/2C_{ox} - \left(qN_D t_{Si}^2\right)/8\varepsilon_{Si}$, where λ_2 is the natural length for DGJLFETs and $\emptyset_0$ is the long-channel central potential, we get a simplified second-order differential equation even for the DGJLFETs as

$$\frac{d^2 \varphi_0(y)}{dy^2} - \frac{1}{\lambda_2^2} \left(\varphi_0(y) - \emptyset_0 \right) = 0 \quad (8.156)$$

This simplified differential equation has the solution of the form:

$$\varphi_0(y) = \emptyset_0 + a_1 e^{\frac{y}{\lambda_2}} + b_1\, e^{-\frac{y}{\lambda_2}} \quad (8.157)$$

where a_1 and b_1 are constants. These constants may be found by utilizing the boundary condition along the y-direction, which are

$$\varphi_0 (y = 0) = V_S = 0 \tag{8.158}$$

$$\varphi_0 \left(y = L_g\right) = V_D = V_{DS} \tag{8.159}$$

Using the boundary conditions given by equations (8.158) and (8.159), we get

$$a_1 = \beta_1 V_G + \delta \tag{8.160}$$

$$b_1 = \gamma V_G + \theta \tag{8.161}$$

where

$$\beta_1 = \frac{e^{-\frac{L_g}{\lambda}} - 1}{2 \sinh\left(\frac{L_g}{\lambda}\right)} \tag{8.162}$$

$$\delta = \frac{V_{DS} - \omega_2 \left(e^{-\frac{L_g}{\lambda}} - 1\right)}{2 \sinh\left(\frac{L_g}{\lambda}\right)} \tag{8.163}$$

$$\gamma = \frac{1 - e^{\frac{L_g}{\lambda}}}{2 \sinh\left(\frac{L_g}{\lambda}\right)} \tag{8.164}$$

$$\theta = \frac{-V_{DS} + \omega_2 \left(e^{\frac{L_g}{\lambda}} - 1\right)}{2 \sinh\left(\frac{L_g}{\lambda}\right)} \tag{8.165}$$

Once the constants a_1 and b_1 are known, the next step is to find an analytical relation for the threshold voltage of DGJLFETs. The location of the minimum central potential can be obtained by differentiating equation (8.157) and using $d\varphi_0 (y) / dy = 0$, which yields that the minimum central potential exists at a location:

$$y_{\min} = \frac{\lambda}{2} \ln\left(\frac{b_1}{a_1}\right) \tag{8.166}$$

The value of the minimum central potential is obtained by putting equation (8.166) in equation (8.157) as

$$\varphi_0 \left(y_{\min}\right) = \sqrt{a_1 b_1} + \emptyset_0 \tag{8.167}$$

Now, at threshold voltage, the neutral channel begins to be uncovered at the center. Therefore, the minimum central potential may be assumed to be zero when the gate voltage equals the threshold voltage V_{Th}[56]. Using this, we find the threshold voltage expression for the short-channel DGJLFETs as

$$V_{Th} = \frac{2(\delta\gamma + \beta\theta) + \omega_2 + \sqrt{\left(2(\delta\gamma + \beta\theta) + \omega_2\right)^2 - (1 - 4\beta\gamma)\left(\omega^2 - 4\delta\theta\right)}}{(1 - 4\beta\gamma)} \tag{8.168}$$

Equation (8.168) presents a closed form analytical solution to the threshold voltage of short-channel DGJLFETs. It may be noted that the constants β and γ cease to be 0 for long-channel lengths and the threshold voltage approaches ω_2, which is the threshold voltage expression obtained for the long-channel DGJLFETs in Section 8.2.1.

8.5.2 Scaling Implication for MOSFETs Using Quasi-2D Approach

Similar to the DGJLFETs, the threshold voltage of the DGMOSFETs can be obtained using the quasi-2D approach. Since the current flows at the surface in the MOSFETs, the surface potential is considered for DGMOSFETs, which is given as equation (8.152):

$$\varphi_S(y) = \emptyset_S + a\, e^{\frac{y}{\lambda_1}} + b\, e^{-\frac{y}{\lambda_1}}$$

The boundary conditions in the y-direction in the case for MOSFETs are different from that of JLFETs as given below:

$$\varphi_S(y = 0) = V_{bi} \tag{8.169}$$

$$\varphi_S\left(y = L_{eff}\right) = V_{bi} + V_{DS} \tag{8.170}$$

where V_{bi} is the built-in potential at the source–channel and the channel-drain interface given by $V_{bi} = V_t \ln\left(N_{S/D}/N_A\right)$, where $N_{S/D}$ is the doping of the source and drain regions and N_A is the channel doping. It may be noted that for the DGMOSFET, the gate length L_g has been replaced by the effective channel length L_{eff} which is given by $L_{eff} = L_g - L_S - L_D + 2L_d$ and takes into account the encroachment of the source/drain depletion regions denoted by L_S and L_D, respectively, in the channel length defined as

$$L_S = \frac{2\left(V_{bi} - \varphi_S\left(y_{min}\right)\right)}{\dfrac{d\varphi_S(y = 0)}{dy}} \tag{8.171}$$

$$L_D = \frac{2\left(V_{DS} + V_{bi} - \varphi_S\left(y_{min}\right)\right)}{\dfrac{d\varphi_S\left(y = L_g\right)}{dy}} \tag{8.172}$$

and also the contribution of the Debye length, which accounts for the transition between the drift and diffusion regions in the channel given as

$$L_{\rm d} = \sqrt{\frac{2\varepsilon_{\rm Si}kT}{q^2N_{\rm A}}} \tag{8.173}$$

After obtaining the constants a and b using the boundary conditions, the minimum surface potential can be obtained as:

$$\varphi_S\left(y_{\min}\right) = \sqrt{ab} + \emptyset_S \tag{8.174}$$

Now, for DGMOSFETs, at the threshold voltage, the minimum surface potential should be equal to twice the Fermi-potential, i.e. $\varphi_S\left(y_{\min}\right) = 2\varphi_{\rm f} = 2V_{\rm t}\ln\left(N_{\rm A}/n_{\rm i}\right)$. The threshold voltage for DGMOSFET may be obtained in this manner.

Now, we can draw few conclusions right away. The effective channel length is larger in DGJLFETs as compared to the DGMOSFET. Therefore, JLFETs are expected to perform better as compared to the MOSFETs with respect to short-channel effects such as threshold voltage roll-off (Fig. 8.16) and DIBL (Fig. 8.17). The DIBL has been defined as the difference in the threshold voltage at a drain voltage of 0.1 V and a higher drain voltage specified in Fig. 8.17. Also, it may be concluded that a lower gate oxide thickness and a smaller channel thickness leads to reduced short-channel effects and facilitates the scaling of the JLFETs [56].

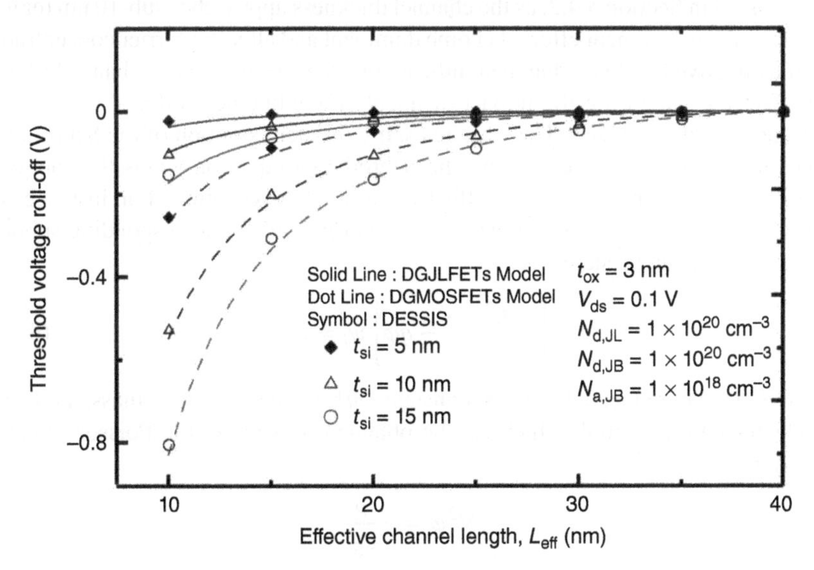

FIGURE 8.16 Threshold voltage roll-off for the DGJLFETs and the DGMOSFETs for different silicon film thicknesses [56].

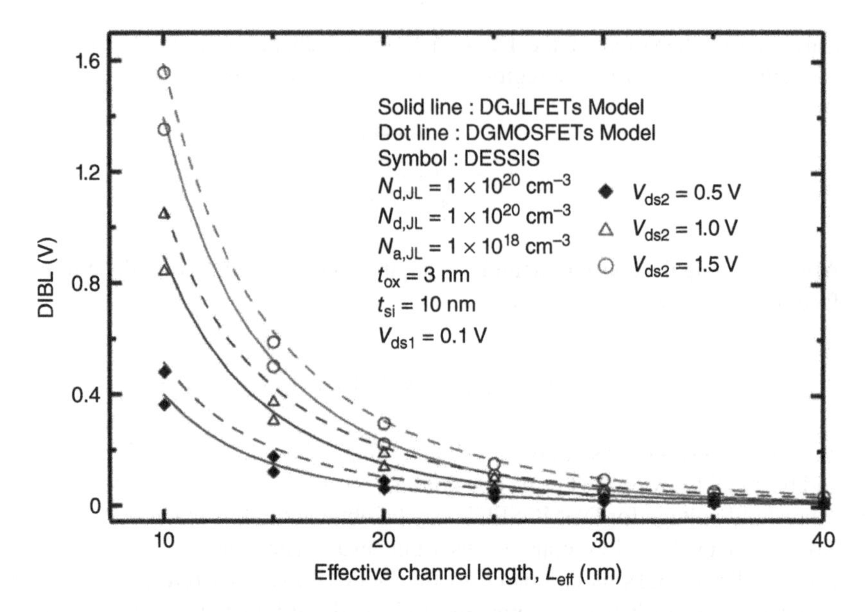

FIGURE 8.17 Drain-induced barrier lowering for the DGJLFETs and the DGMOSFETs for different drain voltage [56].

8.6 MODELING QUANTUM CONFINEMENT

As discussed in Section 5.4.2, as the channel thickness approaches sub-10 nm regime, the quantum confinement effects become dominant and alter the carrier concentration and the effective band gap due to significant discretization of energy bands [57–62]. For effectively capturing the impact of quantum confinement effects, each energy subband must be analyzed. This requires self-consistent solution of the Schrödinger equation and the Poisson equation. The Schrödinger equation yields the energy of the different subbands using the effective mass approximation. For instance, the Schrödinger equation for the ith wave function (ψ_i) and the corresponding subband energy level E_i can be stated as

$$\left[-\frac{\hbar^2}{2m^*}\nabla^2 - q\varphi\right]\psi_i = E_i\psi_i \tag{8.175}$$

where $\hbar$ is the reduced Planck's constant, m^* is the effective mass, and φ is the electrostatic potential, which can be obtained by solving the Poisson equation (equation 8.2):

$$\nabla^2\varphi = -\frac{\rho}{\varepsilon_{\text{Si}}}$$

However, self-consistent solution of the Schrödinger and Poisson equations cannot be obtained analytically.

In case of MOSFETs, in the strong inversion regime, the inversion carriers at the Si–SiO_2 lie in a triangular potential well formed by the gate oxide potential barrier on one side and the heavily bent conduction band acting as a barrier on the other side. Because of a significant quantization in the direction along the channel thickness, i.e. x-direction, the inversion carriers must be treated as a 2D gas which is free to move in the direction along the channel length (y-direction) or the direction along the channel width (z-direction) [57]. Therefore, the quantum confined electrons should be treated with discrete energy subbands with 2D density of states formalism rather than with the conventional continuous 3D density of state formalism taught in the freshman semiconductor physics books and courses. First, we discuss a method to find the carrier density in a particular energy subband for the 2D electron gas, which is essentially the behavior of the quantized electrons. The number of electronic states at energy can be simply thought of as the number of possible momentum values that an electron at that energy level may possess. According to the Heisenberg principle, there is only one electronic state in a phase space of volume $\left(\Delta y \Delta p_y\right)\left(\Delta z \Delta p_z\right) = h^2$, where p_y and $p_z$are the electron momentum along the y-direction and z-direction, respectively, and h is the Planck's constant. Now, the number of electronic states per unit area (density of states ($N(E)$)) in an energy subband at an energy level between E and $E + dE$ can be expressed as

$$N(E)\,dE = \frac{2g\,dp_y dp_z}{h^2} \tag{8.176}$$

where g is the degeneracy of the energy subband, $dp_y dp_z$ is the area in the momentum space where the electron lies within the energy level E and $E + dE$. Now, using the effective mass approximation, the energy–momentum relationship at the bottom of this energy subband with a ground state energy level E_0 can be represented as

$$E - E_0 = \frac{p_y^2}{2m_y} + \frac{p_z^2}{2m_z} \tag{8.177}$$

where $E - E_0$ is the kinetic energy and m_y and m_z are the effective mass of the electrons in the y- and z-directions, respectively. Equation (8.177) represents an ellipse in the momentum space whose area can be given as $2\pi\sqrt{m_y m_z}\left(E - E_0\right)$. Therefore, the area $dp_y dp_z$ in the momentum space where the electron lies within the energy level E and $E + dE$ can also be written as $2\pi\sqrt{m_y m_z}\,dE$. Using this expression in equation (8.176), the 2D density of states in the subband can be expressed as

$$N(E)\,dE = \frac{4g\pi\sqrt{m_y m_z}\,dE}{h^2} \tag{8.178}$$

Now, the number of electrons in this energy subband can be obtained by multiplying the density of states at this energy level by the probability of finding electrons at this energy, which is given by the Fermi–Dirac distribution. Therefore, the electron density in this particular subband is given as

$$n = \int_{E_0}^{\infty} N(E) f(E)\, dE = \frac{4gkT\pi\sqrt{m_y m_z}}{h^2} \ln\left(1 + e^{\frac{E_f - E_0}{kT}}\right) \tag{8.179}$$

The total electron density in the conduction band may be obtained by summing the contributions of all the energy subbands. For this, we need to find the location of the energy subbands, which cannot be obtained analytically owing to the self-consistent solution of the Schrödinger and Poisson equations [57].

However, in the subthreshold regime of operation, the mobile charge density is negligible and the energy band profile is governed by the depletion charges. Therefore, in this regime, the Schrödinger equation can be decoupled from the Poisson equation and solved independently to find the location of the different energy subbands [57]. Let us consider the case of the DGJLFETs. The solution of the Poisson equation for the DGJLFETs, as discussed in Section 8.4.1, can be simplified utilizing equations (8.86) and (8.115) as

$$\varphi(x) = V_G - V_{Th} - \frac{qN_D}{2\varepsilon_{Si}} x^2 \tag{8.180}$$

where V_{Th} is the threshold voltage given by equation (8.39). Now, we also solve the Schrödinger equation independently. Numerical simulations utilizing TCAD ATLAS reveal that for large channel thickness and channel doping, the energy subbands lie enclosed within the boundaries of the bent classical conduction band similar to the DGMOSFETs as shown in Fig. 8.18. However, for the case of ultrathin channel, the energy subbands lie within the potential well dictated by the potential barriers of the gate oxide and the potential at the bottom is governed by the bent classical conduction band as shown in Fig. 8.19. Therefore, for the two extreme cases, the energy subbands show different behavior and boundary conditions [62].

The first case of heavy channel doping and large channel thickness with energy bands confined within a parabolic barrier is similar to the standard quantum harmonic oscillator problem with a potential energy given by

$$V_{QM} = \frac{q^2 N_D}{2\varepsilon_{Si}} x^2 \tag{8.181}$$

Utilizing the effective mass approximation, the solution of this standard problem can be obtained as

$$E_{k,n}(\text{QHO}) = \frac{\left(n + \frac{1}{2}\right) h}{2\pi} \sqrt{\frac{q^2 N_D}{m_k^* \varepsilon_{Si}}} \tag{8.182}$$

where $E_{k,n}$ is the quantized energy of the kth energy subband at the nth orbital, n is the principal quantum number, and m_k^* is the effective mass in the kth energy subband.

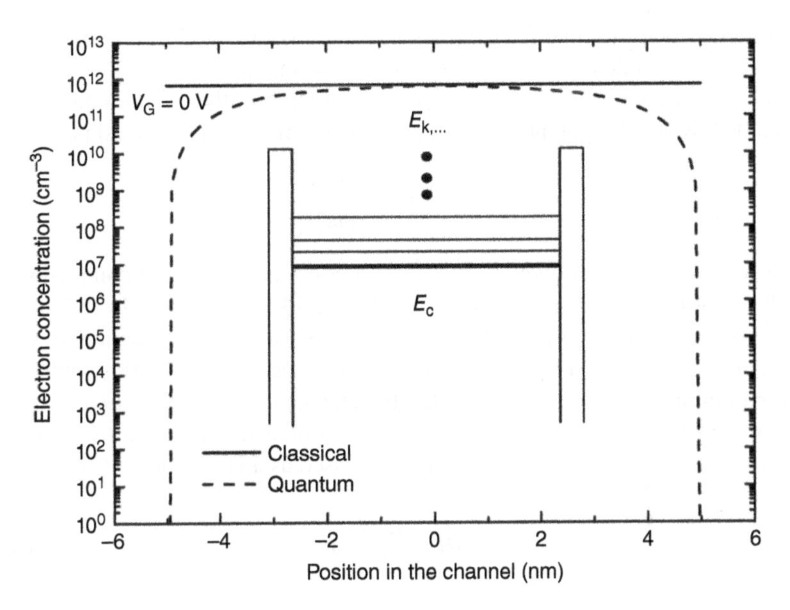

FIGURE 8.18 Comparison of the quantum and classical electron concentration along the channel thickness in DGMOSFETs for a channel thickness of 10 nm [62].

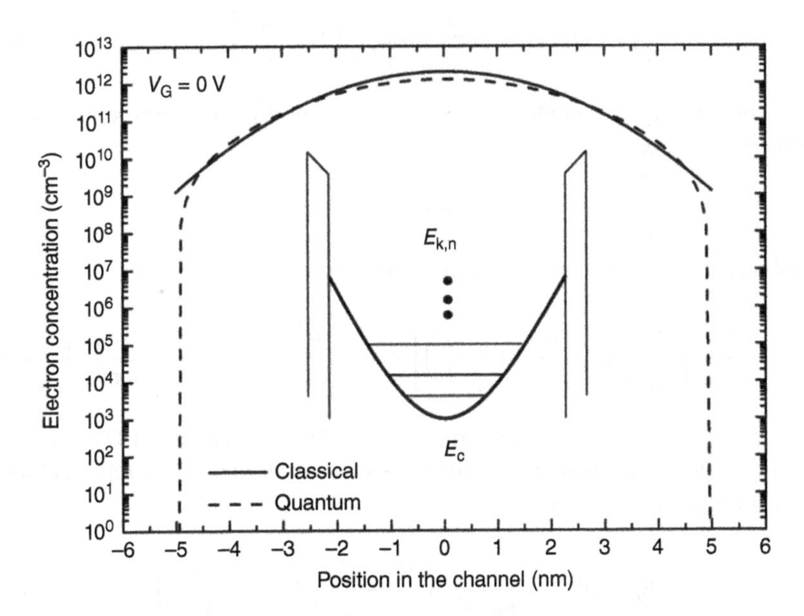

FIGURE 8.19 Comparison of the quantum and classical electron concentration along cutline B-B' (Fig. 8.1) in DGJLFET for a channel thickness of 10 nm [62].

An important point to note from equation (8.182) is that the location of the energy subband is independent of the thickness of the channel [62].

Now, the case with small-channel thickness resembles the standard quantum well surrounded by infinite potential barriers. The solution for this standard problem is

$$E_{k,n}(\mathrm{QW}) = \frac{h^2 n^2}{8 m_k^* t_{\mathrm{Si}}^2} \tag{8.183}$$

However, the potential well at the bottom consists of the bent conduction band and does not have a flat surface in DGJLFETs unlike DGMOSFETs [57–61]. This bottom potential energy is given by equation (8.181) and its contribution to the location of the energy subbands must be included for accuracy. The time-independent perturbation theory enables us to predict the changes in the energy Eigen values by treating the small perturbation energy as an additional Hamiltonian [61]. Therefore, the perturbation potential energy may be expressed as a Hamiltonian (δH) as

$$\delta H = \frac{q^2 N_{\mathrm{D}}}{2\varepsilon_{\mathrm{Si}}} x^2 \tag{8.184}$$

According to the first-order time-independent perturbation theory, the expectation value of the Hamiltonian for a sinusoidal wave function $w(x)$, where

$$w(x) = \sqrt{\frac{2}{t_{\mathrm{Si}}}} \sin\left[\frac{n\pi}{t_{\mathrm{Si}}}\left(x + \frac{t_{\mathrm{Si}}}{2}\right)\right] \tag{8.185}$$

yields an energy correction term for the energy Eigen values of the energy subbands as

$$\delta E = \langle w(x) \, |\delta H| \, w(x) \rangle \tag{8.186}$$

Solving for the expectation value of energy Eigen values, we get

$$\delta E = \frac{q^2 N_{\mathrm{D}}}{\varepsilon_{\mathrm{Si}} t_{\mathrm{Si}}} \int_{-t_{\mathrm{Si}}/2}^{t_{\mathrm{Si}}/2} \sin^2\left[\frac{n\pi}{t_{\mathrm{Si}}}\left(x + \frac{t_{\mathrm{Si}}}{2}\right)\right] x^2 dx = \frac{q^2 N_{\mathrm{D}} t_{\mathrm{Si}}^2}{24\varepsilon_{\mathrm{Si}}}\left(1 - 6\left(\frac{1}{n\pi}\right)^2\right) \tag{8.187}$$

Adding the energy correction term obtained from equation (8.187) to equation (8.183), we get an accurate expression for the energy Eigen values representing the different energy subbands as

$$E_{k,n}(\mathrm{QW}) = \frac{h^2 n^2}{8 m_k^* t_{\mathrm{Si}}^2} + \frac{q^2 N_{\mathrm{D}} t_{\mathrm{Si}}^2}{24\varepsilon_{\mathrm{Si}}}\left(1 - 6\left(\frac{1}{n\pi}\right)^2\right) \tag{8.188}$$

The location of the lowest energy subband has been compared for the DGJLFETs and the DGMOSFETs for different channel thicknesses in Fig. 8.20. The energy of

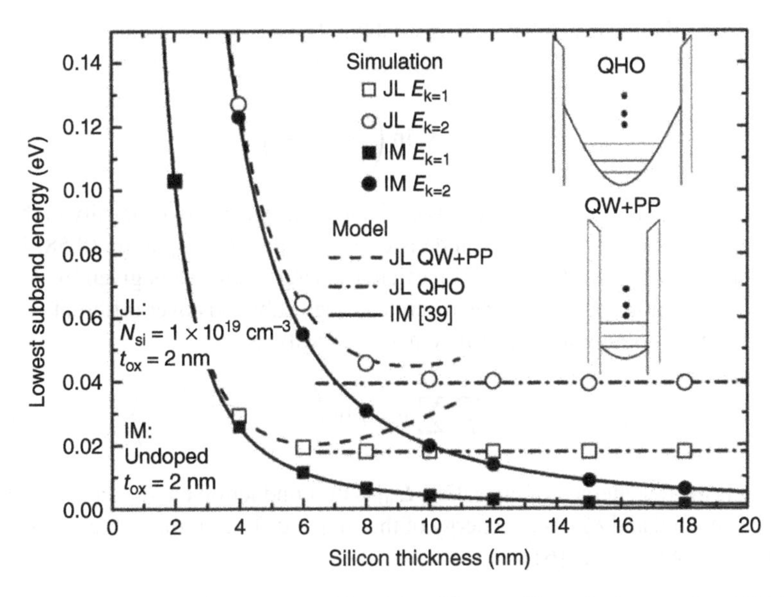

FIGURE 8.20 Energy of the lowest subband for different silicon thicknesses for DGJLFETs [62] and DGMOSFETs [58].

the first subband is higher for the DGJLFETs as compared to the DGMOSFETs for $t_{Si} > 4$ nm and becomes nearly equal as the channel thickness reduces below 4 nm. Also, it can be observed that the energy Eigen values given by the solution for quantum well (with the perturbation potential correction term included) approaches the value obtained from the solution of the quantum harmonic oscillator for a channel thickness of ~8 nm [62]. Moreover, at this channel thickness, the Eigen value obtained from the solution of the quantum well problem is minimum. Therefore, the channel thickness at which the transition from the quantum well problem to the quantum harmonic oscillator problem takes place can be obtained by finding the minima of equation (8.188). Differentiating equation (8.188) and equating it to zero, we get

$$t_{Si}\ (\text{transition}) = \left[\frac{\left(\frac{h^2 n^2}{m_k^*}\right)}{\frac{q^2 N_D}{3\varepsilon_{Si}}\left(1 - 6\left(\frac{1}{n\pi}\right)^2\right)}\right]^{\frac{1}{4}} \tag{8.189}$$

Since the potential distribution inside the channel in JLFETs is a strong function of both channel thickness and doping, the channel doping at which the transition takes place can also obtained at a particular channel thickness and quantum number (n) using equation (8.189).

Now, once the location of the energy subbands have been obtained, the next step is to find the shift in the threshold voltage (ΔV_{Th}), which can be expressed as

$$\Delta V_{\mathrm{Th}} = V_{\mathrm{t}} \ln\left(\frac{n_{\mathrm{Classical}}}{n_{\mathrm{quantum}}}\right) \tag{8.190}$$

where $n_{\mathrm{classical}}$ and n_{quantum} are the classical and quantum electron density in the conduction band. The classical electron density was obtained as equation (8.58) in Section 8.4.1. The electron density in a particular energy subband is given by equation (8.179). Now, summing over all the discrete energy subbands over different quantum numbers, we can obtain the quantum electron density as

$$n_{\mathrm{quantum}} = \frac{4kT\pi}{h^2}\sum_{k}\sum_{n} g_k m_k^* \ln\left(1 + e^{\frac{E_f - E_{\mathrm{c}0} - E_{k,n}}{kT}}\right) \tag{8.191}$$

where g_k is the degeneracy of the valley, E_{c0} is the conduction band energy at the center of the channel and $E_{k,n}$ is the energy of the subband. The minima of the conduction band may be obtained as [61]

$$E_{\mathrm{c}0} = \frac{E_g}{2} + \frac{V_{\mathrm{t}}}{2}\ln\left(\frac{N_{\mathrm{c}}}{N_{\mathrm{v}}}\right) - q\left(V_{\mathrm{G}} - V_{\mathrm{Th}} - \emptyset_{\mathrm{B}}\right) + E_{\mathrm{f}} \tag{8.192}$$

where E_{g} is the band gap, N_{C} and N_{V} are the density of states in the conduction and valence bands, respectively, E_{f} is the Fermi level, and $\emptyset_{\mathrm{B}} = -V_{\mathrm{t}}\ln\left(N_{\mathrm{D}}/n_{\mathrm{i}}\right)$, where n_{i} is the intrinsic carrier concentration.

Now, $\log(1 + ax)$ may be simplified as ax when $ax < 1$. In the subthreshold regime, $e^{(E_{\mathrm{f}} - E_{\mathrm{C}0} - E_{k,n})/kT} < 1$. Therefore, the logarithmic term can be simplified and equation (8.191) may be rewritten as

$$\begin{aligned} n_{\mathrm{quantum}} &= \frac{4kT\pi}{h^2}\sum_{k}\sum_{n} g_k m_k^* e^{\frac{E_f - E_{\mathrm{C}0} - E_{k,n}}{kT}} \\ &= \frac{4kT\pi N_{\mathrm{D}}}{N_{\mathrm{C}}h^2} e^{\frac{V_{\mathrm{G}} - V_{\mathrm{Th}}}{V_t}}\left[2m_{d,1}^*\sum_{n} e^{-\frac{E_{1,n}}{kT}} + 4m_{d,2}^*\sum_{n} e^{-\frac{E_{2,n}}{kT}}\right] \end{aligned} \tag{8.193}$$

where $m_{d,1}^*$ and $m_{d,2}^*$ are the effective masses of the electrons in the conduction band valleys with degeneracy 2 and 4, respectively.

The shift in the threshold voltage can be obtained by putting the value obtained from equations (8.193) and (8.58) in equation (8.190) as

$$\Delta V_{\mathrm{Th}} = V_{\mathrm{t}} \ln\left(\frac{\frac{\alpha}{2}\sqrt{\frac{\pi}{\beta}}\operatorname{erf}\left(\frac{\sqrt{\beta}t_{\mathrm{Si}}}{2}\right)}{\frac{4kT\pi N_{\mathrm{D}}}{N_{\mathrm{C}}h^2} e^{\frac{V_{\mathrm{G}} - V_{\mathrm{Th}}}{V_t}}\left[2m_{d,1}^*\sum_{n} e^{-\frac{E_{1,n}}{kT}} + 4m_{d,2}^*\sum_{n} e^{-\frac{E_{2,n}}{kT}}\right]}\right) \tag{8.194}$$

This threshold voltage shift may be used as a quantum correction term for compact modeling purpose. First, the classical simulations using the drift–diffusion approach may be used for obtaining the transfer characteristics. Then to obtain the quantum transfer characteristics, the classical transfer characteristics may be shifted by the value obtained by equation (8.194).

8.7 CONCLUSION

In this chapter, starting with the fundamentals of FET modeling, we discussed the different approaches used for analytical modeling of surface potential, semiconductor charge density, and the drain current of the long-channel JLFETs. We considered DG architecture as it yields an additional boundary condition due to its symmetric structure and eases the modeling approach. After the discussion on the long-channel JLFETs, the quasi-2D scaling technique to model short-channel JLFETs was also covered and a comparison with MOSFETs was done based on the developed model. In addition, the modeling of quantum confinement effects pertaining to the ultrathin channel JLFETs was also analyzed in depth. The approach discussed in the chapter may be extended to model other architectures such as ultrathin body SOI, TG, or gate-all-around nanowire JLFETs [63, 74]. Apart from the methods discussed in this chapter, several other techniques such as the conformal mapping technique [75], evanescent mode analysis [76], and so on are also used to model JLFETs yielding accurate results. Also, the source/drain depletion regions [77], dynamic boundary conditions [78], and so on should be included to improve the accuracy of the analytical models. This chapter serves as a comprehensive guide to the beginners in the field of analytical modeling of JLFETs, and the readers are directed to [21–29] for gaining further insights.

REFERENCES

[1] S. Sahay and M. J. Kumar, "Physical insights into the nature of gate-induced drain leakage in ultrashort channel nanowire FETs," *IEEE Trans. Electron Devices*, vol. 64, no. 6, pp. 2604–2610, June 2017.

[2] S. Sahay and M. J. Kumar, "A novel gate-stack-engineered nanowire FET for scaling to the sub-10-nm regime," *IEEE Trans. Electron Devices*, vol. 63, no. 12, pp. 5055–5059, Dec. 2016.

[3] S. Sahay and M. J. Kumar, "Spacer design guidelines for nanowire FETs from gate-induced drain leakage perspective," *IEEE Trans. Electron Devices*, vol. 64, no. 7, pp. 3007–3015, July 2017.

[4] S. Sahay and M.J. Kumar, "Insight into lateral band-to-band-tunneling in nanowire junctionless FETs," *IEEE Trans. Electron Devices*, vol. 63, no. 10, pp. 4138–4142, Oct. 2016.

[5] S. Sahay and M. J. Kumar, "Controlling L-BTBT and volume depletion in nanowire JLFETs using core-shell architecture," *IEEE Trans. Electron Devices*, vol. 63, no. 9, pp. 3790–3794, Sept. 2016.

[6] S. Sahay and M. J. Kumar, "Diameter dependency of leakage current in nanowire junctionless field-effect transistors," *IEEE Trans. Electron Devices*, vol. 64, no. 3, pp. 1330–1335, Mar. 2017.

[7] S. Sahay and M. J. Kumar, "Nanotube junctionless FET: Proposal, design, and investigation," *IEEE Trans. Electron Devices*, vol. 64, no. 4, pp. 1851–1856, Apr. 2017.

[8] M. J. Kumar and S. Sahay, "Controlling BTBT induced parasitic BJT action in junctionless FETs using a hybrid channel," *IEEE Trans. Electron Devices*, vol. 63, no. 8, pp. 3350–3353, Aug. 2016.

[9] S. Sahay and M. J. Kumar, "Realizing efficient volume depletion in SOI junctionless FETs," *IEEE J. Electron Devices Soc.*, vol. 4, no. 3, pp. 110–115, May 2016.

[10] S. Sahay and M. J. Kumar, "Controlling the drain side tunneling width to reduce ambipolar current in tunnel FETs using heterodielectric BOX," *IEEE Trans. Electron Devices*, Vol. 62, no. 11, pp. 3882–3886, Nov. 2015.

[11] S. Sahay and M. J. Kumar, "Comprehensive analysis of gate-induced drain leakage in emerging FET architectures: Nanotube FETs vs. nanowire FETs," *IEEE Access*, vol. 5, pp. 18918–18926, Dec. 2017.

[12] S. Sahay and M. J. Kumar, "Symmetric operation in an extended back gate JLFET for scaling to the 5 nm regime considering quantum confinement effects," *IEEE Trans. Electron Devices*, vol. 64, no. 1, pp. 21–27, Jan. 2017.

[13] A. K. Jain, S. Sahay, and M. J. Kumar, "Controlling L-BTBT in emerging nanotube FETs using dual-material gate," *IEEE J. Electron Dev. Soc.*, vol. 6, pp. 611–621, June 2018.

[14] S. Datta, *Electronic Transport in Mesoscopic Systems*, Cambridge University Press, New York, 1997.

[15] S. Datta, *Quantum Transport: Atom to Transistor*, Cambridge University Press, New York, 2005.

[16] S. Datta, *Lessons from Nanoelectronics: A New Perspective on Transport* (vol. 1), World Scientific, Singapore, 2012.

[17] M. Lundstrom and J. Guo, *Nanoscale Transistors: Device Physics, Modeling and Simulation*, Springer Science & Business Media, Boston, MA, 2006.

[18] M. Lundstrom, *Fundamentals of Carrier Transport*, Cambridge University Press, New York, 2009.

[19] M. Lundstrom and C. Jeong, *Near-Equilibrium Transport: Fundamentals and Applications* (Vol. 2), World Scientific, Singapore, 2012.

[20] S. Khandelwal, Y. S. Chauhan, D. D. Lu, S. Venugopalan, M. A. Karim, A. B. Sachid, B. Y. Nguyen, O. Rozeau, O. Faynot, A. M. Niknejad, and C. Hu, "BSIM-IMG: A compact model for ultrathin-body SOI MOSFETs with back-gate control," *IEEE Trans. Electron Devices*, vol. 59, no. 8, pp. 2019–2026, Aug. 2012.

[21] Y. S. Chauhan, D. D. Lu, V. Sriramkumar, S. Khandelwal, J. P. Duarte, N. Payvadosi, A. Niknejad, and C. Hu, *FinFET Modeling for IC Simulation and Design: Using the BSIM-CMG Standard*, Academic Press, New York, 2015.

[22] R. van Langevelde, A. J. Scholten, and D. B. M. Klaassen, "MOS Model 11, Level 1101," Unclassified Report NL-UR 2002/802, Philips Electronics N.V., 2002.

[23] Compact transistor models. Philips Semiconductors. [online], Available: https://www.nxp.com/support/developer-resources/models-and-test-data/simkit/compact-models:COMPACT-MODELS, Accessed: Aug. 10, 2018.

[24] R. van Langevelde and F. M. Klaassen, "An explicit surface-potential based MOSFET model for circuit simulation," *Solid State Electron.*, vol. 44, pp. 409–418, 2000.

[25] F. Jazaeri and J. -M. Sallese, *Modeling Nanowire and Double-Gate Junctionless Field-Effect Transistors*, Cambridge University Press, New York, 2018.

[26] S. K. Saha, *Compact Models for Integrated Circuit Design: Conventional Transistors and Beyond*, CRC Press, Boca Raton, FL, 2016.

[27] A. B. Bhattacharyya, *Compact MOSFET Models for VLSI Design*, John Wiley & Sons, Inc., Hoboken, NJ, 2009.

[28] M. J. Kumar, H. Batwani, and M. Gaur, "Approaches to nanoscale MOSFET compact modeling using surface potential based models," in *IWSPD*, Dec. 2007.

[29] M. J. Kumar and A. Chaudhry, "Two-dimensional analytical modeling of fully depleted dual-material gate (DMG) SOI MOSFET and evidence for diminished short-channel effects," *IEEE Trans. Electron Devices*, vol. 51, pp. 569–574, Apr. 2004.

[30] D. B. Abdi and M. J. Kumar, "Two-dimensional threshold voltage model for the double gate PNPN TFET with localized charges," *IEEE Trans. Electron Devices*, vol. 63, no. 9, pp. 3663–3668, Sept. 2016.

[31] R. Vishnoi and M. J. Kumar, "Two dimensional analytical model for the threshold voltage of a tunneling FET with localized charges," *IEEE Trans. Electron Devices*, vol. 61, pp. 3054–3059, Sept. 2014.

[32] Y. Pratap, S. Haldar, R. S. Gupta, and M. Gupta, "Localized charge-dependent threshold voltage analysis of gate-material-engineered junctionless nanowire transistor," *IEEE Trans. Electron Devices*, vol. 62, no. 8, pp. 2598–2605, 2015.

[33] Z. Chen, Y. Xiao, M. Tang, Y. Xiong, J. Huang, J. Li, X. Gu, and Y. Zhou, "Surface-potential based drain current model for long-channel junctionless double-gate MOSFETs," *IEEE Trans. Electron Devices*, vol. 59, pp. 3292–3298, Dec. 2012.

[34] R. M. Corless, G. H. Gonnet, D. E. G. Hare, D. J. Jeffrey, and D. E. Knuth, "On the Lambert W function," *Adv. Comput. Math.*, vol. 5, pp. 329–359, 1996.

[35] T. C. Banwell, "Bipolar transistor circuit analysis using the Lambert W-function," *IEEE Trans. Circuits Syst. I, Fundam. Theory Appl.*, vol. 47, no. 11, pp. 1621–1633, Nov. 2000.

[36] K. K. Young, "Short-channel effect in fully depleted SOI-MOSFETs," *IEEE Trans. Electron Devices*, vol. 40, no. 10, pp. 1812–1817, Oct. 1993.

[37] M. J. Kumar and G. V. Reddy, "Diminished short channel effects in nanoscale double-gate silicon-on-insulator metal oxide field-effect transistors due to induced back-gate step potential," *Jap. J. Appl. Phys.*, vol. 44, no. 9A, pp. 6508–6509, Sept. 2005.

[38] A. Chaudhry and M. J. Kumar, "Controlling short-channel effects in deep submicron SOI MOSFETs for improved reliability: A review," *IEEE Trans. Dev. Mater. Rel.*, vol. 4, pp. 99–109, Mar. 2004.

[39] J. Wan, C. L. Royer, A. Zaslavsky, and S. Cristoloveanu, "A tunneling field-effect transistor model combining interband tunneling with channel transport," *J. Appl. Phys.*, vol. 110, no. 10, 104503, 2011.

[40] A. Pan and C. O. Chui, "A quasi-analytical model for double-gate tunneling field-effect transistors", *IEEE Electron Device Lett.*, vol. 33, pp. 1468–1470, Oct. 2012.

[41] L. Zhang, X. Lin, J. He, and M. Chan, "An analytical charge model for double-gate tunnel FETs," *IEEE Trans. Electron Devices*, vol. 59, no. 12, pp. 3217–3223, Dec. 2012.

[42] J. P. Duarte, S. -J. Choi, and Y. -K. Choi, "A full-range drain current model for double-gate junctionless transistors," *IEEE Trans. Electron Devices*, vol. 58, pp. 4219–4225, Dec. 2011.

[43] Y. Taur, "An analytical solution to a double-gate MOSFET with undoped body," *IEEE Electron Device Lett.*, vol. 21, no. 5, pp. 245–247.

[44] Y. Taur, X. Liang, W. Wang, and H. Lu, "A continuous, analytic drain-current model for DG MOSFETs," *IEEE Electron Device Lett.*, vol. 25, pp. 107–109, Feb. 2004.

[45] H. Lu and Y. Taur, "An analytic potential model for symmetric and asymmetric DG MOSFETs," *IEEE Trans. Electron Devices*, vol. 53, no. 5, pp. 1161–1168, 2006.

[46] X. Jin, X. Liu, J. -H. Lee, and J. -H. Lee, "A continuous current model of fully-depleted symmetric double-gate MOSFETs considering a wide range of body doping concentrations," *Semiconduc. Sci. Technol.*, vol. 25, 055018, May 2010.

[47] X. Jin, X. Liu, M. Wu, R. Chuai, J.-H. Lee, and J.-H. Lee, "A unified analytical continuous current model applicable to accumulation mode (junctionless) and inversion mode {MOSFETs} with symmetric and asymmetric double-gate structures," *Solid-State Electron.*, vol. 79, pp. 206–209, 2013.

[48] J. -M. Sallese, N. Chevillon, C. Lallement, B. Iñiguez, and F. Prégaldiny, "Charge-based modeling of junctionless double-gate field-effect transistors," *IEEE Trans. Electron Devices*, vol. 58, pp. 2628–2637, Aug. 2011.

[49] J. Singh, V. Gadi, and M. J. Kumar, "Modeling a dual-material-gate junctionless FET under full and partial depletion conditions using finite-differentiation method," *IEEE Trans. Electron Devices*, vol. 63, no. 6, pp. 2282–2287, June 2016.

[50] J. P. Duarte, S. -J. Choi, D. -I. Moon, and Y. -K. Choi, "Simple analytical bulk current model for long-channel double-gate junctionless transistors," *IEEE Trans. Electron Devices*, vol. 32, pp. 704–706, June 2011.

[51] Z. Zhengfan, L. Zhaoji, T. Kaizhou, and Z. Jiabin, "Investigation into sub-threshold performance of double-gate accumulation-mode SOI PMOSFET," in *Proc. ASICON*, pp. 1150–1153, Oct. 2007.

[52] N. Arora, *MOSFET Modeling for VLSI Simulation: Theory and Practice*, World Scientific, Singapore, 2007.

[53] H. C. Pao and C. T. Sah, "Effects of diffusion current on characteristics of metal-oxide (insulator)-semiconductor transistors," *Solid-State Electron.*, vol. 9, no. 10, pp. 927–937, 1966.

[54] E. Gnani, A. Gnudi, S. Reggiani, and G. Baccaranicd, "Theory of the junctionless nanowire FET," *IEEE Trans. Electron Devices*, vol. 58, no. 9, pp. 2903–2910, Sept. 2011.

[55] T. K. Chiang, "A scaling theory for fully-depleted, SOI double-gate MOSFETs: Including effective conducting path effect," *Solid State Electron.*, vol. 49, no. 3, pp. 317–322, Mar. 2005.

[56] T.-K. Chiang, "A quasi-two-dimensional threshold voltage model for short-channel junctionless double-gate MOSFETS," *IEEE Trans. Electron Devices*, vol. 59, pp. 2284–2289, Sept. 2012.

[57] Y. Taur and T. H. Ning, *Modern VLSI Devices*, 2nd ed., Cambridge University Press, Cambridge, UK, 2009.

[58] R. Granzner, F. Schwierz, and V. Polyakov, "An analytical model for the threshold voltage shift caused by two-dimensional quantum confinement in undoped

multiple-gate MOSFETs," *IEEE Trans. Electron Devices*, vol. 54, no. 9, pp. 2562–2565, Sept. 2007.

[59] W. Wang, H. Lu, J. Song, S.-H. Lo, and Y. Taur, "Compact modeling of quantum effects in symmetric double-gate MOSFETs," *Microelectron. J.*, vol. 41, no. 10, pp. 688–692, Oct. 2010.

[60] V. P. Trivedi and J. G. Fossum, "Quantum–mechanical effects on the threshold voltage of undoped double-gate MOSFETs," *IEEE Electron Device Lett.*, vol. 26, no. 8, pp. 579–582, Aug. 2005.

[61] G. Baccarani and S. Reggiani, "A compact double-gate MOSFET model comprising quantum–mechanical and nonstatic effects," *IEEE Trans. Electron Devices*, vol. 46, no. 8, pp. 1656–1666, Aug. 1999.

[62] J. Duarte, M. -S. Kim, S. -J. Choi, and Y. -K. Choi, "A compact model of quantum electron density at the subthreshold region for double-gate junctionless transistors," *IEEE Trans. Electron Devices*, vol. 59, pp. 1008–1012, April 2012.

[63] E. Gnani, A. Gnudi, S. Reggiani, and G. Baccarani, "Physical model of the junctionless UTB SOI-FET," *IEEE Trans. Electron Devices*, vol. 59, pp. 941–948, April 2012.

[64] R. D. Trevisoli, R. T. Doria, M. de Souza, and M. A. Pavanello, "A physically-based threshold voltage definition, extraction and analytical model for junctionless nanowire transistors," *Solid-State Electron.*, vol. 90, pp. 12–17, 2013

[65] F. Lime, O. Moldovan, and B. Iniguez, "A compact explicit model for long-channel gate-all-around junctionless MOSFETs. Part I: DC characteristics," *IEEE Trans. Electron Devices*, vol. 61, pp. 3036–3041, Sept. 2014.

[66] O. Moldovan, F. Lime, and B. Iniguez, "A compact explicit model for long-channel gate-all-around junctionless MOSFETs. Part II: Total charges and intrinsic capacitance characteristics," *IEEE Trans. Electron Devices*, vol. 61, pp. 3042–3046, Sept. 2014.

[67] C. Li, Y. Zhuang, S. Di, and R. Han, "Subthreshold behavior models for nanoscale short channel junctionless cylindrical surrounding-gate MOSFETs," *IEEE Trans. Electron Devices*, vol. 60, pp. 3655–3662, Nov. 2013.

[68] T. Holtij, A. Kloes, and B. Iñíguez, "3-D compact model for nanoscale junctionless triple-gate nanowire MOSFETs, including simple treatment of quantization effects," *Solid-State Electron.*, vol. 112, pp. 85–98, Oct. 2015.

[69] Z. Guo, J. Zhang, Z. Ye, and Y. Wang, "3-D analytical model for short-channel triple-gate junctionless MOSFETs," *IEEE Trans. Electron Devices*, vol. 63, no. 10, pp. 3857–3863, Oct. 2016.

[70] R. D. Trevisoli, R. T. Doria, M. de Souza, S. Das, I. Ferain, and M. A. Pavanello, "Surface-potential-based drain current analytical model for triple-gate junctionless nanowire transistors," *IEEE Trans. Electron Devices*, vol. 59, no. 12, pp. 3510–3518, Dec. 2012.

[71] T. K. Chiang, "A new subthreshold current model for junctionless trigate MOSFETs to examine interface-trapped charge effects," *IEEE Trans. Electron Devices*, vol. 62, no. 9, pp. 2745–2750, Sept. 2015.

[72] R. Trevisoli, R. T. Doria, M. de Souza, S. Barraud, M. Vinet, and M. A. Pavanello, "Analytical model for the dynamic behavior of triple-gate junctionless nanowire transistors," *IEEE Trans. Electron Devices*, vol. 63, no. 2, pp. 856–863, Feb. 2016.

[73] B. Singh, D. Gola, K. Singh, E. Goel, S. Kumar, and S. Jit, "Analytical modeling of channel potential and threshold voltage of double-gate junctionless FETs with a vertical

Gaussian-like doping profile," *IEEE Trans. Electron Devices*, vol. 63, no. 6, pp. 2299–2305, June 2016.

[74] J. -M. Sallese, F. Jazaeri, L. Barbut, N. Chevillon, and C. Lallement, "A common core model for junctionless nanowires and symmetric double gate FETs," *IEEE Trans. Electron Devices*, vol. 60, pp. 4277–4280, Dec. 2013.

[75] T. Holtij, M. Graef, F. Hain, A. Kloes, and B. Iniguez, "Compact model for short-channel junctionless accumulation mode double gatemosfets," *IEEE Trans. Electron Devices*, vol. 61, pp. 288–299, Feb. 2014.

[76] X. Jin, X. Liu, H. -I. Kwon, J. -H. Lee, and J. -H. Lee, "A subthreshold current model for nanoscale short channel junctionless MOSFETs applicable to symmetric and asymmetric double-gate structure," *Solid-State Electron.*, vol. 82, pp. 77–81, Apr. 2013.

[77] A. Gnudi, S. Reggiani, E. Gnani, and G. Baccarani, "Semianalytical model of the subthreshold current in short-channel junctionless symmetric double-gate field-effect transistors," *IEEE Trans. Electron Devices*, vol. 60, pp. 1342–1348, Apr. 2013.

[78] Y. Xiao, X. Lin, H. Lou, B. Zhang, L. Zhang, and M. Chan, "A short channel double-gate junctionless transistor model including the dynamic channel boundary effect," *IEEE Trans. Electron Devices*, vol. 63, no. 12, pp. 4661–4667, Dec. 2016.

9

SIMULATION OF JLFETS USING SENTAURUS TCAD

In Chapter 3, we discussed the fundamentals of the junctionless field-effect transistors (JLFETs) and the different parameters such as mobility, carrier ballisticity, temperature dependence, strain, and so on that affect the performance of the JLFETs. While the experimental measurements are necessary to analyze the macroscopic behavior of the field-effect transistors (FETs) such as the transfer/output characteristics or capacitance–voltage (C–V) relationships, they do not offer any insight into the microscopic device physics of FETs. For instance, the experimental transfer characteristics may relate the drain current and gate voltage of the JLFETs, but no physical insight into the independence of the drain current for negative gate voltage can be obtained using the experimental data [1–4]. Similarly, no physical explanation for the reduction in the drain current of nanowire junctionless field-effect transistor (NWJLFETs) with decreasing nanowire diameter can be inferred using the experimental results of [1]. For obtaining a deeper physical insight into the FET behavior, it becomes necessary to analyze the electric field profile, the potential distribution profile, the electron and hole concentration profile at different gate voltages, etc. However, the experimental measurements fail to give any of these relevant microscopic data for explaining the device physics.

Technology computer aided design (TCAD) tools facilitate the analysis of the microscopic properties of the FETs. Unlike the experimental setups, TCAD tools are not limited to the measurement of terminal voltages/currents/capacitances and

Junctionless Field-Effect Transistors: Design, Modeling, and Simulation, First Edition.
Shubham Sahay and Mamidala Jagadesh Kumar.

enable us to analyze the electric field profile, potential profile, and carrier distribution profile at any point within the device. Therefore, they are indispensable tools for device physicists, experimentalists, and even the device modelers to understand and explain the underlying physics behind the operation of the FETs. TCAD simulations can also be used for predicting the performance of new architectures such as those with new geometry or inherent device physics. Since the TCAD simulations are fast and inexpensive compared to the experimental setup, the performance of any new FET architecture should be first analyzed using simulations. Once the simulation results appear promising, then only the device should be fabricated and tested experimentally. However, it may be noted that the accuracy of the TCAD tools while analyzing the behavior of any device depends on the set of physical models used for the simulations. Therefore, all the physical models that account for the device physics of a particular FET must be plugged in the simulations to obtain accurate results. Also, most physical models available in the simulator consist of parameters calibrated to some experimental data. However, using the default parameters does not ensure that the simulation results are precise. Therefore, the practice of tuning the model parameters of different physical models to reproduce an experimental data utilizing the same experimental FET structure in the simulations, known as calibration, is necessary to validate the simulation setup. Calibration may be considered a prerequisite before predicting the performance of a new architecture using TCAD simulations.

Many commercial TCAD simulators are available in the market. Sentaurus TCAD [5] and Silvaco TCAD [6] are the most widely used simulators. Apart from these, Cogenda visual TCAD [7] is also gaining popularity. Noncommercial TCAD tools also exist, which are proprietary of the university or the organization where it is developed. For instance, IBM has its own simulator FIELDAY [8]. Many universities such as Purdue have also developed online TCAD simulators, which are freely available on nanohub.org [9]. However, in this chapter, we deal with only Sentaurus TCAD. After introducing the basics of how the Sentaurus TCAD works, we would discuss the simulation of JLFETs in Sentaurus.

9.1 INTRODUCTION TO TCAD

TCAD mimics the electrical behavior of the FETs by numerical calculation of the physical equations governing the carrier profile and the conduction mechanisms. Any real FET is approximated as a virtual device whose physical property such as doping is discretized into a number of mesh points (also called nodes or grids). Therefore, the actual continuous device is represented as a finite number of sparse mesh points. The physical equations such as the Poisson equation and the electron and hole current continuity equations are self-consistently solved at these discrete mesh points in an iterative manner. In each iteration, an error is calculated and the tool tries to converge upon a solution which produces an error which is lower than an acceptable value (which may be predefined). Therefore, the TCAD essentially emulates the electrical behavior of any FET by numerical solution of some differential equations. You may wonder whether it is possible to compute the terminal charge, voltage, or current relation of any electrical device via solution of few mathematical equations. In fact, even the analytical modeling approach discussed in Chapter 8 utilizes the solution of mathematical equations such as the Poisson equation to extract the FET behavior. However, it is different from a TCAD in sense that the TCAD uses mathematical solvers to find the exact solution of the Poisson equation and the current continuity equations rather than relying on assumptions or approximations as in the case of analytical or compact modeling. Therefore, a TCAD-based modeling approach is accurate than the analytical/compact modeling although it is computationally not as efficient and fast.

In the next section, we discuss the input files required by the Sentaurus TCAD for simulating the FET behavior and the output files that it generates after simulations from which we may determine the FET properties.

9.2 TOOL FLOW

The typical flow of the Sentaurus TCAD is shown in Fig. 9.1. The device structure is created in either Sentaurus structure editor (SDE) by defining materials with appropriate geometry and boundary conditions or created from the process flow to emulate the experimental conditions using the Sentaurus process (Sprocess). The structure created using SDE or Sprocess is then discretized into mesh points utilizing either SDE itself or Sentaurus mesh (Smesh). The SDE requires a command file as input which specifies the different regions in the device, the material used for that region, the doping, the distance between the consecutive mesh points in a particular region, as well as the definition of electrical contacts and their location. Based on this information, SDE generates a TDR file.

The electrical characteristics of the device are simulated using the Sentaurus device (Sdevice). In addition to the TDR file, which contains information about the device structure, Sdevice requires a command file that contains all the relevant

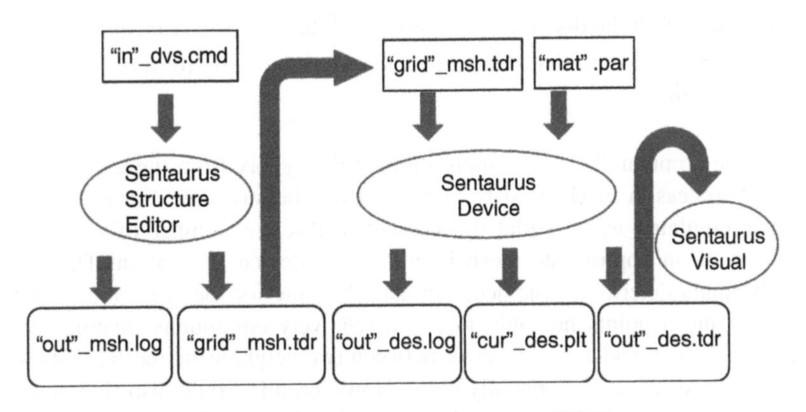

FIGURE 9.1 Tool flow of Sentaurus TCAD.

information about the physical models to be used for device simulations, the initial boundary condition of the contacts, the mathematical solver to be used for numerical simulations, the voltage or current values to which the contact electrodes need to be ramped, the microscopic properties that need to be analyzed such as electric field, potential profile, and so on. Apart from the command file, Sdevice also requires a parameter file as input. The parameter file contains the default values of all the parameters used in the physical models governing the FET behavior. These parameters may be tuned to match the experimental results which is known as model calibration. Sdevice generates a final TDR file, which contains the spatial solution of the different equations, and a PLT file, which contains the relationship between the terminal voltages, currents, and charges.

Sentaurus visual (Svisual) utilizes the final TDR file as input and can be used to visualize the different microscopic properties such as the electric field, carrier concentration, potential profile, and so on, which were specified in the command file used for Sdevice. These properties are obtained from the spatial solution of the different differential equations in Sdevice. Moreover, Svisual can also be used to perform mathematical operations on the obtained microscopic properties such as integration, multiplication, differentiation, and so on. The information about the terminal charges, voltages, or currents at different contacts is embedded in the PLT file. Svisual can also be used for analyzing data from the PLT file.

After a brief discussion on the flow of the Sentaurus TCAD, let us discuss the major steps in detail.

9.2.1 Structure Specification

The structure file specification can be done in a command file input to SDE. The command file must contain the following information:

a) region wise geometry and material specifications,
b) electrical contact definition and contact locations,
c) location of all the discrete mesh points and their connectivity, and
d) description of data fields such as doping profile in the form of value at each mesh point.

After the command file is written utilizing the syntax from the user manual of SDE or Sprocess, a mesh must be generated for the structure. Since the original device is emulated as a virtual device with a discrete number of mesh points, generation of an appropriate mesh is crucial for device simulation. The number of mesh points determines the accuracy and the time for the simulations. A large number of mesh points not only means a relatively continuous treatment of the discretized virtual device but the error between two neighboring mesh points would also be reduced if they are spatially close. However, this would also slow down the simulation time as the physical equations need to be solved for each mesh point.

Therefore, to achieve an acceptable simulation time without compromising with the accuracy, a dense mesh should be constructed only in the regions where high current density (metal-oxide-semiconductor field-effect transistors [MOSFET] channel region, bipolar junction transistor [BJT] base region), high electric field (depletion regions, tunneling interfaces), or high-charge generation is expected. A general practice is to construct a dense mesh at any interface (semiconductor/insulator or semiconductor/semiconductor) where a large variation in the potential or electric field or current is expected and to generate a coarse mesh in the bulk or regions where limited variation in the physical parameters is expected. Moreover, an extremely dense mesh is required for accurate analysis of complex quantum mechanical phenomenon such as band-to-band tunneling (BTBT) or energy band quantization due to quantum confinement effects in TCAD. Therefore, designing an appropriate mesh is the key to an accurate TCAD-based analysis.

SDE allows us to specify maximum and minimum distance between two mesh points in different directions. Close to the interface, the mesh points are kept with the specified minimum distance while in the bulk (far away from the interface), the mesh points are positioned with the maximum specified distance. The delaunizer in the meshing engine of the tool then connects all the mesh points in a triangular fashion, i.e. by constructing acute-angled triangles. Only if the mesh points are located such that it cannot be covered by constructing acute-angled triangles, the mesh engine connects it using the obtuse-angled triangles. It may be noted that is difficult to obtain convergence utilizing obtuse-angled triangles. Therefore, the meshing must be done appropriately.

9.2.2 Model Parameter Specification

Most of the parameters used in the physical models are material specific and fitted to the experimental data. The model parameter file for any material with default parameters can be obtained by using the command:

```
sdevice -P: <material>
```

Although the different model parameters are already fitted to some experimental data, utilizing the default parameter values does not ensure a match between the experimental characteristics and the simulation data for every new device. For this purpose, considering device physics, some model parameters may be tuned. This is an essential step called model calibration and is necessary while utilizing TCAD for predicting the performance of new devices. However, an altogether absurd value of the model parameters should not be chosen to fit the experimental data.

9.2.3 Device Simulation

The device simulations are carried out in Sdevice. A command file needs to be written for running Sdevice. The syntax can be obtained from the Sdevice manual. The command file is not case or order sensitive (apart from mixed-mode simulations), and most keywords or syntax used for simulations can be abbreviated. The structure TDR file and the parameter file are specified as input for performing numerical calculations of the specified physical equations. The output files such as the final TDR file and the plot files are also specified. Then the name and the initial boundary conditions of all the electrodes are defined. Care must be taken while specifying the name of the contact electrodes in the Sdevice command file. They must match with the names of the contact electrode defined while specifying the structure in SDE.

Once the information about the electrodes is given, the next step is to include all the physical models relevant for simulating the particular device. This is the most crucial part. All the aspects of the device physics must be considered while choosing the models. It is recommended that the readers browse through the entire set of physical models available for including mobility degradation, band structure, quantization, carrier transport, generation and recombination (including tunneling), impact ionization, and so on in the manual for taking an informed decision while choosing the models for any device simulation.

For instance, if there are regions in the device with doping greater than the density of states, band gap narrowing phenomenon is indispensable. Therefore, the models for band gap narrowing must be activated. But then there are several models available for band gap narrowing such as Del Alamo, Slotboom, and so on. This is where you need to have a clear understanding of the models and their validity. Similarly, if there are impurity atoms present in the channel, impurity scattering, carrier–carrier scattering, or shielding effect discussed in Chapter 3 are also inevitable. To include the mobility degradation due to these effects, we need to search for appropriate mobility models. If we browse through the manual, we find that the Philips unified mobility model is the most complete model, which accounts for all these phenomena at once. Therefore, we should think logically and then select the most complete set of models from those available in the manual for simulating the device at hand. This will become clearer once we discuss the sample codes for JLFETs.

The next step is to include the microscopic parameters that we want to visualize in the final output. For instance, electric field profiles, potential profiles, carrier density, current density, band-to-band generation rate, Shockley–Read–Hall (SRH) rate, and so on are some of the key parameters that one needs to analyze for clear understanding of device behavior. Unlike experiments, all these parameters may be analyzed at any point within the device. This is the major advantage of TCAD simulations.

We want to observe the terminal voltage/charge/current relationships apart from the microscopic parameters at different operating conditions. For this, the voltage or current needs to be ramped in a quasi-stationary way. The simulator translates this electrical domain behavior into a mathematical domain solution of the physical equations governing the device behavior such as the Poisson equation and continuity equations. The numerical solver specifications are plugged in the maths section of the command file. In general, the fully coupled or Newton method is used to self-consistently solve the Poisson equation along with the continuity equations. The Newton method requires an initial guess before each iteration. The information about the methods to find the initial solution, number of iterations per bias step, the relative error control during each iteration, and so on are specified in the maths section. The information about the different physical equations that need to be solved while ramping the voltages or current values at the different electrodes to appropriate values is then given in the solve section of the command file for Sdevice.

Having addressed the entire flow of the Sentaurus TCAD, let us take a sample input deck for the long-channel JLFETs and explain each step for enhanced clarity to the readers.

9.3 SAMPLE INPUT DECK FOR LONG-CHANNEL JLFETS

The detailed explanation of the command files used for generating the structure file is given in Section 9.3.1. Every step has been explained in a comprehensive manner. The command file for simulating long-channel JLFETs generated in Section 9.3.1 is given in Section 9.3.2.

9.3.1 Structure Specification

The first step while carrying out simulations using any TCAD tool is to specify the structure. A general practice is to first draw a rough sketch of the structure specifying all the dimensions clearly. The Si–SiO_2 interface is normally taken as the *x*-axis, and the origin is usually defined at the beginning of the source region. Then the coordinates of all the regions such as the semiconductor, the gate oxide, the contacts, and so on are specified in the sketch. We have shown the sketch of the long-channel DGJLFET used for simulations in Fig. 9.2. Once the coordinates and regions are clearly specified in the rough sketch, it may be translated into the simulator compatible form by writing the command file which is shown below using the syntax provided in the user manual of SDE.

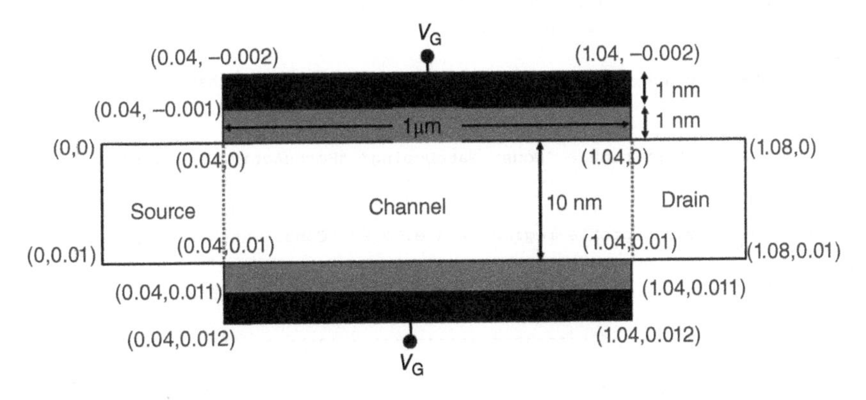

FIGURE 9.2 Rough sketch of a long-channel DGJLFET showing all the coordinates.

```
;-----Section 1--------------------------------------------------------------

; Overlap resolution: New replaces Old

(sdegeo:set-default-boolean "ABA")

;-----Section 2--------------------------------------------------------------

; Creating silicon region

(sdegeo:create-rectangle (position 0.0 0.0 0) (position 1.08 0.01 0) "Silicon"
"region_1" )

; Creating Bottom gate oxide layer

(sdegeo:create-rectangle (position 0.04 0.01 0.0 ) (position   1.04    0.011 0.0
) "SiO2" "region_2")

; Creating Top gate oxide layer

(sdegeo:create-rectangle (position 0.04 -0.001 0.0 ) (position  1.04 0.0 0.0 )
"SiO2" "region_3")

; Creating Top GatePolySilicon layer

(sdegeo:create-rectangle (position 0.04 -0.002 0.0 ) (position 1.04 -0.001 0.0 )
"GatePolySilicon" "region_4")

; Creating Bottom GatePolySilicon layer

(sdegeo:create-rectangle (position 0.04 0.011 0.0 ) (position 1.04 0.012 0.0 )
"GatePolySilicon" "region_5")

;--------Section 3-----------------------------------------------------------

; Doping specification

(sdedr:define-constant-profile "Const.ChannelDoping"
"PhosphorusActiveConcentration" 1e+19 )

(sdedr:define-constant-profile-region "Place.Channel" "Const.ChannelDoping"
"region_1")

(sdedr:define-constant-profile "Const.GateDoping" "BoronActiveConcentration"
1e+20)

(sdedr:define-constant-profile-region "Place.Gate" "Const.GateDoping" "region_4")

(sdedr:define-constant-profile-region "Place.Gate1" "Const.GateDoping" "region_5")

;-------Section 4------------------------------------------------------------

; Defining the mesh spaces
```

```
(sdedr:define-refinement-size "Ch_refinement" 0.0008 0.0008 0.0001 0.0001 )

(sdedr:define-refinement-region "Place.Ch_refine" "Ch_refinement" "region_1" )

(sdedr:define-refinement-size "oxide_refinement" 0.005 0.005 0.001 0.001 )

(sdedr:define-refinement-region "Place.Sioxide" "oxide_refinement" "region_2" )

(sdedr:define-refinement-region "Place.Sioxide_1" "oxide_refinement" "region_3" )

(sdedr:define-refinement-size "Gate_refinement" 0.005 0.005 0.001 0.001 )

(sdedr:define-refinement-region "Place.Gate_refine" "Gate_refinement" "region_4" )

(sdedr:define-refinement-region "Place.Gate_refine1" "Gate_refinement" "region_5")

;------Section 5------------------------------------------------------------

; inserting vertex for defining contact boundaries

(sdegeo:insert-vertex (position 0.025 0 0))

(sdegeo:insert-vertex (position 1.065 0 0))

(sdegeo:insert-vertex (position 0.025 0.01 0))

(sdegeo:insert-vertex (position 1.065 0.01 0))

;-------Section 6-----------------------------------------------------------

; Contact declarations

(sdegeo:define-contact-set "source" 4.0  (color:rgb 1.0 0.0 0.0 ) "##")

(sdegeo:define-contact-set "drain"  4.0  (color:rgb 0.0 1.0 0.0 ) "##")

(sdegeo:define-contact-set "gate" 4.0 (color:rgb 0.0 0.0 1.0 ) "##")

;-------Section 7-----------------------------------------------------------

; Contact positioning

(sdegeo:define-2d-contact (find-edge-id (position 0.01 0.0 0.0)) "source" )

(sdegeo:define-2d-contact  (find-edge-id (position 0.01 0.01 0.0)) "source" )

(sdegeo:define-2d-contact  (find-edge-id (position 1.075 0.0 0.0)) "drain")

(sdegeo:define-2d-contact (find-edge-id (position 1.075 0.01 0.0)) "drain")

(sdegeo:define-2d-contact (find-edge-id (position 0.055 -0.002 0.0)) "gate")

(sdegeo:define-2d-contact (find-edge-id (position 0.055 0.012 0.0)) "gate")
```

```
;---------Section 8-----------------------------------------------------------

; Build mesh

(sde:build-mesh "snmesh" "-offset" "sdemodel_msh")

;-------------------------------------------------------------------------
```

We have divided the command file into various sections to enable the readers to comprehend it easily. The lines beginning with ";" are considered as comments by SDE. No action is performed by SDE even on a valid code if it is written after ";". This notation is generally used for documentation of the code.

The JLFETs consists of multiple regions such as the active silicon layer, gate dielectric, gate material, spacer, and so on. Therefore, there may be an overlap or intersection between the existing regions and the regions defined subsequently. Section 1 specifies the scheme used to resolve such overlaps, i.e. the action that SDE will perform in case of an overlap between two defined regions. There are several schemes available which are represented as codes such as "ABA" (new replaces old), "BAB" (old replaces new), "AB" (merge), and so on. For instance, we have used "ABA," which simply conveys that if region 2 is defined after region 1 in the command file and there exists an overlap between region 1 and region 2, the overlapping region will be considered as a part of region 2 and will follow its properties such as material, mesh spacing, doping, etc. The generalized statement for specifying the scheme used for overlap resolution is `(sdegeo:set-default-boolean "code for the scheme")`

The different regions defined in the structure are mentioned in Section 2. The regions with different material/doping/properties should be defined as different regions for SDE to distinguish them. For region specification, the proper geometric shape of the region along with the material must be defined. Each region must be identified by a unique name, which would be used for defining mesh spacing or doping. The SDE offers several two-dimensional (2D) and three-dimensional (3D) geometric shapes such as rectangle, any regular polygon, square, circle, cylinder, cuboid, etc. It also offers application of Boolean operators to perform union, intersection, etc. of two different regions.

In this sample code, we have first defined the active silicon region as a rectangle with appropriate coordinates. SDE requires the specification of the coordinates of the diagonally opposite corners of the rectangle and the material name (Silicon) for this specification. Then we have defined the bottom and top gate oxides with a thickness of 1 nm and a length of 1 μm (gate length) by specifying rectangle with appropriate coordinates and SiO_2 as the material. After this, we have defined the gate region of 1 μm utilizing a polysilicon material with appropriate coordinates. Note that the unit for coordinates specified in the code is micrometer. The command for creating a rectangle in SDE is

```
(sdegeo:create-rectangle position position material-name region-name)
```

It may be noted that even the graphical user interface (GUI) feature of SDE may be used to create the shapes with exact coordinates. However, it becomes tedious to specify each region via GUI when using multiple regions.

Once the regions have been defined, the next step is to define the doping. First, a doping profile is defined and then it is placed appropriately in the specific region. The statement

```
(sdedr:define-constant-profile ConstProfDef-name species concentration)
```

is used for defining the doping profile. SDE allows us to define different doping profiles such as uniform, Gaussian, etc. For instance, in Section 3, we have defined a uniform (constant) doping profile with the name "`Const.ChannelDoping`" and a 10^{19} cm^{-3} concentration of the phosphorus dopants. Then in the subsequent statement, we have placed this "`Const.ChannelDoping`" in region 1, which is the active silicon layer. The statement for placing the specified doping profile is

```
(sdedr:define-constant-profile-region ConstProfPlace-name

ConstProfDef-name region-name)
```

It may be noted that the doping profile name is a crucial parameter, and the same name should be used while placing the doping profile in any region. Similarly, we define and place a uniform doping profile "`Const.GateDoping`" for the polysilicon top and bottom gates with a boron doping of 10^{20} cm^{-3}.

Once the material and the doping profile of the different regions have been defined, we need to define the mesh spacing, which is done in Section 4. We already discussed in Section 9.1 that the TCAD simulator is a numerical solver, which treats the continuous structures as discrete mesh points and solves the physical equations at each mesh point known as node to obtain meaningful results. The importance of defining an adequate mesh with appropriate mesh spacing has already been discussed in detail in Section 9.2.1. The mesh spacing not only dictates the accuracy of the results but the simulation time and the resources such as memory allocation, number of cores of the processors, etc. to be used for simulation also depends on the mesh spacing. It also governs the convergence of the physical equations. Therefore, to obtain accurate results while consuming a small simulation time and resource such as memory, it is advised to optimize the mesh spacing.

The process of defining the mesh grid follows the same procedure as the doping. First, a refinement size is defined with a refinement name. The maximum and the minimum spacing between the consecutive mesh points along the *x*-, *y*-, and *z*-directions for a particular region is also specified. The statement

```
(sdedr:define-refinement-size definition-name max-x max-y max-z min-x min-
y min-z)
```

is used for this purpose. While generating the mesh, SDE places the mesh points at a separation given by the minimum spacing at the interfaces. However, in the bulk, the mesh points are separated by the maximum specified spacing. Since most of the physical phenomena such as tunneling in JLFETs or inversion layer formation in MOSFETs occur at the interface, a dense mesh is required to effectively capture these effects. For applying different refinement size in different areas of the same region, a ref/eval window may be used or multiple box placement techniques may be utilized. For instance, we have used a maximum mesh spacing of 0.8 nm and a minimum separation of 0.1 nm between the mesh points for defining the refinement size to be placed in the active silicon region. For placing the refinement size in a particular region, the statement

```
(sdedr:define-refinement-region placement-name definition-name region-
name)
```

may be used. Care must be taken to ensure that the same refinement size name is used while placing the regionwise refinement. It may be noted here that contacts do not contain any mesh points.

Once the regions have been appropriately defined, we need to define the contacts and place them adequately. Since we are performing 2D simulations, the contacts appear as a line. The name of the line contacts (source/drain/gate, etc.) need to be defined first and then they can be placed on the structure by specifying an appropriate edge. However, during this contact placement approach, SDE searches for the nearest vertex on the left- and right-hand side of the edge on which the contact has been placed. SDE then extends the line contact from the nearest left-hand side vertex to the nearest right-hand side vertex. Since the vertex exists at only four corners of the rectangle, if a contact is placed at an edge anywhere on the one side of the rectangle, it will cover the entire side. Similarly, if a vertex of the rectangle is chosen for placing the contact, the contact will extend on both sides of the rectangle, which have this common corner (vertex). In a fabricated JLFET, the source/drain contacts are isolated from the gate electrode by means of a spacer. Therefore, the source/drain contacts must be separated from the channel by the spacer length. For this, extra vertex need to be defined which act as the boundary for the source and drain contact definition. Section 4 shows how the new vertex may be inserted by simply specifying the appropriate coordinate where we want to restrict the source/drain contact using the statement:

```
(sdegeo:insert-vertex vertex-list)
```

The syntax to declare the contacts has been given in Section 6. The contact name, the line thickness of the contact, and the color of the contact as a function of the RGB components must be specified for contact declaration as

```
(sdegeo:define-contact-set name [edgeThickness Color] | [Color
facePattern])
```

Additionally, the type of line (solid, dotted, etc.) can also be defined. The ## face pattern is the code for defining solid contacts. The name of the contact is a crucial parameter since the same contact name must be used to specify the different voltages while performing device simulations in Sdevice.

Section 7 deals with positioning the contacts. As discussed earlier, the contacts are placed by identifying an appropriate edge, which is restricted between the boundary points (vertex) defined in Section 5. For instance, defining an edge at (0.01,0,0) restricts the source contact from (0,0,0) [corner of the rectangle] to the vertex (0.015,0,0) defined in Section 5. Therefore, the source/drain contact length is restricted to 15 nm and a separation of 25 nm is defined between the gate and the source/drain contacts. The generalized statement used for placing the contacts is

```
(sdegeo:define-2d-contact entity-list name)
```

After the contacts have been placed, the next step is to build the mesh. The statement used for this purpose is

```
(sde:build-mesh mesher options file-basename)
```

This command generates a TDR file consisting of the information about the region geometry, the materials used, the region boundaries, the contact placement, and the doping profile. A mesh command file is also generated, and the meshing engine "snmesh" builds the mesh. During the mesh construction, the different mesh points are connected by the means of triangles. This process is called delaunization. The delaunizer in the meshing engine of the tool connects all the mesh points by constructing acute-angled triangles. Only if the mesh points are located such that it cannot be covered by constructing acute-angled triangles, the mesh engine connects it using the obtuse-angled triangles. Convergence is difficult to obtain when obtuse-angled triangles are present in the mesh. Here also, caution must be taken while specifying the file-basename since the same file name will be used in Sdevice for device simulations.

The structure may be generated by first opening SDE in GUI mode and then copying the above command and pasting it in the command line window of SDE as shown in Fig. 9.3. The final structure is also shown in Fig. 9.3. To further analyze the structure, we may use Svisual and open the structure to visualize the doping, dimensions or the mesh spacing.

9.3.2 Device Simulation

Once the TDR file is generated by SDE, the device simulations may be performed using Sdevice. The readers are advised to go through Section 9.2.3 once they read the command file to get an overview of the entire code at once.

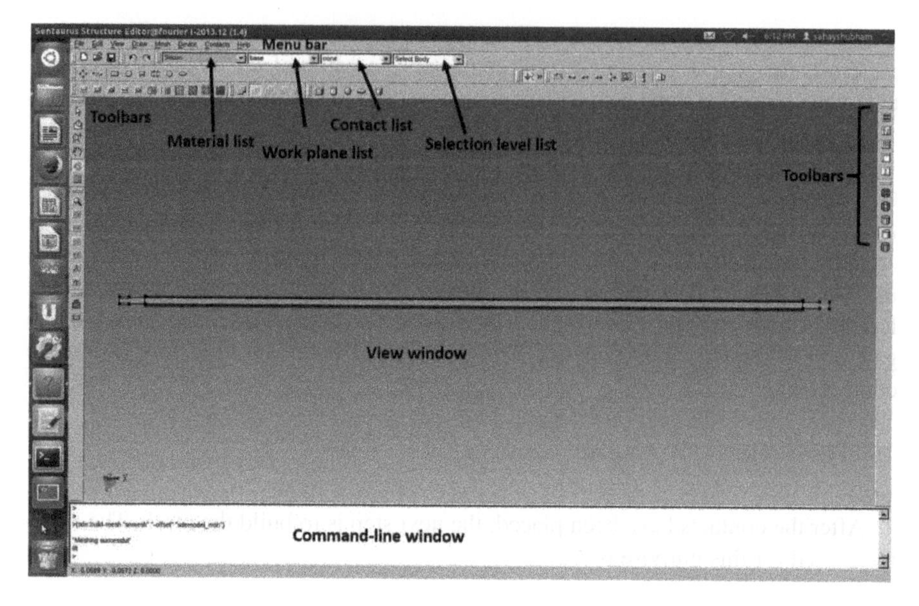

FIGURE 9.3 Final device structure in SDE obtained using the command file in Section 9.3.1.

The command file for simulating the electrical behavior of the structure generated in Section 9.3.1 is given below:

```
#---------File section-----------------------------------------------------------#

File {

# input file

   Grid     = "sdemodel_msh.tdr"

   Param    = "Silicon.par"

# output file

   Plot     = "@tdrdat@"

   Current  = "@plot@"

   Output   = "@log@"

}

#---------Electrode section-----------------------------------------------------#

Electrode {

   { Name="source"     Voltage= 0.0   }
```

```
   { Name="drain"     Voltage= 0.0 }
   { Name="gate"      Voltage= 0.0 }
}
#---------Physical model specification--------------------------------------#
Physics
{
Mobility( PhuMob Enormal)
EffectiveIntrinsicDensity(BandGapNarrowing (oldSlotboom))
Recombination( SRH Auger)
}
Physics (Material = "GatePolySilicon") {
       MetalWorkfunction ( Workfunction= 5.1 )
}
#---------Plot section-------------------------------------------------------#
Plot
{
eDensity hDensity eCurrent hCurrent equasiFermi
hquasiFermi AvalancheGeneration eGradQuasiFermi
ConductionBandEnergy ValenceBandEnergy ConductionCurrent
DisplacementCurrent eAlphaAvalanche hAlphaAvalanche
EffectiveBandGap EquilibriumPotential eQuasiFermi
EffectiveBandGap hGradQuasiFermi
hQuasiFermi IntrinsicDensity
EffectiveIntrinsicDensity eTemperature eIonIntegral Auger
ElectricField/Vector eEparallel hEparallel Potential Doping
SpaceCharge SRHRecombination eMobility hMobility
eVelocity hVelocity
DonorConcentration AcceptorConcentration
```

```
eIonIntegral hIonIntegral

MeanIonIntegral BuiltinPotential

eDriftVelocity Band2Band BandGap

HotElectronInjection HotHoleInjection

eBarrierTunneling hBarrierTunneling

}

#----------Numerical solver specification-------------------------------------#

Math

{

Extrapolate

Derivatives

RelErrcontrol

Iterations=20

Method=Blocked SubMethod=ParDiso

number_of_threads = 8

wallclock

stacksize = 20000000

CNormPrint

}

#-----------Solve section-----------------------------------------------------#

Solve {

NewCurrentPrefix="init_1"

Coupled(Iterations= 100 LineSearchDamping= 1e-4){ Poisson }

Coupled { Poisson Electron Hole }

Plot ( FilePrefix="initdg00_n@node@" )

quasistationary ( Goal {name="drain" voltage=1.0}

Initialstep=0.001 Maxstep=0.1 Minstep=1e-7)

{coupled {poisson Electron Hole }}
```

```
CurrentPlot( Time=(Range=(0 1) Intervals= 30)  )

Plot ( FilePrefix="Vgs-0_n@node@" Time=
(0;0.1;0.2;0.3;0.4;0.5;0.6;0.7;0.8;0.9;1.0) NoOverWrite)

NewCurrentPrefix="init_3"

quasistationary ( Goal {name="gate" voltage= 1}

Initialstep=0.001 Maxstep=0.01 Minstep=1e-7)

{

coupled {poisson electron hole }}

CurrentPlot( Time=(Range=(0 1) Intervals= 30)  )

Plot ( FilePrefix="Vgs-1.1_n@node@" Time=
(0;0.1;0.2;0.3;0.4;0.5;0.6;0.7;0.8;0.9;1.0) NoOverWrite)

}

#----------------------------------------------------------------------#
```

Similar to our approach in Section 9.3.1, we have divided the entire command file for performing device simulations in Sdevice into several sections for enhanced clarity. The lines beginning with "#" are considered comments by Sdevice. It is worth mentioning that the commands/statements for Sdevice can be in any order.

The file section specifies the input files that Sdevice requires for device simulations and the output files it generates. The mesh TDR file generated in SDE, which contains the information about the geometry, regions, materials, electrodes, doping, etc. is given as an input to the Sdevice via the "`Grid`" statement. For 2D simulations, Sdevice automatically assumes a thickness of 1 μm. In addition, the parameter file, which consists of materialwise parameters of the physical models used in the simulation, is also fed as an input to Sdevice using the "`param`" statement.

Three types of output files are generated by Sdevice: (a) the "`Plot`" output file, which contains the spatial solution of the physical equations in the form of different physical quantities defined in the plot section such as electric field, potential, electron concentration, etc. in the structure. This is also saved as a TDR file. (b) the "`Current`" file, which contains the information about the currents, voltages, charges, etc. at the different electrodes, and (c) "`Output`" file, which is the output log file for each run of Sdevice.

The electrode section contains the information about the name of the electrodes defined in SDE and their boundary conditions and initial bias conditions. As defined in Section 9.3.1, the electrodes do not contain any mesh and are treated with only electrical boundary conditions. Caution must be taken to ensure that the same electrode name is used in the electrode section, which was defined in Section 9.3.1 while specifying the structure in SDE.

The most important part of the Sdevice command file is the physical model specification. This is the heart of the simulations, and all the physical models that capture the essential physics behind the device operation are specified in this section. Missing even one of these physical models guarantees an inaccurate result. All the aspects of the device physics must be considered while choosing the models. It is recommended that the readers browse through the entire set of physical models available for including mobility degradation, band structure, quantization, carrier transport, generation and recombination (including tunneling), impact ionization, etc. in the manual for taking an informed decision while choosing the models for any device simulation. Now, we discuss all the relevant device physics involved in the operation of JLFETs and the corresponding physical model used to account for that physical phenomenon in JLFETs.

The silicon film in JLFETs is heavily doped. Therefore, the mobility degradation due to the impurity scattering mechanism is expected to be severe. If you browse through the manual, you will encounter several mobility models capturing impurity scattering. However, we have used the Philips unified mobility model since it is the most comprehensive model, which takes into account (a) temperature dependence of mobility (due to phonon scattering), (b) ionized impurity scattering, (c) carrier–carrier scattering, as well as (d) clustering of impurities and screening of the ionized impurity scattering centers by the charge carriers [5, 10].

Also, we know from our discussion in Section 3.4.1, in the subthreshold regime (full depletion mode) and the accumulation mode, the vertical electric field is very high in JLFETs. Therefore, we have also used the Lombardi mobility model to take into account the mobility degradation due to the electric field and acoustic surface phonons and the surface roughness [11]. The model parameters for these mobility models are material specific and well calibrated for silicon. Therefore, we have used the default parameter values. These models have been activated using the following statement in the physics section: `Mobility(PhuMob Enormal)`.

Also, the silicon film doping in JLFETs is comparable to the density of states of electrons. Therefore, band gap narrowing is inevitable in JLFETs. The Slotboom band gap narrowing model is used for considering the reduction in the band gap due to the heavily doped silicon film in JLFETs. This is done using the following statement: `EffectiveIntrinsicDensity(BandGapNarrowing (oldSlotboom))`.

SRH and Auger recombination model have also been used to take into account the different recombination mechanisms. The SRH rates increase with doping. Therefore, the carrier lifetimes need to be changed to 10^{-7} s for both electrons and holes corresponding to a doping of 10^{19} cm^{-3} [12]. The procedure to change the carrier lifetime value is discussed in Section 9.3.3. These models are activated using the statement: `Recombination(SRH Auger)`

The consecutive statement `(Material="GatePolySilicon") {Metal-Workfunction (Workfunction=5.1)` defines a gate polysilicon material as a metal with a work function of 5.1 eV required to deplete the heavily doped channel in JLFETs. We could have directly used a platinum gate electrode. Another method to specify a gate electrode of work function 5.1 eV is to define a line electrode on top of SiO_2 and specify the work function of 5.1 directly in the electrode statement in Sdevice.

The microscopic parameters such as electric field, carrier concentration, recombination rate, electrostatic potential, etc. that we want to analyze after the device simulation can be specified in the plot section. Sdevice saves the value of these parameters at each mesh point within the device structure in the output TDR file. These parameters may be visualized as a contour plot in the structure using Svisual.

As discussed in Section 9.2.3, Sdevice basically solves a set of physical equations such as the Poisson equation and the electron and hole continuity equations, etc. at each mesh point using some mathematical methods in an iterative manner. The use of coupled command in the solve section activates a numerical solver for the partial differential equations based on the Newton method. The settings for the solver based on the Newton method are specified in the numerical solver section. In the Newton method, an initial guess is required. In the subsequent iterations, the value of the function referred as the RHS is analyzed and the new updated guesses and the relative errors are also obtained. If the RHS vector is less than a predefined value, it is said that the solver converged and that particular guess becomes the solution.

The `Extrapolate` statement simply means that the initial guess for a particular step is obtained by extrapolating the solutions of the two previous steps.

The `Derivatives` statement implies that the full derivatives are calculated while implementing the Newton method. In addition to calculating the RHS value, i.e. the value of the function after each update, Sdevice also determines the value of the updated error during a solve statement. The solver also converges to a solution if the updated error value is below a predefined constant.

`RelErrcontrol` states that the convergence of the updated error depends on a relative error criterion. `Iterations=20` means that a total of 20 iterations would be performed until convergence is achieved for a particular bias point.

The keyword `Method` selects the linear solver to be used, and the `Blocked` statement specifies that the block-decomposition method has been used. The keyword `Sub-method` specifies the inner method for the block decomposition. `ParDiso`, the inner method used is software package used for solving a parallel set of linear equations with high accuracy. To know more about the solvers available, the readers are instructed to follow the solver manual of Sentaurus [13].

The `number_of_threads` simply specifies the number of multiple threads to be used for the linear solver and assembly. The multithreading option available in Sentaurus to increase the computation speed at the expense of utilizing more parallel licenses and processing capabilities can be exploited using this statement.

The `Stacksize` statement is used to specify the stack size (memory) for each thread. Some simulations may take more than the default stack size (1 MB) for simulations. For this simulation, we specify a stack size of 20 MB.

The `Wallclock` statement allows Sdevice to report wallclock times rather than the CPU time after each simulation step.

`CNormPrint` is a powerful tool to analyze the error in case the simulations are not converging. If it is activated, Sdevice specifies (a) largest error, (b) the coordinate where this error occurs, and (c) the value of the solution variable for that coordinate after each iteration and for all the physical equations that are being solved.

Once the solver settings are provided, the different physical equations to be solved and the final values to which the different electrodes of the FET are to be ramped are specified in the Solve section. The first statement `NewCurrentPrefix="init_1"` generates a new output plot file init_1, which would save the information about the voltage, charge, and current at the electrodes. To obtain an initial guess, only the `Poisson` equation is solved using the keyword `Coupled`, which invokes a fully coupled Newton solver. To achieve fast convergence to the initial guess based on the initial boundary conditions of the electrodes, the number of iterations is increased to 100. Moreover, `LineSearchDamping`, which is a damping method, is also used. During the damped Newton iterations, the solver tries to achieve convergence of the coupled Poisson equation far from the actual solution by changing it in smaller steps (the minimum factor by which the Newton update can be damped is mentioned as the value) as compared to a normal Newton solver. This method is appropriate for getting initial guess and fast convergence. The next step is to solve the `Poisson` equation, `Electron` and `Hole` continuity equations in a self-consistent manner using the `Coupled` Newton solver. This step provides the solution for the JLFET in the equilibrium state, i.e. when no external voltages have been applied. At this stage, to visualize the details of the structure, we have also saved the output TDR file by using the statement:

```
Plot ( FilePrefix="initdg00_n@node@" )
```

We want to obtain the transfer characteristics of the JLFET. For this, we have to ramp the drain voltage to the desired value and then sweep the gate voltage. The statement `quasistationary` implies that quasi-static or near steady-state equilibrium solutions are to be obtained for each value of drain voltage while `(Goal {name="drain" voltage=1.0}` simply specifies the final value of the voltage to which the drain electrode has to be ramped. Now, the drain voltage can only be increased in a particular step with a step size, which is defined in the next statement: `(Initialstep=0.001 Maxstep=0.1 Minstep=1e-7)`

This statement specifies the constraint on the initial step size for the increment of the drain voltage as a ratio compared to the final value, i.e. one. A small initial size is taken to achieve convergence easily in the beginning of the voltage ramp. The Sdevice continues to increase the step size up to the maximum step size in the subsequent iterations. However, if the simulations do not converge for any bias point even after the maximum number of specified iterations (20), Sdevice automatically reduces the step size by half of the present value and continues to reduce it by half until convergence is achieved or the minimum step size is reached. If the simulations do not converge even if the minimum step size is reached, it terminates the simulation for that particular `Goal`. The next statement: `{coupled {poisson Electron Hole}}` simply conveys that Poisson equation and the electron and hole continuity equations are solved in a self-consistent manner using the Newton solver for each drain voltage value as it is ramped to 1.0 V. The statement

`CurrentPlot(Time=(Range=(0 1) Intervals=30))` directs Sdevice to save the solutions of the current, charges, or voltages at each electrode only at specific values of the drain voltage between the range 0.0–1.0 V. In this case, the solutions are saved at 31 equally spaced value of the voltage between 0.0 and 1.0 V. The solutions are obtained even at other drain voltages but not saved in the present current output file. Similarly, the statement: `Plot (FilePrefix="Vgs-0_n@node@" Time=(0;0.1;0.2;0.3;0.4;0.5;0.6;0.7;0.8;0.9;1.0) NoOverWrite)` directs Sdevice to save the spatial solutions of the equations in the form of different parameters specified in the plot section at specific times only. Time = 0 corresponds to drain voltage 0.0 V, and time = 1 corresponds to drain voltage = 1.0 V. The final output TDR file is saved at 11 instances as the drain voltage is ramped between 0.0 and 1.0 V so that the structure may be analyzed later.

Once the drain voltage is ramped to the final value, the next step in obtaining the transfer characteristics is to sweep the gate voltage till the desired value. First, a new current plot file is specified which would store the electrode current, voltage, and charge values corresponding to the gate voltage sweep by using: `NewCurrentPrefix= "init_3"`

Then following a similar methodology as done for the drain voltage, the gate voltage is also ramped to 1.0 V while obtaining quasi-static solutions. The different step sizes for the ramp are also given similar to the drain voltage. The same fully coupled or Newton solver is used to solve the Poisson and the current continuity equations. Similar to the drain voltage sweeps, the solutions are saved in the plot file only at 31 instances between gate voltage 0.0 and 1.0 V in the current plot file. The spatial solutions are also saved in the output TDR file at 11 instances between gate voltage 0.0 and 1.0 V.

9.3.3 Model Parameter Specification

The parameter file with extension ".par," which is taken as an input in Sdevice contains the model parameters of the different physical models used in Sdevice. Some model parameters may need to be changed from the default values according to the device to be simulated. For instance, as discussed in Section 9.3.2, the carrier lifetimes need to be changed to 10^{-7} s for both electrons and holes corresponding to a doping of 10^{19} cm^{-3} [12]. The modified values of the carrier lifetimes may be obtained in the simulations by changing the taumax and Nref parameter under the SRH model parameters in the silicon.par file. Changing taumax from the default value of 10^{-5} s for electrons and 3×10^{-6} s for holes to 2×10^{-7} s and Nref to 10^{19} cm^{-3} ensures that tau, i.e. carrier lifetime is equal to 10^{-7} s for both electrons and holes.

```
;----------------Changes in PARAMETER FILE --------------

* Material = "Silicon" {

Scharfetter * relation and trap level for SRH recombination:
```

```
{ * tau = taumin + ( taumax - taumin ) / ( 1 + ( N/Nref )^gamma)

* tau(T) = tau * ( (T/300)^Talpha )          (TempDep)

* tau(T) = tau * exp( Tcoeff * ((T/300)-1) ) (ExpTempDep)

taumin = 0.0000e+00 , 0.0000e+00 # [s]

* taumax = 1.0000e-15 , 1.0000e-15 # [s]
taumax  = 2e-7 ,   2e-7      # [s]
Nref = 1.0000e+19 , 1.0000e+19 # [cm^(-3)]
# Change the carrier lifetimes according to the doping in SRH

gamma = 1 , 1 # [1]

Talpha = -1.5000e+00 , -1.5000e+00 # [1]

Tcoeff = 2.55 , 2.55 # [1]

Etrap = 0.0000e+00 # [eV]

}

;--------------------------------------------------------------
```

9.3.4 Visualizing the Outputs

The transfer characteristics may be visualized using Svisual. After opening the plot file, the gate voltage and the drain current may be plotted on the X and Y axis, respectively. Similarly, the different output TDR files may also be opened in Svisual and the different microscopic parameters saved in the plot file may be visualized. Figure 9.4 shows the transfer characteristics. The different profiles such as the carrier concentration and electric field may also be visualized. Svisual also provides the option of analyzing the different parameters along different cut lines. For instance, we have plotted the energy band profiles along the cutline C_1 (Fig. 9.5(a)) as shown in Fig. 9.5(b). It can be observed that there exists a significant spatial proximity between the valence band of the channel region and the conduction band of the drain region. Such a band overlap would facilitate tunneling of electrons from the channel region to the drain region and lead to the formation of a parasitic BJT as discussed in Section 5.2. However, for long-channel JLFETs, owing to a large-channel length (effective base width), the gain of the parasitic BJT is very small and its impact on the transfer characteristics is negligible. However, for the short-channel JLFETs, the BTBT must be included to simulate JLFETs, which has been ignored in Section 9.3.2. Therefore, in Section 9.5, we would discuss the methodology to simulate short-channel JLFETs

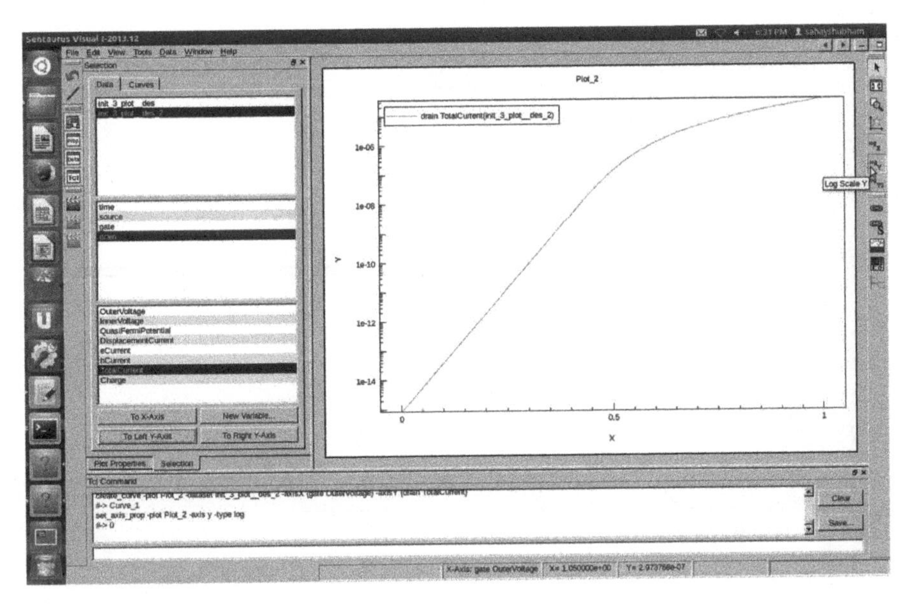

FIGURE 9.4 Transfer characteristics of the long-channel double-gate JLFET.

while using the calibrated BTBT models. Before moving to Section 9.5, we discuss the importance of model calibration and how it is performed.

9.4 MODEL CALIBRATION

Like a computer, a TCAD software is also a numerical solver. It does not have a brain of its own. It gives an output based on the inputs that we provide. In this regard, several questions may arise: Can we validate if the results given by the TCAD simulator is accurate? If yes, how do we ensure that our simulation setup is valid? The answer to these inevitable questions is model calibration. Model calibration is the first step to be carried out before starting the actual device simulations and involves reproducing the experimental results. The same structure reported in the experiments should be used for simulations and the model parameters of the appropriate physical models should be tuned to match the experimental characteristics.

However, this is very tricky and the most difficult part of any TCAD simulation-based study. Sometimes it is difficult to obtain all the dimensions by analyzing only the TEM image, which most experimental papers provide. Also, often the important parameters such as doping or stress in the semiconductors or the fixed oxide charge density or the trap density is not measured or reported in the experimental papers. The lack of information about all the dimensions or the properties of the experimental structure may complicate the task. This is where we need to make intelligent approximations. For instance, if upon simulating the experimental structure, we get transfer characteristics which are shifted as compared to the experimental transfer

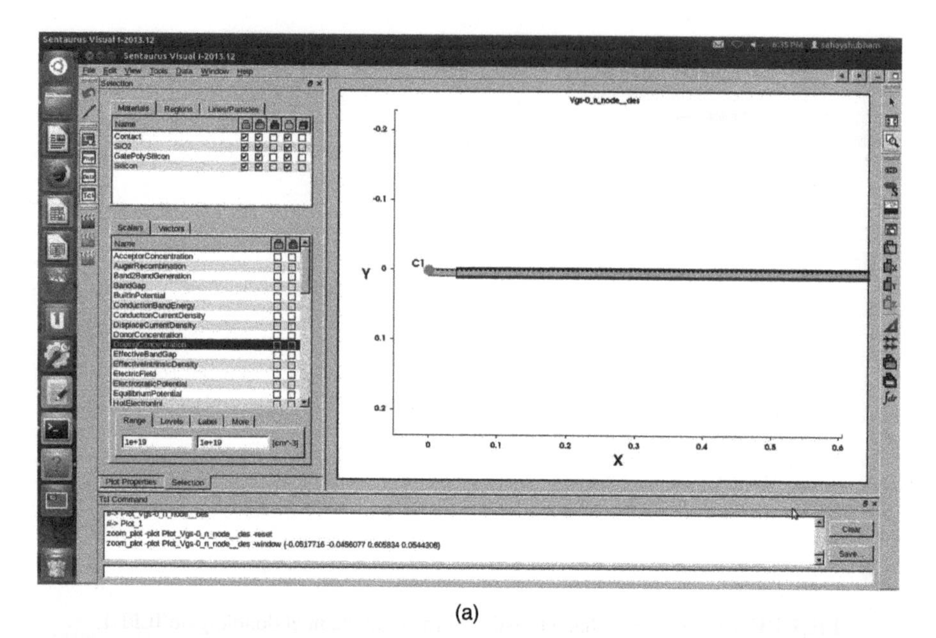

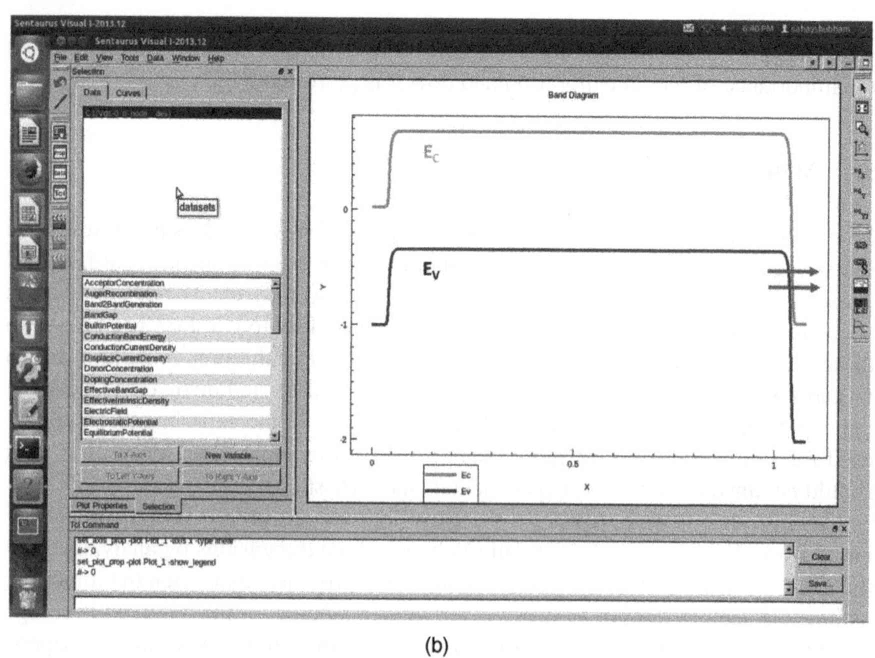

FIGURE 9.5 (a) The cutline C_1 at a distance 1 nm below the top Si–SiO_2 interface and (b) energy band profile along C_1.

characteristics, we should consider matching the threshold voltage by utilizing fixed oxide charges or traps. Oxide charges and traps are inevitable while fabricating any FET. Therefore, this is a valid approximation. Sometimes, the experimental structures may contain some irregular shapes such as oval shape or trapezoidal shape that are difficult to reproduce in the simulator and may cause numerical convergence issues. For these structures, it is better to approximate them to the nearest regular geometric shape such as a circle for an oval shape, a rectangle for a pyramid shape, etc. This may cause a slight difference between the simulation results and the experimental results. However, it should be kept in mind that no matter how hard we try, we can only approximate the experimental conditions during model calibration in TCAD software, i.e. a mismatch between the simulation results and the experimental results is inevitable due to lack of the knowledge about the entire experimental structure, setup etc. If a decent match is found between the experimental results and the simulation results especially in the region which the study focusses, the simulation setup can be said to be calibrated and may be used to predict the device performance of similar architectures. For instance, if we are interested in a study of gate-induced drain leakage (GIDL) of FETs, the parameters in the BTBT model must be tuned to match the drain current close to zero gate voltage where GIDL is dominant. Therefore, we need to identify the regions where a particular physical model governs the results and tune it to match the experimental result.

At this juncture, we would also like to mention that reproducing the simulation results from another simulation-based study, which has its models calibrated to any experimental data, is also a smart and valid practice. It is easier to reproduce the regular geometric shape used in a simulation-based study than to reproduce complicated experimental structures. We have followed the same approach in our model calibration for the short-channel JLFETs discussed in Section 9.5. We would calibrate our simulation setup by reproducing the results reported in [14]. The simulation setup in [14] was first calibrated with the experimental results of a long-channel (1 μm) JLFET [15], in which the BTBT-induced parasitic BJT action is not dominant. Then, a calibrated nonlocal tunneling model (used earlier for tunnel FETs in [16]) was used to account for tunneling in JLFETs.

9.5 MODEL CALIBRATION FOR SHORT-CHANNEL JLFETs

As discussed in Section 9.4, we would calibrate our simulation setup by reproducing the simulation results of [14] which is already calibrated to experimental results of [15] and tunneling models of [16]. However, we consider the following assumptions for our simulations: (a) The direct tunneling model for considering the gate leakage current will not be included assuming high-κ/metal gate stack as done in [14, 17–28]. (b) The atomic structure of interfaces and dopant fluctuations will not be taken into account in our simulations as done in [26, 27] since the main objective of our work is to demonstrate the relative impact of new architectures on the electrostatics of the JLFET rather than showing the exact values of the currents.

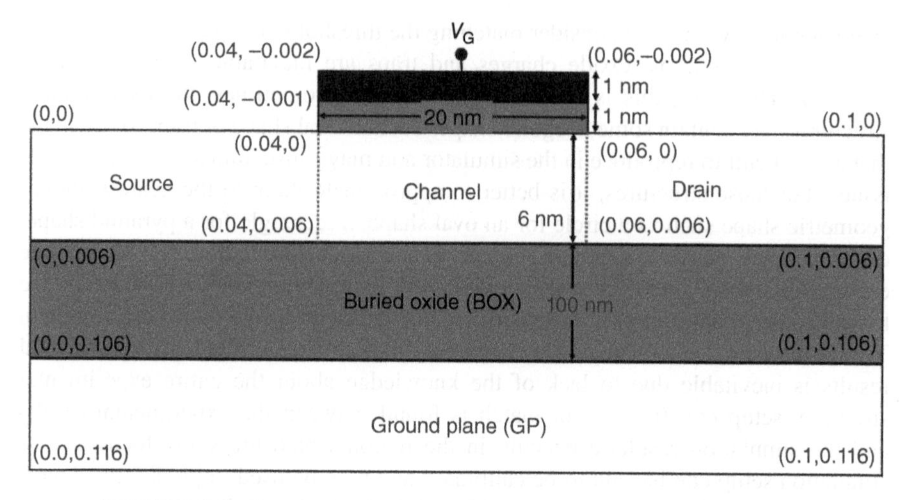

FIGURE 9.6 Rough sketch of the SOI JLFET used for model calibration.

With these assumptions, the single-gate silicon-on-insulator (SOI) structure whose results are to be reproduced is shown in Fig. 9.6. The data of the transfer characteristics of this structure may be extracted from [14] utilizing any tool such as web plot digitizer [29], etc. The simulation results can then be compared against this data by overlapping the plots. This is a standard practice while performing model calibration.

9.5.1 Structure Specification

The command file for the short-channel SOI JLFET shown in Fig. 9.6 with a channel length of 20 nm, and a gate oxide thickness of 1 nm is given below:

```
;-------------------------------------------------------------------

; Overlap resolution: New replaces Old

(sdegeo:set-default-boolean "ABA")

;-------------------------------------------------------------------

; Creating source region
(sdegeo:create-rectangle (position 0 0 0)(position 0.04 0.006 0) "Silicon"
"region_Epi1" )

; Creating channel region

(sdegeo:create-rectangle (position 0.04 0.0 0) (position 0.06 0.006 0) "Silicon"
"region_Epi2" )

; Creating drain region
```

```
(sdegeo:create-rectangle (position 0.06 0.0 0) (position 0.1 0.006 0) "Silicon"
"region_Epi3" )

; Creating Buried oxide (BOX) region

(sdegeo:create-rectangle (position 0.0 0.006 0.0 ) (position 0.1 0.106 0.0 )"SiO2"
"region_2")

; Creating Gate oxide (SiO2) layer

(sdegeo:create-rectangle (position 0.04 -0.001 0.0 ) (position 0.06 0.0 0.0 )
"SiO2" "region_3")

; Creating Gate Polysilicon layer

(sdegeo:create-rectangle (position 0.04 -0.002 0.0 ) (position 0.06 -0.001 0.0 )
"GatePolySilicon" "region_5")

; Creating ground plane (GP)

(sdegeo:create-rectangle (position 0.0 0.106 0.0 ) (position 0.1 0.116 0.0 )
"Silicon" "region_6")

;-------------------------------------------------------------------

; Defining refinement size

(sdedr:define-refinement-size "Ch_refinement" 0.0008 0.0008 0.0001 0.0001 )

; Placing refinement size

(sdedr:define-refinement-region "Place.Ch_refine" "Ch_refinement" "region_Epi1" )

(sdedr:define-refinement-size "Ch_refinement_1" 0.001 0.001 0.0001 0.0001 )

(sdedr:define-refinement-region "Place.Ch_refine_1" "Ch_refinement_1" "region_Epi2" )

(sdedr:define-refinement-size "Ch_refinement_2" 0.001 0.001 0.0001 0.0001 )

(sdedr:define-refinement-region "Place.Ch_refine_2" "Ch_refinement_2" "region_Epi3" )

(sdedr:define-refinement-size "Siooxide_refinement" 0.001 0.001 0.0005 0.0005 )

(sdedr:define-refinement-region "Place.Siooxide" "Siooxide_refinement" "region_3")

(sdedr:define-refinement-size "Gate_refinement" 0.001 0.001 0.0005 0.0005 )

(sdedr:define-refinement-region "Place.Gate_refine" "Gate_refinement" "region_5" )

(sdedr:define-refinement-size "GP_refinement" 0.001 0.001 0.0005 0.0005 )

(sdedr:define-refinement-region "Place.GP_refine" "GP_refinement" "region_6" )

;-------------------------------------------------------------------
```

```
; Defining doping in source, channel and drain regions

(sdedr:define-constant-profile "Const.ChannelDoping"
"PhosphorusActiveConcentration" 1e+19 )

(sdedr:define-constant-profile-region "Place.Channel" "Const.ChannelDoping"
"region_Epi1")

(sdedr:define-constant-profile-region "Place.Channel1" "Const.ChannelDoping"
"region_Epi2")

(sdedr:define-constant-profile-region "Place.Channel2" "Const.ChannelDoping"
"region_Epi3")

; Defining polysilicon gate doping

(sdedr:define-constant-profile "Const.GateDoping" "BoronActiveConcentration"
1e+20)

(sdedr:define-constant-profile-region "Place.Gate" "Const.GateDoping" "region_5")

; Defining ground plane (GP) doping

(sdedr:define-constant-profile "Const.GPDoping" "BoronActiveConcentration" 5e+18 )

(sdedr:define-constant-profile-region "Place.GP" "Const.ChannelDoping" "region_6")

;-----------------------------------------------------------------------

; Insert vertex for S/D contact boundaries

(sdegeo:insert-vertex (position 0.015 0 0))

(sdegeo:insert-vertex (position 0.085 0 0))

;-----------------------------------------------------------------------

; Contact declarations

(sdegeo:define-contact-set "source" 4.0  (color:rgb 1.0 0.0 0.0 ) "##")

(sdegeo:define-contact-set "drain" 4.0  (color:rgb 0.0 1.0 0.0 ) "##")

(sdegeo:define-contact-set "gate" 4.0 (color:rgb 0.0 0.0 1.0 ) "##")

;-----------------------------------------------------------------------

; Contact placement

(sdegeo:define-2d-contact (find-edge-id (position 0.01 0.0 0.0)) "source" )

(sdegeo:define-2d-contact (find-edge-id (position 0.09 0.0 0.0)) "drain")

(sdegeo:define-2d-contact (find-edge-id (position 0.05 -0.002 0.0)) "gate")
```

```
;-------------------------------------------------------------

; Generating the mesh using SNMesh

(sde:build-mesh "snmesh" "-offset" "junctionless_msh")

;-------------------------------------------------------------
```

We first begin by specifying the overlap resolution as discussed in Section 9.3.1. Then we define the different regions sketched in Fig. 9.6 as rectangles with appropriate dimensions. It may be noted that we have defined source, channel, and drain regions as three different regions unlike in Section 9.3.1 where they were represented as a single silicon film. Since we have to use the nonlocal tunneling model to take BTBT into account, it becomes convenient to construct the nonlocal mesh if the interface between the source–channel and channel–drain regions is defined. The nonlocal mesh definition based on interface between two regions can then be easily utilized.

The next step as discussed in Section 9.3.1 is to define the refinement size and to place the refinement for specifying the mesh spacing. It may be noted that the maximum mesh spacing is kept lower in the source region as compared to the channel and drain regions. This is done to ensure convergence, and this set of numbers is attained by the hit and trial method. Once the mesh spacing has been defined for each region, we define the doping concentration of the regions.

After defining the various regions along with the meshing and properties such as doping, we insert the vertex to restrict the source/drain contact length and to separate gate electrode from the source/drain electrodes. Following this, we declare the source, drain, and gate electrodes and then place them in the structure by identifying the appropriate edges.

Once the contacts are also defined, we call SNmesh to run the delaunizer and construct the mesh joining the mesh points in a triangular fashion. Figure 9.7 shows the final structure of short-channel SOI JLFET in Svisual created using the above command file.

9.5.2 Device Simulations

The Sdevice command file used for simulating the short-channel SOI JLFET structure defines in Section 9.5.1 is given below:

```
#-------Initialising electrode voltages-----------------------------------------#

Electrode {

   { Name="source"    Voltage= 0.0  }

   { Name="drain"     Voltage= 0.0 }

   { Name="gate"      Voltage= 0.0 } }
```

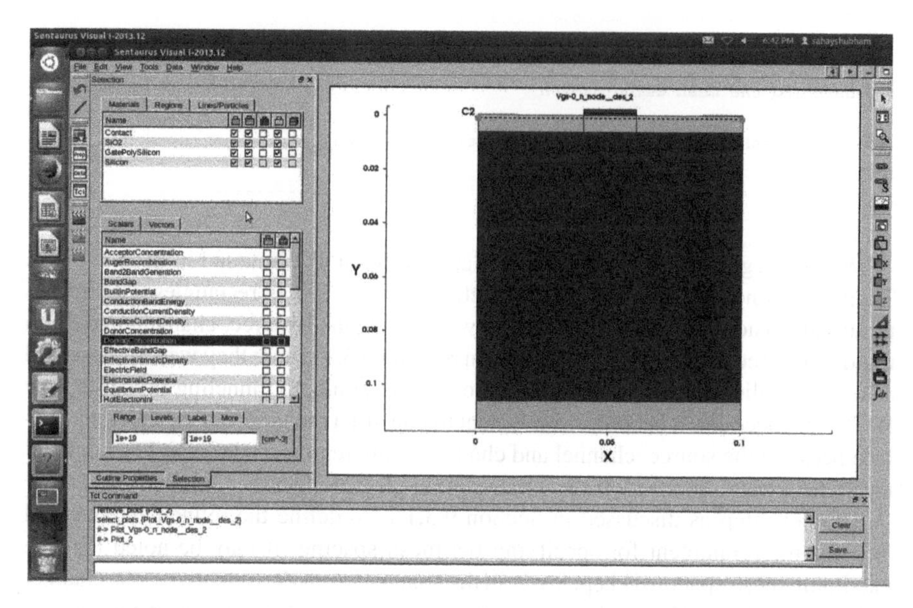

FIGURE 9.7 Final structure of the short-channel SOI JLFET in Svisual.

```
#-------Defining Input and output files-----------------------------------------#

File {

   Grid    = " junctionless_msh.tdr"

   Plot    = "@tdrdat@"

   Current = "@plot@"

   Output  = "@log@"

   Param   = "Silicon.par"

}

#-------Specifying physical models----------------------------------------#

Physics

{

Mobility( PhuMob HighFieldSat Enormal)

EffectiveIntrinsicDensity(BandGapNarrowing (oldSlotboom))

Recombination( SRH Auger

eBarrierTunneling(Band2Band TwoBand Transmission)
```

```
hBarrierTunneling(Band2Band TwoBand Transmission))

}

Physics (Material = "GatePolySilicon") { MetalWorkfunction ( Workfunction= 5.1 )}

#--------Specifying the parameters to be plotted--------------------------------#

Plot

{

eDensity hDensity eCurrent hCurrent equasiFermi

hquasiFermi AvalancheGeneration eGradQuasiFermi

ConductionBandEnergy ValenceBandEnergy ConductionCurrent

DisplacementCurrent eAlphaAvalanche hAlphaAvalanche

EffectiveBandGap EquilibriumPotential eQuasiFermi

EffectiveBandGap hGradQuasiFermi

hQuasiFermi IntrinsicDensity

EffectiveIntrinsicDensity eTemperature eIonIntegral Auger

ElectricField eEparallel hEparallel Potential Doping

SpaceCharge SRHRecombination eMobility hMobility

eVelocity hVelocity

DonorConcentration AcceptorConcentration

eIonIntegral hIonIntegral

MeanIonIntegral BuiltinPotential

eDriftVelocity Band2Band BandGap

HotElectronInjection HotHoleInjection

eBarrierTunneling hBarrierTunneling

}

#-------Defining solver settings------------------------------------------#

Math

{CNormPrint
```

```
Extrapolate

Derivatives

RelErrcontrol

Iterations=20

Method=Blocked SubMethod=ParDiso

number_of_threads = 8

stacksize = 2000000

wallclock

}

#-------Defining non-local mesh for tunneling----------------------------------#

Math(regionInterface="region_Epi2/region_Epi3")

{ NonLocal(Length=2.5e-6 Permeable Permeation=2.5e-6)

Digits(NonLocal)=3

EnergyResolution(NonLocal)=0.001

}

Math(regionInterface="region_Epi1/region_Epi2")

{NonLocal(Length=2.5e-6 Permeable Permeation=2.5e-6)

Digits(NonLocal)=3

EnergyResolution(NonLocal)=0.001

}

#-------Solve section-----------------------------------------#

Solve {

NewCurrentPrefix="init_1"

Coupled(Iterations= 100 LineSearchDamping= 1e-4){ Poisson }

Coupled {Poisson Electron Hole}

quasistationary ( Goal {name="drain" voltage=1.0}

Initialstep=0.001 Maxstep=0.1 Minstep=1e-7)

{coupled {poisson Electron Hole }}
```

```
CurrentPlot( Time=(Range=(0 1) Intervals= 30)  )

Plot ( FilePrefix="Vgs-0_n@node@" Time=
(0;0.1;0.2;0.3;0.4;0.5;0.6;0.7;0.8;0.9;1.0) NoOverWrite)

NewCurrentPrefix="init_2"

quasistationary ( Goal {name="gate" voltage= -0.5}

Initialstep=0.001 Maxstep=0.01 Minstep=1e-7)

{coupled {poisson electron hole }}

CurrentPlot( Time=(Range=(0 1) Intervals= 30)  )

Plot ( FilePrefix="Vgs-0.5_n@node@" Time=
(0;0.1;0.2;0.3;0.4;0.5;0.6;0.7;0.8;0.9;1.0) NoOverWrite)

NewCurrentPrefix="init_3"

quasistationary ( Goal {name="gate" voltage=1.0}

Initialstep=0.001 Maxstep=0.01 Minstep=1e-7)

{coupled {poisson electron hole }}

}

#-------------------------------------------------------------------------#
```

As discussed in Section 9.3.2, first of all, we define the initial voltages at the different electrodes. Then, we specify the name of the input TDR file (“`junctionless_msh.tdr`”) and the parameter file and define the extension for the output TDR file, plot files, and the log file. Once the proper input and output files have been specified, the next step is to define the various physical models to be used for simulation of short-channel JLFETs.

As discussed in Section 9.3.2, the most complete model for taking into account the impurity scattering, i.e. the Philips unified mobility model, is used. Also, the Lombardi mobility model is used to take into account the mobility degradation due to the high vertical electric field similar to the long-channel JLFETs. However, in short-channel JLFETs, due to the high lateral electric field, the linear relationship between the velocity of electrons and the electric field is not valid owing to the reduction in the mobility. To take into account the velocity saturation, a unique property of short-channel FETs, the Canali mobility model [5] is specified as `HighFieldSat`. `SRH` and Auger recombination models have also been used.

Since there exists a spatial proximity between the valence band of the channel region and the conduction band of the drain region in JLFETs as shown in Fig. 9.8, BTBT of electrons from the channel to the drain region is inevitable. Therefore, we must include the BTBT model in our simulations.

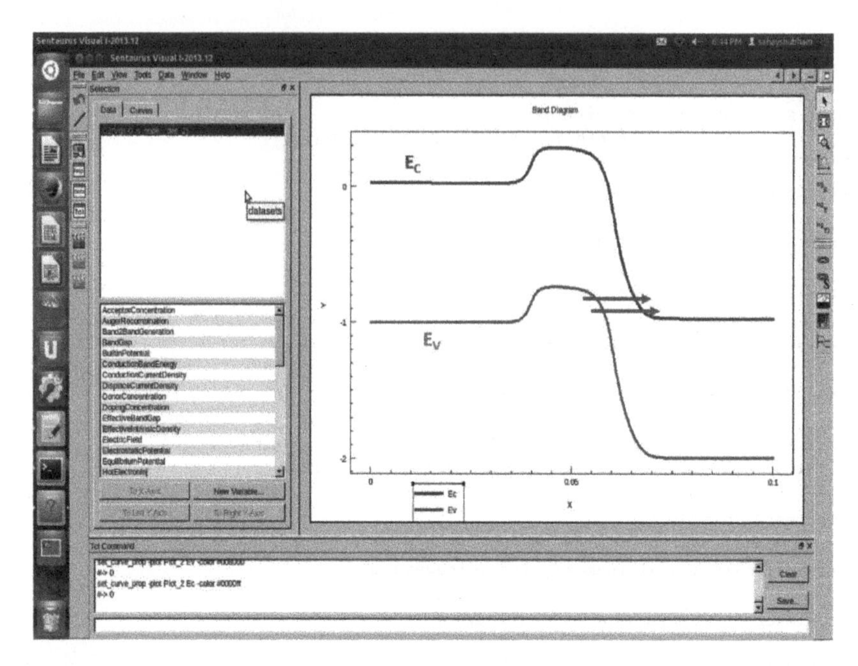

FIGURE 9.8 Energy band profiles of the short-channel JLFET at a cutline 1 nm below the Si–SiO_2 interface.

BTBT is a complex quantum mechanical phenomenon. However, the TCAD simplifies this complex quantum phenomenon by treating it as a classical generation–recombination process whereby the holes are generated at the beginning of the tunnel path and electrons are generated at the end of the tunneling path. For TCAD simulations, two different types of BTBT models have been formulated:

a) *Local BTBT models*: The local BTBT models assume a constant electric field in the forbidden gap to calculate the generation rate of electrons and holes. These models tend to overestimate the generation rate and the tunneling current value. Also, the generation rate is non-zero even at equilibrium, i.e. when no bias is applied. Moreover, instead of considering the electron and hole generation at the end and starting of the tunneling path, respectively, these models do not resolve the spatial generation of the electrons and holes. Therefore, these models are often not employed in TCAD-based simulation studies. However, owing to the numerical efficiency of these simple models, they are frequently employed for analytical and compact modeling studies [30–35].

b) *Nonlocal BTBT models*: The nonlocal BTBT models take into account the spatial variation of the electric field within the forbidden gap to calculate the generation rates according to the Wentzel–Kramers–Brillouin (WKB) approximation. They are accurate as compared to the local models and consider the spatial generation of electrons at the end of the tunneling path and holes at the beginning of the tunneling path. Two flavors of the nonlocal tunneling model are available in SDevice [5]: the basic nonlocal BTBT model where the tunneling path has to be predefined with the help of a nonlocal tunneling mesh and the dynamic nonlocal BTBT model where the simulator automatically finds the tunneling path based on the gradient of the valence band energy at the beginning of the tunneling path. In the basic nonlocal BTBT model, a specialized nonlocal mesh needs to be defined in the Maths section. The nonlocal mesh connects the points on the left of the tunneling junction to those at the right of the tunneling junction via nonlocal lines. Therefore, the entire tunneling junction is divided into a number of one-dimensional (1D) slices. The BTBT generation rate is then calculated along each 1D slice by considering the variation in the 1D electric field along the slice (horizontal direction). The net tunneling current is then obtained by integrating the generation rates over the individual slices.

The dynamic nonlocal BTBT model, on the other hand, does not require the specification of a nonlocal path or a predefined tunneling mesh. It calculates the generation rate by dynamically identifying the tunneling path based on the gradient of the valence band energy at the starting point of the tunneling path. Therefore, the tunneling path taken into account by the dynamic nonlocal model can be in any arbitrary direction and need not be always horizontal. The dynamic nonlocal tunneling model considers the net electric field rather than the horizontal component of the electric field as done by the basic nonlocal tunneling model. Therefore, the basic nonlocal tunneling model rather underestimates the value of the tunneling current as compared to the dynamic nonlocal model since it computes the generation rate due to only the horizontal (or vertical) component of the electric field [36, 37]. However, even the nonlocal BTBT may give accurate results if it is calibrated to experimental data.

We have employed the basic nonlocal BTBT model by constructing a predefined horizontal tunneling path to consider the BTBT in JLFETs by using the statement: `eBarrierTunneling(Band2Band TwoBand Transmission)`. This statement specifies that the BTBT generation rate for electrons has to be calculated at the tunneling barriers, i.e. source–channel or channel–drain junctions where the tunneling mesh has been defined. The keyword `TwoBand` conveys that the dispersion relation for both the conduction band, and the valence band has to be considered while calculating the generation rate. Similarly, to compute the BTBT hole generation rate, the following statements are used: `hBarrierTunneling(Band2Band TwoBand Transmission)`. It may be noted that BTBT is classified as a generation–recombination process by TCAD and defined under the Recombination section as discussed earlier.

Once the appropriate physical models are specified, the microscopic parameters to be analyzed later in the output TDR file are defined in the plot section. The different settings for the numerical coupled Newton solver are defined in the Maths section followed by the Plot section. The details of the different keywords are already specified in Section 9.3.2.

The nonlocal mesh for calculating the BTBT current is also specified in the Maths section. The keyword `regionInterface="region_Epi2/region_Epi3")` conveys to Sdevice that a nonlocal mesh has to be constructed at the interface of region defined as `region_Epi2` (channel) and `region_Epi3` (drain). To ease the nonlocal mesh specification, we purposely defined source, channel and the drain regions as three separate regions rather than a single silicon film as done in Section 9.3.1.

The keyword `Length` specifies the distance from the interface (in cm) up to which the nonlocal mesh lines are to be constructed. In this case, we have taken it as 25 nm. The keyword `Permeable` indicates that the nonlocal mesh lines can be extended across the interface by a length (in cm) defined by the keyword `Permeation`. The permeation is also taken as 25 nm. The nonlocal mesh lines connect the vertices on either side of the interface up to a distance given by length and permeation by shortest geometric paths.

The keyword `Digits(NonLocal)=3` defines the accuracy limit for the calculation of the tunneling current.

Similarly, the keyword `EnergyResolution(NonLocal)=0.001` specifies the minimum resolution of the energy level to be used for integration while calculating the tunneling current. We have used a resolution of 0.001 eV. The nonlocal mesh lines are also constructed at the source–channel interface for symmetry.

Once the settings of the solvers are defined and the nonlocal mesh lines are constructed, the different voltages to which the electrodes need to be ramped may be specified in the Solve section. The detail about the steps is already discussed in detail in Section 9.3.2. A fully coupled Newton solver is used, and first a damped solution for only the Poisson equation is obtained for initial guess. The Poisson equation along with the electron and hole continuity equation is then solved as the drain voltage is ramped to 1.0 V using the predefined step size. The output TDR files and the current plot files are then saved, and similar steps are followed for ramping the gate voltage.

9.5.3 Model Parameters for Calibration

For simulating the short-channel JLFETs, in addition to modifying the `taumax` parameter of the electrons and holes to 10^{-7} s for the SRH model as done in Section 9.3.3, the parameters of the BTBT model should also be tuned to match the results of [14], which are calibrated to experimental results. Therefore, for model calibration, the parameters of the BTBT model such as tunneling mass for electrons and holes and the density of states for electrons and holes may be tuned. However, it should be kept in mind that any arbitrary value cannot be assigned to any parameter to forcefully match the simulation results to the experimental data or data extracted from the reference paper based on the simulation study. For instance, the density of states is well defined for silicon. Therefore, it should not be altered. The tunneling mass on the other hand is a hypothetical concept like the effective mass and

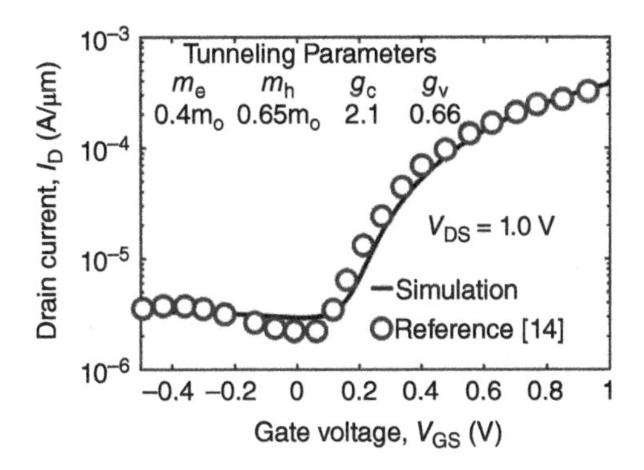

FIGURE 9.9 Model calibration by reproducing the results reported in [14].

simply conveys the ease of motion of electrons following the tunneling mechanism. Therefore, tunneling mass may be tuned to match the simulation results to the data extracted from [14]. However, even the tunneling mass should be handled with care and it should be ensured that its value is taken in the range used in the previous works utilizing the same model. For instance, the basic nonlocal BTBT model used here was used for simulations of TFETs after calibrating the tunneling mass as $0.5m_0$ for electrons and $0.65m_0$ for holes in [38]. Therefore, we should only tweak the value of the tunneling mass between $0.1m_0$ and $0.9m_0$ without affecting the physical meaning of the tunneling mass. A very high or very low tunneling mass denotes impractical systems.

Now, for model calibration, first the region in the transfer characteristic which is affected by a particular model should be identified. For instance, the drain current below zero gate voltage depends on the BTBT. Therefore, first of all, we need to tune the model parameters of BTBT, i.e. the tunneling mass of electrons and holes to match the simulation results to the results reported in [14]. This requires a series of hit and trial simulation experiments with a different set of tunneling mass of electrons and holes. After an exhaustive set of simulations, for an electron tunneling mass of $0.4m_0$ and a hole tunneling mass of $0.65m_0$, the simulation results match with the results reported in [14] as shown in Fig. 9.9. Therefore, these values of tunneling mass and default values of density of states for electrons and holes need to be given as input to the Sdevice in the parameter file for carrying out simulations of short-channel JLFETs with the BTBT model. This can be done by using the following statements in the parameter file:

```
;-------------------------------------------------------------

BarrierTunneling

{ * Non Local Barrier Tunneling

* G(r) = g*A*T/kB*F(r)*Pt(r)*ln[(1+exp((E(r)-Es)/kB/T))/(1+exp((E(r)-Em)/kB/T))]
```

```
* where:

*       Pt(r) is WKB approximation for the tunneling probability

*       g = As/A, As is the Richardson constant for carriers in semiconductor

*       A is the Richardson constant for free electrons

*       F(r) is the electric field

*       E(r) is carrier energy

*       Es is carrier quasi fermi energy in semiconductor

*       Em is carrier fermi energy in metal

*       alpha is the prefactor for quantum potential correction

*       eoffset and hoffset are lists of band offsets
g       = 2.1 ,      0.66     # [1]
mt      = 0.4 ,      0.65     # [1]
# Calibrate the tunneling masses to match the results reported in [14]

alpha = 0.0000e+00 , 0.0000e+00 # [1]

eoffset = () # eV

hoffset = () # eV

;--------------------------------------------------------------
```

9.6 MODEL CALIBRATION FOR NWFETS

In Section 9.5, we discussed the model calibration methodology for carrying out 2D simulation of short-channel JLFETs including the BTBT model. In this section, we present an insight into the simulation and calibration methodology for 3D architectures such the gate-all-around NWFETs (GAA NWFETs). As discussed in Section 2.6, the GAA NWFETs are hailed as the most promising candidates for the ultimate scaling of silicon FETs. However, they suffer from a lateral BTBT GIDL, which hinders their scaling to the sub-10 nm technology nodes. Therefore, it becomes necessary to understand the impact of lateral BTBT (L-BTBT) in NWFETs and propose architectures to mitigate L-BTBT. The calibrated simulation setup provided in this section, and the discussions about the simulation methodology may provide an incentive to the readers interested in working on NWFETs. For carrying out the model calibration for NWFETs, we chose [39] as the reference where L-BTBT was discussed for the first time. The experimental results reported in [39] need to be reproduced through the simulation setup for model calibration.

9.6.1 Assumptions

As discussed in Section 9.4, the knowledge about every parameter needed for simulation of experimental structure is generally not available in the experimental papers. Even in [40], where the details about the structures fabricated and studied in [39] are given, the information about the doping dose, the doping profile, the exact cross-sectional view of the nanowire along with the dimensions including the gate oxide thickness are not explicitly available. Also, since the nanowires are defined using self-limited oxidation, they are bound to be strained. The stress measurements are also not reported in [40]. Also, the gate work function or the information about the gate material is not mentioned in [40].

Therefore, we have to take few intelligent approximations to consider the experimental NWFET [39] [shown in Fig. 3.10(h) of Chapter 3] in our simulations. First, a source/drain doping of $N_D = 5 \times 10^{19}$ cm^{-3}, a channel doping of $N_A = 10^{16}$ cm^{-3}, an effective oxide thickness (EOT) of 1 nm were assumed for the experimental NWMOSFETs as per the discussions given in [39]. Furthermore, an abrupt doping profile was assumed at the source–channel and the channel–drain interfaces although the experimental structure may have a gradient in the doping profile. Even while doping the source/drain regions in the experimental nanowires, the mask used had an offset of 16 nm for reducing the lateral encroachment of dopants in the channel region [40]. Therefore, an abrupt doping is a valid approximation. Also, the main aim of the simulations is to calibrate the tunneling models to account L-BTBT, which is dominant at low gate voltages. The T-BTBT, which originates due to the transverse BTBT in the gate overlapped drain region owing to the dopant encroachment inside the channel region, occurs at extremely negative gate voltages (< -1.5 V). Therefore, due to the use of abrupt doping profile, our simulations would capture the effect of L-BTBT efficiently though the simulation setup would fail to capture the impact of transverse BTBT (T-BTBT). However, it would not make much difference since it is the L-BTBT, which increases the OFF-state current significantly and is of utmost importance to the device designers. T-BTBT originates at gate voltages lower than -1.5 V and falls outside the operating regime of the FETs with a supply voltage lower than 1.0 V.

The schematic view of the NWMOSFET used for the simulation is shown in Fig. 9.10.

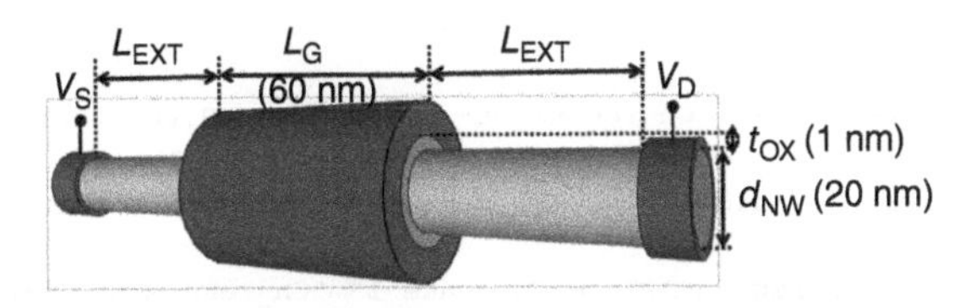

FIGURE 9.10 Rough sketch of the NWMOSFET to be created in Section 9.6.2.

9.6.2 Structure Specification

The structure of the experimental nanowire is specified using the command file given below:

```
;----------------------------------------------------------------------

; Overlap resolution: New replaces Old

(sdegeo:set-default-boolean "ABA")

;----------------------------------------------------------------------

; Creating source contact

(sdegeo:create-cylinder (position 0 0 0)  (position 0.035 0 0)  0.012 "Platinum"
"region_1" )

; Creating drain contact

(sdegeo:create-cylinder (position 0.105 0 0)  (position 0.14 0 0)  0.012
"Platinum" "region_2" )

; Creating gate contact

(sdegeo:create-cylinder (position 0.04 0 0)  (position 0.1 0 0)  0.012
"GatePolySilicon" "region_3" )

; Creating gate oxide region

(sdegeo:create-cylinder (position 0.04 0 0)  (position 0.1 0 0)  0.011 "SiO2"
"region_4" )

; Creating cylindrical silicon nanowire region

(sdegeo:create-cylinder (position 0 0 0)  (position 0.04 0 0)  0.01 "Silicon"
"region_Epi1" )

(sdegeo:create-cylinder (position 0.04 0 0)  (position 0.1 0 0)  0.01 "Silicon"
"region_Epi2" )

(sdegeo:create-cylinder (position 0.1 0 0)  (position 0.14 0 0)  0.01 "Silicon"
"region_Epi3" )

;---------Defining refinement size----------------------------------------------

(sdedr:define-refinement-size "Ch_refinement" 0.002 0.002 0.002 0.001 0.001 0.001 )

(sdedr:define-refinement-region "Place.Ch_refine" "Ch_refinement" "region_Epi1" )

(sdedr:define-refinement-region "Place.Ch_refine_1" "Ch_refinement" "region_Epi2")

(sdedr:define-refinement-region "Place.Ch_refine_2" "Ch_refinement" "region_Epi3")
```

```
(sdedr:define-refinement-size "oxide_refinement" 0.002 0.002 0.002 0.001 0.001
0.001)

(sdedr:define-refinement-region "Place.Siooxide" "oxide_refinement" "region_4" )

(sdedr:define-refinement-size "Gate_refinement" 0.005 0.005 0.005 0.002 0.002
0.002)

(sdedr:define-refinement-region "Place.Gate_refine" "Gate_refinement" "region_1" )

(sdedr:define-refinement-region "Place.Gate_refine1" "Gate_refinement" "region_2")

(sdedr:define-refinement-region "Place.Gate_refine2" "Gate_refinement" "region_3")

;----------Defining Doping------------------------------------------

; Defining Source/drain doping

(sdedr:define-constant-profile "Const.sdDoping" "PhosphorusActiveConcentration"
5e+19 )

(sdedr:define-constant-profile-region "Place.Channel" "Const.sdDoping"
"region_Epi1")

(sdedr:define-constant-profile-region "Place.Channel1" "Const.sdDoping"
"region_Epi3")

; Defining channel doping

(sdedr:define-constant-profile "Const.ChannelDoping" "BoronActiveConcentration"
1e+16 )

(sdedr:define-constant-profile-region "Place.Channel2" "Const.ChannelDoping"
"region_Epi2")

; Defining Gate polysilicon doping

(sdedr:define-constant-profile "Const.GateDoping" "BoronActiveConcentration"
1e+20)

(sdedr:define-constant-profile-region "Place.Gate" "Const.GateDoping" "region_1")

;----------Defining contacts--------------------------------------------------------

; Contact declarations

(sdegeo:define-contact-set "source" 4.0  (color:rgb 1.0 0.0 0.0 ) "##")

(sdegeo:define-contact-set "drain"  4.0  (color:rgb 0.0 1.0 0.0 ) "##")

(sdegeo:define-contact-set "gate" 4.0 (color:rgb 0.0 0.0 1.0 ) "##")

;---------Contact placement----------------------------------------------------------
```

```
; Contact placement to be done via GUI. Steps:

; Select the contact in the contact dialogue box and chose whether it is to be
defined as a face contact or as the body contact in the selection pane. We chose
the face contact for source and drain regions.

; Select the face/body to be defined as contact by right clicking on it
(Fig. 9.11(a))

; Place the contact by selecting "Set contact" option from the Contact drop down
box.

; Repeat these steps for selecting and defining source, drain and gate contacts
to obtain the structure shown in Fig. 9.11(b).

;---------Generating the mesh using SNMesh------------------------------------

(sde:build-mesh "snmesh" "-offset" "nwmosfet_msh")

; or by using GUI. Steps: Mesh drop down box -> Build mesh -> Give appropriate
mesh name -> build mesh

;-----------------------------------------------------------------------------
```

First, the overlap resolution was set to new replaces old. This is a very crucial step.

The statement `(sdegeo:create-cylinder (position 0 0 0) (position 0.035 0 0) 0.012 "Platinum" "region_1")` creates a solid platinum cylinder whose central axis starts from the origin and end at 35 nm along the x-axis and whose radius is specified as 12 nm, i.e. it spans from $z = -12$ nm to $z = 12$ nm. Similarly, a solid platinum cylinder is constructed for a drain contact.

After this, a solid gate polysilicon cylinder is created for gate contact. Then, a solid SiO_2 cylinder with a radius of 11 nm is constructed to form the gate oxide. Owing to the new replaces old overlap resolution, since the diameter of gate oxide cylinder is 2 nm lower than the gate polysilicon, the SiO_2 replaces the gate polysilicon leaving only a hollow cylindrical region. Now, you may think how a solid cylinder can act as the gate oxide. This would become clear in this discussion. The definition of the active silicon film with a radius of 10 nm, i.e. 1 nm lower than the gate oxide leads to replacement of the gate oxide by silicon solid cylinder. The gate oxide is limited to a hollow cylinder of thickness 1 nm encompassing the solid silicon channel region and covering it from all sides.

It may be noted that here also we have defined the source, channel, and drain regions as different silicon regions rather than a single silicon region to ease the specification of the nonlocal mesh. Once the regions have been defined, we define the refinement size and place them adequately. Then the doping profile is defined and placed in source, drain, channel, and gate polysilicon regions.

The contact declaration is the next step followed by the placement of the contacts. We use a simple GUI approach for placing the contacts. We simply copy and paste the command file in the SDE window and then use GUI functions to define

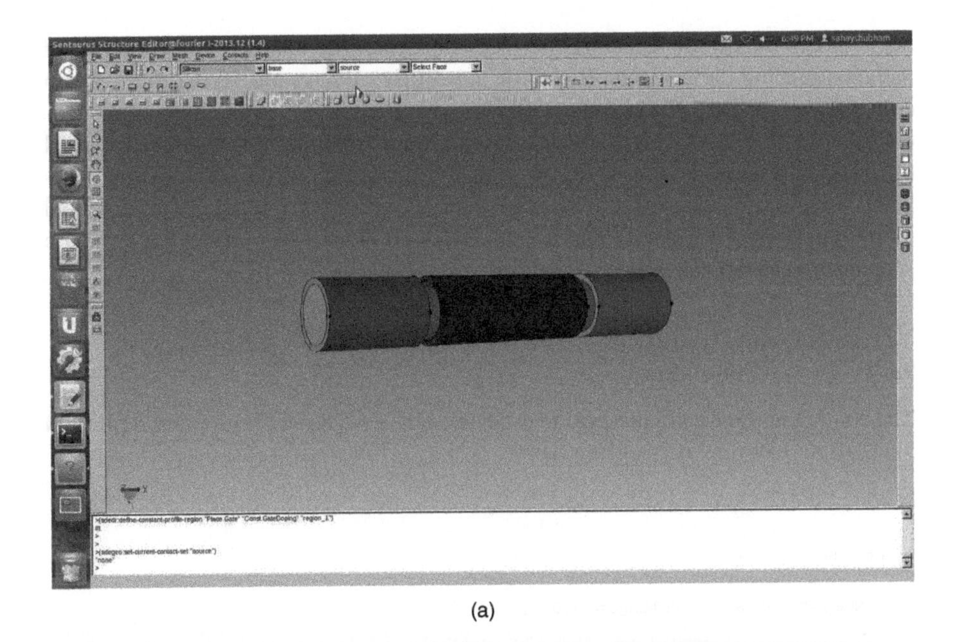

(a)

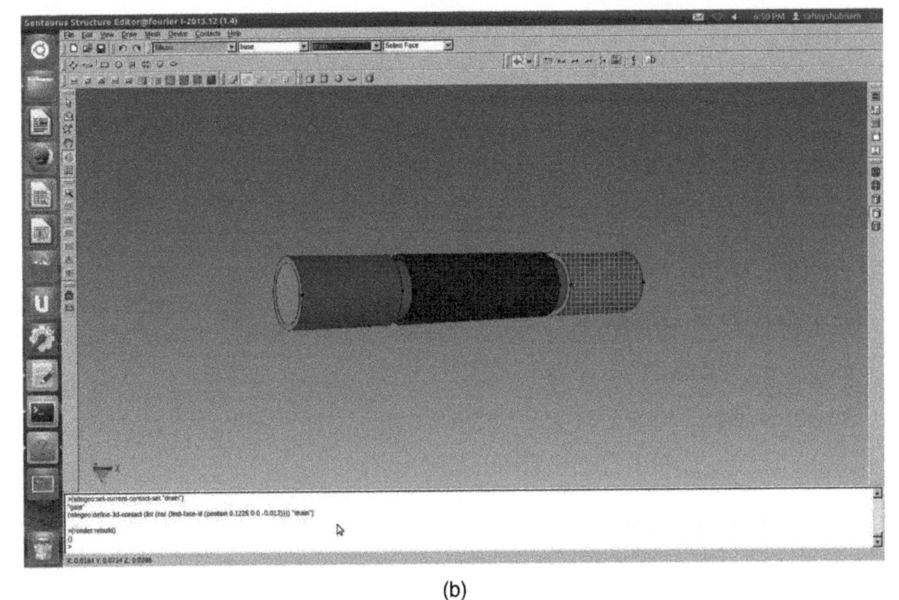

(b)

FIGURE 9.11 Contact placement via the GUI interface (a) selecting the contact by right click and (b) the final structure with all the contacts placed using GUI.

the contacts by simply selecting the faces on which the contacts are to be placed and then placing the contacts. It may be noted that we have placed the source/drain contacts on the face of the NWMOSFET, which is similar to the case of a vertically stacked NWMOSFET. However, in [40], the source/drain contacts were placed on top of the large silicon area far from the nanowire which was defined by e-beam lithography. The mesh should be constructed either by using the command or by using the GUI itself. The use of GUI + command file makes the structure specification easier.

9.6.3 Device Simulations

The command file for simulating the experimental NWMOSFET structure replicated via SDE in Section 9.6.2 is given below:

```
#----------Electrode specification-------------------------------------------------#

Electrode {

   { Name="source"    Voltage= 0.0  }

   { Name="drain"     Voltage= 0.0 }

   { Name="gate"      Voltage= 0.0 }

}

#-----------File section-----------------------------------------------------------#

File {

#-Predefined SWB paranmeters

   Grid    = "nwmosfet_msh.tdr"

   Plot    = "@tdrdat@"

   Current = "@plot@"

   Output  = "@log@"

   Param   = "Silicon.par"

}

#-----------Physical model specification-----------------------------------------#

Physics

{

Mobility( PhuMob HighFieldSat Enormal)
```

```
EffectiveIntrinsicDensity(BandGapNarrowing (oldSlotboom))

Recombination( SRH Auger

eBarrierTunneling(Band2Band TwoBand Transmission)

hBarrierTunneling(Band2Band TwoBand Transmission))

}

Physics (Material = "GatePolySilicon") {MetalWorkfunction ( Workfunction= 4.4 )}

#----------Plot section--------------------------------------------------------#

Plot

{

eDensity hDensity eCurrent hCurrent equasiFermi

hquasiFermi AvalancheGeneration eGradQuasiFermi

ConductionBandEnergy ValenceBandEnergy ConductionCurrent

DisplacementCurrent eAlphaAvalanche hAlphaAvalanche

EffectiveBandGap EquilibriumPotential eQuasiFermi

EffectiveBandGap hGradQuasiFermi

hQuasiFermi IntrinsicDensity

EffectiveIntrinsicDensity eTemperature eIonIntegral Auger

ElectricField eEparallel hEparallel Potential Doping

SpaceCharge SRHRecombination eMobility hMobility

eVelocity hVelocity

DonorConcentration AcceptorConcentration

eIonIntegral hIonIntegral

MeanIonIntegral BuiltinPotential

eDriftVelocity Band2Band BandGap

HotElectronInjection HotHoleInjection

eBarrierTunneling hBarrierTunneling

}

#-----------Maths section-------------------------------------------------------#
```

```
Math
{
Extrapolate
Derivatives
Digits=6
ErrRef(Electron) = 1e8
ErrRef(Hole)     = 1e8
RelErrcontrol
Iterations=25
Method=Blocked SubMethod=ParDiso
number_of_threads = 8
stacksize = 20000000
wallclock
}
Math(regionInterface="region_Epi1/region_Epi2")
{ NonLocal(Length=2.5e-6 Permeable Permeation=2.5e-6)
Digits(NonLocal)=3
EnergyResolution(NonLocal)=0.001
}
Math(regionInterface="region_Epi2/region_Epi3")
{NonLocal(Length=2.5e-6 Permeable Permeation=2.5e-6)
Digits(NonLocal)=3
EnergyResolution(NonLocal)=0.001
}
#----------Solve section------------------------------------------------------#
Solve {
NewCurrentPrefix="init_1"
Coupled(Iterations= 100 LineSearchDamping= 1e-4){ Poisson }
```

```
Coupled {Poisson Electron Hole}

quasistationary ( Goal {name="drain" voltage=1.0}

Initialstep=0.001 Maxstep=0.05 Minstep=1e-7)

{coupled {poisson Electron Hole }}

CurrentPlot( Time=(Range=(0 1) Intervals= 30)  )

Plot ( FilePrefix="Vgs-0_n@node@" Time=(0;0.1;0.2;0.3;0.4;0.5;0.6;0.7;0.8;0.9;1.0)
NoOverWrite)

NewCurrentPrefix="init_2"

Coupled (LineSearchDamping=0.01){Poisson}

Coupled {Poisson Electron Hole}

quasistationary ( Goal {name="gate" voltage=1.0}

Initialstep=0.001 Maxstep=0.01 Minstep=1e-7)

{coupled {poisson electron hole }}

CurrentPlot( Time=(Range=(0 1) Intervals= 30)  )

Plot ( FilePrefix="Vgs1.0_n@node@" Time=
(0;0.1;0.2;0.3;0.4;0.5;0.6;0.7;0.8;0.9;1.0) NoOverWrite)

NewCurrentPrefix="init_3"

Coupled (LineSearchDamping=0.01){Poisson}

Coupled {Poisson Electron Hole}

quasistationary ( Goal {name="gate" voltage=-1.5}

Initialstep=0.001 Maxstep=0.01 Minstep=1e-7)

{coupled {poisson electron hole }}

CurrentPlot( Time=(Range=(0 1) Intervals= 30)  )

Plot ( FilePrefix="Vgs-1.0_n@node@" Time=
(0;0.1;0.2;0.3;0.4;0.5;0.6;0.7;0.8;0.9;1.0) NoOverWrite)

}

#----------Solve section-----------------------------------------------------#
```

The Sdevice command file for simulating NWMOSFETs is almost similar to that of the SOI JLFET in Section 9.5.2. We first specify the initial boundary conditions of the electrodes followed by definition of the different input and output file names. The different physical models are then specified.

For NWMOSFET, the Philips unified mobility model has been used to take into account phonon scattering and carrier–carrier scattering in the inversion layer. The Lombardi mobility model has also been used to take into account the surface roughness since inversion layer formation takes place at the surface. Moreover, the Canali model has been used to consider velocity saturation for the short-channel (60 nm) NWMOSFET [5]. To consider the band gap narrowing in the heavily doped source/drain regions of the MOSFET, the Slotboom band gap narrowing model was used. For taking the recombination processes into account, the SRH recombination model with default carrier lifetimes and the Auger recombination model was also used.

Furthermore, the basic nonlocal tunneling model based on user-specified tunneling path and nonlocal mesh was used to take L-BTBT into account. The gate polysilicon was defined as a metal with a midgap work function of 4.4 eV. The physical parameters to be visualized later in the output TDR file have been mentioned in the plot section.

The different settings for the coupled Newton solver are specified in the Maths section. This section is almost similar to Section 9.5.2. The additional statements `Digits` and `ErrRef` have been added for tackling the convergence issues. The statement `Digits` specifies the number of digits after the decimal point to which the physical equations must be solved for convergence. `ErrRef` statement is used to define the value of relative error for electrons and holes. As discussed in Section 9.3.2, the guess for any particular step is considered as solution and the equation is said to be converged if either the error is lower than a predefined absolute value or the relative value defined using `ErrRef`.

The nonlocal mesh is then defined at the source–channel and the channel–drain interface to take L-BTBT into account. The values of `Permeation` and `Length` of 25 nm are similar to those used in Section 9.5.2. After the settings for the numerical solver have been defined, the solve statements are specified to ramp the drain voltage and the gate voltage to the desired values and to save the voltage and current values at the different electrodes using the current plot and to save the TDR structure file at different instances.

The output TDR structure file for gate voltage = 0.0 V obtained after Sdevice simulation named as `Vgs-0_n@node@_des.tdr` is shown in Fig. 9.12. The energy band profiles along a cutline at a location 1 nm below the Si–SiO_2 interface is also shown in Fig. 9.12. It can be clearly seen that there exists a considerable band overlap in NWMOSFETs in the OFF-state which facilitates L-BTBT similar to the JLFETs.

9.6.4 Model Parameters

As discussed in Section 9.5.4, the model parameters must be tuned to match the experimental transfer characteristics of [39]. First, we identify that for gate voltages below 0.0 V, the current is entirely due to the L-BTBT mechanism. Therefore, we need to

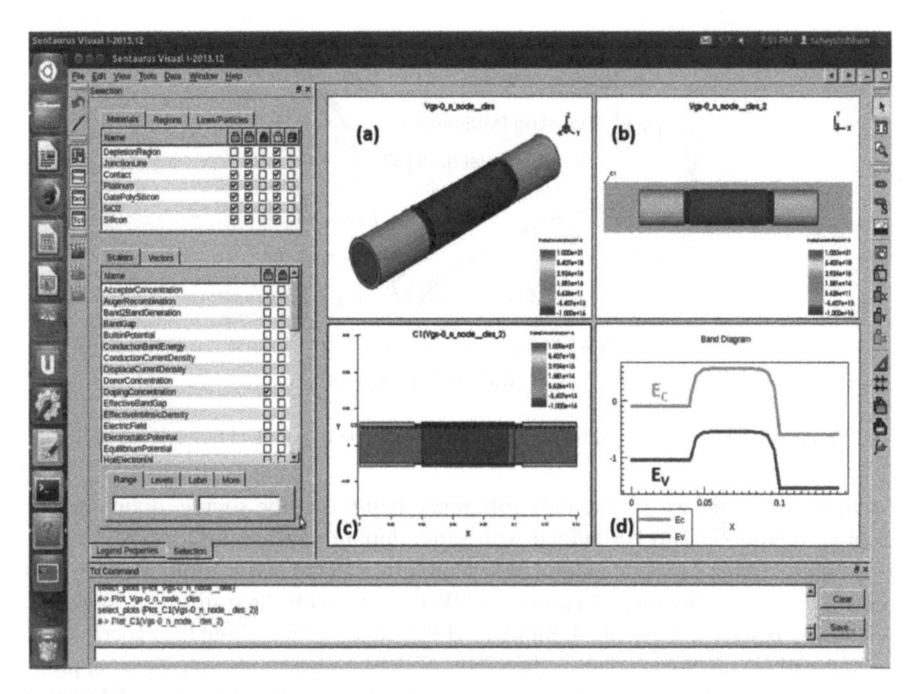

FIGURE 9.12 The output TDR file `Vgs-0_n@node@_des.tdr` in SVisual (a) 3D view, (b) the cutline along the *z*-axis, (c) the doping profile, and (d) the energy band profile at a cutline 1 nm below the Si–SiO_2 interface.

tune the tunneling model parameters such as tunneling mass to match the simulation results with the experimental results. This requires a lot of hit and trial simulations with different set of tunneling mass for electrons and holes. However, fortunately, the simulation results matched with the experimental data using the value of the tunneling mass for electrons and holes specified in the parameter file used in Section 9.5.4 as shown in Fig. 9.13. However, as can be observed from Fig. 9.13, there is a slight mismatch between the ON-state current of the simulated NWMOSFET as compared to the experimental characteristics. This is attributed to two reasons: (a) The fabricated nanowires are bound to be strained and, therefore, may have a larger mobility as compared to the simulated structure where we have not defined any strain. (b) The area for source/drain contact was larger in the experimental device as compared to the simulation resulting in a smaller series resistance of the source and drain regions. The reasons for the mismatch, if any, between the experimental data and the simulated results should also be logically analyzed while model calibration.

9.6.5 Simulation of NWJLFETs

Once the models for simulating NWFETs are calibrated, they may be utilized for simulation of NWJLFETs in the same way utilizing the command file given in

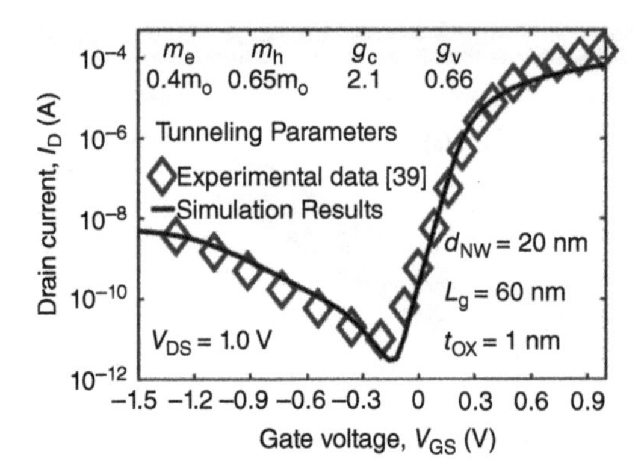

FIGURE 9.13 Model calibration by reproducing the experimental results of [39].

Sections 9.6.2 and 9.6.3. The only difference being that the source, drain and the channel region need to be defined with same doping concentration of 10^{19} cm^{-3}. Now, utilizing the calibrated simulation setup given in the last section, we have also reproduced the experimental results of NWJLFETs with three different nanowire widths [1] as shown in Fig. 9.14. Similar to [39], all the details about the experimental structure of [1] are not available. Therefore, we have to take some intelligent approximations. For instance, the cross-sectional structure of the nanowires fabricated in [1] (shown in Fig. 3.10(g)) appears somewhat oval shaped. Reproducing such an oval-shaped structure in Sentaurus TCAD is itself a challenge. Therefore, we assumed a circular cross section for these nanowires and assumed their diameter to be equal to the nanowire widths reported.

However, simulating these NWFETs utilizing the calibrated simulation deck given in Sections 9.6.2 and 9.6.3 would yield transfer characteristics which are shifted

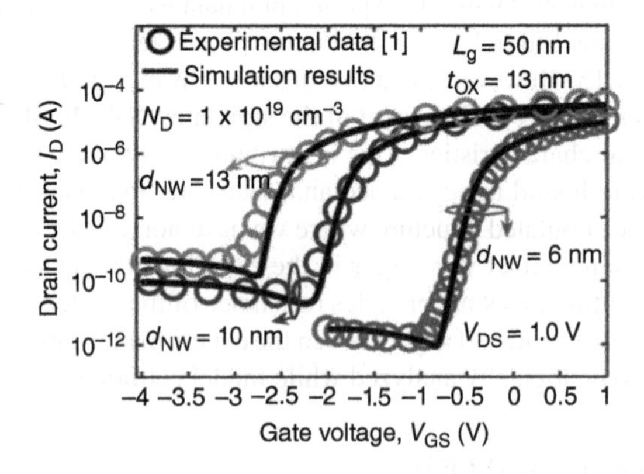

FIGURE 9.14 The simulation of experimental NWJLFETs of [1] using the calibrated simulation deck provided in Section 9.6.3. The simulation results match with the experimental results validating the simulation setup.

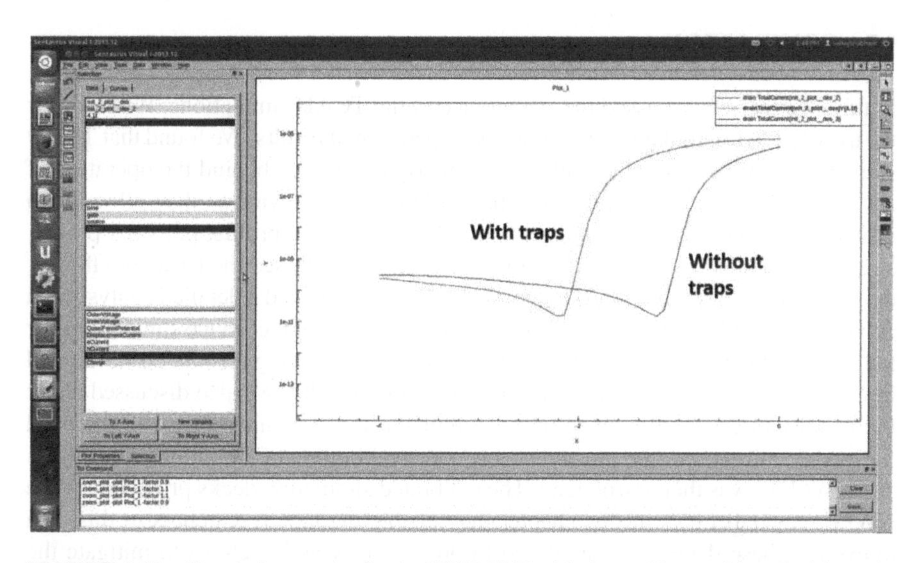

FIGURE 9.15 Simulation results for the NWJLFET with $d_{NW} = 10$ nm with and without the traps.

toward the positive gate voltage by some amount. Even after using the gate electrode with minimum possible work function, i.e. 3.9 eV, the transfer characteristics obtained by simulating the simplified experimental nanowire with circular cross section and nanowire diameter 10 nm would appear as shown in Fig. 9.15. The threshold voltage of the simulated transfer characteristics appears shifted by a particular value as compared to the experimental transfer characteristics.

Upon careful investigation, we can notice that the NWJLFETs fabricated in [1] were reported for application as silicon-oxide-nitride-oxide-silicon (SONOS) memory. Therefore, the oxide charges or charges in the nitride layer are inevitable. Using this logical assumption, if we use a fixed charge value in the simulations, we can shift the simulated transfer characteristics to match the experimental data. However, this shift also requires a lot of hit and trial experiments with the simulator. After a large number of efforts, we found that for a fixed charge of 3.4×10^{12} cm^{-2} at the Si–SiO_2 interface, the simulated transfer characteristics finally matched with the experimental data as shown Fig. 9.14. The statements used in Sdevice for specifying the interface charge is given below:

```
Physics (MaterialInterface="Silicon/Oxide") {Traps((FixedCharge Conc=3.4e+12))}
```

The simulation deck provided in Sections 9.5.2 and 9.5.3 has been used extensively to analyze the physics of the JLFETs and NWFETs and to predict the performance of different architectures proposed to mitigate the challenges faced by these emerging FET architectures in [17–25, 28, 41]. Interested readers are directed to [42] for further information.

9.7 CONCLUSION

In this chapter, we discussed the importance of the TCAD simulations. We saw how the use of TCAD simulation augments the experimental results. We found that TCAD simulations are necessary to explain the underlying physics behind the operation of any FET. They give an insight into the working of the device as they allow us to visualize the microscopic parameters such as electric field profile, potential profile, etc. which cannot be done via experiments. Then we discussed the basic tool flow of the most commonly used TCAD Sentaurus. We also provided a detailed analysis of a demo command file for both creating structure and running the device simulations for long-channel JLFETs. This code may be utilized for validating the models developed in Chapter 8 by matching them with the simulation results. We also discussed about the validity of the results given by the simulator and the importance and the need for model calibration. The procedure for model calibration of short-channel JLFETs and NWFETs was then introduced. The calibrated simulation decks provided for the NWFETs and JLFETs in this chapter may enable the new researchers in this area to explore these devices in detail and to propose new architectures to mitigate the challenges for these FETs.

In the next chapter, we present a brief overview of the road ahead for JLFETs and the various emerging fields where junctionless architecture may be efficiently exploited.

REFERENCES

[1] S.-J. Choi, D.-I. Moon, S. Kim, J. P. Duarte, and Y.-K. Choi, "Sensitivity of threshold voltage to nanowire width variation in junctionless transistors," *IEEE Electron Device Lett.*, vol. 32, no. 2, pp. 125–127, Feb. 2011.

[2] C.-H. Park, M.-D. Ko, K.-H. Kim, R.-H. Baek, C.-W. Sohn, C. K. Baek, S. Park, M. J. Deen, Y.-H. Jeong, and J.-S. Lee, "Electrical characteristics of 20-nm junctionless Si nanowire transistors," *Solid-State Electron.*, vol. 73, pp. 7–10, July 2012.

[3] I. Wong, Y. Chen, S. Huang, W. Tu, Y. Chen, and C. W. Liu, "Junctionless gate-all-around PFETs using in-situ boron doped Ge channel on Si," *IEEE Trans. Nanotech.*, vol. 14, no. 5, pp. 878–882, Sept. 2015.

[4] S. Migita, Y. Morita, T. Matsukawa, M. Masahara, and H. Ota, "Experimental demonstration of ultrashort-channel (3 nm) junctionless FETs utilizing atomically sharp V-grooves on SOI," *IEEE Trans. Nanotechnol.*, vol. 13, no. 2, pp. 208–215, Mar. 2014.

[5] *Sentaurus Device User Guide*, Synopsys, Mountain View, CA, 2017.

[6] *ATLAS Device Simulation Software*, Silvaco, Santa Clara, CA, 2017.

[7] *Genius*, 3-D Device Simulator, Version1.9.0, Reference Manual, Cogenda Pvt. Ltd., Singapore, 2008.

[8] E. M. Buturla, P. E. Cottrell, B. M. Grossman, and K. A. Salsburg, "Finite-element analysis of semiconductor devices: The FIELDAY program," *IBM J, Res. Dev.*, vol. 25, no. 4, pp. 218–231, July 1981.

[9] Nanohuh [online], Available: www.nanohub.org, Accessed: Dec. 23, 2017.

[10] D. B. M. Klaassen, "A unified mobility model for device simulation—I. Model equations and concentration dependence," *Solid-State Electron.*, vol. 35, no. 7, pp. 953–959, 1992.

[11] C. Lombardi, S. Manzini, A. Saporito, and M. Vanzi, "A physically based mobility model for numerical simulation of nonplanar devices," *IEEE Trans. Comput.-Aided Design*, vol. 7, no. 11, pp. 1164–1171, 1988.

[12] S. Sahay and M. J. Kumar, "Realizing efficient volume depletion in SOI junctionless FETs," *IEEE J. Electron Devices Soc.*, vol. 4, no. 3, pp. 110–115, May 2016.

[13] *Sentaurus Solvers User Guide*, Synopsys, Mountain View, CA, 2017

[14] S. Gundapaneni, M. Bajaj, R. K. Pandey, K. V. R. M. Murali, S. Ganguly, and A. Kottantharayil, "Effect of band-to-band tunneling on junctionless transistors," *IEEE Trans. Electron Devices*, vol. 59, no. 4, pp. 1023–1029, Apr. 2012.

[15] J.-P. Colinge, C.-W. Lee, A. Afzalian, N. D. Akhavan, R. Yan, I. Ferain, P. Razavi, B. O'Neill, A. Blake, M. White, A.-M. Kelleher, B. McCarthy, and R. Murphy, "Nanowire transistors without junctions," *Nature Nanotechnol.*, vol. 5, no. 3, pp. 225–229, Mar. 2010.

[16] H. G. Virani, A. R. B. Rao, and A. Kottantharayil, "Dual-k spacer device architectures for the improvement of performance of hetero structure n-channel tunnel FETs," *IEEE Trans. Electron Devices*, vol. 57, no. 10, pp. 2410–2417, Oct. 2010.

[17] S. Sahay and M. J. Kumar, "Physical insights into the nature of gate-induced drain leakage in ultrashort channel nanowire FETs," *IEEE Trans. Electron Devices*, vol. 64, no. 6, pp. 2604–2610, Jun. 2017.

[18] S. Sahay and M. J. Kumar, "A novel gate-stack-engineered nanowire FET for scaling to the sub-10-nm regime," *IEEE Trans. Electron Devices*, vol. 63, no. 12, pp. 5055–5059, Dec. 2016.

[19] S. Sahay and M. J. Kumar, "Spacer design guidelines for nanowire FETs from gate-induced drain leakage perspective," *IEEE Trans. Electron Devices*, vol. 64, no. 7, pp. 3007–3015, July 2017.

[20] S. Sahay and M. J. Kumar, "Insight into lateral band-to-band-tunneling in nanowire junctionless FETs," *IEEE Trans. Electron Devices*, vol. 63, no. 10, pp. 4138–4142, Oct. 2016.

[21] S. Sahay and M. J. Kumar, "Controlling L-BTBT and volume depletion in nanowire JLFETs using core-shell architecture," *IEEE Trans. Electron Devices*, vol. 63, no. 9, pp. 3790–3794, Sept. 2016.

[22] S. Sahay and M. J. Kumar, "Diameter dependency of leakage current in nanowire junctionless field-effect transistors," *IEEE Trans. Electron Devices*, vol. 64, no. 3, pp. 1330–1335, Mar. 2017.

[23] S. Sahay and M. J. Kumar, "Nanotube junctionless FET: Proposal, design, and investigation," *IEEE Trans. Electron Devices*, vol. 64, no. 4, pp. 1851–1856, Apr. 2017.

[24] M. J. Kumar and S. Sahay, "Controlling BTBT induced parasitic BJT action in junctionless FETs using a hybrid channel," *IEEE Trans. Electron Devices*, vol. 63, no. 8, pp. 3350–3353, Aug. 2016.

[25] S. Sahay and M. J. Kumar, "Symmetric operation in an extended back gate JLFET for scaling to the 5 nm regime considering quantum confinement effects," *IEEE Trans. Electron Devices*, vol. 64, no. 1, pp. 21–27, Jan. 2017.

[26] V. Thirunavukkarasu, Y.-R. Jhan, Y.-B. Liu, and Y.-C. Wu, "Performance of inversion, accumulation, and junctionless mode n-type and p-type bulk silicon FinFETs with 3-nm gate length," *IEEE Electron Device Lett.*, vol. 36, no. 7, pp. 645–647, July 2015.

[27] Y. R. Jhan, V. Thirunavukkarasu, C. P. Wang, and Y. C. Wu, "Performance evaluation of silicon and germanium ultrathin body (1 nm) junctionless field-effect transistor with ultrashort gate length (1 nm and 3 nm)," *IEEE Electron Device Lett.*, vol. 36, no. 7, pp. 545–656, July 2015.

[28] S. Sahay and M. J. Kumar, "Comprehensive analysis of gate-induced drain leakage in emerging FET architectures: Nanotube FETs vs. nanowire FETs," *IEEE Access*, vol. 5, pp. 18918–18926, Dec. 2017.

[29] Web plot digitizer [online], Available: http://arohatgi.info/WebPlotDigitizer/app3_12/, Accessed: Dec. 23, 2017.

[30] A. Schenk, "Rigorous theory and simplified model of the band-to-band tunneling in silicon," *Solid-State Electron.*, vol. 36, no. 1, pp. 19–34, 1993.

[31] J. J. Liou, "Modeling the tunnelling current in reverse-biased p/n junctions," *Solid-State Electron.*, vol. 33, no. 7, pp. 971–972, 1990.

[32] G. A. M. Hurkx, D. B. M. Klaassen, and M. P. G. Knuvers, "A new recombination model for device simulation including tunneling," *IEEE Trans. Electron Devices*, vol. 39, no. 2, pp. 331–338, 1992.

[33] E. O. Kane, "Theory of tunneling," *J. Appl. Phys.*, vol. 32, no. 1, pp. 83–91, 1961.

[34] M. J. Kumar, R. Vishnoi, and P. Pandey, *Tunnel Field-effect Transistors (TFET): Modelling and Simulation*, John Wiley and Sons, Ltd, Ltd., West Sussex, UK, 2016.

[35] S. Sahay and M. J. Kumar, "Controlling the drain side tunneling width to reduce ambipolar current in tunnel FETs using heterodielectric BOX," *IEEE Trans. Electron Devices*, vol. 62, no. 11, pp. 3882–3886, Nov. 2015.

[36] L. D. Michielis, M. Iellina, P. Palestri, A. M. Ionescu, and L. Selmi, "Effect of the choice of the tunnelling path on semi-classical numerical simulations of TFET devices," *Solid-State Electron.*, vol. 71, pp. 7–12, 2012.

[37] D. Esseni, M. Pala, P. Palestri, C. Alper and T. Rollo, "A review of selected topics in physics based modeling for tunnel field-effect transistors," *Semicond. Sci. Technol.*, vol. 32, pp. 083005–083031, July 2017.

[38] S. Mookerjea, R. Krishnan, S. Datta, and V. Narayanan, "On enhanced Miller capacitance effect in interband tunnel transistors," *IEEE Electron Device Lett.*, vol. 30, no. 10, pp. 1102–1104, Oct. 2009.

[39] J. Fan, M. Li, X. Xu, Y. Yang, H. Xuan, and R. Huang, "Insight into gate-induced drain leakage in silicon nanowire transistors," *IEEE Trans. Electron Devices*, vol. 62, no. 1, pp. 213–219, Jan. 2015.

[40] J. Fan, M. Li, X. Xu, and R. Huang, "New observation on gate induced drain leakage in silicon nanowire transistors with epi-free CMOS compatible technology on SOI substrate," in *Proc. IEEE SOI-3D-Subthreshold Microelectron. Technol. Unified Conf.* (S3S), Oct. 2013, pp. 1–2.

[41] A. K. Jain, S. Sahay, and M. J. Kumar, "Controlling L-BTBT in emerging nanotube FETs using dual-material gate," *IEEE J. Electron Dev. Soc.*, vol. 6, pp. 611–621, June 2018.

[42] S. Sahay, "Design and analysis of emerging nanoscale junctionless FETs from gate-induced drain leakage perspective," Ph.D. Thesis, IIT Delhi, New Delhi, India, Mar. 2018.

10

CONCLUSION AND PERSPECTIVES

In the previous chapters, we discussed the promising potential of the junctionless field-effect transistors (JLFETs) and the distinct advantages that they offer over the conventional metal–oxide–semiconductor field-effect transistors (MOSFETs). We saw how the absence of metallurgical junction not only reduces the fabrication complexity and cost but also provides flexibility while choosing the materials for the gate stack due to a lower thermal budget [1–15]. We discussed how the junctionless architecture leads to an altogether new conduction mechanism whereby the field-effect transistor (FET) operates under the flat band condition in the ON-state with a significantly reduced vertical electric field. The low electric field alleviates the reliability issues in JLFETs which is a menace for the conventional MOSFETs [16]. Also, the bulk conduction feature enables the integration of other materials like Ge, III-V, SiC, etc. in JLFETs as the performance is not severely degraded due to the interface traps and defects [16]. We also discussed how the electrostatic squeezing of carriers in the OFF-state leads to an unintentional underlap in JLFETs and helps to realize a larger effective channel length than the drawn gate length [15]. This unique feature of JLFETs not only increases their immunity to the short-channel effects as compared to the MOSFETs but also enhances their scalability. Moving further, we saw how the impact ionization induced steep subthreshold swing occurs at a lower drain voltage in JLFETs as compared to MOSFETs [17]. We also discussed the device architectures, which harness impact ionization phenomenon in JLFETs to achieve

Junctionless Field-Effect Transistors: Design, Modeling, and Simulation, First Edition.
Shubham Sahay and Mamidala Jagadesh Kumar.

steep subthreshold swing at subunity drain voltages. These attributes strengthen the candidature of the JLFETs to replace the MOSFETs in the future complementary metal–oxide–semiconductor (CMOS) technology for logic applications.

However, the high source/drain series resistance in JLFETs reduces their ON-state current and degrades their dynamic performance [1–15]. Moreover, achieving volume depletion requires costly ultrathin SOI films. Also, the OFF-state leakage current is significantly increased in JLFETs due to the band-to-band tunneling (BTBT) induced parasitic bipolar junction transistor (BJT) action [1–15]. Therefore, we also discussed in detail the different device architectures proposed to boost the ON-state current, achieve volume depletion, and mitigate the BTBT gate-induced drain leakage (GIDL) in the previous chapters. We also discussed the specific challenges faced by each of these device architectures to evaluate their potential for commercial applications.

The high sensitivity of JLFETs to random dopant fluctuations (RDF) and process variations as compared to MOSFETs limit their applications. However, the chemical doping-free electrostatic doping techniques such as charge plasma (CP) doping discussed in Chapter 7 may mitigate RDF and sensitivity to process variations in JLFETs. This may pave the way for utilization of field-induced source/drain doping in charge plasma JLFETs (CPJLFETs) for application in the mainstream CMOS technology.

Moreover, we also discussed the specific challenges associated with different emerging FET architectures such as tunnel FETs (TFETs), impact ionization-MOS (I-MOS), negative capacitance FETs (NCFETs), nanowire FETs (NWFETs), nanotube FETs (NTFETs) to replace the conventional MOSFETs in Chapter 1. One of the major issues bothering all these FETs with metallurgical junctions is the stringent requirement of ultrasteep doping profile at the junctions. The chemical doping-free field-induced doping of source/drain regions on an intrinsic or lightly doped ultrathin silicon film utilizing electrostatic doping techniques such as CP may enable the realization of junctionless version of these emerging FETs [18–22]. The junctionless versions of these emerging FET architectures such as JLTFETs, JL-IMOS, discussed in Chapter 7, are not only immune to RDF but also exhibit comparable performance to their doped counterparts. Therefore, junctionless versions of the emerging FETs are also lucrative alternative to the conventional CMOS.

The major challenge associated with CP doping is the metal-induced gap states (MIGS), which lead to Fermi-level pinning at the semiconductor–metal interface and degrade the performance of CP-based FETs [23]. With the advancements in the fabrication technology in future and the introduction of novel techniques such as ultrathin insulator between the metal and the semiconductor, MIGS may be mitigated and junctionless versions of the emerging FETs may become viable alternative to the MOSFETs in the mainstream CMOS technology for logic applications.

So far, we have only discussed the advantages and limitations of the JLFETs for logic applications. However, we are living in an era of Internet of Things (IoT) and big data where almost all the facets of our life are connected to the Internet. A new ecosystem of connected devices including a swarm of sensors, actuators, memories, etc. under the umbrella of IoT is touching our daily lives. Low-cost and low-power sensors and memories have become inevitable to sustain this era of IoT. Junctionless architecture with lower cost, low fabrication complexity, and a better immunity to the short-channel effects may enable realization of ultralow power and low-cost FET-based biosensors and memories. In the subsequent sections, we shall see the application of JLFET for biosensing and memory applications.

10.1 JLFETS AS A LABEL-FREE BIOSENSOR

MOSFET-based biosensors are gaining popularity owing to their low cost and the ability to detect unlabeled biomolecules when functionalized with appropriate receptors [24–39]. In this approach, the gate oxide is etched forming a nanocavity as shown in Fig. 10.1 and the target biomolecules are allowed to flow in the cavity. The silicon surface is functionalized with appropriate receptors, which bind the target biomolecules once they contact the receptors. The change in the FET characteristics in the presence and absence of the biomolecules allows for their detection. The target biomolecules are characterized by their equivalent electrical property such as dielectric constant or charge.

However, the MOSFET-based biosensors offer poor selectivity and the signal-to-noise ratio is very low due to the intrinsic variations. To overcome these limitations, other variants of FETs such as TFETs and I-MOS have been utilized for biosensing applications [32–35]. The steep subthreshold swing offered by these emerging FETs enhances the sensing margin. However, these FETs especially I-MOS have a p–i–n asymmetric structure and is not scalable. The size of the proteins lies in the sub-10 nm range, and the most commonly used proteins such as biotin or streptavidin are ~3–5 nm in size. Therefore, for efficient detection, the nanogaps should be designed in the range of ~10 nm so that they allow the biomolecules to flow while exhibiting enhanced sensitivity. Therefore, the FETs used as biosensors should also be scalable

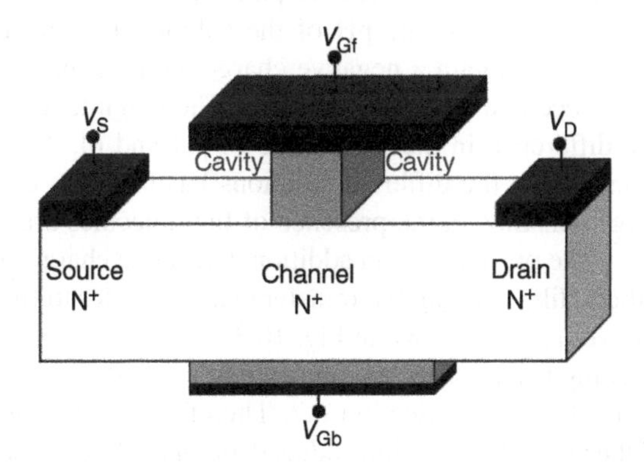

FIGURE 10.1 Three-dimensional (3D) view of the DGJLFET with nanocavities for biosensing application.

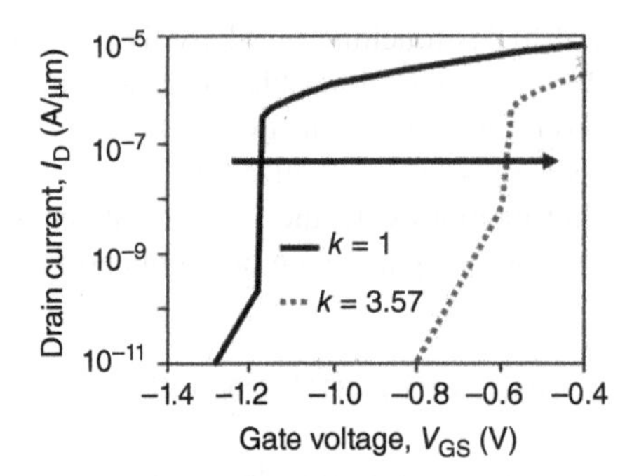

FIGURE 10.2 Transfer characteristics of the DGJLFET biosensor in the presence of different biomolecules [39].

to the sub-100 nm range to offer increased packing density along with enhanced sensitivity.

The JLFETs can offer a low-cost, scalable, and highly sensitive biosensor when operating in the impact ionization dominant regime exhibiting steep subthreshold slopes like I-MOS as discussed in Chapter 6. The performance of the JLFET-based biosensor in the impact ionization regime is shown in Fig. 10.2. The efficacy of the JLFET-based biosensor has been tested on three target biomolecules: biotin, streptavidin, and (3-aminopropyl)triethoxysilane (APTES) modeled with a dielectric constant of 2.63, 2.1, and 3.57, respectively [39]. The shift in the threshold voltage in the presence and absence of biomolecules is taken as the sensing margin.

The presence of the biomolecule increases the effective gate capacitance leading to a larger electric field coupling. This suppresses the lateral electric field reducing the impact ionization rate. Therefore, a larger gate voltage has to be supplied to attain the steep switching behavior.

In addition to the dielectric constant, the biomolecules also exhibit an inherent charge depending on their isoelectronic point (pI), pH of the solution, and the concentration of the solution. If the pH of the solution is more than the isoelectric point, the biomolecules gain a negative charge while if the pH is less than the pI, the biomolecules become positively charged. The magnitude of the charge also depends upon the difference in the magnitude of pH and pI. Although the charge density of biomolecules in the different solutions has not been characterized well, the observed charge densities in the presence of biomolecules are in the range of -10^{11} to -10^{12} cm^{-2}. The presence of an additional negative charge helps in depleting the underneath silicon film, forcing the transfer characteristics to further shift toward more positive gate voltages as shown in Fig. 10.3.

The steep switching behavior can be obtained at lower drain voltages with asymmetric operation as observed in Section 6.7. Therefore, the asymmetric operation, which enhances the impact ionization-induced bipolar effects, may be utilized to further enhance the sensitivity of the JLFET-based biosensor in the impact ionization dominant regime.

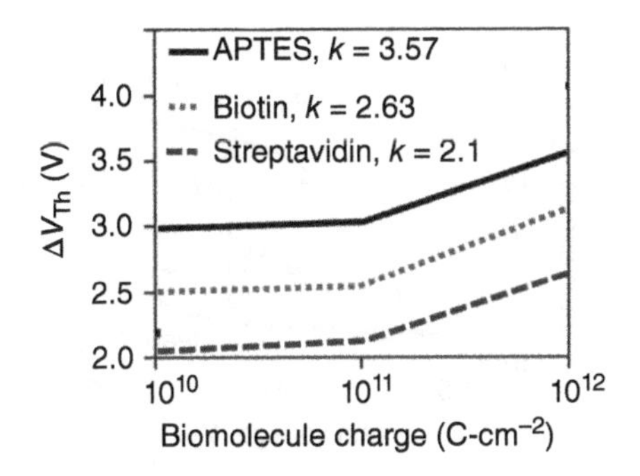

FIGURE 10.3 The shift in the threshold voltage (ΔV_{Th}) for different biomolecules with different charge [39].

10.2 JLFETS AS CAPACITORLESS DRAM

The snapback property in the output characteristics of JLFETs may be utilized for capacitorless dynamic random access memory (DRAM) application [40–53]. The DRAM, a memory element, as you may know consists of an array of capacitors in a 1T-1C (1 transistor, 1 capacitor) configuration. The information is stored in the form of charge in the capacitor. However, the capacitor needs to be refreshed continuously, which consumes a lot of power, and scaling the capacitor is also a technological challenge. Therefore, the field of capacitorless DRAM where the silicon body in the MOSFET itself stores the charge carriers has attracted the attention of the researchers. However, the snapback characteristics are obtained at a large drain voltage >8 V in a SOI MOSFET [40–53] and the read current margin (i.e., difference between read currents in state "0" and "1") is also low. Therefore, the quest for capacitorless DRAM devices operating at low supply voltages is underway.

In JLFET-based capacitorless DRAM, the charges would be stored in the highly doped silicon channel region. The snapback window allows to distinguish between the states "0" or "1" during read cycle once the JLFET has been written (programmed) to state "0" or "1" by applying an appropriate drain voltage. For instance, we may select a drain voltage, $V_{DS} = 1.75$ V for writing state "0" and a $V_{DS} = 2.25$ V for writing state "1" and chose a drain voltage of 2.1 V for reading the state of the JLFET according to the output characteristics of the JLFET shown in Fig. 10.4. For this case, the read current obtained after programming the memory cell (JLFET) in state "1" is three orders of magnitude more than that obtained after programming the memory cell in state "0." The two states may be easily distinguished without the need of an ultrasensitive sense amplifier. Therefore, the JLFETs offer a large memory window, i.e. snapback window along with a large read margin at a low drain voltage and may be a lucrative alternative to the MOSFETs for low static power capacitorless DRAM.

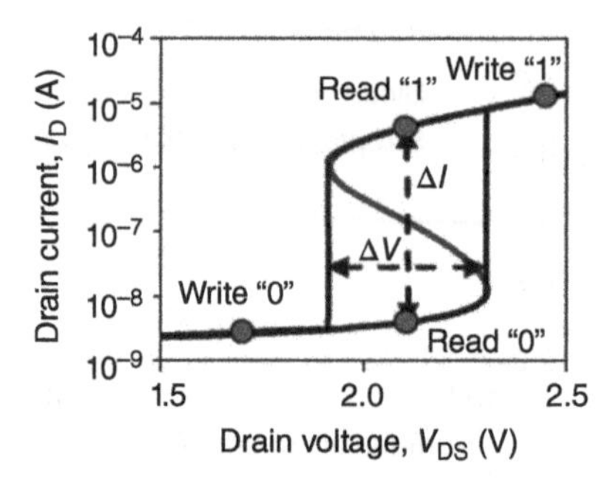

FIGURE 10.4 The hysteresis in the output characteristics of the DGJLFET obtained during voltage-controlled measurements and the different programming voltages for the application of DGJLFETs as capacitorless DRAM [40].

10.3 NANOWIRE JUNCTIONLESS NAND FLASH MEMORY

The introduction of flash memory changed the storage landscape. We can find flash memory in the external storage card of every smartphone owing to the high storage density, low program/erase time, and high endurance and retention. The flash memory exploits the floating gate MOSFETs to store data. The floating gate MOSFET has two gates as shown in Fig. 10.5(a): a floating gate (FG), which is not connected to any external bias and stores charge (in the form of trapped electrons), and a control gate (CG), which controls the charge transfer between the MOSFET channel and the floating gate. Generally, the floating gate is made of polysilicon. In the program mode, a large bias is applied to CG. A portion of the voltage applied to CG couples to the FG through capacitive coupling. FG acts as the gate of the underlying MOSFET and depending on the capacitively coupled voltage, electrons are injected from the channel region of the MOSFET to FG via Fowler–Nordheim (FN) tunneling or channel hot electron injection. Since the FG is surrounded by insulators, the electrons are trapped on FG. The stored negative charge on FG increases the threshold voltage of the underlying MOSFET and the flash memory is said to store bit "1." The application of a CG voltage of opposite polarity enables the stored electrons to tunnel back into the MOSFET channel region restoring the original threshold voltage. This is erase operation, and the flash memory is said to store bit "0." During the read operation, a small voltage is applied to the gate and the threshold voltage shift is evaluated by measurement of the drain current. On the other hand, the voltages required for program/erase operation is very high (in excess of ~12 V), which requires an additional supply or a charge pump circuitry increasing area overhead and routing complexity. Also, to increase the storage density, the floating gate MOSFETs must be scaled. However, the CG–FG coupling ratio degrades significantly with scaling of the floating gate MOSFETs [54]. Therefore, flash memory architectures with a different charge trap layer such as nitride layer, high-κ dielectric, or silicon nanocrystals were explored [55–62].

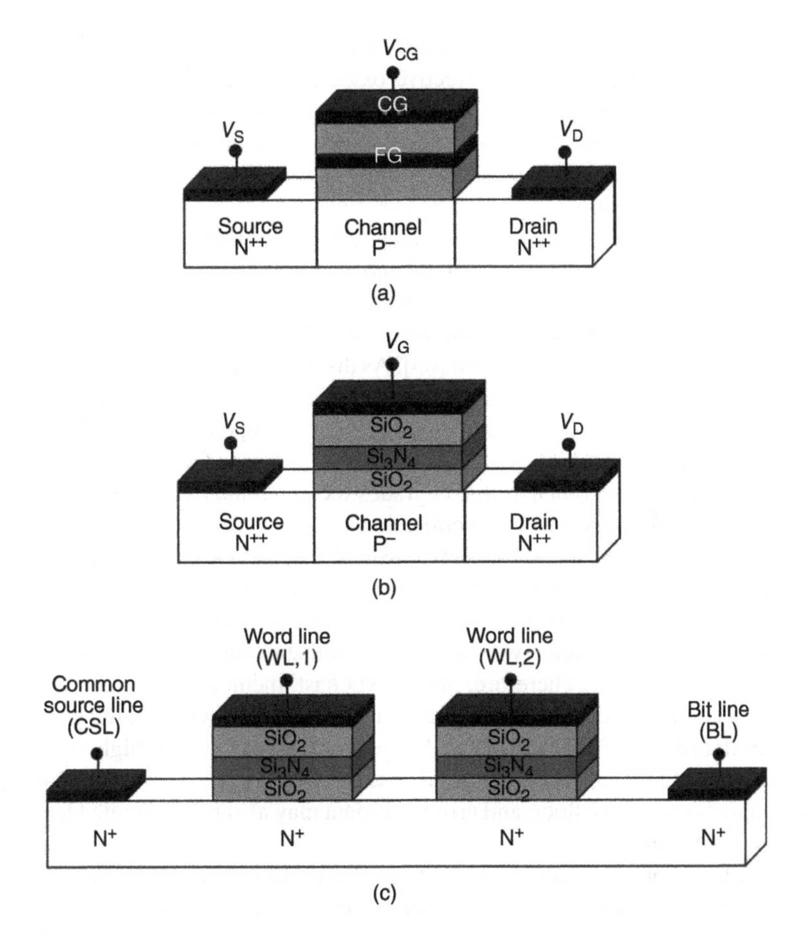

FIGURE 10.5 Three-dimensional view of (a) floating gate MOSFET, (b) conventional SONOS memory cell, and (c) NAND flash memory based on JLFETs.

The flash memory utilizing the silicon nitride layer for charge trapping is known as silicon-oxide-nitride-oxide-silicon SONOS as the gate stack consists of a silicon nitride sandwiched between an ultrathin SiO_2 tunneling layer and a relatively thick SiO_2 blocking layer as shown in Fig. 10.5(b). In the program mode, the application of a voltage on the gate results in a large electric field leading to tunneling of electrons (FN tunneling or direct tunneling depending on tunnel layer thickness) through ultrathin SiO_2 layer into the charge trapping nitride layer. The storage of charge on the nitride layer increases the threshold voltage of the SONOS memory and stores bit “1.” Similarly, the application of a gate voltage of opposite polarity leads to tunneling of holes from the channel region to the nitride layer and bit “0” is stored in the SONOS memory cell. The SONOS memory cell offers a better scalability as compared to the floating gate MOSFETs [55–61]. However, the short-channel effects hinder the scaling of SONOS memory cell. The multi-layered gate stack consisting of O-N-O restricts the scaling of effective oxide thickness (EOT) since the charge stored depends on the thickness of the nitride layer. The blocking layer also cannot be scaled as it blocks the tunneling of the charges from the nitride layer to the gate electrode. A higher EOT reduces the immunity of the SONOS memory cell against the short-channel effects.

The use of ultrasteep doping profiles may reduce the short-channel effects. Also, holes tunnel from the channel region to the nitride layer during the erase operation. If somehow the holes are increased in the channel region before the erase operation, the efficiency of the erase operation would increase and the distribution of the threshold voltage after erase would be narrowed [63]. As discussed in Chapter 5, GIDL leads to an accumulation of holes in the channel region [1–13]. Therefore, if somehow GIDL is triggered before the erase operation, an efficient erase operation would be ensured. An ultrasteep doping profile is required even from a higher GIDL perspective. You may wonder that a phenomenon that degrades the performance of devices for logic operation is beneficial for memory application.

To increase the flash storage density, the SONOS memory cells may be stacked vertically via three-dimensional (3D) integration [64–71]. However, realizing ultrasteep doping profiles for reducing short-channel effects and achieving efficient erasure of memory cell is extremely difficult in 3D multistacking integration owing to the complex thermal budget. Therefore, 3D NAND flash memories with a junction-free polysilicon channel and virtual source/drain regions (created via fringing gate fields) were introduced [72, 73]. The lower mobility of polysilicon and the high series resistance of the virtual source/drain regions lead to a very low read current. The read current may hit the noise floor, and erroneous data may also be interpreted leading to failure of read operation.

The JLFETs with a uniformly doped silicon film may circumvent all the above challenges owing to their low thermal budget, ease in fabrication, reduced short-channel effects, and reduction in surface scattering due to bulk conduction. Moreover, the GIDL in JLFETs which is detrimental for logic application may enable efficient erasure and aid its memory applications. Therefore, SONOS NAND flash memory based on JLFETs (Fig. 10.5(c)) was proposed and experimentally realized [63, 71, 74–76]. The cross-sectional TEM image of the fabricated NWJLFETs with SONOS stack is also shown in Fig. 3.10(g). The JLFETs based 3D NAND flash memory exhibits a high read current, a large programming window, i.e. the difference in the threshold voltage between the program and erase operations, a narrow distribution of threshold voltage after erase operation and an improved endurance (number of program/erase cycles which the device can sustain without failure). Therefore, the JLFETs-based flash memory overcomes all the challenges faced by the 3D flash memory.

So far, we have discussed the potential of JLFETs for application as biosensors and memory devices owing to their low cost. Since the JLFETs do not have any metallurgical junctions, there are no stringent constraints on realizing ultrasteep doping profiles. As a result, JLFETs do not require rapid annealing processes and complex thermal budgets. The lower thermal budget for JLFETs and junctionless versions of emerging FETs facilitates their realization on single-crystal silicon-on-glass substrates. This may lead to a new domain of exciting possibilities whereby these devices could be employed in display devices and for biocompatible and optoelectronic applications. Therefore, polysilicon junctionless thin film transistors (TFTs) have been experimentally demonstrated [78, 79]. However, the large source/drain series resistance in JLTFTs result in a very low ON-state current. To reduce the source/drain series resistance and improve the performance of JLTFTs, a hybrid channel JLTFT was proposed [79, 80]. We discuss the JLTFTs with a hybrid channel in the next section.

10.4 JUNCTIONLESS POLYSILICON TFTS WITH A HYBRID CHANNEL

The structure of the junctionless polysilicon TFT with a hybrid channel is shown in Figs. 10.6(a) and 6(b). The structure is similar to the HSJLFET discussed in Section 5.8.1 and consists of a n^+ substrate below the p^+ active polysilicon channel region. Although there is a vertical p–n junction, the JLTFT with a hybrid channel is still junctionless in the direction of current flow, i.e. in the lateral direction. The vertical p–n junction induced depletion helps to realize a smaller effective polysilicon p^+-channel thickness as compared to the actual physical thickness. This helps to realize efficient volume depletion even when the p^+-polysilicon film thickness is large. A large p^+-polysilicon film thickness also reduces the source/drain series resistance increasing the ON-state current of the JLTFT with the hybrid channel.

The JLTFT with a hybrid channel offers a better immunity to the short-channel effects due to the enhanced gate control owing to a lower effective channel thickness. The low OFF-state leakage current and the improved subthreshold swing of the JLTFT with a hybrid channel are attributed to these factors. Moreover, the threshold voltage may be tuned by changing the doping of the underlying n^+-substrate allowing multithreshold voltage circuit design.

JLTFTs with multiple hybrid channels (p^+-channel-n^+ channel) stacked on top of each other as shown in Figs. 10.6(c) and 6(d) have also been proposed and experimentally demonstrated [80]. The bottom p^+ channel in the JLTFT with a stacked hybrid channel is sandwiched between the top and bottom n^+-substrates. As a result, it is depleted from both top and bottom leading to an effective channel thickness, which is significantly small as compared to the actual physical thickness. Therefore, the gate controllability is further improved in the JLTFT with the stacked hybrid channel, resulting in a significantly reduced OFF-state current as shown in Fig. 10.7. Moreover, the ON-state current is also larger in the JLTFT with a stacked hybrid channel owing to the reduced source/drain series resistance and increased number of channels for current flow. Also, the JLTFT with the stacked hybrid channel exhibits better immunity to the negative bias instability and a reduced low-frequency noise [80].

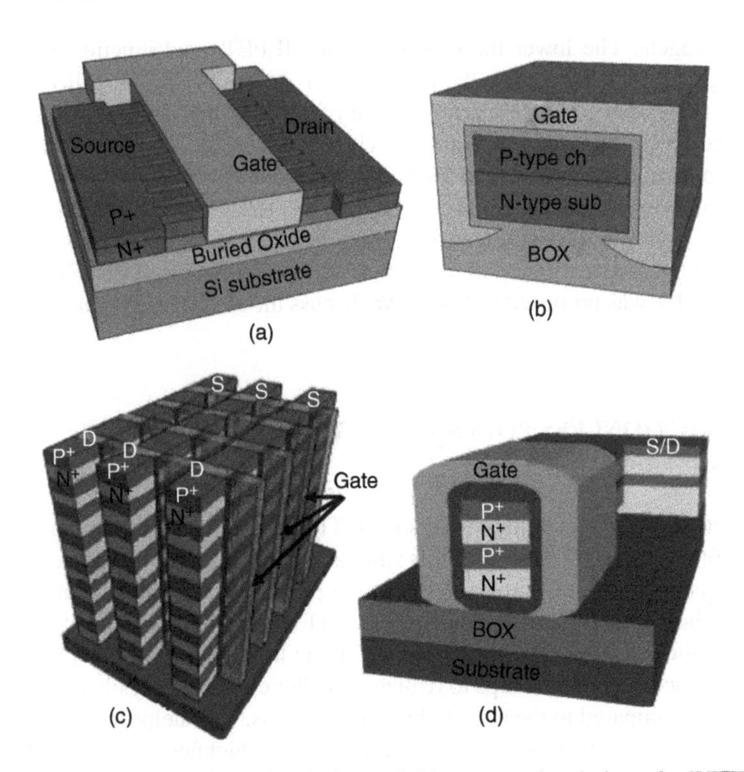

FIGURE 10.6 (a) Three-dimensional view and (b) cross-sectional view of a JLTFT with a hybrid channel [79] and (c) 3D view and (d) cross-sectional view of the JLTFT with vertically stacked hybrid channels [80].

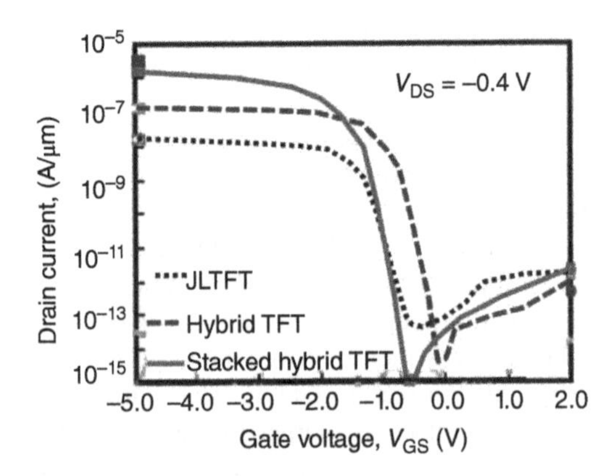

FIGURE 10.7 Transfer characteristics of the JLTFT, JLTFT with a hybrid channel, and the JLTFT with a stacked hybrid channel [80].

10.5 JLFETS FOR 3D INTEGRATED CIRCUITS

As discussed in Chapter 1, gate length scaling of the conventional MOSFETs is limited due to the short-channel effects, GIDL, direct source-to-drain-tunneling, and inability to realize ultrasteep doping profile. Therefore, to sustain the Moore's law, researchers have started exploring the vertical dimension and stacking devices on top of each other to increase the area efficiency without scaling the lateral dimensions such as gate length. Vertical stacking of the front end of line FETs on top of the back end of line (BEOL) interconnects is an integral part of these sequential 3D integrated circuits. This has triggered the research and development in the field of thru-silicon via and other 3D integration techniques.

In addition, it has also led to a quest for BEOL compatible transistors. The BEOL consists of a dense network of copper interconnects with low-κ dielectric. Therefore, the BEOL transistors must be processed at a lower temperature (< 430°C) without affecting the thermal stability of the copper interconnects with low-κ dielectric. Moreover, the BEOL compatible transistors should exhibit high performance and high scalability. In view of these requirements, the JLFETs owing to their low thermal budget could be a lucrative alternative for "cold" processed BEOL compatible transistors. Recently, BEOL-compatible JLFETs for 3D sequential integration have been experimentally demonstrated in [81, 82]. Double-gate JLFETs may be fabricated on top of a processed wafer at temperature below 430°C using the process flow demonstrated in [81] and shown in Fig. 10.8.

First, the silicon-on-insulator (SOI) wafer containing the active silicon film is ion implanted and annealed at high temperature to achieve the required uniform doping for realizing JLFETs. A gate stack may then be deposited over the doped crystalline SOI layer. This gate stack would act as the back gate once this wafer is transferred on to the processed wafer containing the BEOL interconnects. A low-temperature direct wafer–wafer dielectric bonding process (using oxide–oxide or SiCN–SiCN) may then be used to transfer the SOI wafer with active JLFET to the processed wafer. The substrate and BOX of the donor SOI wafer are then etched. Once the wafer has been transferred, high-temperature processes are not permissible due to the thermal instability of BEOL at temperatures above ~450°C. Therefore, a gate first approach is used and HfO_2 gate dielectric is deposited using atomic layer deposition at 300°C followed by deposition of TiN metal gate using chemical vapor deposition at 430°C. The replacement metal gate process commonly used in fabricating Fin field-effect transistors requires a temperature greater than 500°C and cannot be used for this process [81]. Furthermore, the selective epitaxial growth of source/drain regions to obtain a lower source/drain series resistance also requires a temperature more than 500°C. Therefore, direct tungsten (W) contacts are taken at the source and drain regions using a Ti/TiN barrier layer. As a result, the source/drain series resistance dominates and the ON-state current is low [81].

The stacked hybrid-channel nanowire junctionless polysilicon TFTs discussed in Section 10.4 may also be used for monolithic sequential 3D integrated circuits.

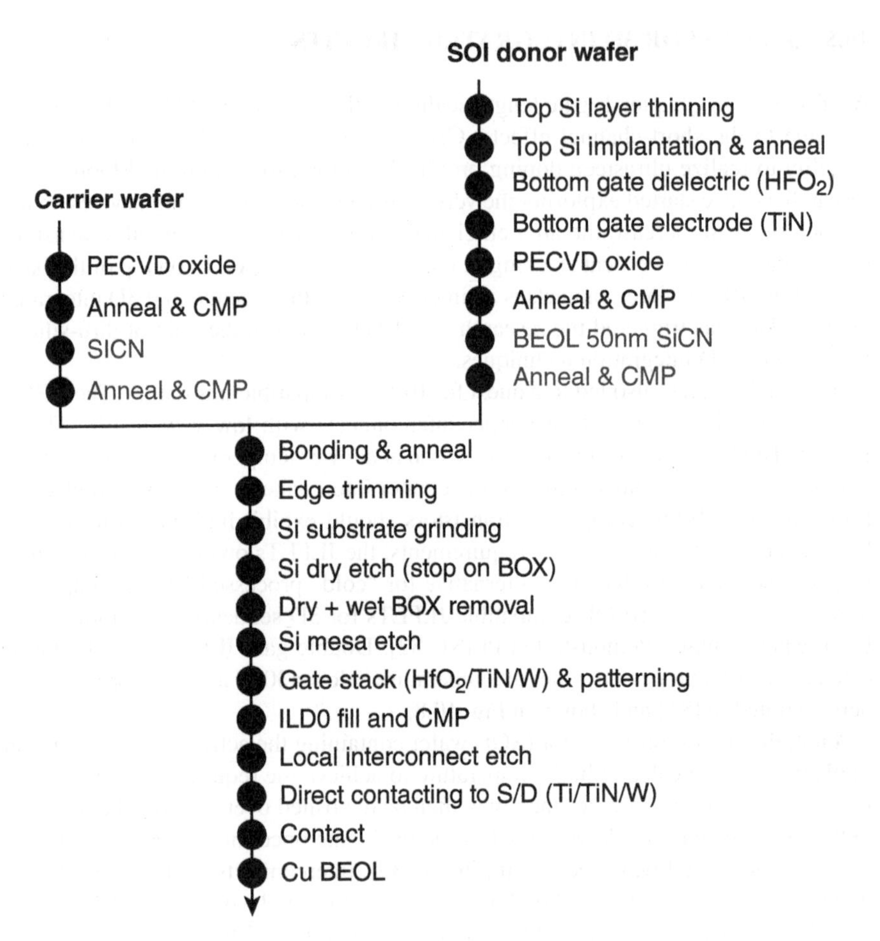

FIGURE 10.8 Fabrication flow for realizing BEOL-compatible DGJLFET on an already processed carrier wafer [81].

10.6 SUMMARY

In this chapter, we discussed the novel attributes of JLFETs and examined the competitive advantages of JLFETs over the conventional MOSFETs. It is quite possible that JLFETs or the junctionless version of the emerging FETs such as JLTFETs may be employed in the mainstream CMOS technology in future. In addition, the junctionless architecture, owing to its low cost, low fabrication complexity, and lower thermal budget may open up a new domain of exciting possibilities whereby these devices could be employed as sensors, memories, such as capacitor-less DRAM, NAND flash memory, display devices, and for biocompatible, optoelectronic, and 3D sequential integrated circuit applications apart from logic applications. The enormous possibilities offered by the junctionless architecture offers exciting opportunities to the researchers to explore and invent novel JLFET architectures for a variety of applications ranging from logic circuits to memory, sensor, 3D integration, and display technology. Chapter 8 on analytical modeling of JLFETs and Chapter 9 on the simulation of JLFETs using technology computer-aided design (TCAD) will definitely aid the researchers especially the beginners in the field and provide them with an effective tool to analyze, evaluate, and invent new junctionless architectures for different applications. We hope that this book covering the fundamentals of the JLFET along with their analytical modeling and simulation using TCAD would encourage the beginners to pursue research on JLFETs and augment the efforts of the existing researchers to realize a power-efficient JLFET for "green" electronics, which would eventually lead to a better society.

REFERENCES

[1] S. Sahay and M. J. Kumar, "Physical insights into the nature of gate-induced drain leakage in ultrashort channel nanowire FETs," *IEEE Trans. Electron Devices*, vol. 64, no. 6, pp. 2604–2610, June 2017.

[2] S. Sahay and M. J. Kumar, "A novel gate-stack-engineered nanowire FET for scaling to the sub-10-nm regime," *IEEE Trans. Electron Devices*, vol. 63, no. 12, pp. 5055–5059, Dec. 2016.

[3] S. Sahay and M. J. Kumar, "Spacer design guidelines for nanowire FETs from gate-induced drain leakage perspective," *IEEE Trans. Electron Devices*, vol. 64, no. 7, pp. 3007–3015, July 2017.

[4] S. Sahay and M. J. Kumar, "Insight into lateral band-to-band-tunneling in nanowire junctionless FETs," *IEEE Trans. Electron Devices*, vol. 63, no. 10, pp. 4138–4142, Oct. 2016.

[5] S. Sahay and M. J. Kumar, "Controlling L-BTBT and volume depletion in nanowire JLFETs using core-shell architecture," *IEEE Trans. Electron Devices*, vol. 63, no. 9, pp. 3790–3794, Sept. 2016.

[6] S. Sahay and M. J. Kumar, "Diameter dependency of leakage current in nanowire junctionless field-effect transistors," *IEEE Trans. Electron Devices*, vol. 64, no. 3, pp. 1330–1335, Mar. 2017.

[7] S. Sahay and M. J. Kumar, "Nanotube junctionless FET: Proposal, design, and investigation," *IEEE Trans. Electron Devices*, vol. 64, no. 4, pp. 1851–1856, Apr. 2017.

[8] M. J. Kumar and S. Sahay, "Controlling BTBT induced parasitic BJT action in junctionless FETs using a hybrid channel," *IEEE Trans. Electron Devices*, vol. 63, no. 8, pp. 3350–3353, Aug. 2016.

[9] S. Sahay, and M. J. Kumar, "Realizing efficient volume depletion in SOI junctionless FETs," *IEEE J. Electron Devices Soc.*, vol. 4, no. 3, pp. 110–115, May 2016.

[10] S. Sahay and M. J. Kumar, "Controlling the drain side tunneling width to reduce ambipolar current in tunnel FETs using heterodielectric BOX," *IEEE Trans. Electron Devices*, vol. 62, no. 11, pp. 3882–3886, Nov. 2015.

[11] S. Sahay and M. J. Kumar, "Comprehensive analysis of gate-induced drain leakage in emerging FET architectures: Nanotube FETs vs. nanowire FETs," *IEEE Access*, vol. 5, pp. 18918–18926, Dec. 2017.

[12] S. Sahay and M. J. Kumar, "Symmetric operation in an extended back gate JLFET for scaling to the 5 nm regime considering quantum confinement effects," *IEEE Trans. Electron Devices*, vol. 64, no. 1, pp. 21–27, Jan. 2017.

[13] A. K. Jain, S. Sahay, and M. J. Kumar, "Controlling L-BTBT in emerging nanotube FETs using dual-material gate," *IEEE J. Electron Dev. Soc.*, vol. 6, pp. 611–621, June 2018.

[14] S. Gundapaneni, S. Ganguly, and A. Kottantharayil, "Bulk planar junctionless transistor (BPJLT): An attractive device alternative for scaling," *IEEE Electron Device Lett.*, vol. 32, no. 3, pp. 261–263, Mar. 2011.

[15] J.-P. Colinge, C.-W. Lee, A. Afzalian, N. D. Akhavan, R. Yan, I. Ferain, P. Razavi, B. O'Neill, A. Blake, M. White, A.-M. Kelleher, B. McCarthy, and R. Murphy, "Nanowire transistors without junctions," *Nature Nanotechnol.*, vol. 5, no. 3, pp. 225–229, Mar. 2010.

[16] M. Toledano-Luque, P. Matagne, A. Sibaja-Hernández, T. Chiarella, L. A. Ragnarsson, B. Sorée, M. Cho, A. Mocuta, and A. Thean, "Superior reliability of junctionless pFinFETs by reduced oxide electric field," *IEEE Electron Device Lett.*, vol. 35, no. 12, pp. 1179–1181, Dec. 2014.

[17] C. W. Lee, A. N. Nazarov, I. Ferain, N. D. Akhavan, R. Yan, P. Razavi, R. Yu, R. T. Doria, and J. P. Colinge, "Low subthreshold slope in junctionless multigate transistors," *Appl. Phys. Lett.*, vol. 96, no. 10, 102106, 2010.

[18] B. Rajasekharan, R. J. E. Hueting, C. Salm, T. V. Hemert, R. A. Wolters, and J. Schmitz, "Fabrication and characterization of the charge-plasma diode," *IEEE Electron Device Lett.*, vol. 31, no. 6, pp. 528–530, 2010.

[19] R. J. Hueting, B. Rajasekharan, C. Salm, and J. Schmitz, "The charge plasma PN diode," *IEEE Electron Device Lett.*, vol. 29, no. 12, pp. 1367–1369, 2008.

[20] G. Gupta, B. Rajasekharan, and R. J. Hueting, "Electrostatic doping in semiconductor devices," *IEEE Trans. Electron Devices*, vol. 64, no. 8, pp. 3044–3055, 2017.

[21] S. Ramaswamy and M. J. Kumar, "Junctionless impact ionization MOS: Proposal and investigation," *IEEE Trans. Electron Devices*, vol. 61, no. 12, pp. 4295–4298, Dec. 2014.

[22] M. J. Kumar and S. Janardhanan, "Doping-less tunnel field-effect transistor: Design and investigation," *IEEE Trans. Electron Devices*, vol. 60, no. 10, pp. 3285–3290, 2013.

[23] K. H. Kao and L. Y. Chen, "A dopingless FET with metal–insulator–semiconductor contacts," *IEEE Electron Device Lett.*, vol. 38, no. 1, pp. 5–8, 2017.

[24] P. Bergveld, "The development and application of FET-based biosensors," *Biosensors*, vol. 2, no. 1, pp. 15–33, 1986.

[25] Y. Cui, Q. Q. Wei, H. K. Park, and C. M. Lieber, "Nanowire nanosensors for highly sensitive and selective detection of biological and chemical species," *Science*, vol. 293, no. 5533, pp. 1289–1292, 2001.

[26] K. W. Lee, S.-J. Choi, J.-H. Ahn, D.-I. Moon, T. J. Park, S. Y. Lee, and Y.-K. Choi "An underlap field-effect transistor for electrical detection of influenza," *Appl. Phys. Lett.*, vol. 96, no. 3, 033703, 2010.

[27] J. Y. Kim, J. H. Ahn, D. I. Moon, T. J. Park, S. Y. Lee, and Y. K. Choi, "Multiplex electrical detection of avian influenza and human immunodeficiency virus with an underlap-embedded silicon nanowire field-effect transistor," *Biosensors Bioelectron.*, vol. 55, pp. 162–167, 2014.

[28] H. Im, X. J. Huang, B. Gu, and Y. K. Choi, "A dielectric-modulated field-effect transistor for biosensing," *Nature Nanotechnol.*, vol. 2, pp. 430–434, 2007.

[29] C. H. Kim, C. Jung, K. B. Lee, H. G. Park, and Y. K. Choi, "Label-free DNA detection with a nanogap embedded complementary metal-oxide-semiconductor," *Nanotechnology*, vol. 22, no. 13, 135502, 2011.

[30] M. Im, J. H. Ahn, J. W. Han, T. J. Park, S. Y. Lee, and Y. K. Choi, "Development of a point-of-care testing platform with a nanogap-embedded separated double-gate field-effect transistor array and its readout system for detection of avian influenza," *IEEE Sensors J.*, vol. 11, no. 2, pp. 351–360, Feb. 2011.

[31] X. P. A. Gao, G. Zheng, and C. M. Lieber, "Subthreshold regime has the optimal sensitivity for nanowire FET biosensors," *Nano Lett.*, vol. 10, no. 2, pp. 547–552, 2009.

[32] D. Sarkar and K. Banerjee, "Proposal for tunnel-field-effect-transistor as ultra-sensitive and label-free biosensors," *Appl. Phys. Lett.*, vol. 100, no. 14, 143108, 2012.

[33] D. Sarkar, H. Gossner, W. Hansch, and K. Banerjee, "Impact-ionization field-effect-transistor based biosensors for ultra-sensitive detection of biomolecules," *Appl. Phys. Lett.*, vol. 102, no. 20, 203110, 2013.

[34] N. Kannan and M. J. Kumar, "Dielectric-modulated impact-ionization MOS (DIMOS) transistor as a label-free biosensor," *IEEE Electron Device Lett.*, vol. 34, no. 12, pp. 1575–1577, Dec. 2013.

[35] N. Kannan and M. J. Kumar, "Charge-modulated underlap I-MOS transistor as a label-free biosensor: A simulation study," *IEEE Trans. Electron Devices*, vol. 62, no. 8, pp. 2645–2651, Aug. 2015.

[36] J. Y. Kim, "An underlap channel-embedded field-effect transistor for biosensor application in watery and dry environment," *IEEE Trans. Nanotechnol.*, vol. 11, no. 2, pp. 390–394, Mar. 2012.

[37] S. Kalra, M. J. Kumar, and A. Dhawan, "Dielectric-modulated field-effect transistors for DNA Detection: Impact of DNA orientation," *IEEE Electron Device Lett.*, vol. 37, no. 11, pp. 1485–1488, Nov. 2016.

[38] I. Y. Chung, H. Jang, J. Lee, H. Moon, S. M. Seo, and D. H. Kim, "Simulation study on discrete charge effects of SiNW biosensors according to bound target position using a 3D TCAD simulator," *Nanotechnology*, vol. 23, no. 6, 2012.

[39] M. S. Parihar and A. Kranti, "Enhanced sensitivity of double gate junctionless transistor architecture for biosensing applications," *Nanotechnology*, vol. 26, no. 14, 145201, 2015.

[40] M. S. Parihar, D. Ghosh, G. A. Armstrong, and A. Kranti, "Bipolar snapback in junctionless transistors for capacitorless dynamic random access memory," *Appl. Phys. Lett.*, vol. 101, no. 26, 263503, 2012.

[41] P. G. D. Agopian, M. D. V. Martino, J. A. Martino, R. Rooyackers, D. Leonelli, and C. Claeys, "Experimental analog performance of pTFETs as a function of temperature," in *Proc. IEEE Int. SOI Conf.*, pp. 1–2, Jan. 2012.

[42] T. Tanaka, E. Yoshida, and T. Miyashita, "Scalability study on a capacitorless 1T-DRAM: From single-gate PD-SOI to double-gate FinDRAM," *IEDM Tech. Dig.*, pp. 919–922, Apr. 2004

[43] C. Hu, T.-J. King, and C. Hu, "A capacitorless double-gate DRAM cell," *IEEE Electron Device Lett.*, vol. 23, no. 6, pp. 345–347, June 2002.

[44] A. Biswas and A. M. Ionescu, "1T capacitor-less DRAM cell based on asymmetric tunnel FET design," *IEEE J. Electron Devices Soc.*, vol. 3, no. 3, pp. 217–222, May 2015.

[45] D.-O. Kim, D.-I. Moon, and Y. K. Choi, "Optimization of bias schemes for long-term endurable 1T-DRAM through the use of the biristor mode operation," *IEEE Electron Device Lett.*, vol. 35, no. 2, pp. 220–222, Feb. 2014.

[46] N. Rodriguez, C. Navarro, F. Gamiz, F. Andrieu, O. Faynot, and S. Cristoloveanu, "Experimental demonstration of capacitorless A2RAM cells on silicon-on-insulator," *IEEE Electron Device Lett.*, vol. 33, no. 12, pp. 1717–1719, Dec. 2012.

[47] L. M. Almeida, K. R. A. Sasaki, C. Caillat, M. Aoulaiche, N. Collaert, M. Jurczak, E. Simoen, C. Claeys, and J. A. Martino, "Optimizing the front and back biases for the best sense margin and retention time in UTBOX FBRAM," *Solid-State Electron.*, vol. 90, pp. 149–154, Dec. 2013.

[48] M. Aoulaiche, A. Bravaix, E. Simoen, C. Caillat, M. Cho, L. Witters, P. Blomme, P. Fazan, G. Groeseneken, and M. Jurczak "Endurance of one transistor floating body RAM on UTBOX SOI," *IEEE Trans. Electron Devices*, vol. 61, no. 3, pp. 801–805, Mar. 2014.

[49] J.-T. Lin, P.-H. Lin, S. W. Haga, Y.-C. Wang, and D.-R. Lu, "Transient and thermal analysis on disturbance immunity for 4F 2 surrounding gate 1T-DRAM with wide trenched body," *IEEE Trans. Electron Devices*, vol. 62, no. 1, pp. 61–68, Jan. 2015.

[50] E. Yoshida and T. Tanaka, "A capacitorless 1T-DRAM technology using gate-induced drain-leakage (GIDL) current for low-power and high-speed embedded memory," *IEEE Trans. Electron Devices*, vol. 53, no. 4, pp. 692–697, Apr. 2006.

[51] S. Okhonin, M. Nagoga, E. Carman, R. Beffa, and E. Faraoni, "New generation of Z-RAM," *IEDM Tech. Dig.*, pp. 925–928, Sept. 2007.

[52] N. Navlakha, J. T. Lin, and A. Kranti, "Improved retention time in twin gate 1T DRAM with tunneling based read mechanism," *IEEE Electron Device Lett.*, vol. 37, no. 9, pp. 1127–1130, Sept. 2016.

[53] A. Lahgere and M. J. Kumar, "1-T Capacitorless DRAM using bandgap-engineered silicon-germanium bipolar I-MOS," *IEEE Trans. Electron Devices*, vol. 64, no. 4, pp. 1583–1590, 2017.

[54] J. D. Lee, S. H. Hur, and J. D. Choi, "Effects of floating-gate interference on NAND flash memory cell operation," *IEEE Electron Device Lett.*, vol. 23, no. 5, pp. 264–266, May 2002.

[55] K. Kim, "Technology for sub-50 nm DRAM and NAND flash manufacturing," in *IEDM Tech. Dig.*, 2005, pp. 323–326.

[56] Y. Zhao, X. Wang, H. Shang, and M. H. White, "A low voltage SANOS nonvolatile semiconductor memory (NVSM) device," *Solid State Electron.*, vol. 50, no. 9/10, pp. 1667–1669, Sept./Oct. 2006.

[57] H. H. Hsu, I. Y. Chang, and J. Y. Lee, "Metal–oxide–high-κ dielectric–oxide–semiconductor (MOHOS) capacitors and field-effect transistors for memory application," *IEEE Electron Device Lett.*, vol. 28, no. 11, pp. 964–966, Nov. 2007.

[58] Y. N. Tan, W. K. Chim, W. K. Choi, M. S. Joo, and B. J. Cho, "Hafnium aluminum oxide as charge storage and blocking-oxide layers in SONOS type nonvolatile memory for high-speed operation," *IEEE Trans. Electron Devices*, vol. 53, no. 4, pp. 654–662, Apr. 2006.

[59] S. Jeon, J. H. Han, J. H. Lee, S. Choi, H. Hwang, and C. Kim, "High work-function metal gate and high-k dielectrics for charge trap flash memory device applications," *IEEE Trans. Electron Devices*, vol. 52, no. 12, pp. 2654–2659, Dec. 2005.

[60] Y. H. Shih, H. T. Lue, K. Y. Hsieh, R. Liu, and C. Y. Lu, "A novel 2-bit/cell nitride storage flash memory with greater than 1 M P/E-cycle endurance," in *IEDM Tech. Dig.*, 2004, pp. 881–884.

[61] M. H. White, D. A. Adams, and J. Bu, "On the go with SONOS," *IEEE Circuits Devices Mag.*, vol. 16, no. 4, pp. 22–31, Jul. 2000.

[62] H. I. Hanafi, S. Tiwari, and I. Khan, "Fast and long retention-time nanocrystal memory," *IEEE Trans. Electron Devices*, vol. 43, no. 9, pp. 1553–1558, Sept. 1996.

[63] S. J. Choi, D. I. Moon, J. P. Duarte, S. Kim, and Y. K. Choi, "A novel junctionless all-around-gate SONOS device with a quantum nanowire on a bulk substrate for 3D stack NAND flash memory," in *IEEE VLSI Tech. Symp.*, pp. 74-75, 2011.

[64] H. Tanaka, M. Kido, K. Yahashi, M. Oomura, R. Katsumata, M. Kito, Y. Fukuzumi, M. Sato, Y. Nagata, Y. Matsuoka, Y. Iwata, H. Aochi, and A. Nitayama, "Bit cost scalable technology with punch and plug process for ultra high density Flash memory," in *Proc. VLSI Symp. Tech.*, 2007, pp. 14–15.

[65] Y. Fukuzumi, R. Katsumata, M. Kito, M. Kido, M. Sato, H. Tanaka, Y. Nagata, Y. Matsuoka, Y. Iwata, H. Aochi, and A. Nitayama, "Optimal integration and characteristics of vertical array devices for ultra-high density, bit-cost scalable Flash memory," in *IEDM Tech. Dig.*, Dec. 2007, pp. 449–452.

[66] Y. Komori, M. Kido, M. Kito, R. Katsumata, Y. Fukuzumi, H. Tanaka, Y. Nagata, M. Ishiduki, H. Aochi, and A. Nitayama, "Disturbless Flash memory due to high boost efficiency on BiCS structure and optimal memory film stack for ultra high density storage device," in *IEDM Tech. Dig.*, Dec. 2008, pp. 851–854.

[67] J. Kim, A. J. Hong, S. M. Kim, E. B. Song, J. H. Park, J. Han, S. Choi, D. Jang, J.-T. Moon, and K. L. Wang, "Novel vertical-stacked-array transistor (VSAT) for ultra-high-density and cost-effective NAND Flash memory devices and SSD (solid state drive)," in *VLSI Symp. Tech. Dig.*, 2009, pp. 186–187.

[68] J. Jang, H.-S. Kim, W. Cho, H. Cho, J. Kim, S. I. Shim, Y. Jang, J.-H. Jeong, B.-K. Son, D. W. Kim, K. Kim, J.-J. Shim, J. S. Lim, K.-H. Kim, S. Y. Yi, J.-Y. Lim, D. Chung, H.-C. Moon, S. Hwang, J.-W. Lee, Y.-H. Son, U-I. Chung, and W.-S. Lee, "Vertical cell array using TCAT (terabit cell array transistor) technology for ultra high density NAND flash memory," in *VLSI Symp. Tech. Dig.*, 2009, pp. 192–193.

[69] W. Kim, S. Choi, J. Sung, T. Lee, C. Park, H. Ko, J. Jung, I. Yoo, and Y. Park, "Multi-layered vertical gate NAND Flash overcoming stacking limit for terabit density storage," in *VLSI Symp. Tech. Dig.*, 2009, pp. 188–189.

[70] A. Hubert, E. Nowak, K. Tachi, V. Maffini-Alvaro, C. Vizioz, C. Arvet, J.-P. Colonna, J.-M. Hartmann, V. Loup, L. Baud, S. Pauliac, V. Delaye, C. Carabasse, G. Molas, G. Ghibaudo, B. De Salvo, O. Faynot, and T. Ernst, "A stacked SONOS technology, up to 4 levels and 6 nm crystalline nanowires, with gate-all-around or independent gates (Φ-flash), suitable for full 3-D integration," in *IEDM Tech. Dig.*, Dec. 2009, pp. 637–640.

[71] M. C. Chen, H. Y. Yu, N. Singh, Y. Sun, N. S. Shen, X. H. Yuan, G. Q. Lo, and D. L. Kwong, "Vertical Si nanowire SONOS memory for ultra-high density application," *IEEE Electron Device Lett.*, vol. 30, no. 8, pp. 879–881, Aug. 2009.

[72] A. J. Walker, S. Nallamothu, E.-H. Chen, M. Mahajani, S. B. Herner, M. Clark, J. M. Cleeves, S. V. Dunton, V. L. Eckert, J. Gu, S. Hu, J. Knall, M. Konevecki, C. Petti, S. Radigan, U. Raghuram, J. Vienna, and M. A. Vyvoda, "3-D TFT-SONOS memory cell for ultra-high density file storage applications," *in Proc. VLSI Symp. Tech. Dig.*, 2003, pp. 29–30.

[73] E.-K. Lai, H.-T. Lue, Y.-H. Hsiao, J.-Y. Hsieh, S.-C. Lee, C.-P. Lu, S.-Y. Wang, L.-W. Yang, K.-C. Chen, J. Gong, K.-Y. Hsieh, J. Ku, R. Liu, and C.-Y. Lu, "A highly stackable thin-film transistor (TFT) NAND-type flash memory," *in Proc. VLSI Symp. Tech.* Dig., 2006, pp. 46–47.

[74] S. J. Choi, D. I. Moon, S. Kim, J. H. Ahn, J. S. Lee, J. Y. Kim, and Y. K. Choi, "Nonvolatile memory by all- around-gate junctionless transistor composed of silicon nanowire on bulk substrate," *IEEE Electron Device Lett.*, vol. 32, no. 5, pp. 602–604, May 2011.

[75] Y. Sun, H. Y. Yu, N. Singh, K. C. Leong, G. Q. Lo, and D. L. Kwong, "Junctionless vertical-Si-nanowire-channel-based SONOS memory with 2-bit storage per cell," *IEEE Electron Device Lett.*, vol. 32, no. 6, pp. 725–727, Jun 2011.

[76] Y. H. Lin, M. S. Yeh, Y. R. Jhan, M. H. Chung, C. C. Chung, M. Yen, and Y. C. Wu, "Band-to-band hot hole erase mechanism of p-channel junctionless silicon nanowire nonvolatile memory," *IEEE Trans. Nanotechnology*, vol. 15, no. 1, pp. 80–84, Jan. 2016.

[77] V. Thirunavukkarasu, C. H. Cheng, Y. R. Lin, E. D. Kurniawan, and Y. C. Wu, "Performance of hybrid p-channel trench poly-Si junctionless field-effect gate-all-around transistors," in *proc. SISC*, p. 11.2, 2016.

[78] Y. C. Cheng, H. B. Chen, C. S. Shao, J. J. Su, Y. C. Wu, C. Y. Chang and T. C. Chang, "Performance of a novel P-type junctionless transistor using a hybrid poly-Si fin channel," in *IEDM Tech. Dig.*, pp. 622–625, 2014

[79] Y. C. Cheng, H. B. Chen, J. J. Su, C. S. Shao, V. Thirunavukkarasu, C. Y. Chang and Y. C. Wu, "Characteristics of a novel poly-Si P-channel junctionless transistor with hybrid P/N-substrate," *IEEE Electron Device Lett.*, vol. 36, no. 2, pp. 159–161, 2015.

[80] Y. C. Cheng, H. B. Chen, C. Y. Chang, C. H. Cheng, Y. J. Shih, and Y. C. Wu, "A highly scalable poly-Si junctionless FETs featuring a novel multi-stacking hybrid P/N layer and vertical gate with very high Ion/Ioff for 3D stacked ICs," in *Proc. VLSI Tech. Dig.*, pp 188–189, 2016.

[81] A. Vandooren, L. Witters, E. Vecchio, E. Kunnen, G. Hellings, L. Peng, F. Inoue, W. Li, N. Waldron, D. Mocuta, and N. Collaert, "Double-gate Si junction-less n-type transistor for high performance Cu-BEOL compatible applications using 3D sequential integration," in *IEEE S3S*, pp. 1–2, 2017.

[82] A. Vandooren, "3D sequential stacked planar devices on 300 mm wafers featuring replacement metal gate junctionless top devices processed at 525°C with improved reliability," in *Proc. VLSI Tech. Symp.*, June 18–21, 2018.

INDEX

Junctionless Field-Effect Transistors: Design, Modeling, and Simulation, First Edition.
Shubham Sahay and Mamidala Jagadesh Kumar.

BOOKS IN THE IEEE PRESS SERIES ON MICROELECTRONIC SYSTEMS

The focus of the series is on all aspects of solid-state circuits and systems including: the design, testing, and application of circuits and subsystems, as well as closely related topics in device technology and circuit theory. The series also focuses on scientific, technical and industrial applications, in addition to other activities that contribute to the moving the area of microelectronics forward.

R. Jacob Baker, *Series Editor*

---

1. *CMOS: Circuit Design, Layout, and Simulation*
   R. Jacob Baker, Harry W. Li, and David E. Boyce
2. *Nonvolatile Semiconductor Memory Technology: A Comprehensive Guide to Understanding and Using NVSM Devices*
   William D. Brown and Joe E. Brewer
3. *Advanced Electronic Packaging: With Emphasis on MultiChip Modules*
   William D. Brown
4. *Low-Voltage/Low-Power Integrated Circuits and Systems: Low-Voltage Mixed-Signal Circuits*
   Edgar Sanchez-Sinencio and Andreas G. Andreou
5. *High-Performance System Design: Circuits and Logic*
   Vojin G Oklobdzija
6. *Hardware Description Languages: Concepts and Principles*
   Sumit Ghosh
7. *CMOS: Circuit Design, Layout, and Simulation, Second Edition*
   R. Jacob Baker
8. *Advanced Electronic Packaging, Second Edition*
   William D. Brown and Richard K. Ulrich
9. *DRAM Circuit Design: Fundamental and High-Speed Topics*
   Brent Keeth, R. Jacob Baker, Brian Johnson, and Feng Lin
10. *Nonvolatile Memory Technologies with Emphasis on Flash: A Comprehensive Guide to Understanding and Using NVM Devices*
    Joseph E. Brewer and Manzur Gill
11. *CMOS: Mixed-Signal Circuit Design, Second Edition*
    R. Jacob Baker
12. *Reliability Wearout Mechanisms in Advanced CMOS Technologies*
    Alvin W. Strong, Ernest Y. Wu, Rolf-Peter Vollertsen, Jordi Sune, Giuseppe La Rosa, Timothy D. Sullivan, and Stewart Rauch, III
13. *CMOS: Circuit Design, Layout, and Simulation, Third Edition*
    R. Jacob Baker
14. *Quantum Mechanics for Electrical Engineers*
    Dennis M. Sullivan

15. *Nanometer Frequency Synthesis beyond Phase-Locked Loop: A New Frontier in Electronic Design*
    Liming Xiu
16. *Electrical, Electronics and Digital Hardware Essentials for Scientists and Engineers*
    Ed Lipiansky
17. *Enhanced Phase-Locked Loop Structures for Power and Energy Applications*
    Masoud Karimi-Ghartemani
18. *From Frequency To Time-Average-Frequency: A Paradigm Shift in the Design of Electronic System*
    Liming Xiu
19. *NAND Flash Memory Technologies*
    Seiichi Aritome
20. *Understanding Delta-Sigma Data Converters, Second Edition*
    Shanthi Pavan, Richard Schreier, and Gabor C. Temes
21. *Junctionless Field-Effect Transistors: Design, Modeling, and Simulation*
    Shubham Sahay and Mamidala J. Kumar